HEAT TRANSFER AND THERMAL CONTROL

Edited by
A. L. Crosbie
Thermal Radiative Transfer Group
Department of Mechanical and Aerospace Engineering
University of Missouri-Rolla, Rolla, Missouri

Volume 78
PROGRESS IN
ASTRONAUTICS AND AERONAUTICS
Martin Summerfield, Series Editor-in-Chief
Princeton Combustion Research Laboratories, Inc.,
Princeton, New Jersey

Technical papers from the AIAA 18th Aerospace Sciences Meeting, January 1980, and the AIAA 15th Thermophysics Conference, July 1980, subsequently revised for this volume.

Published by the American Institute of Aeronautics and Astronautics,
1290 Avenue of the Americas, New York, N.Y. 10104.

American Institute of Aeronautics and Astronautics
New York, New York

Library of Congress Cataloging in Publication Data

AIAA Aerospace Sciences Meeting (18th: 1980:
  Los Angeles, Calif.)
  Heat transfer and thermal control.

  (Progress in astronautics and aeronautics; v. 78)
  Includes bibliographical references.
  1. Heat—Transmission—Congresses. 2. Temperature control—
Congresses. I. Crosbie, A. L. (Alfred L.) II. AIAA Thermophysics
Conference (15th: 1980: Snowmass, Colo.) III. American Institute of
Aeronautics and Astronautics. IV. Title. V. Series.
TL507.P75 vol. 78 [TJ260]   621.402'2   81-10795
ISBN 0-915928-53-1   AACR2

Copyright © 1981 by
American Institute of Aeronautics and Astronautics

All rights reserved. No part of this book may be reproduced in any form or by any means, electronic or mechanical, including photocopying, recording, or by any information storage and retrieval system, without permission in writing from the publisher.

Table of Contents

Preface vii

Editorial Committee xi

List of Series Volumes xii

Chapter I.   Radiation Heat Transfer 1

**Radiative Properties of a Painted Layer
Containing Nonspherical Pigment** 3
   H. M. Shafey and T. Kunitomo, *Kyoto University, Kyoto, Japan*

**Bidirectional Reflectance Measurements of Specular and
Diffuse Surfaces with a Simple Spectrometer** 25
   A. I. Funai, *Lockheed Research Laboratories, Palo Alto, Calif.*

Radiative Equilibrium in a General Plane-Parallel Environment ... 49
   A. Sharma and A. C. Cogley, *University of Illinois, Chicago, Ill.*

**Application of Finite-Element Techniques to the Interaction
of Conduction and Radiation in Participating Medium** 61
   S. T. Wu and R. E. Ferguson, *University of Alabama in Huntsville,
Huntsville, Ala.*, and L. L. Altgilbers, *U. S. Army Metrology and
Calibration Center, U. S. Army Missile Command, Redstone Arsenal,
Ala.*

**A Finite-Element Approach to Combined Conductive and
Radiative Heat Transfer in a Planar Medium** 92
   R. Fernandes, J. Francis, and J. N. Reddy, *University of Oklahoma,
Norman, Okla.*

Heat Transfer in Irradiated Shallow Layers of Water 110
   M. Behnia and R. Viskanta, *Purdue University, West Lafayette, Ind.*

**An Analytical and Experimental Investigation
of Temperature Distribution in Laser Heated Gases** 130
   J. P. Schuster, W. O. Li, and W. J. McLean, *Cornell University, Ithaca,
N. Y.*

**Numerical Methods for the Analysis of Laser Annealing
of Doped Semiconductor Wafers** 152
   J. R. Kirkpatrick, G. E. Giles Jr., and R. F. Wood, *Union Carbide
Corporation, Nuclear Division, Oak Ridge, Tenn.*

Chapter II. Conduction Heat Transfer 177

**Thermal Resistance of Two-Dimensional Geometries
Having Isothermal and Adiabatic Boundaries................ 179**
G. E. Schneider and M. M. Yovanovich, *University of Waterloo,
Waterloo, Ontario, Canada*

**Some Basic Three-Dimensional Influence Coefficients
for the Surface Element Method 202**
M. M. Yovanovich and K. A. Martin, *University of Waterloo, Waterloo,
Ontario, Canada*

**Approximate Solutions of Transient Heat Conduction
in a Finite Slab 229**
T. F. Zien, *Naval Surface Weapons Center, Silver Spring, Md.*

**Finite-Element Analysis for Conduction and
Ablation Moving Boundary............................... 249**
J. H. Chin, *Lockheed Missiles & Space Company, Inc., Sunnyvale, Calif.*

Thermal Resistance of Contacts: Influence of Oxide Films 266
F. R. Al-Astrabadi, *British Aerospace Dynamics Group, Stevenage,
Hertforshire, United Kingdom,* and P. W. O'Callaghan and S. D.
Probert, *Cranfield Institute of Technology, Cranfield, Bedfordshire,
United Kingdom*

Thermal Contact Conductance of Coated Multi-layered Sheets 285
J. W. Sheffield, *University of Missouri-Rolla, Rolla, Mo.,* and
T. N. Veziroglu and A. Williams, *University of Miami, Coral Gables,
Fla.*

Chapter III. Heat Pipes............................... 303

Performance Testing of a Hydrogen Heat Pipe 305
J. Alario and R. Kosson, *Grumman Aerospace Corporation, Bethpage,
N. Y.*

**Design, Development, and Test
of a 1000 W Osmotic Heat Pipe 324**
A. Basiulis, G. L. Fleischman, and C. P. Minning, *Hughes Aircraft
Company, Torrance, Calif.*

Heat Pipe Performance with Gravity Assist and Liquid Overfill ... 345
F. C. Prenger Jr. and J. E. Kemme, *Los Alamos National Laboratory, Los Alamos, N. Mex.*

The Heat Pipe Thermal Canister........................... 357
W. Harwell, *Grumman Aerospace Corporation, Bethpage, N. Y.,* and S. Ollendorf, *NASA Goddard Space Flight Center, Greenbelt, Md.*

Vapor Chambers for an Atmospheric Cloud Physics Laboratory... 375
G. L. Fleischman, *Hughes Aircraft Company, Torrance, Calif.,* T. R. Scollon Jr., *General Electric Company, Valley Forge, Pa.,* and J. D. Loose, *NASA George C. Marshall Space Flight Center, Huntsville, Ala.*

A Prototype Heat Pipe Radiator for the German Direct Broadcasting TV Satellite 402
R. Schlitt and R. Meyer, *ERNO Raumfahrttechnik GmbH, Bremen, Federal Republic of Germany*

Chapter IV. Thermal Control 427

Development of a 5 W 70 K Passive Radiator 429
J. P. Wright, *Rockwell International Corporation, Downey, Calif.*

Transient Response of Thermal Louvers with Bimetallic Actuators 452
H. Hwangbo, *MRJ, Inc., Fairfax, Va.,* and W. H. Kelly, *Fairchild Space and Electronics Company, Germantown, Md.*

α_s/ϵ_H Measurements of Thermal Control Coatings on the P78-2 (SCATHA) Spacecraft.................................. 467
D. F. Hall and A. A. Fote, *The Aerospace Corporation, El Segundo, Calif.*

Free Convection in Enclosures Exposed to Compressive Heating... 487
R. P. Bobco, *Hughes Aircraft Company, Los Angeles, Calif.*

Thermal Analysis of a Multipurpose Furnace for Material Processing in Space... 516
G. S. Karp, J. N. Holsen, and S. H. Miesner, *McDonnell Douglas Astronautics Company, St. Louis, Mo.*

Author Index 538

Preface

Heat transfer and thermal control are two extremely important areas of thermophysics. Heat transfer is a fundamental discipline which forms the basic foundation of thermophysics. In the vacuum environment of space, radiation and conduction heat transfer play dominant roles in the thermal balance of spacecraft. In turn, the traditional field of heat transfer has been enriched by thermophysics. Advances in radiation heat transfer closely parallel the achievements of the space program. Developments in thermal contact resistance and heat pipe technology can also be attributed to the space program. Future space exploration and utilization coupled with energy and environmental problems here on Earth will continue to stimulate heat-transfer research. Thermal control is a prime concern in the design of all spacecraft. The continuing demand for more accurate, more reliable, and longer-lived instruments leads to even more stringent temperature control requirements. Increasing power demands will require more efficient heat rejection systems. The Space Shuttle with its large payload capacity will further stimulate thermal control research.

This volume contains a selection of recent studies in heat transfer and thermal control. The papers were drawn from the AIAA 18th Aerospace Sciences Meeting in Los Angeles, California in January 1980 and the AIAA 15th Thermophysics Conference in Snowmass, Colorado in July 1980. The papers were reviewed, revised, and updated especially for this volume. They have been grouped into four chapters: radiation heat transfer, conduction heat transfer, heat pipes, and thermal control.

Besides the obvious appeal of this volume to the heat transfer and thermophysics community, the volume should be of interest to engineers and scientists working in related areas. Specifically, these groups include chemical engineers, nuclear engineers, energy specialists, laser specialists, optical physicists, paint chemists, finite-element specialists, and applied mathematicians. Hopefully, the developments reported here will stimulate further research in heat transfer and thermal control.

The eight papers of Chapter I illustrate the wide range of problems in which radiation heat transfer is important. The radiative properties of coatings are the topic of the first two papers. *Shafey and Kunitomo* develop a theoretical model to predict the reflected radiation from a painted layer containing nonspherical pigment particles in random orientation. The scattering properties of the particles are calculated using a method based on the volume integral form of Maxwell's equations, and the radiation transport equation is solved to find the reflected radiation. This work is unique in considering nonspherical particles and should become a standard reference for future work. *Funai* describes an experimental apparatus for measuring bidirectional reflectance.

The next three papers of Chapter I deal with solution techniques. *Sharma and Cogley* present a generalized solution technique for radiative equilibrium in a planar medium with anisotropic scattering and nongray properties. Independently, *Wu, Ferguson, and Algilbers* and *Fernandes, Francis, and Reddy* apply the finite-element method to the problem of combined conduction and radiation heat transfer. Steady-state results are presented in the first paper, while both steady-state and transient results are presented in the second. These two papers probably signal the beginning of a sustained effort to apply finite-element methods to radiative heat-transfer problems.

In the last three papers of Chapter I, heat transfer induced by collimated radiation is analyzed. *Behnia and Viskanta* report on an experimental and analytical investigation of the temperature structure in a shallow layer of water heated by an external radiation source. *Schuster, Li, and McLean* also conduct an experimental and analytical investigation. They study the temperature and velocity distributions produced in a laser-heated reactor consisting of a finite cylinder containing argon and the selective absorber, sulfur hexafluoride, and irradiated with a cw carbon dioxide laser. Heat and dopant transport in a semiconductor material undergoing a laser annealing process is studied numerically by *Kirkpatrick, Giles, and Wood*.

Techniques suitable for solving various heat-conduction problems are discussed in the first four papers of Chapter II. *Schneider and Yovanovich* apply a series of Schwartz-Christoffel transformations to find the thermal resistance for a general class of steady-state heat-conduction problems without heat generation. The technique is applicable to two-dimensional geometries with isothermal and adiabatic boundaries. *Yovanovich and Martin* consider the surface element method and present several basic integral solutions of Laplace's equation for point and distributed sources. *Zien* presents an approximate integral procedure for transient heat transfer in a finite slab, while a finite-element analysis for an ablation moving boundary is given by *Chin*.

The last two papers in Chapter II deal with thermal contact resistance. A theoretical prediction for the thermal resistance of a contact between oxidized nominally flat randomly rough metallic surfaces is presented by *Al-Astrabadi, O'Callaghan, and Probert*. Steady-state thermal contact conductances of multilayered electrically insulated sheets are examined theoretically and experimentally by *Sheffield, Verziroglu, and Williams*.

Chapter III consists of six papers concerned with heat pipe performance and application. The development and performance of specialized heat pipes is discussed in the first three papers. *Alario and Kossom* report test results for a re-entrant groove heat pipe with hydrogen working fluid. Steady-state performance data is taken over a 19-23 K range. *Basiulis, Fleischman, and Minning* operate an osmotically pumped heat pipe continuously for 500 h with a thermal load of 500 W. A peak thermal transport of 1000 W is demonstrated. Performance limits for gravity-assist heat pipes are investigated by *Prenger and Kemme* with the volume of the working fluid and tilt angle as independent variables.

The next three papers of Chapter III deal with the application of heat pipe technology to the thermal control of spacecraft and their components. *Harwell and Ollendorf* describe the acceptance thermal vacuum test results of the heat pipe thermal canister experiment which will be carried on an early Space Shuttle flight. *Fleischman, Scollon, and Loose* report on a flat-plate heat pipe with methanol working fluid to be used in a spaceborne atmosphere cloud chamber. *Schlitt and Meyer* describe the development, fabrication, and testing of a high-performance heat radiator to cool high dissipating repeater units on future communication satellites. An axially grooved heat pipe with ammonia working fluid is used.

Various aspects of thermal control are discussed in the five papers of Chapter IV. *Wright* reports on the development of a large two-stage heat pipe radiator for ground testing of the thermal performance and structural integrity of large, multiwatt passive cryogenic radiators. The transient response of thermal louvers with bimetallic actuators is investigated analytically by *Hwangbo and Kelly*. *Hall and Fote* analyze data acquired over the first year in orbit of a thermal control coating experiment. *Bobco* describes an experimental study of heat transfer in a vertical annulus and a three-dimensional gap that is used to establish the influence of compressive heating on the convective process in enclosures. The correlations will be applied to the design of a vented descent module parachuting into the Jupiter atmosphere. *Karp, Holsen, and Miesner* report on the thermal design of a multipurpose furnace for material processing in space.

This editor gratefully acknowledges the contributions of the Editorial Committee listed on page *xi*, Miss Ruth F. Bryans, Associate Series Editor, Miss Brenda Hio, Managing Editor of the Series, and Dr. Martin Summerfield, Editor-in-Chief of the AIAA *Progress in Astronautics and Aeronautics* series. The efforts of Dr. Kenneth E. Harwell who organized the thermophysics sessions at the AIAA 18th Aerospace Sciences Meeting, Dr. John E. Francis who chaired the AIAA Thermophysics Technical Committee in 1979 and who served as General Program Chairman of the AIAA 15th Thermophysics Conference, and Mr. Jesse F. Keville who chaired the AIAA Thermophysics Technical Committee in 1980 are also greatly appreciated. Finally, the contributors to this volume are thanked for their patience, cooperation, and care in the preparation of their papers.

Alfred L. Crosbie
March 1981

Editorial Committee for Volume 78

A. L. Crosbie
University of Missouri-Rolla

P. Bauer
McDonnell Douglas Astronautics Co.

Y. Bayazitoglu
Rice University

H. E. Collicott
Bendix Research Laboratories

G. R. Cunnington Jr.
Lockheed Palo Alto Research Laboratory

W. M. Farmer
University of Tennessee Space Institute

K. T. Feldman Jr.
University of New Mexico

J. E. Francis
University of Oklahoma

R. A. Haslett
Grumman Aerospace Corp.

D. C. Look
University of Missouri-Rolla

T. J. Love
University of Oklahoma

H. F. Nelson
University of Missouri-Rolla

J. A. Oren
Vought Corporation

D. W. Ruth
University of Calgary

G. E. Schneider
University of Waterloo

T. F. Smith
University of Iowa

**Progress in
Astronautics and Aeronautics**

Martin Summerfield,
Series Editor-in-Chief
*Princeton Combustion Research
Laboratories, Inc.*

Ruth F. Bryans,
Associate Series Editor
AIAA

Norma J. Brennan,
Director, Editorial Department
AIAA

Brenda J. Hio,
Series Managing Editor
AIAA

VOLUMES	**EDITORS**
*1. **Solid Propellant Rocket Research.** 1960	Martin Summerfield *Princeton University*
2. **Liquid Rockets and Propellants.** 1960	Loren E. Bollinger *The Ohio State University* Martin Goldsmith *The Rand Corporation* Alexis W. Lemmon Jr. *Battelle Memorial Institute*
3. **Energy Conversion for Space Power.** 1961	Nathan W. Snyder *Institute for Defense Analyses*
*4. **Space Power Systems.** 1961	Nathan W. Snyder *Institute for Defense Analyses*
5. **Electrostatic Propulsion.** 1961	David B. Langmuir *Space Technology Laboratories, Inc.* Ernst Stuhlinger *NASA George C. Marshall Space Flight Center* J. M. Sellen Jr. *Space Technology Laboratories, Inc.*

*Now out of print.

*6.	Detonation and Two-Phase Flow. 1962	S. S. Penner *California Institute of Technology* F. A. Williams *Harvard University*
7.	Hypersonic Flow Research. 1962	Frederick R. Riddell *AVCO Corporation*
8.	Guidance and Control. 1962	Robert E. Roberson *Consultant* James S. Farrior *Lockheed Missiles and Space Company*
*9.	Electric Propulsion Development. 1963	Ernst Stuhlinger *NASA George C. Marshall Space Flight Center*
*10.	Technology of Lunar Exploration. 1963	Clifford I. Cummings and Harold R. Lawrence *Jet Propulsion Laboratory*
11.	Power Systems for Space Flight. 1963	Morris A. Zipkin and Russell N. Edwards *General Electric Company*
12.	Ionization in High-Temperature Gases. 1963	Kurt E. Shuler, Editor *National Bureau of Standards* John B. Fenn, Associate Editor *Princeton University*
*13.	Guidance and Control—II. 1964	Robert C. Langford *General Precision Inc.* Charles J. Mundo *Institute of Naval Studies*
*14.	Celestial Mechanics and Astrodynamics. 1964	Victor G. Szebehely *Yale University Observatory*
*15.	Heterogeneous Combustion. 1964	Hans G. Wolfhard *Institute for Defense Analyses* Irvin Glassman *Princeton University*

		Leon Green Jr.
Air Force Systems Command
16.	Space Power Systems Engineering. 1966	George C. Szego
Institute for Defense Analyses
J. Edward Taylor
TRW Inc.
*17.	Methods in Astrodynamics and Celestial Mechanics. 1966	Raynor L. Duncombe
U. S. Naval Observatory
Victor G. Szebehely
Yale University Observatory
18.	Thermophysics and Temperature Control of Spacecraft and Entry Vehicles. 1966	Gerhard B. Heller
NASA George C. Marshall Space Flight Center
*19.	Communication Satellite Systems Technology. 1966	Richard B. Marsten
Radio Corporation of America
20.	Thermophysics of Spacecraft and Planetary Bodies: Radiation Properties of Solids and the Electromagnetic Radiation Environment in Space. 1967	Gerhard B. Heller
NASA George C. Marshall Space Flight Center
21.	Thermal Design Principles of Spacecraft and Entry Bodies. 1969	Jerry T. Bevans
TRW Systems
22.	Stratospheric Circulation. 1969	Willis L. Webb
Atmospheric Sciences Laboratory, White Sands, and University of Texas at El Paso
23.	Thermophysics: Applications to Thermal Design of Spacecraft. 1970	Jerry T. Bevans
TRW Systems
24.	Heat Transfer and Spacecraft Thermal Control. 1971	John W. Lucas
Jet Propulsion Laboratory

25.	Communications Satellites for the 70's: Technology. 1971	Nathaniel E. Feldman *The Rand Corporation* Charles M. Kelly *The Aerospace Corporation*
26.	Communications Satellites for the 70's: Systems. 1971	Nathaniel E. Feldman *The Rand Corporation* Charles M. Kelly *The Aerospace Corporation*
27.	Thermospheric Circulation. 1972	Willis L. Webb *Atmospheric Sciences Laboratory, White Sands, and University of Texas at El Paso*
28.	Thermal Characteristics of the Moon. 1972	John W. Lucas *Jet Propulsion Laboratory*
29.	Fundamentals of Spacecraft Thermal Design. 1972	John W. Lucas *Jet Propulsion Laboratory*
30.	Solar Activity Observations and Predictions. 1972	Patrick S. McIntosh and Murray Dryer *Environmental Research Laboratories, National Oceanic and Atmospheric Administration*
31.	Thermal Control and Radiation. 1973	Chang-Lin Tien *University of California, Berkeley*
32.	Communications Satellite Systems. 1974	P. L. Bargellini *COMSAT Laboratories*
33.	Communications Satellite Technology. 1974	P. L. Bargellini *COMSAT Laboratories*
34.	Instrumentation for Airbreathing Propulsion. 1974	Allen E. Fuhs *Naval Postgraduate School* Marshall Kingery *Arnold Engineering Development Center*

35. Thermophysics and Spacecraft Thermal Control. 1974 — Robert G. Hering, *University of Iowa*

36. Thermal Pollution Analysis. 1975 — Joseph A. Schetz, *Virginia Polytechnic Institute*

37. Aeroacoustics: Jet and Combustion Noise; Duct Acoustics. 1975 — Henry T. Nagamatsu, Editor, *General Electric Research and Development Center*; Jack V. O'Keefe, Associate Editor, *The Boeing Company*; Ira R. Schwartz, Associate Editor, *NASA Ames Research Center*

38. Aeroacoustics: Fan, STOL, and Boundary Layer Noise; Sonic Boom; Aeroacoustics Instrumentation. 1975 — Henry T. Nagamatsu, Editor, *General Electric Research and Development Center*; Jack V. O'Keefe, Associate Editor, *The Boeing Company*; Ira R. Schwartz, Associate Editor, *NASA Ames Research Center*

39. Heat Transfer with Thermal Control Applications. 1975 — M. Michael Yovanovich, *University of Waterloo*

40. Aerodynamics of Base Combustion. 1976 — S. N. B. Murthy, Editor, *Purdue University*; J. R. Osborn, Associate Editor, *Purdue University*; A. W. Barrows and J. R. Ward, Associate Editors, *Ballistics Research Laboratories*

41. Communication Satellite Developments: Systems. 1976 — Gilbert E. LaVean, *Defense Communications Engineering Center*; William G. Schmidt, *CML Satellite Corporation*

42. Communication Satellite Developments: Technology. 1976 — William G. Schmidt, *CML Satellite Corporation*; Gilbert E. LaVean, *Defense Communications Engineering Center*

43. **Aeroacoustics: Jet Noise, Combustion and Core Engine Noise.** 1976

Ira R. Schwartz, Editor
NASA Ames Research Center
Henry T. Nagamatsu, Associate Editor
General Electric Research and Development Center
Warren C. Strahle, Associate Editor
Georgia Institute of Technology

44. **Aeroacoustics: Fan Noise and Control; Duct Acoustics; Rotor Noise.** 1976

Ira R. Schwartz, Editor
NASA Ames Research Center
Henry T. Nagamatsu, Associate Editor
General Electric Research and Development Center
Warren C. Strahle, Associate Editor
Georgia Institute of Technology

45. **Aeroacoustics: STOL Noise; Airframe and Airfoil Noise.** 1976

Ira R. Schwartz, Editor
NASA Ames Research Center
Henry T. Nagamatsu, Associate Editor
General Electric Research and Development Center
Warren C. Strahle, Associate Editor
Georgia Institute of Technology

46. **Aeroacoustics: Acoustic Wave Propagation; Aircraft Noise Prediction; Aeroacoustic Instrumentation.** 1976

Ira R. Schwartz, Editor
NASA Ames Research Center
Henry T. Nagamatsu, Associate Editor
General Electric Research and Development Center
Warren C. Strahle, Associate Editor
Georgia Institute of Technology

47. **Spacecraft Charging by Magnetospheric Plasmas.** 1976

Alan Rosen
TRW Inc.

48. Scientific Investigations on the
    Skylab Satellite. 1976

Marion I. Kent and
Ernst Stuhlinger
*NASA George C. Marshall Space
Flight Center*
Shi-Tsan Wu
The University of Alabama

49. Radiative Transfer and
    Thermal Control. 1976

Allie M. Smith
ARO Inc.

50. Exploration of the
    Outer Solar System. 1977

Eugene W. Greenstadt
TRW Inc.
Murray Dryer
*National Oceanic and
Atmospheric Administration*
Devrie S. Intriligator
University of Southern California

51. Rarefied Gas Dynamics,
    Parts I and II
    (two volumes). 1977

J. Leith Potter
ARO Inc.

52. Materials Sciences in Space
    with Application to Space
    Processing. 1977

Leo Steg
General Electric Company

53. Experimental Diagnostics
    in Gas Phase Combustion
    Systems. 1977

Ben T. Zinn,
Editor
Georgia Institute of Technology
Craig T. Bowman,
Associate Editor
Stanford University
Daniel L. Hartley,
Associate Editor
Sandia Laboratories
Edward W. Price,
Associate Editor
Georgia Institute of Technology
James G. Skifstad,
Associate Editor
Purdue University

54. Satellite Communications:
    Future Systems. 1977

David Jarett
TRW Inc.

55. Satellite Communications: Advanced Technologies. 1977 — David Jarett, *TRW Inc.*

56. Thermophysics of Spacecraft and Outer Planet Entry Probes. 1977 — Allie M. Smith, *ARO Inc.*

57. Space-Based Manufacturing from Nonterrestrial Materials. 1977 — Gerard K. O'Neill, Editor, *Princeton University*; Brian O'Leary, Assistant Editor, *Princeton University*

58. Turbulent Combustion. 1978 — Lawrence A. Kennedy, *State University of New York at Buffalo*

59. Aerodynamic Heating and Thermal Protection Systems. 1978 — Leroy S. Fletcher, *University of Virginia*

60. Heat Transfer and Thermal Control Systems. 1978 — Leroy S. Fletcher, *University of Virginia*

61. Radiation Energy Conversion in Space. 1978 — Kenneth W. Billman, *NASA Ames Research Center*

62. Alternative Hydrocarbon Fuels: Combustion and Chemical Kinetics. 1978 — Craig T. Bowman, *Stanford University*; Jørgen Birkeland, *Department of Energy*

63. Experimental Diagnostics in Combustion of Solids. 1978 — Thomas L. Boggs, *Naval Weapons Center*; Ben T. Zinn, *Georgia Institute of Technology*

64. Outer Planet Entry Heating and Thermal Protection. 1979 — Raymond Viskanta, *Purdue University*

65. Thermophysics and Thermal Control. 1979 — Raymond Viskanta, *Purdue University*

66. Interior Ballistics
    of Guns. 1979

Herman Krier
*University of Illinois
at Urbana-Champaign*
Martin Summerfield
New York University

67. Remote Sensing of Earth
    from Space: Role of
    "Smart Sensors." 1979

Roger A. Breckenridge
NASA Langley Research Center

68. Injection and Mixing in
    Turbulent Flow. 1980

Joseph A. Schetz
*Virginia Polytechnic
Institute and State University*

69. Entry Heating and
    Thermal Protection. 1980

Walter B. Olstad
NASA Headquarters

70. Heat Transfer, Thermal
    Control, and Heat Pipes.
    1980

Walter B. Olstad
NASA Headquarters

71. Space Systems and
    Their Interactions
    with Earth's Space
    Environment. 1980

Henry B. Garrett and
Charles P. Pike
Hanscom Air Force Base

72. Viscous Flow Drag
    Reduction. 1980

Gary R. Hough
*Vought Advanced
Technology Center*

73. Combustion Experiments
    in a Zero-Gravity Laboratory.
    1981

Thomas H. Cochran
NASA Lewis Research Center

74. Rarefied Gas Dynamics,
    Parts I and II
    (two volumes). 1981

Sam S. Fisher
*University of Virginia
at Charlottesville*

75.	**Gasdynamics of Detonations and Explosions.** 1981	J. R. Bowen *University of Wisconsin at Madison* N. Manson *Université de Poitiers* A. K. Oppenheim *University of California at Berkeley* R. I. Soloukhin *Institute of Heat and Mass Transfer, BSSR Academy of Sciences*
76.	**Combustion in Reactive Systems.** 1981	J. R. Bowen *University of Wisconsin at Madison* N. Manson *Université de Poitiers* A. K. Oppenheim *University of California at Berkeley* R. I. Soloukhin *Institute of Heat and Mass Transfer, BSSR Academy of Sciences*
77.	**Aerothermodynamics and Planetary Entry.** 1981	A. L. Crosbie *University of Missouri-Rolla*
78.	**Heat Transfer and Thermal Control.** 1981	A. L. Crosbie *University of Missouri-Rolla*

(Other volumes are planned.)

Chapter I. Radiation Heat Transfer

RADIATIVE PROPERTIES OF A PAINTED LAYER CONTAINING NONSPHERICAL PIGMENT

Hamdy M. Shafey[*] and Takeshi Kunitomo[+]

Kyoto University, Kyoto, Japan

Abstract

The radiative properties of a painted layer containing nonspherical pigment particles are studied theoretically. The scattering properties are calculated using a new method based on the volume integral form of Maxwell's equations. The radiative transfer is treated by Chandrasekhar's theory. The effects of the optical properties of the pigment and the optical thickness are examined. It is found that the assumption for nonspherical pigments having the scattering properties calculated by the Mie theory for equivalent spherical particles leads to a considerable error in predicting the reflectances of an optically thin layer, and a small error in the case of a thick layer.

Nomenclature

c_n = angular distribution coefficient
$\vec{E}, \vec{E}_i, \vec{E}_s$ = total, incident and scattered electric vectors, respectively
$G_v(\vec{r},\vec{r}')$ = Green's function
$g(\vec{r},\vec{r}')$ = tensor Green's function
$\vec{H}, \vec{H}_i, \vec{H}_s$ = total, incident and scattered magnetic vectors, respectively
I = intensity of monochromatic radiation
$\vec{J}_{eq}$ = equivalent current density vector
K_e = extinction coefficient of the painted layer, μm^{-1}

Presented as Paper 80-1521 at the AIAA 15th Thermophysics Conference, Snowmass, Colo., July 14-16, 1980. Copyright© American Institute of Aeronautics and Astronautics, Inc., 1980. All rights reserved.
[*]Graduate Student, Department of Engineering Science (Assistant Lecturer, Assiut University, Assiut, Egypt).
[+]Professor, Department of Engineering Science.

L	=	thickness of the painted layer, μm^{-1}
n_v	=	real refractive index of the surrounding medium
$\hat{O}$	=	unit vector in the scattering direction
P_k	=	Legendre polynomial of order k
$p(\theta)$	=	scattering phase function
R	=	distance, μm
R_D	=	directional-hemispherical diffuse reflectance
R_h	=	directional-hemispherical reflectance
r_{OD}	=	bidirectional relative diffuse reflectance
v	=	pigment volume concentration
X_s, X_e	=	scattering and extinction efficiency factors, respectively
$\hat{\varepsilon}$	=	complex dielectric constant of the particle material
ε_v	=	real dielectric constant of the surrounding medium
θ	=	polar angle
μ	=	$\cos\theta$
λ	=	wavelength in the vacuum, μm
μ_0	=	magnetic permeability of the vacuum
$\vec{\Pi}_s$	=	electric Hertz vector for scattering
σ_d, σ_s	=	differential and total scattering cross sections
σ_e	=	extinction cross section
σ_g	=	geometrical cross section
ρ_1, ρ_2	=	specular reflectivities at the upper (air-layer) and the lower (layer-specular substrate) boundaries, respectively, for an incidence direction of μ_{inc} inside the painted layer
τ	=	normal optical depth
τ_1	=	optical thickness of the layer
ϕ	=	azimuthal angle
ω	=	angular frequency of the light
$\tilde{\omega}$	=	scattering albedo of the painted layer

Introduction

Paint coatings are widely used on the basis of their low cost and ease of application in many engineering applications, for instance, thermal equipment, automotive industry and architectural structutes. When these paints are well and specially formulated they can be used in developing new coatings with spectrally selective radiative properties (e.g. coated surfaces for solar energy collectors and spacecraft thermal control coatings). To achieve these characteristics one must have accurate knowledge of the role of pigment (shape, size and optical properties) in predicting the radiative properties of such paints.

Previous studies[1-3] were carried out on the radiative properties of a painted layer containing spherical pigment.

The pigments selected for these studies were assumed to have spherically shaped particles whose scattering and absorption properties can be calculated by Mie equations. But in many cases the pigment particles are not spherical; and accurate treatment, other than Mie theory, is needed to account for the nonspherical particle. The scattering properties of a nonspherical particle had been treated by Rayleigh scattering[4] for particles with very small sizes compared with the wavelength, or by Rayleigh-Debye scattering (Rayleigh-Gans approximation)[5] for particles with relative refractive indices close to unity. Latimar[6] combined the Rayleigh-Debye scattering and the anomalous diffraction approximations with the exact Lorenz-Mie relations for spheres in a trial to obtain approximate scattering properties of ellipsoids in terms of those for suitable equivalent spheres. Chylek et al.[7] developed an approximate method for calculating the scattering properties of randomly oriented irregular particles by suppressing resonances related to surface waves in Mie theory. The exact solution for scattering of an electromagnetic wave by a spheroidal particle was presented by Asano and Yamamoto[8] and the scattering properties based on this solution were studied by Asano.[9] Barber and Yeh[10] also solved the problem of scattering of electromagnetic waves by arbitrarily shaped dielectric bodies using the exact method of the extended boundary conditions. However, these exact solutions are complicated with much calculation even for one orientation of the particle. Beside the above-mentioned approximations and exact methods, Richmond,[11] Senior and Weil[12] and Shiffer and Thielheim[13] solved the scattering problem in a more general and rigorous treatment using the integral equations for the electromagnetic field. But their results were either analytical with approximate form for the scattering properties[13] or numerical and limited to the special shapes of a nonspherical particle.[11,12] In this study, we develop the method for calculating the scattering properties by an arbitrarily shaped particle of any size, based on the numerical solution of the volume integral equations for the electromagnetic fields inside and outside the particle. The particle is assumed to have any orientation with respect to the incidence and polarization directions of the incident electromagnetic wave. The method includes also the averaging of the scattering properties over all possible orientations for randomly oriented particles of monodispersion. The phase function for scattering of unpolarized light by a randomly oriented particle is calculated and expressed in a series of Legendre polynomials. The scattering properties of a spherical particle calculated by the present method are compared with those calculated by Mie equations. Then, the present method is applied to dielectric and metallic pigments of nonspherical particles to obtain the scattering properties, which are introduced into the same procedure[1-3] for predicting the reflection properties of a painted layer containing such pigments. The selected pigments for the present study are the dehydrated yellow ferric oxide Fe_2O_3 of red color and of acicular particles, the zinc white ZnO of rod-like particles, and the metallic Aluminum pigment of flake-like particles.

The effects of the optical thickness, the optical properties of the pigment and the pigment volume concentration on the bidirectional, directional and hemipherical reflectances of a painted layer obliquely irradiated, are examined. The wavelength dependence of the results is examined in the range from 0.45 to 10 μm.

Scattering Properties for Specified Orientation of the Particle

Particle Divisions and Coordinate Systems

For the purpose of numerical integration over the particle volume, the particle is divided into small N main divisions and M subdivisions for each. The division element is chosen to be suitable for the regular shape of the considered particle. For example, particles of parallelepiped bodies may have their division elements as shown in Fig. 1a, while the cylindrical element shown in Fig. 1b may be adopted for the particles having bodies of revolution as spheroids, circular cylinders and discs. The rectangular coordinate system for a general orientation of a nonspherical particle (three dimensionnal plate is considered for clarity) is shown in Fig. 1c. Ox, Oy and Oz are fixed axes in the space, while OX, OY and OZ are fixed axes in the particle. The orientation of the particle can be described by a polar angle α and an azimuthal angle β for the OZ axis and an angle γ in the X'Y' plane which is parallel to the XY plane. The coordinate systems (X_n, Y_n, Z_n) and (X_{nm}, Y_{nm}, Z_{nm}) for the centers of main divisions and subdivisions respectively, are transformed to the systems (x_n, y_n, z_n) and (x_{nm}, y_{nm}, z_{nm}) by the vector relation.

$$\begin{pmatrix} x \\ y \\ z \end{pmatrix} = [T] \cdot \begin{pmatrix} X \\ Y \\ Z \end{pmatrix} \tag{1}$$

where [T] is the transformation matrix whose elements are trignometric functions of the angles α, β, and γ.

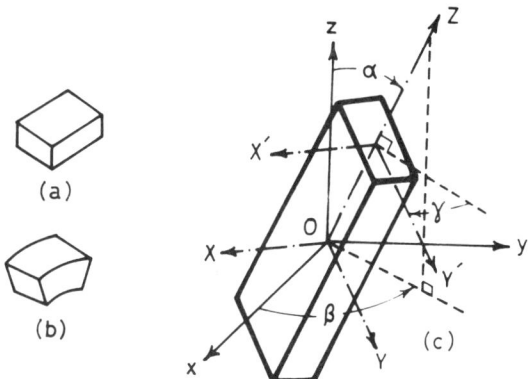

Fig. 1 (a) and (b) Division elements of the nonspherical particle; (c) Coordinate systems and orientation angles of the particle.

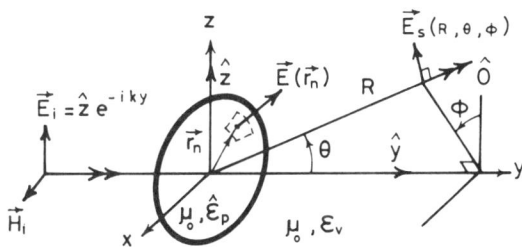

Fig. 2 Configuration of the incidence and scattering of the linearly polarized electromagnetic wave.

General Formulation

The configuration for the incidence and scattering of a linearly polarized electromagnetic wave is shown in Fig. 2. The wave is considered to be incident on the particle in the direction of the y axis, with the incident electric vector of a unit amplitude polarized in the direction of the z axis. For other directions of polarization the z axis (and correspondingly x axis) may be rotated to coincide with the specified polarization direction. The time harmonic function of the incident wave (omitted for simplicity in the following equations) is considered to be $e^{i\omega t}$ type, where ω is the angular frequency. The particle is of a homogeneous and isotropic material with a complex dielectric constant $\hat{\varepsilon}_p$. In the practical application of a painted layer, the pigment particles are dispersed in a vehicle which absorbs electromagnetic radiation. But the imaginary part of the dielectric constant of a vehicle is usually very small. The medium surrounding the particle can thus be considered nonconducting with a real dielectric constant ε_v for scattering calculations. Both the particle material and the surrounding medium are considered nonmagnetic and have the magnetic permeability μ_0 of the vacuum. The propagation constant k is given by $k = 2\pi n_v/\lambda$, where $n_v = \varepsilon_v/\varepsilon_0$ is the real refractive index of the surrounding medium, ε_0 is the electrical permittivity of vacuum and λ is the wavelength in the vacuum. In the following equations it is understood that all time-dependent quantities are of the time harmonic function $e^{i\omega t}$. For the electromagnetic field inside or outside the particle we may write Maxwell's equations as

$$\vec{\nabla} \times \vec{E}(\vec{r}) = -i\omega\mu_0\vec{H}(\vec{r}) \qquad (2)$$

$$\vec{\nabla} \times \vec{H}(\vec{r}) = i\omega\varepsilon_v\vec{E}(\vec{r}) + \vec{J}_{eq}(\vec{r}) \qquad (3)$$

where $\vec{E}(\vec{r})$ and $\vec{H}(\vec{r})$ are the electric and the magnetic field vectors, respectively. $\vec{J}_{eq}(\vec{r})$ is the equivalent current density vector and is given by

$$\vec{J}_{eq}(\vec{r}) = \begin{cases} i\omega\varepsilon_v[\hat{\varepsilon}_p/\varepsilon_v - 1]\vec{E}(\vec{r}) & \text{inside the particle} \\ 0 & \text{outside the particle} \end{cases} \qquad (4)$$

In the absence of the particle the total field at all points of the medium is the incident field. The existance of the particle gives rise to the scattered field and thus the total field $[\vec{E}(\vec{r}), \vec{H}(\vec{r})]$ is expressed by

$$\vec{E}(\vec{r}) = \vec{E}_i(\vec{r}) + \vec{E}_s(\vec{r}) \qquad (5a)$$

and

$$\vec{H}(\vec{r}) = \vec{H}_i(\vec{r}) + \vec{H}_s(\vec{r}) \qquad (5b)$$

where $[\vec{E}_i(\vec{r}), \vec{H}_i(\vec{r})]$ is the incident field, and $[\vec{E}_s(\vec{r}), \vec{H}_s(\vec{r})]$ is the scattered field generated by $\vec{J}_{eq}(\vec{r})$.

Total Electric Field Inside the Particle

To determine the total electric field according to Eq. (5a) one must know the scattered electric field $\vec{E}_s(\vec{r})$ inside the particle. It can be expressed in terms of $\vec{J}_{eq}(\vec{r})$ (Ref. 14)

as
$$\vec{E}_s(\vec{r}) = PV\int_V \vec{J}_{eq}(\vec{r}')\cdot g(\vec{r},\vec{r}')dV' - \vec{J}_{eq}(\vec{r})/(3i\omega\epsilon_v) \qquad (6)$$

where $g(\vec{r},\vec{r}')$ is the tensor Green's function and is defined by

and
$$g(\vec{r},\vec{r}') = -i\omega\mu_0[U + \vec{\nabla}\vec{\nabla}/k^2]\psi(\vec{r},\vec{r}')$$
$$\psi(\vec{r},\vec{r}') = \exp(-ik|\vec{r}-\vec{r}'|/[4\pi|\vec{r}-\vec{r}'|] \qquad (7)$$

where U is the identity tensor. The symbol PV in Eq. (6) means the Principle Value of the integral carried out over the particle volume V.

Substituting Eqs. (4) and (6) in Eq. (5a) we obtain the following integral vector equation for the total electric field vector $\vec{E}(\vec{r})$ inside the particle,

$$\tfrac{1}{3}[n_r^2+2]\vec{E}(\vec{r}) - i\omega\epsilon_v[n_r^2-1]PV\int_V \vec{E}(\vec{r}')\cdot g(\vec{r},\vec{r}')dV' \qquad (8)$$
$$= \vec{E}_i(\vec{r})$$

where
$$n_r^2 = \hat{\epsilon}_p/\hat{\epsilon}_v$$

In Eq. (8) the integration of the inner product $\vec{E}(\vec{r},\vec{r}')$ $\times g(\vec{r},\vec{r}')$ over the particle volume can be carried out numerically by replacing the continuous particle by its N main divisions. Thus, Eq. (8) can be written in the matrix form

$$[G][E] = -[E_i] \qquad (9)$$

where [G] is $3N \times 3N$ matrix whose elements are defined in terms of the complex relative refractive index n_r and the coordinates for the centers of divisions and their volumes. $[E_i]$ and $[E]$ are $3N \times 1$ dimensions representing the x, y, and z components of the incident and the total fields in the main divisions. The dimension [E] can be determined by applying the Gaussian elimination procedure to Eq. (9), where the matrix [G] and the dimension $[E_i]$ are known.

Scattered Field Outside the Particle

The scattered field outside the particle can be expressed by

$$\vec{E}_s(\vec{r}) = \vec{\nabla} \times \vec{\nabla} \times \vec{\Pi}_s(\vec{r}) \qquad (10)$$

and
$$\vec{H}_s(\vec{r}) = i\omega\varepsilon_v \vec{\nabla} \times \vec{\Pi}_s(\vec{r}) \tag{11}$$

where $\vec{\Pi}_s(\vec{r})$ is the electric Hertz vector for scattering and is given by

$$\vec{\Pi}_s(\vec{r}) = (1/i\omega\varepsilon_v)\int_V G_v(\vec{r},\vec{r}')\vec{J}_{eq}(\vec{r}')dV' \tag{12}$$

where $G_v(\vec{r},\vec{r}')$ is the Green function and is given in the far field of the particle by[15]

$$G_v(\vec{r},\vec{r}') = \exp(-ikR+ik\vec{r}'\cdot\hat{O})/4\pi R \; ; \; \vec{r} = R\hat{O} \tag{13}$$

where R is the distance from the particle to the observation point, and $\hat{O}$ is a unit vector in the scattering direction (θ, ϕ). Substituting Eqs. (4, 12, and 13) in Eq. (10), we may express the far scattered field by

$$\vec{E}_s(R,\theta,\phi) = \frac{k^2}{4\pi R}[n_r^2-1]\exp(-ikR)\int_V\{-\hat{O}\times[\hat{O}\times\vec{E}(\vec{r}')]\}\exp(ik\vec{r}'\cdot\hat{O})dV' \tag{14}$$

Noting that $\{-\hat{O}\times[\hat{O}\times\vec{E}]\} = \vec{E}-\hat{O}(\hat{O}\cdot\vec{E})$, we write Eq.(14) in the numerical form

$$\vec{E}_s(R,\theta,\phi) = \frac{k^2}{4\pi R}[n_r^2-1]\exp(-ikR)\sum_{n=1}^{N}[\vec{E}(\vec{r}_n) - \hat{O}\{\hat{O}\cdot\vec{E}(\vec{r}_n)\}]\exp(ik\vec{r}_n\cdot\hat{O})\Delta V_n \tag{15}$$

Equation (15) can be improved by numerical integarion of the function $\exp(ik\vec{r}'\cdot\hat{O})$ over the volume ΔV_n of the main division which is replaced by its M subdivisions.

Scattering and Extinction Cross Sections

The differential scattering cross section $\sigma_d(\theta,\phi)$ is defined by

$$\sigma_d(\theta,\phi) = |Re^{ikR}\vec{E}_s(R,\theta,\phi)|^2 = |\vec{F}(\theta,\phi)|^2 \tag{16}$$

where the complex vector $\vec{F}(\theta,\phi)$, for the case of unit incident electric vector $|\vec{E}_i|=1$, represents the amplitude, phase and polarization of the scattered spherical wave in the direction (θ,ϕ). The total scattering cross section σ_s is given by integrating σ_d over all scattering directions i.e.,

$$\sigma_s = \int_{4\pi} \sigma_d(\theta,\phi) d\Omega \quad ; \quad d\Omega = \sin\theta d\theta d\phi \qquad (17)$$

The extinction cross section σ_e is obtained from the forward scattering theorem[16] by

$$\sigma_e = (-4\pi/k) \text{Im}\{F_z(0,0)\} \qquad (18)$$

where Im means imaginary part of, and $F_z(0,0)$ is the component of $\vec{F}(0,0)$ in the polarization direction of $\vec{E}_i$.

Scattering Properties for Randomly Oriented Particle

The scattering properties for a randomly oriented particle in a monodispersion can be obtained by averaging the various cross sections over all possible orientations of the particle with the assumption of neglecting the effect of interference between the scattered fields by the particles. For a particle that does not have an axis of symmetry the average differential scattering cross section is given by

$$\tilde{\sigma}_d(\theta,\phi) = \frac{1}{8\pi^2} \int_{-\pi}^{\pi} \int_{-\pi}^{\pi} \int_{-1}^{1} \sigma_d(\theta,\phi) d(\cos\alpha) d\beta d\gamma \qquad (19)$$

where $\sigma_d(\theta,\phi)$ is calculated at a specified orientation defined by angles α, β, and γ. The average extinction cross section $\tilde{\sigma}_e$ is given by similar equation. For symmetrical particles (spheroids, circular rods and discs) the integration in Eq. (19) is done on angles α and β only. The azimuthal average differential scattering cross section $\tilde{\sigma}_d(\theta)$ is given by

$$\tilde{\sigma}_d(\theta) = \frac{1}{2\pi} \int_{-\pi}^{\pi} \tilde{\sigma}_d(\theta,\phi) d\phi \qquad (20)$$

Then, the average scattering cross section is given by

$$\tilde{\sigma}_s = 2\pi \int_{-1}^{1} \tilde{\sigma}_d(\theta) d\mu' \qquad (21)$$

where $\mu' = \cos\theta$.

The phase function $p(\theta)$ for scattering of unpolarized incident wave by a randomly oriented particle is given by

$$p(\theta) = 4\pi \tilde{\sigma}_d(\theta) / \tilde{\sigma}_s \qquad (22)$$

Note that $p(\theta)$ is normalized to unity. We can express $p(\theta)$ in a finite series of Legendre polynomials as

$$p(\theta) = \sum_{k=0}^{m'} c_k P_k(\cos\theta) \qquad (23)$$

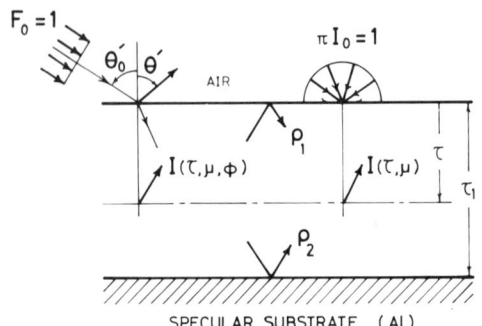

Fig. 3 Radiative model of the painted layer.

where c_k's are angular distribution coefficients, which can be estimated using the orthogonality property of Legendre polynomials. Thus, the coefficient c_n is given by

$$c_n = \left(\frac{2n+1}{1}\right)\int_{-1}^{1} p(\theta) P_n(\cos\theta) d(\cos\theta) \quad (24)$$

The integrals in Eqs.(19) $\sim$ (24) can be calculated numerically by applying the Gaussian quadrature formula.

For a comparison with the Mie theory calculation, it is better to introduce the definition of the average efficiency factor $X_{s,e}$ as

$$X_{s,e} = \tilde{\sigma}_{s,e}/\sigma_g \quad (25)$$

where σ_g is the geometric cross section of the spherical particle having the same volume as the nonspherical particle.

Radiative Transfer in the Painted Layer

Figure 3 shows the model of a plane-parallel painted layer. It represents a monodispersion of absorbing and anisotropically scattering pigment in an absorbing vehicle. The nonspherical particles of the pigment are assumed to be randomly oriented with volume concentration v. The cases of oblique and hemispherical (with uniform irradiation) incidences are shown at once for convenience. A parallel beam of unpolarized radiation of unit net flux ($F_0=1$) is assumed to fall obliquely in a direction defined by a polar angle θ_0' and an azimuthal angle ϕ_0. The interface between the vehicle and air is assumed to be specular. The radiative transfer is treated by Chandrasekhar's theory. So, the transfer equation within a layer on a specular substrate for the case of oblique incidence is expressed by (Ref. 3)

$$\mu \frac{dI(\tau,\mu,\phi)}{d\tau} = I(\tau,\mu,\phi) - \frac{\tilde{\omega}}{4\pi}\int_{4\pi} p(\cos\Theta)I(\tau,\mu',\phi')d\Omega'$$
$$- \frac{\tilde{\omega}}{4\pi\mu_{inc}} \{H \cdot p(\cos\Theta_0) + G \cdot p(\cos\Theta_0')\} \quad (26)$$

where the functions H and G are given by

$$H(\tau) = \frac{(1-\rho_1)\cdot\exp(-\tau/\mu_{inc})}{1-\rho_2\rho_1\cdot\exp(-2\tau_1/\mu_{inc})} \quad (27)$$

$$G(\tau) = \frac{\rho_2(1-\rho_1)\cdot\exp(-[2\tau_1-\tau]/\mu_{inc})}{1-\rho_2\rho_1\cdot\exp(-2\tau_1/\mu_{inc})} \quad (28)$$

In Eq. (26) Θ is the angle between the directions (μ',ϕ') Θ_0 is the angle between the directions (μ,ϕ) and $(-\mu_{inc},\phi_0)$, and Θ_0' is the angle between the directions (μ,ϕ) and (μ_{inc},ϕ_0), where μ_{inc} and $\mu_{inc}' = \cos\theta_0'$ satisfy Snell's law. The boundary conditions linked with Eq. (26) are given by

$$I(0,-\mu,\phi) = \rho_\mu I(0,\mu,\phi) \quad (28a)$$

$$I(\tau_1,\mu,\phi) = \rho_{-\mu}I(\tau_1,-\mu,\phi) \quad (28b)$$

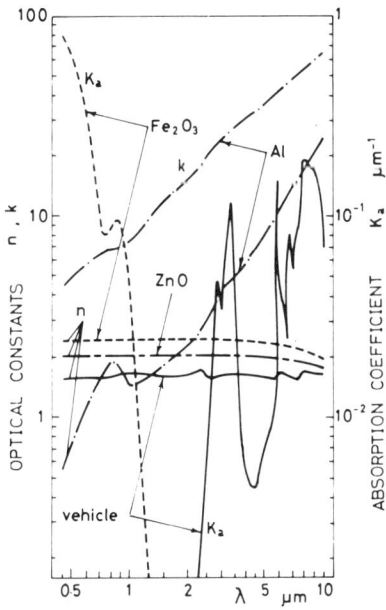

Fig. 4 Optical constants of pigments and vehicle.

The Gaussian quadrature formula is adopted with the Gaussian elimination method to obtain the numerical solution of Eq. (27) together with Eqs. (28a) and (28b). Details of the solution can be found in Ref. 3.

The reflection properties of the layer are obtained in terms of the diffuse radiation intensities $I(0,\mu,\phi)$. The bidirectional relative diffuse reflectance r_{OD} in the reflection direction (μ',ϕ) is defined by

$$r_{OD}(\mu',\phi) = \pi(1-\rho_\mu)I(0,\mu,\phi)/n_v^2 \qquad (29)$$

where μ and μ' satisfy Snell's law.

The directional-hemispherical reflectance R_h and its diffuse component R_D are given by

$$R_h = R_D + R_s \qquad (30)$$

$$R_D = \int_{2\pi}(1-\rho_\mu)I(0,\mu,\phi)\mu d\Omega \qquad (31)$$

The specular component R_s is calculated as in Ref. 3.

Results and Discussion

The dielectric pigments selected for the present study are dehydrated yellow ferric oxide Fe_2O_3 of acicular particles and zinc white ZnO of rod-like particles, as absorbing and nonabsorbing in the visible region, respectively. The Al metal of flake-like particles is selected as an example for the metallic pigment. For Fe_2O_3 particles, an average size of a prolate spheroid is taken to be 0.1 μm (diam) × 2.7 μm length).[17] For ZnO particles an average size of a square rod is taken to be 0.1 μm × 0.1μm × 0.7 μm.[18] For Al pigment particles an average size of a disc is taken to be 1.0 μm (diam) × 0.1μm (thickness).

The optical constants of Fe_2O_3, ZnO, and Al are calculated using the experimental results, the dispersion equations and dispersion parameters.[1,19,20] The refractive indices and the absorption coefficients of the vehicle (alkyd resin) are determined experimentally.[1] The values adopted are shown in Fig. 4.

Comparison with the Mie Theory

The present method for a nonspherical particle is examined for the case of a spherical particle. The results obtained by the present method for a spherical particle of Fe_2O_3 pigment with 0.3 μm diam are compared in Fig. 5 with those obtained by using Mie equations. The angular distributions of the scattered monochromatic intensity $I_s(\theta)$ at different wavelengths are shown. The intensity $I_s(\theta)$ is given by

$$I_s(\theta) = [\tilde{\sigma}_s p(\theta)/4\pi]I_0/R^2 \qquad (32)$$

where the distance R is considered (arbitrarily) 10 μm, and I_0 is the incident monochromatic intensity and assumed unity. The present method shows good agreement with the Mie theory at short wavelengths λ = 0.6 and 1.0 μm, where considerable scattering exists. Noticeable differences are shown at long wavelengths λ = 2.4 and 4 μm, where a small amount of scattering exists. This is due to the exact account for the spherical surface of the particle in the Mie theory, which introduces the exact boundary conditions to the infinite series of the closed form solution of the Maxwell's equations. In the present method, however, the finite divisions of the particle do not exactly have the continuous spherical surface. The surface waves present in scattering by a spherical particle according to the Mie theory yields considerable scattering at large angles, especially at $\theta \sim$ 180 deg.[7] In case of a nonspherical particle, the high symmetry of the spherical shape does not exist. The present method is predicted to give good results and to have better agreement with the exact closed form solution, if such a solution exists.

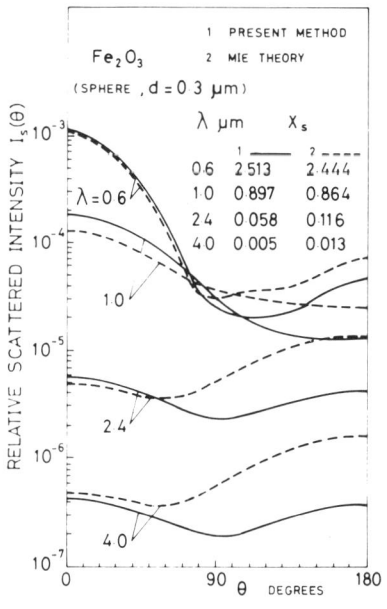

Fig. 5 Comparison with Mie theory.

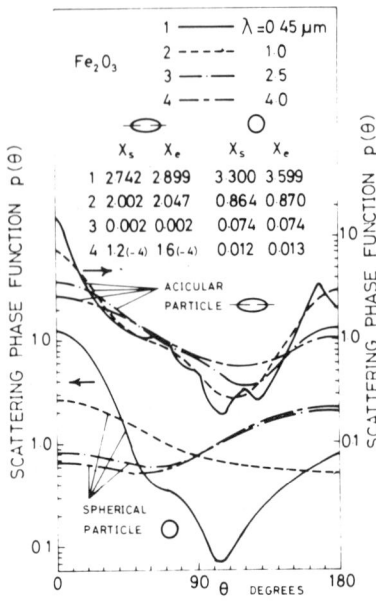

Fig. 6 Wavelength dependence of the phase function for Fe_2O_3 pigment.

In the following discussions the comparison are made between the results for the nonspherical particle and those for the equivalent spherical particle having the same volume.

Scattering Properties

Figures 6 and 7 show the wavelength dependence of the scattering phase function. The results for acicular and equivalent spherical particles of Fe_2O_3 pigment, are shown in Fig. 6. The phase functions for both cases exhibit nearly the same characteristics at a short wavelength $\lambda = 0.45$ μm. But as the wavelength increases the scattering for an equivalent spherical particle changes gradually from dominantly forward to dominantly backward, while the acicular particle shows always the dominant forward scattering for all wavelengths. The values of the scattering efficiency factor X_s indicated in the figure show considerable difference. This is due to the large deviation in the shape and size between the two cases. The results in Fig. 7 show the combined effects of shape, size and optical properties of nonspherical pigment particles. The small size of the rod-like particle of ZnO together with its small relative refractive indices leads to small values of X_s and X_e. Also, the phase function exhibits smooth characteristics and tends to be uniform at long wavelengths. On the other hand the flake-like particle of Al pigment shows large values of X_s and X_e. The corresponding phase function changes

RADIATIVE PROPERTIES OF A PAINTED LAYER

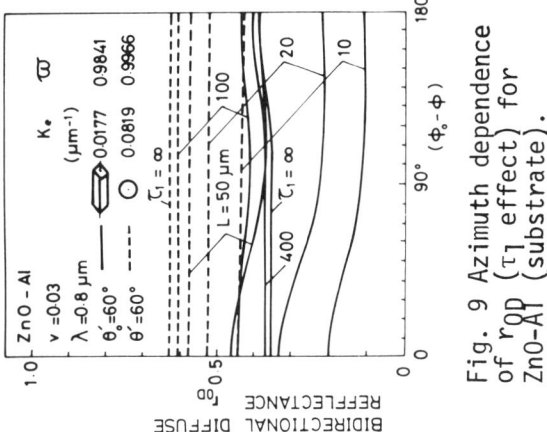

Fig. 9 Azimuth dependence of r_{0D} (τ_1 effect) for ZnO-Al (substrate).

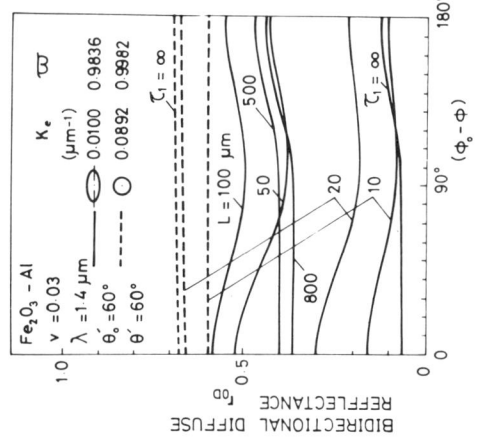

Fig. 8 Azimuth dependence of r_{0D} (τ_1 effect) for Fe_2O_3-Al (substrate).

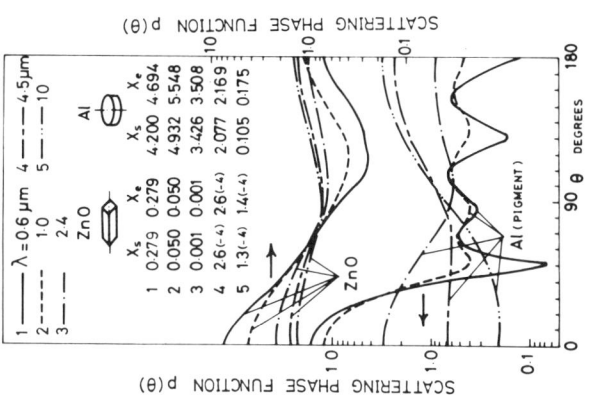

Fig. 7 Wavelength dependence of the phase function for ZnO and Al pigments.

with the wavelength in a manner different from the case of ZnO as shown.

Reflection Properties of the Painted Layer

The scattering properties, as discussed in the previous section, together with the optical constants of the vehicle and Al metal substrate (specular) are introduced into the same programs in Refs. 1-3 for calculating the bidirectional relative diffuse reflectance r_{0D} and, the directional-hemispherical reflectance R_h and its diffuse component R_D. We consider the case of oblique incidence at an azimuthal angle ϕ_0 and a polar angle $\theta_0'=60$ deg. The results for R_D and R_h in this case can be considered nearly the same as for the case of hemispherical incidence with uniform irradiation.[3] Figures 8-10 show the results for the effect of the layer thickenss L, and by turn the optical thickness τ_1, on the azimuth dependence of r_{0D}. The reflection is observed at an azimuthal angle ϕ and polar angle $\theta'=60$ deg. The results for Fe_2O_3 are shown in Fig. 8. In the case of acicular particles, the curves for r_{0D} change gradually from the dominant forward reflection to the dominant backward as the thickness L increases, while they are almost uniform for the case of the equivalent spherical particles. Also, the azimuthal average value of r_{0D} for the case of acicular particles is small compared with that for the case of spherical particles. This is due to the considerable difference in the values of the extinction coefficient K_e of the painted layer as indicated in the figure. The values of the scattering albedo $\tilde{\omega}$ and K_e are much affected by the scattering and extinction of the pigment particles.[1] The results in Fig. 9 for ZnO show a similar tendency. The results in Fig. 10 for

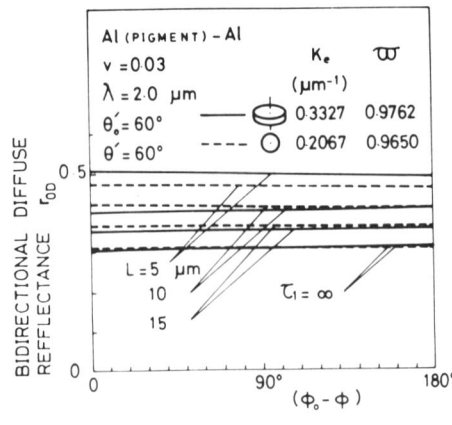

Fig. 10 Azimuth dependence of r_{0D} (τ_1 effect) for Al (pigment)-Al (substrate).

Al pigment show very small differences between the case of flake like particles and that of spherical particles. The curves of r_{0D} for the two cases coincide with each other at $\tau_1 = \infty$.

Figures 11-16 show the wavelength dependence of reflectances R_D and R_h. Shown in Fig. 11 are the results of an optically thin layer containing Fe_2O_3 pigment for both cases of acicular and equivalent spherical particles. Considerable difference is shown in the VIS (visible) and NIR (near infrared) regions, while clearly less or no difference exists in the IR (infrared) region. This is due to the wavelength dependence of scattering efficiency factors X_s as indicated in Fig. 6. The difference between the two cases is also affected by the pigment volume concentration v, especially for the results of R_D. The same tendency for the results of R_h of an optically thick layer ($\tau_1 = \infty$) can be seen in Fig. 12. According to these results, the wavelength dependence for the case of acicular particles in the NIR region differs from that for the spherical particles. Figures 13 and 14 show the corresponding results in the case of Al pigment. The results of an optically thin layer in Fig. 13 show noticeable difference between the two cases of particles for all wavelengths. But, those of an optically thick alyer ($\tau_1 = \infty$) in Fig. 14 show only a little difference. Figures 15 and 16 show the results in the case of the ZnO pigment. The results of an optically thin layer in Fig. 15 show a tendency similar to the case of the Fe_2O_3 pigment (Fig. 11). The results of an optically thick layer ($\tau_1 = \infty$) in Fig. 16 indicate that the R_h for the case of

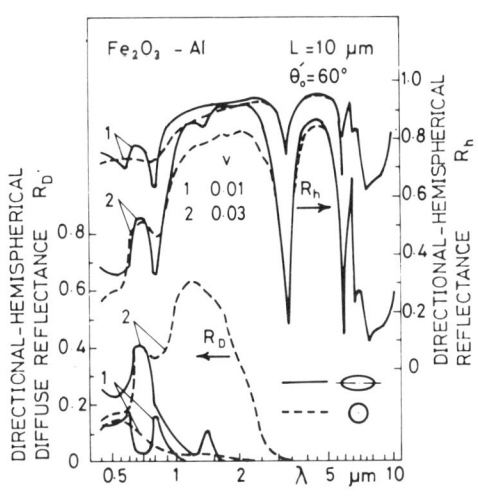

Fig. 11 Wavelength dependence of R_D and R_h for Fe_2O_3-Al system.

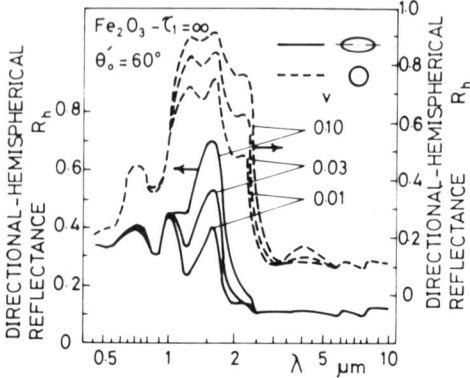

Fig. 12 Wavelength dependence of R_h for Fe_2O_3-($\tau_1=\infty$).

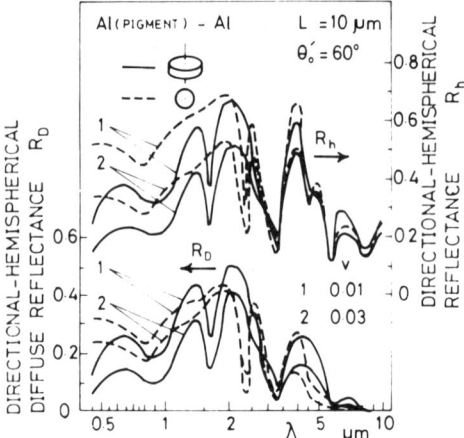

Fig. 13 Wavelength dependence of R_D and R_h for Al (pigment)-Al system.

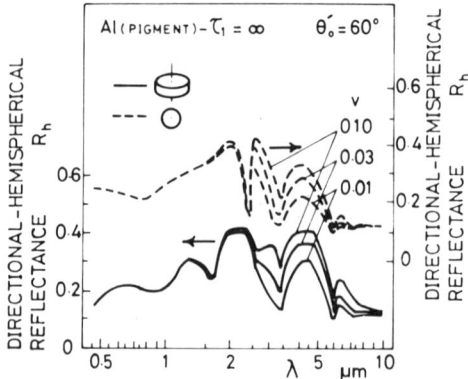

Fig. 14 Wavelength dependence of R_h for Al (pigment)-($\tau_1=\infty$).

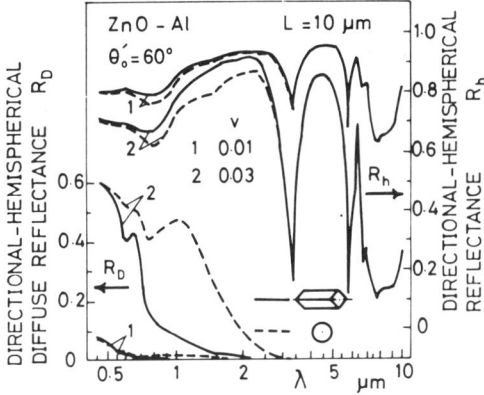

Fig. 15 Wavelength dependence of R_D and R_h for ZnO-Al system.

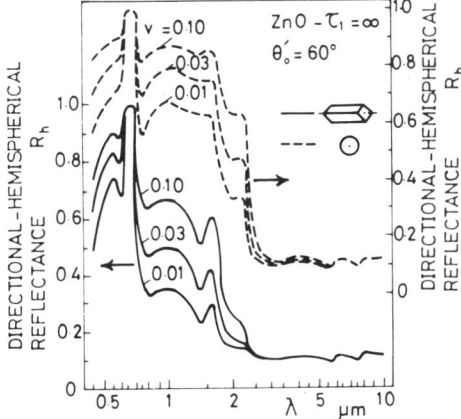

Fig. 16 Wavelength dependence of R_h for ZnO-($\tau_1=\infty$).

rod-like particles decreases more with the increase of the wavelength, compared with those for the case of spherical particles.

Conclusions

The reflection properties of a painted layer containing nonspherical pigment particles in random orientation have been studied theoretically. The scattering properties for a single particle of arbitrary shape was calculated using a developed method in the present sudy, based on the numerical solution for the volume integral form of Maxwell's equations. The summary of the results is shown below.

1) The scattering properties for a randomly oriented nonspherical particle, depending on its shape and size, have mark-

ed differences from those for an equivalent spherical particle of the same volume and material

2) The monochromatic bidirectional relative diffuse reflectance r_{0D} of an optically thick painted layer containing nonspherical particles of metallic pigment may be approximated by that for the case of equivalent spherical particles, provided that the values for the albedo of scattering are small.

3) For an optically thin layer, the assumption that nonspherical pigment particle has the same scattering and absorption properties as for the equivalent spherical particle leads to a considerable error in predicting the monochromatic directional-hemispherical reflectances R_D and R_h, especially in the VIS and NIR regions.

4) In the wavelength region of very small values of scattering efficiency factors, the reflection properties of a painted layer with any thickness containing nonspherical pigment particles can be well approximated by those for the case of equivalent spherical particles.

Acknowledgment

A part of this research was supported by the Grant-in-Aid for Developmental Scientific Research from the Ministry of Education of Japan.

References

[1] Kunitomo, T., Shafey, H. M., and Teramoto, T., "Theoretical Study on Radiative Properties of a Painted Layer Containing Spherical Pigment: Case of Normal Incidence," Bulletin of the Japan Society of Mechanical Engineers, Vol. 22, No. 173, November 1979, pp. 1587-1594.

[2] Shafey, H. M., and Kunitomo, T., "Theoretical Study on Radiative Properties of an Optically Thick Painted Layer Containing Spherical Pigment: Case of Normal Incidence," Bulletin of the Japan Society of Mechanical Engineers, Vol. 23, No. 182, August 1980, pp. 1366-1373.

[3] Shafey, H. M., and Kunitomo, T., "Theoretical Study on Radiative Properties of a Painted Layer Containing Spherical Pigment: Cases of Oblique and Hemispherical Incidences," Bulletin of the Japan Society of Mechanical Engineers, Vol. 23, No. 185, November 1980, pp. 1842-1848.

[4] Van de Hulst, H. C., *Light Scattering by Small Particles*, John Wiley & Sons, New York, 1957, Chap. 7.

[5] Khlebtsov, N. G., and Shchegolev, S. Yu., "Allowance for Nonspherical Particles in Determining the Parameters of Dispersed Systems by the Turbidity-Spectrum Method. 1: Characteristic Scattering Functions for Systems of Nonspherical Particles with Random Orientation in the Rayleigh-Gans Approximation," *Optics and Spectroscopy*, Vol. 42, No. 5, May 1977, pp. 547-552.

[6] Latimer, P., "Light Scattering by Ellipsoids," *Journal of Colloid and Interface Science*, Vol. 53, No. 1, October 1975, pp. 102-109.

[7] Chylek, P., Grams, G. W., and Pinnick, R. G., "Light Scattering by Irregular Randomly Oriented Particles," *Science*, Vol. 193, August 1976, pp. 480-482.

[8] Asano, S., and Yamamoto, G., "Light Scattering by a Spheroidal Particle," *Applied Optics*, Vol. 14, No. 1, January 1975, pp. 29-49.

[9] Asano, S., "Light Scattering Properties of Spheroidal Particle," *Applied Optics*, Vol. 18, No. 5, March 1979, pp. 712-723.

[10] Barber, P., and Yeh, C., "Scattering of Electromagnetic Waves by Arbitrarily Shaped Dielectric Bodies," *Applied Optics*, Vol. 14, No. 12, December 1975, pp. 2864-2872.

[11] Richmond, J. H., "Scattering by a Dielectric Cylinder of Arbitrary Cross Section Shape," *IEEE Transactions on Antennas and Propagation*, Vol. AP-13, PP. 334-341.

[12] Senior, T. B. A., and Weil, H., "Electromagnetic Scattering and Absorption by Thin Walled Dielectric Cylinders With Application to Ice Crystals," *Applied Optics*, Vol. 16, No. 11, November 1977, pp. 2979-2985.

[13] Schiffer, R., and Thielheim, K., O., "Light Scattering by Dielectric Needles and Disks," *Journal of Applied Physics*, Vol. 50, No. 4, April 1979, pp. 2476-2483.

[14] Livesay, D., E., and Chen, M., "Electromagnetic Fields Induced Incide Arbitrarily Shaped Biological Bodies," *IEEE Transactions on Microwave Theory and Techniques*, Vol. MTT-22, No. 12, December 1974, pp. 1273-1280.

[15] Ishimaru, A., Wave Propagation and Scattering in Random Media, Vol. 1, Academic Press, New York, 1978, Chap. 2.

[16] Van de Hulst, H., C., Light Scattering by Small Particles, John Wiley & Sons, New York, 1957, Chap. 4.

[17] Considine, D., M., Chemical and Process Technology Encyclopedia, McGraw-Hill, 1974, pp. 642.

[18] Dunn, E., J., et al., "Agglomeration of Pigment Particles in Dried Paint Films," Journal of Paint Technology, Vol. 40, No. 518, March 1968, pp. 112-122.

[19] Collins, R., J., and Kleinman, D., A., "Infrared Reflectivity of Zinc Oxide," Journal of Physics and Chemistry of Solids, Vol. 11, 1959, pp. 190-194.

[20] Bennett, J. M., and Booty, M., J., "Computational Method for Determining n and k for a Thin Film from the Measured Reflectance, Transmittance, and Film Thickness," Applied Optics, Vol. 5, No. 1, January 1966, pp. 41-43; Beattie, J., R., "Optical Constants of Metals in the Infrared-Experimental Methods,' The Phylosophical Magazine, Vol. 46, No. 373, February 1955, pp. 235-245; Lenham, A., P., "Optical Constants of Single Crystals of Mg, Zn, Cd, Al, Ga, In, and White Sn," Journal of The Optical Society of America, Vol. 56, No. 6, June 1966, pp. 752-756.

BIDIRECTIONAL REFLECTANCE MEASUREMENTS OF SPECULAR AND DIFFUSE SURFACES WITH A SIMPLE SPECTROMETER

A. I. Funai[*]

Lockheed Research Laboratories, Palo Alto, Calif.

Abstract

An experimental apparatus is described for determining the spatial distribution of reflected solar energy from the surfaces of specular and diffusely-reflecting materials and coatings. A simple spectrometer with a photomultiplier attachment is utilized in conjunction with a lockin amplifier/voltmeter and optical chopper. Angles of incidence can be varied between 0 and 85 deg. Solar energy wavelengths are selected using spectral filters. Reflected beam profiles at several angles of incidence are presented and discussed for a specular mirror, three diffuse surfaces, and several other materials and coatings with different scattering characteristics. The results obtained by this comparatively simple method are shown to agree with those reported in the literature by previous investigators.

Nomenclature

D_T	=	diameter of collimator and telescope objective lenses
I_o	=	reflected beam intensity in the direction of the surface normal
$I(\theta)$	=	reflected beam intensity in the direction θ
S	=	distance from axis of rotation to telescope objective

Presented as Paper 80-1529 at the AIAA 15th Thermophysics Conference, Snowmass, Colo., July 14-16, 1980. Copyright © 1980 by A. I. Funai. Published by the American Institute of Aeronautics and Astronautics with permission.
　*Research Scientist, Applied Materials and Technology Laboratory.

V_i	=	detector signal for incident beam
$V(\theta)$	=	detector signal for reflected beam at viewing angle θ
$V(\theta)_{Max}$	=	maximum $V(\theta)$ value for a reflected beam profile
W	=	width of sample surface area irradiated by W_i
W_i	=	width of incident beam
W_r	=	width of reflected beam from a specular surface
W_T	=	width of sample surface area viewed by the telescope
θ	=	angle of reflection or viewing angle
θ_{Max}	=	reflection angle where $V(\theta)_{Max}$ is observed
$\Delta\theta_s$	=	angular width of specularly reflected profile peak
λ	=	wavelength
ψ	=	angle of incidence
ω_r	=	solid angle of diffusely scattered energy collected by the telescope objective

Introduction

As the size of structures being proposed for future space applications gets larger, a better understanding of the spatial distribution of reflected sunlight from their surfaces is required for the thermal design and analysis of the systems. In certain orientations, even low intensity scattered components of reflection from the large surface areas associated with an unfurled antenna, solar sail, or high power solar cell panel can contribute an undesirable level of solar energy to a sensitive payload component. A similar problem occurs in the design of tube and baffle systems for large telescopes and detector arrays. Black diffuse coatings are often selected to minimize stray light effects, but at grazing angles of incidence the reflectance of these coatings may be significantly higher and more specular than they are at near normal angles of incidence.

One common laboratory method for determining the directional reflectance characteristics of surfaces and coatings is with an integrating sphere reflectometer.[1] This method measures the directional hemispherical reflectance of either specular or diffuse surfaces equally well, and for a wide range of incidence angles, but is not suitable for determining the spatial distribution of the reflected energy. To determine these reflected energy distributions an apparatus is required which can measure a wide range of energy levels over a wide range of small, well defined viewing angles for each of a number of prescribed angles of incidence. Such an apparatus is usually referred to as a bidirectional reflectance apparatus, or biangular if it is simplified by restricting the viewing angles to the plane containing the incident beam and surface normal vector.

Several bidirectional measurement facilities have been built by different investigators during the past 15 years to study the directional and wavelength dependency of radiant energy distributions from roughened surfaces. A biangular apparatus described by Torrance and Sparrow.[2,3] has been used to study roughened metal and ceramic surfaces, and the results have been analyzed to develop theories for the location and polarization of the off-specular peaks they observed.[4,5] Another facility has been described by Smith et al.[6,7] for studying roughened surfaces of glass, and aluminum and steel alloys. Each of these specimen surfaces was carefully characterized in terms of its root-mean-square (rms) roughness and slope to establish correlations between these parameters and the reflected energy distributions for a number of wavelengths and angles of incidence.

Another apparatus of somewhat different optical design is described by Look[8] for his study of roughened brass surfaces. The results are analyzed to obtain specular and diffuse shape factors for an emperical expression to fit the reflected power distributions for several angles of incidence at a wavelength of 0.55 μm. Two experimental situations that apply to all bidirectional reflectance measurements, depending on whether the irradiated sample area is larger (overilluminated) or smaller (overdetected) than the area viewed by the detector, are also analyzed. A study of the directional reflected power from roughened sphere surfaces has also been made with this apparatus.[9]

Descriptions of two additional bidirectional apparatus designs are contained in the reports of studies made by Love and Francis[10] and by Miller and VunKannon[11]. In the first of these, results are presented for roughened surfaces of type 302 stainless steel at angles of incidence of 10, 30, and 60 deg and for wavelengths between 0.6 and 10 μm. The second reports measurements which show the bidirectional reflectance characteristics of smoked MgO surfaces to deviate significantly from those for a theoretically diffuse surface as the angle of incidence increases from 0 to 60 deg.

The bidirectional apparatus described in this paper differs significantly from the ones described in the references above. Its principal optical component is a simple research spectrometer of a type commonly found in most optics laboratories and easily adaptable for measurements of the kind described. Experimental data in the form of reflected beam profiles are presented for a variety of materials and coatings that are being considered for space system structures or as thermal control surfaces, as well as for specular and diffuse reference surfaces. The profiles are for angles of incidence between 0 and 75 deg and for a representative solar energy wavelength of 0.60 μm. The term 'reflected beam profile' is used to differentiate the measurements from the more familiar bidirectional measurement formats used in the references. Additional analysis of the method is needed to determine how well it satisifies the requirements for accurate bidirectional measurements, but good agreement with the referenced results is indicated.

Bidirectional Reflectance

The relationship between the direction of an incident beam of radiant energy on a surface and the spatial distribution of the reflected energy from the surface is determined experimentally from bidirectional reflectance measurements. The energy may be incident on the surface from any possible direction and may be scattered into any or all of the direction angles contained in the 2π sr of hemispherical space in front of the surface. With a rectangular coordinate system therefore, four angles are generally required to define the incident and reflected beam directions with respect to the surface plane. If the reflection angles are confined to those that lie in the plane of incidence, the coordinate system is simplified to two dimensions and only one angle is required to define each beam. In this case, the incident beam direction is described by a polar angle ψ, measured from the direction normal to the surface, and each reflection angle is described by another polar angle θ, also measured from the surface normal, as shown in Fig. 1. Measurements of this type are referred to as biangular reflectance measurements[2,3], and can be considered to be a special, or limited class of the more general bidirectional type measurement.

For smooth or isotropically rough surfaces, a set of biangular reflectance measurements at the angles of incidence and wavelengths of interest will usually provide most of the information needed for analysis. For rough or textured surfaces with nonisotropic features, such as machined or embossed surfaces or woven fabrics, the biangular reflectance characteristics obtained for one sample orientation may differe significantly from those obtained for a different sample orientation.[7] These other orientations can easily be examined by rotating the sample around its surface normal.

For perfectly specular surfaces the spatial distribution of reflected energy is completely determined by the geometry of the incident beam and by the law of reflection. The intensity and polarization of the reflected energy for different angles of incidence are determined by the optical constants of the surface and Fresnel's equations. For perfectly diffuse surfaces, the spatial distribution of reflected energy should be symmetrical with respect to the surface normal for any and all angles of incidence, and the intensity at non-normal reflection angles should change in accordance with Lambert's cosine law. The surfaces of most engineering materials however, are neither perfectly specular nor perfectly diffuse; and those that do obey Lambert's law at near normal angles of incidence usually show significant departures from the cos θ distribution at grazing angles of incidence. The location of maximum intensity for energy reflected from roughened surfaces is not always at the specular reflection angle but instead may be located as much as 10 to 15 deg away. [2,8,10] Hence, the need and justification for experimental measurements of bidirectional reflectance.

Apparatus and Method

A schematic sketch of the experimental apparatus used to make the bidirectional reflectance measurements in this paper

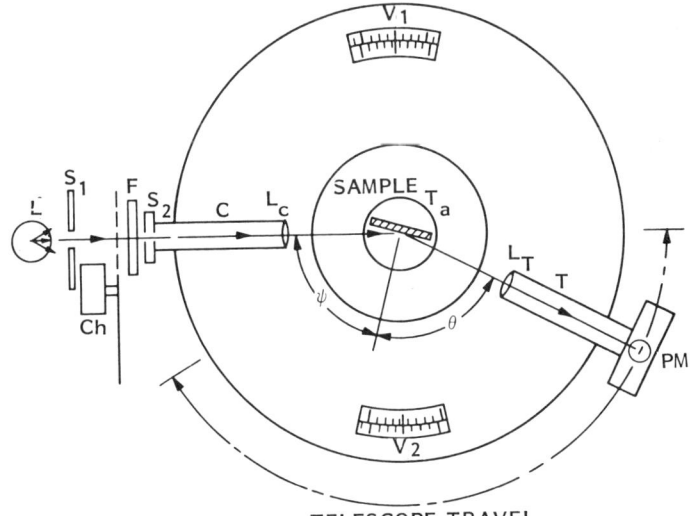

Fig. 1 Schematic of spectrometer setup for bidirectional reflectance measurements.

is shown in Fig. 1. The principal optical component is a simple spectrometer with a photomultiplier attachment in place of the eyepiece normally used for visual observations and alignment. The light source, L, is a standard 15 W frosted light bulb. Light from the source passes through a fixed aperture slit, S_1, a 13 cycles/s chopper, Ch, a spectrally selective filter, F, and uniformly illuminates a bilateral slit, S_2, located at the focal plane of the spectrometers' collimator lens. A collimated beam of light emerges from the collimator lens, L_c, and is incident on the sample surface located at the center of the spectrometer table, T_a. The angle of incidence, ψ, can be set to any angle between 0 and 85 deg by rotating the table whose axis of rotation coincides with that of the spectrometer's telescope and passes through the surface plane of the sample. Light reflected from the sample in the direction angle θ that falls within the solid collecting angle of the telescope lens, L_T, is then focused on the photo-multiplier detector, PM. The detector signal is amplified and displayed by a lockin voltmeter that is synchronized in phase with the 13 cycles/s chopper.

The construction of the spectrometer is such that the telescope cannot be rotated closer than 30 deg to the collimator; consequently the range of viewing angles or reflection angles is limited and depends on the angle of incidence setting. With the sample oriented for $\psi = 0$ deg, the range of θ settings is limited to between 30 and 90 deg. When $\psi = 30$ deg, the range is from 0 to 90 deg, and when $\psi = 75$ deg, the range is from -45 deg to 90 deg. Negative values of θ designate reflection angles on the opposite side of the surface normal from the side the specular reflection angle is directed to. Angular settings of the telescope are readable to 10 minutes of arc on the spectrometer's divided circle and vernier plates (V_1 and V_2). Higher resolution is possible - to 6 seconds of arc - but is not required for these types of measurements. Other features of the spectrometer that are of interest are the following:

1) Diameter of the divided circle for ψ and θ readings: 190 mm.

2) Aperture and focal length of the collimator and telescope objectives: 28.4 mm and 250 mm, respectively.

3) Distance from axis of rotation to collimator and telescope objectives: 140 mm.

For most of the measurements the height and width of slits S_1 and S_2 were fixed at 12 x 6 mm and 10 x 2.6 mm, respectively with S_1 located 40 cm in front of S_2. Slit S_1 serves to control the width of the collimated beam that emerges from the collimator lens and slit S_2 controls the intensity of the beam. With these settings, the height and width of the collimated beam incident on a sample surface at $\psi = 0$ deg is 16 x 6 mm. This irradiated area is more than adequate for the study of smooth or polished surfaces, and is also large enough for good measurements of roughened or coarse textured surfaces such as fabrics or embossed plastic films.

Various measurement wavelengths are obtained by inserting spectral filters into the holder, F, located in front of the collimator slits. The photomultiplier detector is a type 931A with an S-4 response that extends from 0.30 μm to 0.62 μm, consequently measurements are presently restricted to this spectral region. It is anticipated that the range could be extended fairly easily to the cutoff wavelength for the glass optics of the spectrometer (≈ 2.5 μm), by replacing the photomultiplier with a PbS detector attachment. All of the measurements in this paper were made using a red filter (Wratten #25), with a band pass between 0.56 μm and 0.81 μm and a peak transmittance at $\lambda = 0.61$ μm. The peak responsivity for the apparatus is therefore between 0.57 and 0.61 μm, a representative bandwidth for the short wavelength side of the solar energy peak.

In addition to the red spectral filter, a neutral density filter (ND = 2) is placed in the filter holder for measurements of the incident beam profile and reflected beam profiles from specular samples. This filter reduces the intensity of the incident beam by a factor of 1.4×10^{-2} at $\lambda = 0.6$ μm and is used to keep the photomultiplier signal below 1.5 mV to guard against saturation and a nonlinear detector output-vs-light intensity response. For the weaker scattered light intensities from diffuse samples, the ND = 1 filter is used with a transmittance of 0.107 at $\lambda = 0.60$ μm. The gain of the lockin amplifier/voltmeter can also be increased by 2 to 3 orders of magnitude without introducing excessive noise. Consequently, low-level scattered light intensities that are 5 to 6 orders of magnitude lower than that of the incident beam are easily measured. Except for two of the diffuse samples noted later, all of the measurements were made with the photomultiplier operated at 500V. The two exceptions were measured with the voltage increased to 600V. This resulted in higher values of the V_i and $V(\theta)$ signals but did not effect the linearity of the detector response.

Samples are described in the results section of this paper along with the discussion of their reflected beam profiles.

All of the samples were at least 5 cm wide and 4 cm high so that none of the incident beam energy overlapped the sample edges at angle-of-incidence settings less than 83 deg. The thin, flexible plastic film samples were examined by mounting them to a flat support block with double-backed tape.

Incident and Reflected Beam Geometry

From Fig. 1 it is apparent that as the sample table is rotated to increase the angle of incidence, the width of the sample surface area irradiated by the incident beam, W, increases in accordance with $W = W_i \sec \psi$, where W_i is the width of the incident beam.

For the measurements in this paper, W_i was a constant (6 mm), therefore W ranged from this width at ψ = 0 deg to a maximum of 23.2 mm at ψ = 75 deg. The height of the incident beam and irradiated surface area was also constant (16 mm), and independent of the angle of incidence setting. The location and width of the irradiated sample area was checked by removing the filters and observing the light pattern on a flat piece of white paper held at the sample location. Since the diameter of the telescope objective lens D_T, used to view the irradiated sample area is more than $4 \times W_i$, all of the measurements reported in this paper were made under conditions corresponding to the 'overdetect' case described by Look.[8]

The geometry pertinent to the case for a reflected beam off of a specular surface is shown in Fig. 2a. Note that the reflected beam is collimated and has the same width as the incident beam for all angles of incidence, i.e. $W_r = W_i$ = 6 mm. All of the energy in this beam is collected by the telescope objective and focused at the detector plane. The ratio of the signals $V(\theta)/V_i$ is therefore a measure of the absolute specular reflectance of the surface. For a small range of viewing angles near $\theta = \psi$ a portion of the reflected beam is collected by the detector optics, consequently the measured beam profile has a characteristic angular beam width $\Delta\theta_s$, that is determined by W_r and the radius S, of the circle traversed by the telescope objective. For this apparatus $\Delta\theta_s$ = 2 arctan (3/140), or 2.4 deg. This calculated value agrees well with the value indicated by experimental measurements for a specular reference mirror surface described later. For all viewing angles outside this range, the detector signal drops to zero. Analytical expressions corresponding to the incident and reflected beam power functions described qualitatively above are given in the references by Torrance and Sparrow,[2,3] Smith et al.,[6,7] and Look.[8]

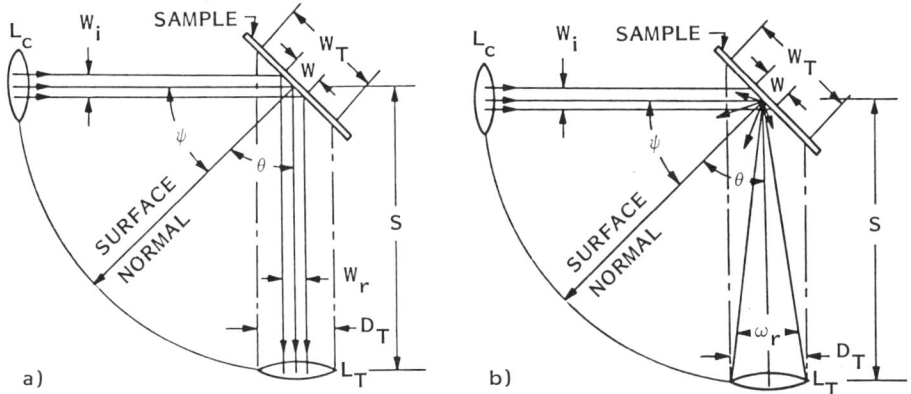

Fig. 2 Incident and reflected beam geometry for: a) specular surface and b) diffuse surface.

In the case of a perfectly diffuse surface, reflected energy is no longer directed to one specific reflection angle in the form of a collimated beam, as in the specular case, but rather is scattered in all possible directions with an intensity distribution that obeys Lambert's cosine law, i.e., $I(\theta) = I_o \cos\theta$. Here, $I(\theta)$ is the intensity of the radiation reflected by the surface in the direction θ, and I_o is the intensity in the direction $\theta = 0°$, and for an ideal lambertian surface, I_o is the same for all angles of incidence. The telescope now collects reflected energy from the irradiated surface area contained in the solid angle ω_r, as shown in Fig. 2b. The value for ω_r in this case is determined by the diameter of the telescope (D_T = 28.4 mm), and by the distance S (140 mm), which gives ω_r = 0.032 sr. Although larger than the solid angles reported for most of the referenced measuring systems, this solid angle appears to be satisfactory for resolving all the principal features of reflected beam profiles for most types of diffuse or scattering surfaces.

Since the energy reflected by diffuse surfaces is no longer collimated, it does not arrive at the detector plane in sharp focus, as in the specular case, but instead is contained in a circular bundle of rays with image points that extend from the focal point of the telescope objective to a distance about 68 mm behind this point. The diameter of this circular bundle of rays at the detector plane is calculated to be about 6 mm, which is considerably smaller than the dimensions of the receiver plate of the photomultiplier. Therefore, all of the reflected energy contained in ω_r is effectively detected, focused or not.

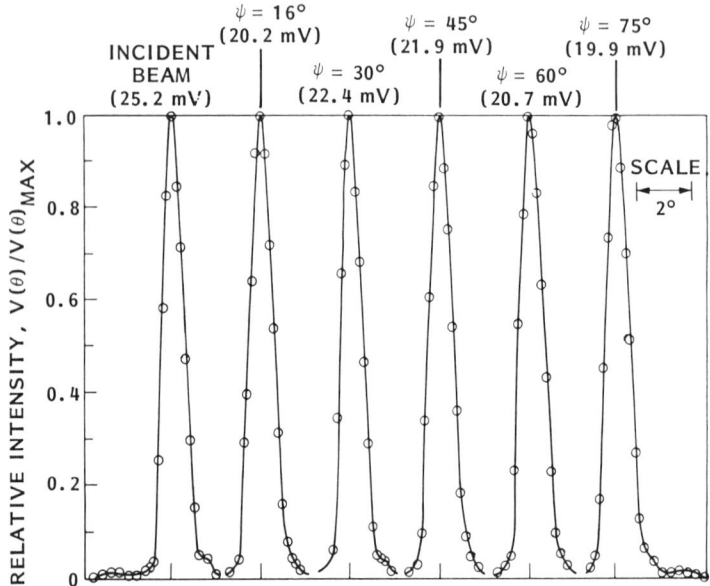

Fig. 3 Incident and reflected beam profiles for a specular mirror at five angles of incidence.

As noted earlier, the telescope overviews the irradiated area of the sample surface at all viewing angles for angles of incidence between 0 and 75 deg. Consequently the directional intensity of reflected energy from a diffuse surface, as measured by $V(\theta)$, should be a maximum at $\theta = 0$ deg and should decrease with $\cos \theta$ as θ increases to either side of the surface normal. This behavior is confirmed by experimental measurements described in the next section for three diffuse surfaces at angles of incidence out to $\psi = 45$ deg. The deviations noted at larger angles of incidence are attributed to real departures from lambertian behavior.

Results and Discussion

Using the spectrometer bidirectional apparatus described in the preceding sections of this paper, a series of reflected beam profiles were obtained for a variety of material and coating surfaces and are shown in Figs. 3-9. For comparative purposes each profile is normalized to its peak intensity value, as indicated by the maximum detector signal $V(\theta)_{max}$ measured for the profile at the referenced angle of incidence setting. The ordinate for each figure therefore, indicates the relative intensity of the profile as a function of the viewing or reflection angle shown along the abscissa. Absolute values of $V(\theta)_{max}$ are tabulated along with an index to the angle of incidence settings for each set of profiles. Each set of profiles can be replotted and compared to one another in terms of absolute signal or intensity levels by multiplying the relative intensity values obtained from the ordinate of the figures by the appropriate $V(\theta)_{max}$ values. Unless noted otherwise, all of the beam profiles were measured with a constant incident beam intensity corresponding to V_i = 25.5 ± 0.5 mv. To compare these beam profile intensities with those that have been reported using an overilluminated measurement system,[2-5] they must be multiplied by the appropriate sec θ factor.

Specular Reference Mirror

Normalized reflected beam profiles for a specular mirror are shown in Fig. 3 to illustrate the profile shape for a perfectly specular surface, as measured with this apparatus. The sample is a first surface aluminized mirror on glass, 7.5 cm wide by 5 cm high with a measured hemispherical reflectance at λ = 0.60 μm of 0.895. The profiles shown are for five angles of incidence between ψ = 16 deg and ψ = 75 deg which are indicated above each peak along with the measured value of $V(\theta)_{max}$ for each peak. The profile at the left of the figure is for the incident beam and is obtained by removing the sample from the spectrometer table and sighting the telescope directly at the collimator. Data points at the base of the profiles are from measurements at 0.33 deg intervals, and the profile peaks are defined by measurements made at 0.17 deg intervals. Within these limits of resolution, all of the peaks were detected at their appropriate reflection angles, i.e., $\theta = \psi$. The scale bar on the figure designates a 2-deg angular spread for each of the profiles.

The shapes of the profiles are seen to be essentially identical to one another and the angular width at their base agrees with the 2.4 deg calculation for $\Delta\theta_s$ made earlier. Two secondary peaks located about 2.2 deg to either side of the main peak and with intensities about 1% of $V(\theta)_{max}$ were observed for each profile. They are shown in Fig. 3 on the left side of the incident beam profile and to the right of the ψ = 75 deg profile. Initially these peaks were attributed to a diffraction effect from the spectrometer slits, but subsequently were found to be caused by reflections off the side walls of the telescope tube when the reflected beam hits either side of the objective lens. They can be eliminated from future measurements either by recoating the interior of the telescope tube wall or by installing a baffle behind the telescope objective.

The profiles in Fig. 3 were measured using the No. 2 neutral density filter mentioned earlier to avoid detector saturation. Background measurements were made by removing the filter and increasing the detector-amplifier gain by a factor of 10^3. These measurements indicated the background level was less than 10^{-6} $V(\theta)_{max}$ at all reflection angles more than 3 degs away from the peaks. This range of detection capability makes the apparatus useful for evaluating the quality of specular surfaces from measurements of their low-level scattered light characteristics.

One other feature of the apparatus indicated by the data in Fig. 3 is its capability for making absolute specular reflectance measurements. The ratio $V(\theta)_{max}/V_i$ for the $\psi = 30°$ profile gives a reflectance value of 0.889 for the mirror, which agrees with the independently measured near normal value of 0.895 at $\lambda = 0.60$ μm obtained with an integrating sphere reflectometer. The low $V(\theta)_{max}$ value for the $\lambda = 16$ deg profile is attributed to the fact that the telescope was within 1.5 deg of its limit of travel at this θ setting and as a result was not collecting all of the reflected beam energy that lies to the left of the peak. The lower $V(\theta)_{max}$ values at larger angles of incidence agree with the directional reflectance changes predicted for an aluminum surface by the Fresnel equations.

Diffuse Surfaces

Normalized reflected beam profiles for three different diffuse surfaces are shown in Fig. 4. The upper set of profiles is for a commercially available reflectance standard of barium sulfate paint on aluminum. The sample evidently was old because an integrating sphere measurement of its near-normal ($\psi = 7$ deg), reflectance at $\lambda = 0.60$ μm was 0.890, which is about 7% lower than its advertised reflectance when freshly prepared. The middle set of profiles is for a 5 cm cube of USP grade $MgCO_3$ with a measured reflectance of 0.915 at $\lambda = 0.60$ μm; and the lower set of profiles is for a 10 x 10 x 2.5 cm block of silica fiber insulation used in the thermal protection system for the space shuttle orbiter. The measured reflectance of this latter sample was 0.895 at $\lambda = 0.60$ μm. The higher $V(\theta)_{max}$ values for the $BaSO_4$ and $MgCO_3$ profiles are due to the photomultiplier being operated in excess of the 500 V setting used for the other sample measurements in this study. Consequently the absolute signal values for these profiles should be reduced by a factor of ≈ 2 for comparison with the insulation block profiles, but the detector linearity and the normalized shape of the profiles was not

BIDIRECTIONAL REFLECTANCE MEASUREMENTS OF SURFACES 37

affected. The measurements for all three samples were made with the neutral density filter removed from the incident beam path since the scattered light intensities from these surfaces are about 4 orders of magnitude lower than the incident beam intensity.

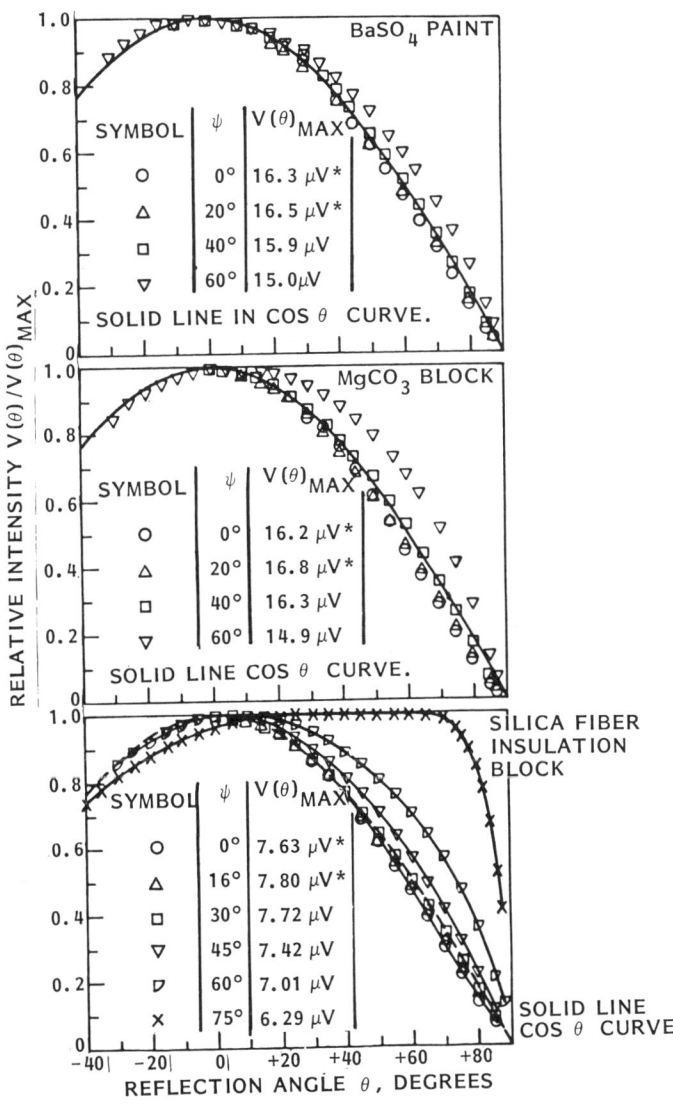

*Estimated values, see text.

Fig. 4 Reflected beam profiles for three diffuse surfaces.

All three sets of profiles are compared to the normalized cos θ intensity distribution for a diffuse surface represented by the solid curve with each set. Since measurements at $\theta = 0$ deg cannot be made for settings of ψ less than 30 deg, the $V(\theta)_{max}$ values listed for the $\psi = 0$ deg and $\psi = 20$ deg profiles are estimates based on extrapolations from the $V(\theta)$ values closest to $\theta = 0$ deg. These values are starred (*) to indicate they are estimates. The profiles for all three surfaces closely follow the cos θ distribution pattern for angles of incidence out to 40 deg. Note that when the cos θ distribution pattern is multiplied by sec θ, the relative intensity values are converted to a constant equal to 1.0 at all reflection angles. This if the format for the biangular reflectance results reported for diffuse surfaces by Torrance and Sparrow,[2-5] therefore the results obtained by either method are the same.

When ψ is increased to 60 deg, the reflected beam profiles for all three samples in Fig. 4 begin to depart from the cos θ distribution pattern. The profiles flatten out in the vicinity of $\theta = 0$ deg which correlates with the lower values for $V(\theta)_{max}$, and bulge towards the forward scattering direction at θ settings greater than 30 deg. A comparison of the $\psi = 60$ deg profiles indicates that the $BaSO_4$ surface retains its diffuse reflecting characteristics better than the $MgCO_3$ and insulation block surfaces, but none of the three is perfectly diffuse when $\psi \geq 60$ deg.

The profile obtained for the insulation block at $\psi = 75$ deg shows an even greater departure from cos θ behavior. The maximum profile intensity is about 20% lower than those measured for near normal angles of incidence but extends over a broad range of reflection angles from $\theta = 10$ deg to $\theta = 70$ deg. The energy backscattered to the negative θ side of the surface normal also appears to be lower. This profile corresponds to one which, if measured with the apparatus described in the references above, would have a pronounced off specular peak located near $\theta = 88$ deg with an intensity about 12 x higher than in the direction $\theta = 0$ deg. The off specular shift in the location of the intensity maxima is not readily apparent from the reflected beam profiles in Fig. 4 until they are multiplied by the sec θ factor for the 'overdetect' measurement case.

Black Paint Surfaces

Normalized reflected beam profiles for three types of black painted surfaces are shown in Fig. 5. The upper set of profiles is for a diffuse flat black polyurethane paint applied

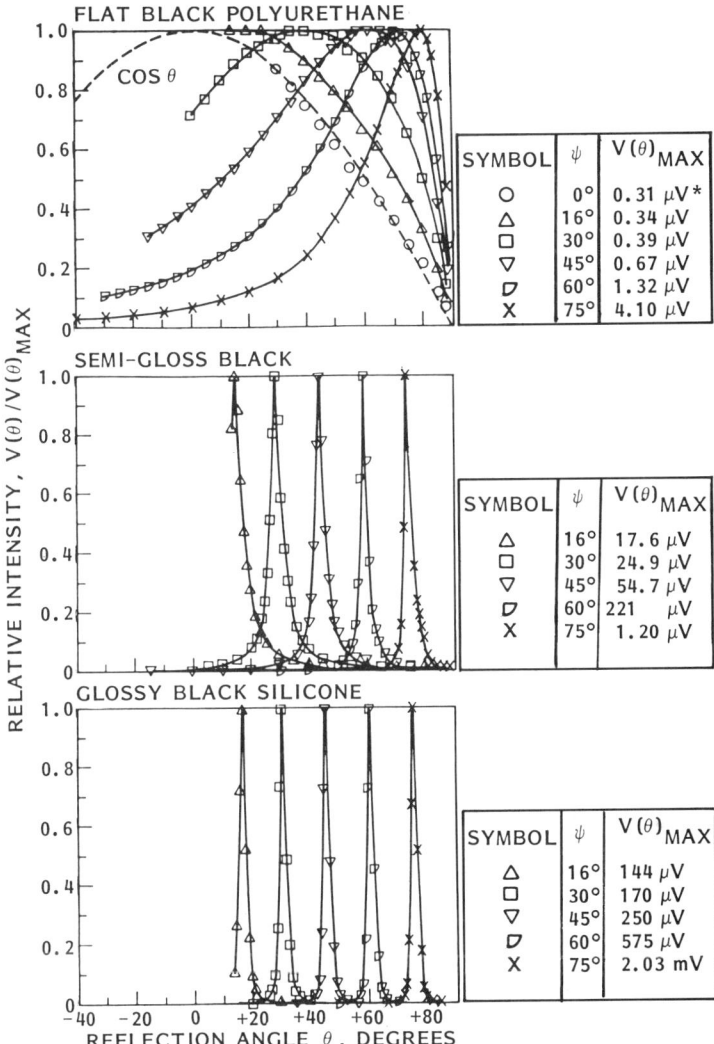

Fig. 5 Reflected beam profiles for three black paint coatings.

to a fiberboard substrate. The middle set of profiles is for a semigloss paint on aluminum, and the lower set is for a glossy black silicone paint, also on aluminum. The thickness of the coatings was not determined but is estimated to be between 0.05 and 0.10 mm for all three samples. Supplemental spectral directional reflectance curves for these coatings, measured with an integrating sphere reflectometer at angles of incidence out to 85 deg, are shown in Fig. 6 At near normal angles of incidence the reflectance of all three coatings is between 4% and 5% for all solar energy wavelengths between 0.28 μm and 2.4 μm. At grazing angles of incidence, the reflectance of the flat black coating increases to between 15% and 20%, while the reflectance of the other two black coatings increases more dramatically to between 55% and 60%. These characteristics correlate nicely with the reflected beam profile characteristics shown in Fig. 5.

Inspection of the reflected beam profiles for the flat black coating indicates that when $\psi = 0$ deg, the distribution of reflected energy corresponds closely to the $\cos \theta$ distribution. (Note that $V(\theta)_{max}$ for this case is estimated in the same manner as described for the diffuse surfaces.) At all the non-normal angle of incidence settings, the shape of the profiles becomes increasingly specular looking as ψ is increased. Significant amounts of scattered light are measured, however, over the full range of reflection angles for each angle of incidence setting. Note also that the location of the profile peaks shifts to the right of the specular reflection angle that corresponds to each ψ-setting, and the shift appears to be maximum (≈ 17 deg) at $\psi = 45$ deg. If the profile intensities are modified by the $\sec \theta$ factor to convert them to the overilluminated measurement format discussed earlier, the off-specular shift appears to be even greater.

Since the reflected energy from this paint sample and from the insulation block sample described earlier appears to be distributed in the same diffuse pattern when $\psi = 0$ deg, the ratio of their peak intensity values should be about the same as the ratio of their absolute reflectance values at $\lambda = 0.60$ μm. The estimated values of $V(\theta)_{max}$ from Figs. 4 and 5 give a ratio of 0.041, and the ratio of their near normal reflectance values is 0.046. This close agreement lends confidence to the photometric consistency of the bidirectional measurements.

Reflected beam profiles for the semigloss and glossy black coatings are seen to be predominantly specular in appearance at all angles of incidence, but with broader peaks than are obtained from the specular mirror. Most of the scattered light from these surfaces appears to be concentrated in a small angle "halo" centered around the specular peak. The profiles tend to narrow as ψ is increased, and the peak intensity values increase by a factor of 1 to 2 orders of magnitude as ψ is increased from 16 deg to 75 deg. Some of this increase can be attributed to the changes in absolute reflectance shown in Fig. 6, but the greater portion appears due to the increas-

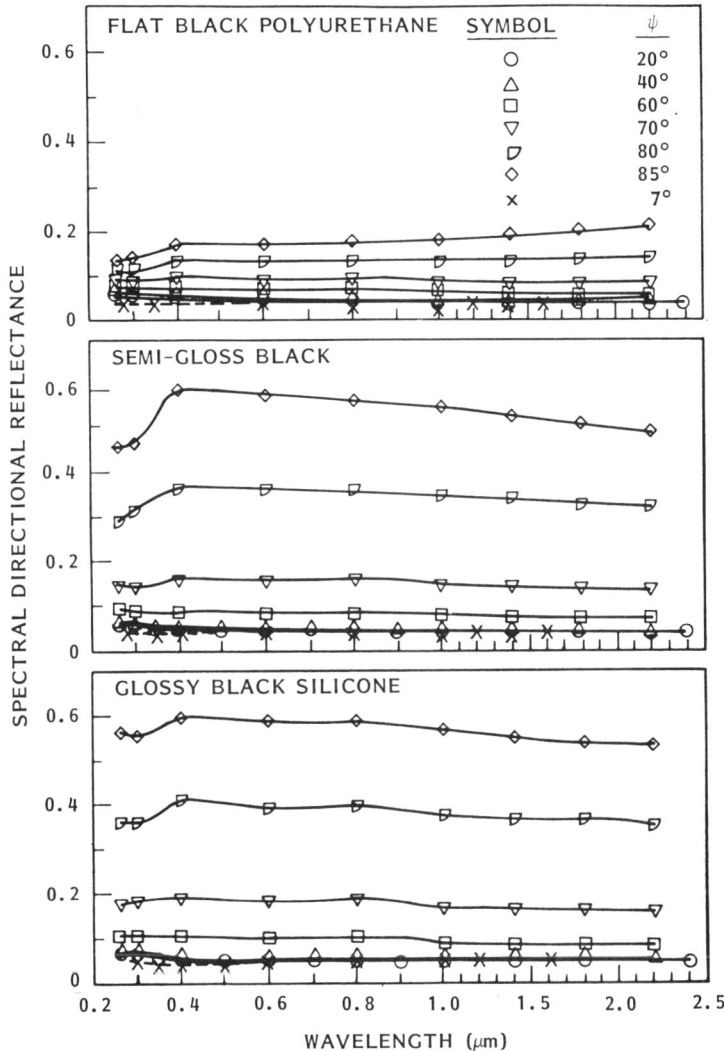

Fig. 6 Spectral directional reflectance curves for the three black paint coatings in Fig. 5.

ing sharpness of the profile peaks as ψ increases. The locations of the profile peaks for these two surfaces are within ± 1 deg of the specular reflection angle for all five angle of incidence settings, and these small deviations can be attributed to the uncertainty in the angle of incidence settings. The influence of the different substrate materials on these bidirectional reflectance characteristics needs to be investigated further, but is believed to be small since the coatings were optically opaque.

White Paint and Plastic Surfaces

Normalized reflected beam profiles for a white paint on aluminum and a sheet of white plastic are shown in Fig. 7 to illustrate the changes in spatial distribution of scattered light from these surfaces for various angles of incidence. The paint is a white, low outgassing, silicone base paint with the trade name S13-GLO. It is used as a thermal control coating for spacecraft surfaces and has a near normal hemispherical reflectance at $\lambda = 0.60$ μm of 0.890. The plastic is a 0.1 mm thick film, identified as cloud white Tedlar, and has a reflectance of 0.885 at near normal angles of incidence which increases to 0.945 at $\psi = 85$ deg.

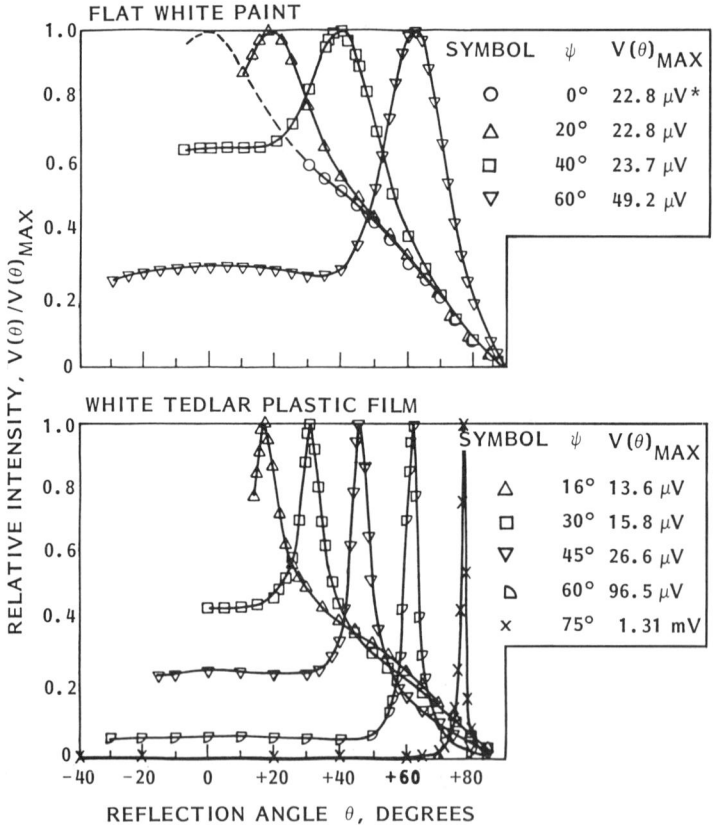

Fig. 7 Reflected beam profiles for two white surfaces.

Profiles for the white paint surface show that at near normal angles of incidence the reflected light is scattered into a nearly diffuse distribution pattern but with a slight specular hump. At high angles of incidence the profiles become increasingly specular in appearance, but with significant levels of scattered light detected at all viewing angles. Scattered light from the plastic film surface is also detected at all viewing angles, but the shape of the specular-type peaks is narrower than those from the white paint surface. At $\psi =$ 75 deg the width of the reflection peak approaches that obtained from the specular mirror surface, and the $V(\theta)_{max}$ value is over 1.3 mV. The ratio of this signal to that for the incident beam (25.5 mV), indicates that 5% to 6% of the incident energy is reflected into this specular peak. Most of the remaining 85% of reflected energy from this surface appears to be directed by small angle scattering into a "halo" around the peak with a half-angle on the order of 10 deg. A low level of scattered light is detected, however, at all viewing angles of the surface. These scattering characteristics are most likely related to the chemical composition, size, and distribution of the pigments used in these materials. The results suggest that bidirectional reflectance could be a useful measurement method for studying the effects of these parameters on the optical properties of materials and coatings.

Embossed Metallized Film

Normalized reflected beam profiles are shown for this surface in Fig. 8 to illustrate yet another type of distribution pattern between the specular and diffuse type patterns described earlier. The sample is a flexible, second-surface metallized film of vacuum deposited silver on a 0.13 mm thick clear plastic film with an adhesive backing. The surface of the film is embossed with a pattern of small dimples on the order of 0.5 mm in diam on 0.75 mm centers in an attempt to eliminate the specularity of the surface but still retain a low solar absorptance or high reflectance. The hemispherical reflectance of the film at $\lambda = 0.60$ μm is 0.94 at near normal angles of incidence and drops to 0.91 at $\psi = 80$ deg.

At $\psi = 20$ deg the reflected energy is scattered into a rather broad specular-type peak that extends 20 to 30 degs on either side of $\theta = \psi$. At $\psi = 60$ deg, the width of the peak appears to broaden which is the reverse of the trend noted for the previously described surfaces. The measured values of $V(\theta)_{max}$ for these two profile peaks were 298 μV at $\psi = 20$ deg and 101 μV at $\psi = 60$ deg, which correlates with the shape of the peaks.

Quartz Cloth Surfaces

As a final example, some rather complex scattered light distribution patterns for four types of white quartz-cloth samples are shown in Fig. 9. The cloth is marketed under the trade name Astroquartz, and the identifying numbers shown in the figure are style numbers that describe the weave pattern and texture of the cloth. Style #503 is a thin, lightweight cloth with a somewhat open weave pattern; style #527 is medium-weight cloth with a close weave pattern; and styles #581 and #593 are coarse textured, heavy cloths that are tightly woven with no open spaces between strands. The materials are not entirely opaque so that measured reflectance values vary depending on the substrate surface behind them. With a black substrate surface, hemispherical reflectance values at $\lambda = 0.60$ μm ranged from 0.45 for style #503 to 0.66 for style #581. With an aluminized Kapton backing surface (first surface Al), the reflectance value for style #503 was 0.83 and for the other three styles was 0.81. All of the profiles shown in Fig. 9 were measured with the aluminized Kapton backing and are compared to the specular-type profile for that surface without any covering. The near normal reflectance of the bare Al/Kapton was 0.905.

The data shown in Fig. 9 indicates that each cloth style reflects with a characteristic scatter pattern which changes with angle of incidence. In all cases significant levels of scattered light are detected at all viewing angles. The pro-

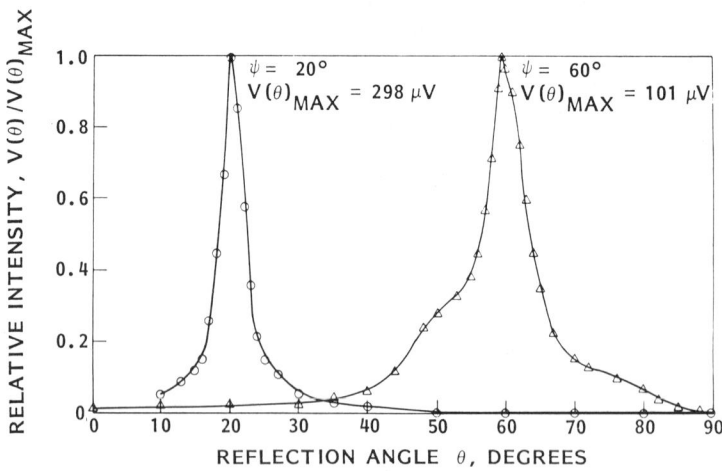

Fig. 8 Reflected beam profiles for an embossed, metallized plastic film.

BIDIRECTIONAL REFLECTANCE MEASUREMENTS OF SURFACES 45

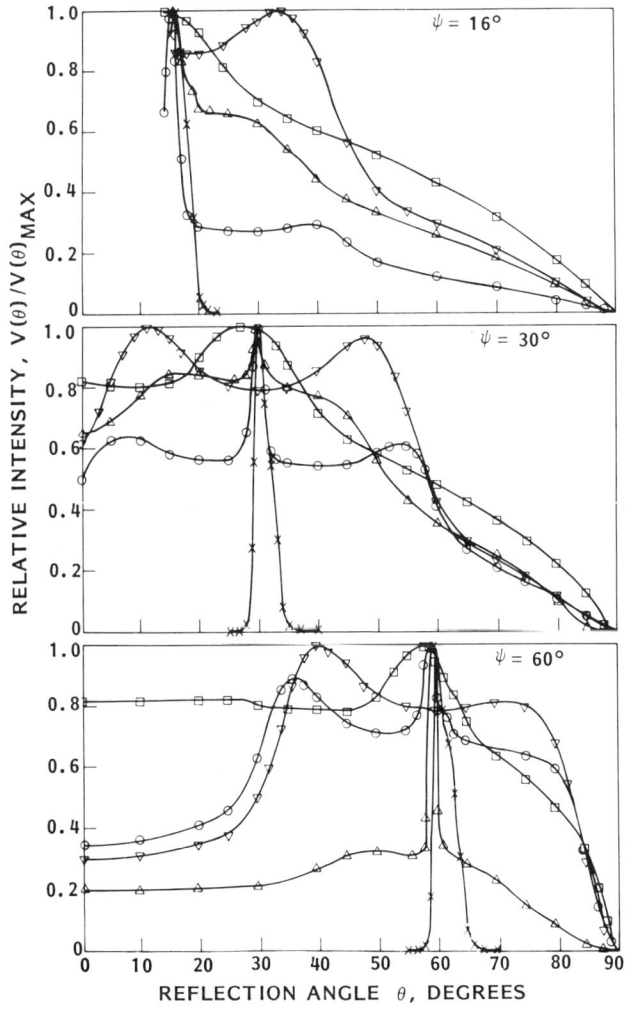

Fig. 9 Reflected beam profiles for four quartz cloth samples over aluminized kapton, at three angles of incidence.

files for cloth styles #503 and #527 show a specular "glint" from the substrate surface that is reflected through the more open weave patterns of these styles at all three angles of incidence. The three profiles for style #581 all have broad peaks centered around $\theta = \psi$ which are characteristic of the cloth material rather than a substrate effect. Profiles for the style #593 sample show two broad diffracted peaks that are located 18 to 20 deg on either side of $\theta = \psi$.

The specular peak profiles obtained for the bare aluminized Kapton surface are seen to be broader than those obtained from the mirror surface and are somewhat asymmetric in shape. This is attributed to the fact that the surface was not perfectly flat but, instead, had a slight waviness which tended to smear out the specular beam over a region of ± 5 deg around $\theta = \psi$.

Summary

A comparatively simple experimental apparatus is described for determining the specular and bidirectional reflectance characteristics of most material surfaces. A spectrometer, commonly available to many laboratories, is utilized to measure reflected beam profiles for surfaces at angles of incidence out to 75 deg. Significant changes in the spatial distribution of reflected energy from most diffuse or scattering surfaces are shown to occur as the angle of incidence increases.

Off-specular peak shifts similar to those reported in the referenced literature on studies of roughened, non-specular surfaces are observed. The reflected beam profiles for diffusely-reflecting surfaces also agree with those reported in the literature when the $\cos \theta$ difference between the overilluminate and overdetect conditions of measurement are accounted for.

The principal limitation of the apparatus is its restricted range of viewing angles at near-normal angle of incidence settings. Improvements and added measurement capabilities which are believed to be easily attainable with minor modifications include:

1) A reduction of the rather wide solid angle of reflected energy collected by the telescope through the use of smaller apertures in front of the telescope objective.

2) An expansion of the spectral range capability out to ≈ 2.5 μm through the use of a different detector attachment

(e.g., PbS), interchangeable with the photomultiplier attachment.

3) An extension of the angle of incidence measurement range out to 85 deg through the use of narrower slit settings, or slit apertures in front of the collimator lens, to reduce the width of the incident beam.

Measurements of polarization effects on the reflected beam profiles could also be obtained by placing a suitable film or screen polarizer in the incident beam.

References

[1] Edwards, D. K., Nelson, K. E., Roddick, R. D., and Gier, J. T., "Integrating Sphere for Imperfectly Diffuse Samples," Journal of the Optical Society of America, Vol. 51, Nov. 1961, pp. 1279-1288.

[2] Torrance, K. E. and Sparrow, E. M., "Off-Specular Peaks in the Directional Distribution of Reflected Thermal Radiation," Journal of Heat Transfer, Vol. 88, May 1966, pp. 223-230.

[3] Torrance, K. E., Sparrow, E. M., and Birkebak, R. C., "Polarization, Directional Distribution, and Off-Specular Peak Phenomena in Light Reflected from Roughened Surfaces," Journal of the Optical Society of America, Vol. 56, No. 7, July 1966, pp. 916-925.

[4] Torrance, K. E. and Sparrow, E. M., "Theory for Off-Specular Reflection from Roughened Surfaces," Journal of the Optical Society of America, Vol. 57, Sept. 1967, pp. 1105-1114.

[5] Torrance, K. E., "Theoretical Polarization of Off-Specular Reflection Peaks," Journal of Heat Transfer, Vol. 91, May 1969, pp. 287-290.

[6] Smith T. F. and Hering, R. G., "Surface Roughness Effects on Bidirectional Reflectance," Thermophysics and Spacecraft Thermal Control, Progress in Astronautics and Aeronautics, Vol. 35, edited by R. G. Hering, AIAA, New York, 1974, pp. 145-165.

[7] Smith, T. F., Suiter, R. L., and Kanayama, K., "Bidirectional Reflectance Measurements for One-Dimensional, Randomly Rough Surfaces," Heat Transfer, Thermal Control, and Heat Pipes, Progress in Astronautics and Aeronautics, Vol. 20, edited by Walter B. Olstad, AIAA, New York, 1980, pp. 189-208.

[8] Look, D. C., "Analysis of Thermal Radiation Bidirectionally Reflected from Roughened Brass," AIAA Journal, Vol. 14, No. 12, Dec. 1976, pp. 1772-1774.

[9] Look, D. C. and Rau, B. W., "Angular Distribution of Radiation Reflected from Roughened Spheres," AIAA Journal, Vol.14, No. 12, Dec. 1976, pp. 1772-1774.

[10] Love, T. J. and Francis, R. E., "Experimental Determination of Reflectance Function for Type 302 Stainless Steel," Thermophysics of Spacecraft and Planetary Bodies, Progress in Astronautics and Aeronautics, Vol. 20, edited by Gerhard Heller, Academic Press, New York, 1967, pp. 115-135.

[11] Miller, E. R. and VunKannon, R. S., "Development and Use of a Bidirectional Spectroreflectometer," Thermophysics of Spacecraft and Planetary Bodies, Progress in Astronautics and Aeronautics, Vol. 20, edited by Gerhard Heller, Academic Press, New York, 1967, pp. 219-233.

RADIATIVE EQUILIBRIUM IN A GENERAL PLANE-PARALLEL ENVIRONMENT

A. Sharma* and A.C. Cogley+

University of Illinois, Chicago, Ill.

Abstract

A new formulation for plane-parallel radiative transfer is used to obtain the first general solution for radiative equilibrium in nongray, anisotropically scattering, inhomogeneous media with arbitrary surfaces. The effects of scattering in the medium and on the surfaces are given by scattering functions (Green's functions) that represent the response of the environment to an unitary-type illumination. These functions are computed by the adding/doubling method and then used in the energy equation. Temperature profiles show the strong effect of inhomogeneous media and scattering for both gray and nongray example problems covering a range of radiative properties. The code is verified by showing that it satisfies certain theoretically predicted results and reproduces the available exact solution for an isotropically scattering medium with black walls. The accuracy of the solutions can ke kept within 1% for computation times of several CPU seconds on an IBM 370/158 computer.

Introduction

Many radiative heat transfer problems in plane-parallel media have been solved in the past using a wide variety of simplifying assumptions. A pedagogical discussion of some of these solutions and their formulation is given by Sparrow and Cess[1]. Most of these solutions are for gray, isotropically scattering media with diffuse surfaces[1-4], although certain special cases with speculary reflecting surfaces[5] and nongray,

---

Presented as Paper 80-1520 at the AIAA 15th Thermophysics Conference, Snowmass, Colo., July 14-16, 1980. Copyright © American Institute of Aeronautics and Astronautics, Inc.,1980. All rights reserved.

*Research Scientist, Department of Energy Engineering.
+Professor, Department of Energy Engineering.

but nonscattering, media with narrow spectral lines[6] have also been solved. These simplifying assumptions are necessary because the usual formulations result in a coupled system of nonlinear, integral equations that are extremely difficult to solve.

The formulation[7] introduced here leads to a conceptually different solution procedure. The linearity of radiative transfer, with respect to radiative sources, is used to handle the scattering phenomena separately by introducing scattering functions that depend only on the radiative properties of the medium and surfaces. This approach removes the coupling of the governing equations and enables one to represent all scattering events by scattering functions that can be computed by a numerically efficient matrix-algebra adding/doubling method[8]. The resulting energy equation which contains these scattering functions is then the only equation solved to obtain the required temperature field and/or heat fluxes. The structure of the formulation leads naturally to general and automated numerical solutions.

As a particular application, this paper presents the first unrestricted solution for radiative equilibrium with anisotropically scattering, inhomogeneous media surrounded by arbitrary reflecting surfaces. The procedure should work equally well for problems including conduction and convection.

Radiative Equilibrium

When radiation is the only (or predominant) mode of energy transfer and the system is in steady state, the energy equation for one-dimensional media is simply

$$\frac{dq_{RT}}{dx} = 0 \tag{1}$$

Here x is the space dimension measured into the medium from the top surface and q_{RT} is the total (frequency integrated) radiative heat flux. Since the radiative transfer equation can be solved generally only for the monochromatic intensity, Eq. (1) is rewritten as

$$\int_0^\infty \frac{\partial q_R(\lambda)}{\partial x} d\lambda = \int_0^\infty K_T(\lambda,x) \frac{\partial q_R}{\partial t}(\lambda,x) d\lambda = 0 \tag{2}$$

where λ is the wavelength of radiation. The optical depth t is based on the volumetric extinction coefficient K_T by

$$t(\lambda,x) = \int_0^x K_T(\lambda,x') \, dx' \qquad (3)$$

where K_T is the sum of the scattering K_s and absorption K_a coefficients. The monochromatic radiative transfer equation driven only by thermal sources can be written as

$$\mu \frac{\partial I(t,s)}{\partial t} = I(t,s) - [1 - \omega(t)] \, \bar{B}[T(t)] - J(t,s) \qquad (4)$$

The direction of radiative propagation is represented by $s(\theta, \phi)$, the set of zeneth θ and azimuthal ϕ angles, with $\mu = \cos \theta$. The specific intensity is I, $\bar{B}$ the Planck function, and ω the single scattering albedo. The scattering source function J is defined by

$$J(t,s) = \frac{\omega(t)}{4\pi} \int_{-1}^{1} d\mu' \int_{0}^{2\pi} d\phi' \, P(t;s,s') \, I(t,s') \qquad (5)$$

where P is the scattering phase function. In this paper the analytic phase function

$$p(\Theta,g) = \frac{1 - g^2}{(1 + g^2 - 2g \cos\Theta)^{3/2}} \qquad (6)$$

is used to simulate areosol scattering. Here Θ is the included angle between incident and scattered rays while the parameter g is the asymmetry factor. The value of g can vary between ±1 with g = 0.0 being isotropic scattering. Positive g values represent particles that scatter preferential in the forward direction. The formulation, however, accepts any phase function information. The present expression (6) is used only for convenience.

The radiative transfer equation is solved formally and the results used in Eq. (5) to form an integral equation which governs J. The usual approach is to solve the governing equation for J which is coupled to both the energy equation and integral equations expressing intensity balances at the reflecting walls. This is not the approach used here. The scattering effects in the present formulation are obtained through scattering functions by an adding/doubling method that is equivalent to solving for the scattering source function. These scattering functions then appear as extra functions in the energy equation. There is also a Fourier expansion in ϕ that is required[8], but thermally driven problems depend only on the zeroth component of such an expansion. Therefore, only the direction cosine μ appears in the following work.

The details of the manipulations described above can be found in Ref. 9 and lead to the following governing equation for radiative equilibrium with gray medium and surfaces:

$$2 \phi(t) = F_{Gs}(t) + \int_0^{t_0} G(t,t') \phi(t') dt \quad (7)$$

For gray problems the Planck function can be integrated as

$$\int_0^\infty \bar{B}(\lambda, T) d\lambda = \frac{\sigma T^4}{\pi}$$

where σ is the Stefan-Boltzman constant. The dimensionless temperature function ϕ is defined by

$$\phi(t) = \frac{T^4(t) - T_T^4}{T_B^4 - T_T^4} \quad (8)$$

where T_T and T_B are surface temperatures at the top and bottom, respectively. The Kernel G is given by

$$G(t,t') = \left[1 - \omega(t')\right] \left\{ E_1(|t-t'|) + f^+(t,t') \right\} \quad (9)$$

and the inhomogeneous function F_{Gs} is

$$F_{Gs}(t) = \frac{1}{1-R} \left[\int_0^1 \varepsilon_B(\mu) \left\{ \exp\left(\frac{t - t_0}{\mu}\right) + \frac{c^-(t,\mu)}{4\pi} \right\} d\mu \right.$$

$$+ R \left(\int_0^1 \varepsilon_T(\mu) \left\{ \exp\left(-\frac{t}{\mu}\right) + \frac{c^+(t,\mu)}{4\pi} \right\} d\mu \right.$$

$$\left. + \int_0^{t_0} G(t,t') dt' - 2 \right) \right] \quad (10)$$

Wall directional emissivities are ε_T and ε_B and $R = T_B^4/T_T^4$.

Equation (7) has the same form as the corresponding governing equation, restricted to isotropic scattering, given in Ref. 1 as Eq. (8-4), and reduces to that equation when scattering is isotropic and surfaces are diffuse. Anisotropic scattering introduces the c scattering functions in F_{GS} and the f^+ scattering function in the Kernel G. Reference 9 gives a complete discussion of the physical meaning of scattering functions $c^{\pm}$ and f^+ and how they are computed. Here we use them as known functions of the radiative properties of the environment. When the radiative properties become temperature dependent, the following solution procedures must include an iterative loop to update these scattering functions.

Equation (7) for gray radiative equilibrium is a Fredholm integral equation of the second kind. Since the $c^{\pm}$ and f^+ functions are numerically calculated at discrete t locations, Eq. (7) must be solved numerically. The only difficulty is that a special quadrature scheme must be used to accurately handle the singular E_1 function present in the integrand. This scheme involves quadrature weights that are functions of the moments of the E_1 function[9]. Converting Eq. (7) to discrete form leads to a set of simultaneous, linear, algebraic equations that are then solved in a standard manner. The details of the special quadrature and numerical method are given in Ref. 9, which also contains a complete discussion of the solutions accuracy.

The results in Fig. 1 show the effects of anisotropic scattering and reflecting surfaces on radiative equilibrium temperature profiles in gray media of total optical thickness t_o = 2.0 with gray surfaces and R = 0.01. Curves 1-5 are for various media with black surfaces. Curve 1 is for an homogeneous, isotropic medium of ω = 0.4 and g = 0.0. This curve reproduces the solution for isotropic scattering media given in Ref. 1 and therefore acts as a strong check for the present code. An isotropic but inhomogeneous medium with ω = 0.4t is solved next and presented as curve 2. Theory says that solutions for isotropically scattering media cannot depend on ω even if inhomogeneously distributed. Curves 1 and 2 are identical to three significant figures and these two cases together prove that the code is valid and computationally accurate since all the ω information is being correctly cancelled out.

The temperature profiles for anisotropically scattering media with ω = 0.4 are given as curves 3(g = 0.4), 4(g = 0.8), and 5(g = 0.4t) in Fig. 1. These curves show that the form of the scattering phase function can significantly perturb

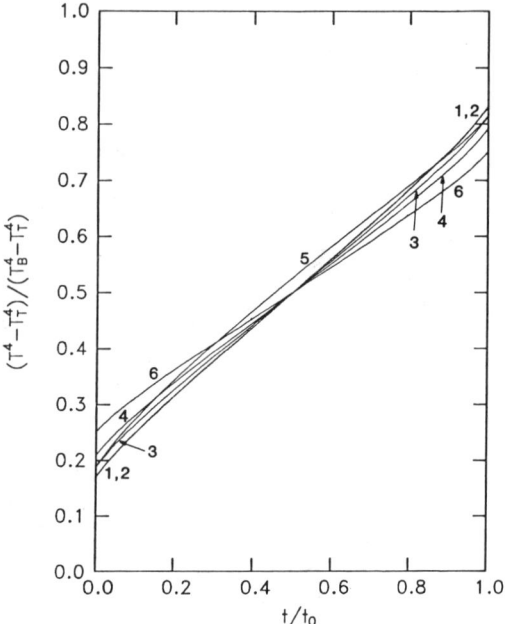

Fig. 1 Radiative equilibrium solutions for grey media with R = 0.01. All curves are for black surfaces except curve 6 which has reflecting surfaces with properties given in Fig. 2. The media properties are 1 (g = 0.0, ω = 0.4), 2 (g = 0.0, ω = 0.4t), 3 (g = 0.4, ω = 0.4), 4 (g = 0.8, ω = 0.4), 5 (g = 0.4t, ω = 0.4), and 6 (g = 0.0, ω = 0.4).

the temperature profile for radiative equilibrium. Forward scatters (g > 0) cause increased temperature differences (radiation slip) between the surfaces and medium next to them (compare the end points of curves 1, 3, and 4). All media with symmetry about the centerline have the same centerline temperature (ϕ_{center} = 0.5). Although this symmetry is not disturbed by variations in ω for isotropic scatters, the case of curve 5 for inhomogeneously distributed anisotropic scatters (g = 0.4t) does not produce symmetric results and ϕ_{center} ≠ 0.5. The final curve, numbered 6 in Fig. 1, is for a homogeneous, isotropic (ω = 0.4, g = 0.0) medium with identical reflecting surfaces having the emissivity and reflectivity shown in Fig. 2 (the surfaces are opaque and these properties are related by Kirchhoff's law). Surface scattering (reflection) is seen to have effects similar to those for anisotropic scattering media and such changes are similar to those that occur when the medium optical thickness t_o is reduced.

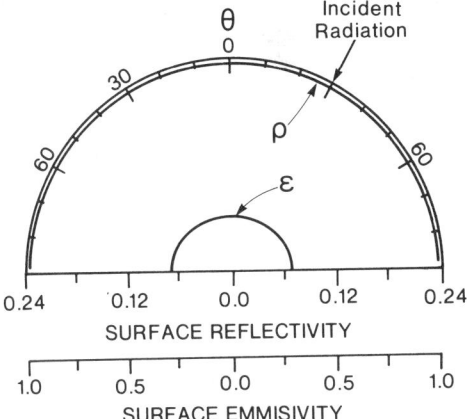

Fig. 2 The bidirectional reflectivity ρ for incidence angle shown and the corresponding directional emissivity ε for the opague surfaces used in Fig. 1, curve 6.

The accuracy of the solutions presented in Fig. 1 are about 0.1% and each case took approximately 2 s of CPU time on an IBM 370/158 computer. This time includes that used to obtain the scattering functions $c^{\pm}$ and f^{+}.

The above gray problem is mostly of academic interest and was carried out as a learning step to solve the general nongray problem. When all properties are wavelength dependent the governing equation for radiative equilibrium becomes

$$\int_0^\infty K_a(\lambda,x) \left\{ -2\overline{B}[\lambda, T(x)] + F_{Ns}(\lambda,x) \right. $$
$$\left. + \int_0^{x_0} G_N(\lambda,x,x') \overline{B}[\lambda, T(x')] \, dx' \right\} d\lambda = 0 \qquad (11)$$

The dependent variable is the temperature that appears in the Planck function $\overline{B}$, and the driving function F_{Ns} is given by

$$F_{Ns}(\lambda,x) = \int_0^1 d\mu \left[I_{t_0}^{0+} \left\{ \exp\left(\frac{t-t_0}{\mu}\right) + \frac{c^-(t,\mu)}{4\pi} \right\} \right.$$
$$\left. + I_0^{0-} \left\{ \exp\left(-\frac{t}{\mu}\right) + \frac{c^+(t,\mu)}{4\pi} \right\} \right] \qquad (12)$$

The optical depth t is mapped to the physical depth x by Eq. (3). The zeroth Fourier component of the boundary intensities $I_{t_0}^{0+}$ and I_0^{0-} for surfaces are obtained from

$$I_{t_0}^{0+}(\lambda,\mu) = \varepsilon_B(\lambda,\mu) \bar{B}(\lambda,T_B)$$

and

$$I_0^{0-}(\lambda,\mu) = \varepsilon_T(\lambda,\mu) \bar{B}(\lambda,T_T) \quad (13)$$

where ε is the surface directional, spectral emissivity. The nongray kernal function is

$$G_N(\lambda,x,x') = K_a(\lambda,x') \left[E_1(|t - t'|) + f^+(t,t') \right] \quad (14)$$

where again the mapping from x to t coordinates must be performed.

Equation (11) expresses conservation of the total radiant energy for nongray media and is an integral over λ of nonlinear Fredholm integral equations in T. No direct analytical or numerical solution of this equation is possible. However, an iterative solution method can be used and one such example of an efficient scheme is developed in Ref. 9. Equation (11) is first put into discrete form as discussed for the gray problem, using the same quadrature method. The governing equation takes the form

$$\int_0^\infty \underset{\sim}{K_a}(\lambda) \left[-2\vec{b}(\lambda) + \vec{f}_{Ns}(\lambda) + \underset{\sim}{A}\vec{b}(\lambda) \right] d\lambda = 0 \quad (15)$$

where the following definitions are needed:

$\underset{\sim}{K_a}(\lambda)$: a diagonal matrix of absorption coefficients $K_a(\lambda, x_i)$

$\vec{b}(\lambda)$: the Planck function vector of the unknown temperature $\bar{B}(\lambda, T_i)$

$\vec{f}_{Ns}(\lambda)$: the vector $F_{Ns}(\lambda, x_i)$

$\underset{\sim}{A}(\lambda)$: the matrix $A_{ik}(\lambda)$

where

$$A_{ik}(\lambda) = K_a(\lambda,x_k) \left[W_{s_{ik}}(\lambda) + W_{ik} f_{ik}^+(\lambda) \right] \quad (16)$$

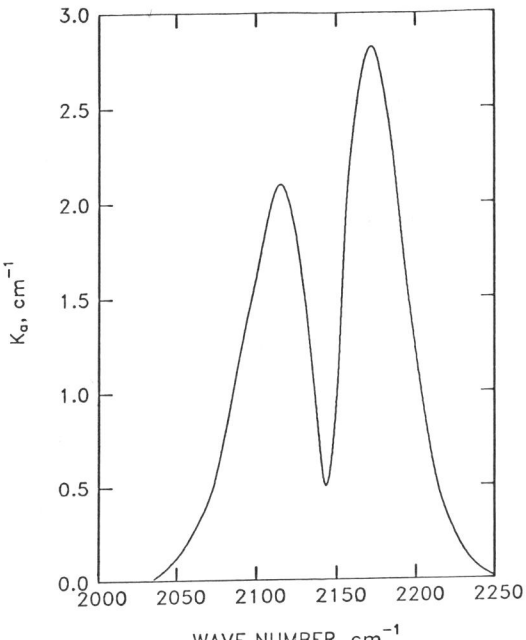

Fig. 3 The CO fundamental absorption band as represented by spline curve-fits used in calculations.

The quadrature weights W_{ik} are the standard (for example Simpson's rule) weights, and W_{sik} are the special weights discussed earlier and given in Ref. 9. The wavelength integration is done using a normal cubic spline integrator to accurately account for the absorption band structure. The Newton-Raphson iteration scheme is then applied to the system of equations represented by Eq. (15). This scheme coverages rapidly to the limit of machine accuracy, especially when a simple approximate solution[9] is used to start the iterations.

To show the capability of the code, example nongray problems are constructed. The medium is taken to be homogeneous with constant scattering of $K_s = 1$ cm^{-1}. The absorption coefficient $K_a(\lambda)$ is shown in Fig. 3 and represents the smoothed fundamental vibration-rotation band of CO. The physical thickness is $X_0 = 1.0$ cm. A small temperature difference between the surfaces is choosen ($T_B = 600$ K and $T_T = 650$ K) such that the assumption of temperature independent properties is valid.

Figure 4 shows five temperature profiles for nongray problems. Curve 1 is the solution when the walls are black and the asymmetry parameter g is 0.8. Curve 2 is the same conditions with g = 0.0 (isotropic scattering) and curve 3 represents results with no scattering, K_S = 0. Comparing curves 1 and 2 demonstrates that forward scatters increase the wall radiation slip (same effect as decreasing t_0) as was also noted for the gray solutions. In curve 3 the scattering coefficient is set to zero but the optical depths are kept constant by adding a constant absorption to the band. This curve falls between curves 1 and 2. These scattering effects are not small in that they are equivalent to substantial changes in t_0.

The last two curves in Fig. 4 are for g = 0.8 but with a reflecting bottom for curve 4 (other wall black) and two reflecting surfaces for curve 5. The surface reflectivity and emissivity used for these surfaces are given in Fig. 5.

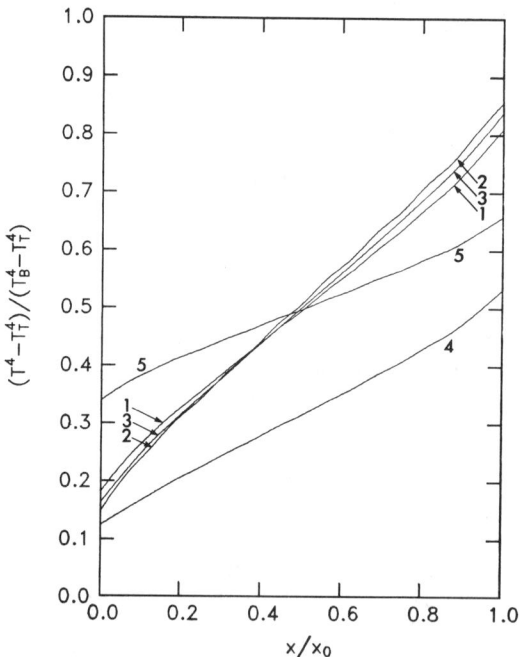

Fig. 4 Radiative equilibrium in nongray media with gaseous absorption of CO and nonabsorbing scatters with K_S = 1 cm^{-1}. The media are 1 cm thick with surface temperatures T_B = 600 K and T_T = 650 K. Curve 1 (g = 0.8), 2 (g = 0.0), and 3 (K_S = 0) have black surfaces. Curve 4 (g = 0.8) has a black top and reflecting bottom and curve 5 (g = 0.8) has identical reflecting surfaces. Surface reflecting properties are given in Fig. 5.

RADIATIVE EQUILIBRIUM

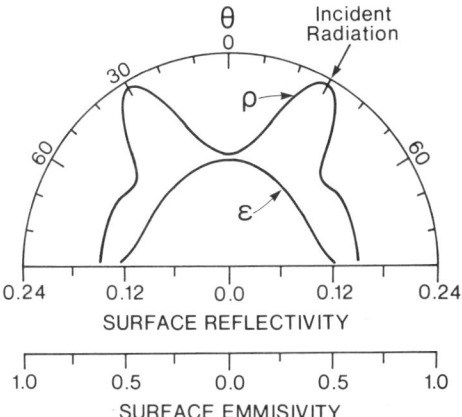

Fig. 5 The bidirectional reflectivity ρ for incidence angle shown and the corresponding directional emissivity ε for the opague surfaces used in Fig. 4, curves 4 and 5.

Notice the dramatic effect that surface reflectivity has on the temperature profiles for this nongray problem and the greatly increased radiation slip.

Conclusion

The purpose of this paper was to present a developed method to solve general radiative transfer problems in plane-parallel media. Although other solution procedures have been developed for plane-parallel radiative equilibrium,[1] none can handle the complete problem of inhomogeneous, anisotropic media with arbitrary reflecting surfaces. The present code has reproduced the available results for certain special subcases, but comparisons have not been made to all the other related problems in the literature. Such an exercise has limited use since there are no other approaches that are general. There is no doubt that this exact code will reproduce results for any special subcase. Although the code is somewhat complex to construct, it is computationally fast and available from the authors for further parametric studies.

References

[1] Sparrow, E.M. and Cess, R.D., Radiation Heat Transfer, Augmented Edition, McGraw-Hill, New York, 1978, pp. 203-271.

[2] Viskanta, R. and Grosh, R.J., "Heat Transfer in a Thermal Radiation Absorbing and Scattering Medium", *International Heat Transfer Conference*, Boulder, Colorado, 1961, p. 820.

[3] Lick, W., "Transient Energy Transfer by Radiation and Conduction", *International Journal of Heat and Mass Transfer*, Vol. 8, 1965, pp. 119-127.

[4] Heaslet, M.A. and Warming, R.F., "Radiative Transport and Wall Temperature Slip in an Absorbing Planar Medium", *International Journal of Heat and Mass Transfer*, Vol. 8, 1965, pp. 979-994.

[5] Cess, R.D. and Sotak, A.E., "Radiation Heat Transfer in an Absorbing Medium Bounded by a Specular Reflector", *Zeitschrift füer Angewandte Mathematik und Physik*, Vol. 15, 1964, pp. 642-647.

[6] Crosbie, A.L. and Viskanta, R., "Nongray Radiative Transfer in a Planar Medium Exposed to a Collimated Flux", *Journal of Quantitative Spectroscopy & Radiative Transfer*, Vol. 10, 1970, pp. 465-485.

[7] Domanus, H.M. and Cogley, A.C., "A Fundamental-Source-Function Formulation of Radiative Transfer and the Resulting Fundamental Reciprocity Relations", *Journal of Quantitative Spectroscopy & Radiative Transfer*, Vol. 14, 1974, p. 705.

[8] Cogley, A.C., "Adding and Invariant Imbedding Equations in Matrix Notation for All the Scattering Functions", *Journal of Quantitative Spectroscopy & Radiative Transfer*, Vol. 19, 1978, pp. 113-126.

[9] Sharma, A., "An Accurate and Computationally Fast Formulation for Radiative Fields and Heat Transfer in General, Plane-Parallel, Non-Grey Media with Anisotropic Scattering", Ph.D. Dissertation, University of Illinois at Chicago Circle, Chicago, 1980.

APPLICATION OF FINITE-ELEMENT TECHNIQUES TO THE INTERACTION OF CONDUCTION AND RADIATION IN A PARTICIPATING MEDIUM

S. T. Wu[*] and R. E. Ferguson[+]

The University of Alabama in Huntsville, Huntsville, Ala.

and

L. L. Altgilbers[≠]

U. S. Army Metrology and Calibration Center, U. S. Army Missile Command, Redstone Arsenal, Ala.

Abstract

In this paper, the authors demonstrate that a Galerkin finite-element method of analysis, utilizing isoparametric elements, offers a viable means of solving continuum thermal radiation problems with conduction in a participating medium. The participating medium was considered to be a gray radiation medium exhibiting isotropic absorption, emission, and scattering characteristics and optical properties that are independent of temperature. The medium was considered to be bounded by infinite parallel opaque, gray surfaces with diffuse emission and reflection characteristics. In solving this problem, a finite-element formulation was developed to describe a system in radiative equilibrium. Then the results of this first analysis were linked with a second finite-element model which incorporated conduction into the analysis. The results of this study were found to be in good agreement with existing published data. The model implies the following advantageous features: geometric generality, a computational algorithm which is "convenient" and "computable," and a functional basis for extension of the radiation model to higher order approximation.

Presented as Paper 80-1486 at the AIAA 15th Thermophysics Conference, Snowmass, Colo., July 14-16, 1980. Copyright © American Institute of Aeronautics and Astronautics, Inc., 1980. All rights reserved.
    *Professor, Department of Mechanics Engineering.
    +Graduate Student, Department of Mechanics Engineering.
    ≠Metrology Engineer.

Introduction

Analytical solutions to engineering problems involving a medium participating in radiant energy interchange have proved mathematically formidable due to the nonlinear mechanism of radiation processes, geometric complexities, and simultaneous interaction of all parts of the participating continuum. In this study a finite-element method (FEM) is used to develop a radiation model which can provide accurate solutions to absorbing, scattering, and emitting systems with conduction considered.

As has been shown by numerous investigators, the FEM method "is sufficiently general to treat a variety of unsteady and nonlinear phenomena, which by patching together a number of purely 'local' approximations of the phenomena under consideration, eliminates the traditional difficulties associated with irregular geometries, multi-connected domains, and mixer boundary conditions."[1]

The Radiation Problem

Until the early 1920s, the participating medium radiation analysis in engineering applications existed merely as a correction to convective processes. But as physicists began to demand more accurate models of stellar and galactic energy transport processes and engineers were faced with the problems of rocket nozzle and ablative system design, more accurate representations of the radiative processes within a participating medium were required. However, when the thermal radiation exchange mechanism is examined, the investigator is immediately faced with the following problems:
1) Radiation processes follow a nonlinear mechanism.
2) Except for certain limiting cases, all parts of the participating continuum interact simultaneously, in sharp contrast to the conduction-convection energy transfer processes which follow a linear diffusion mechanism.
3) Exact formulation of the radiation exchange processes in the absorbing, emitting, and scattering medium, even under ideal and simplified conditions, exists in the form of simultaneous nonlinear integrodifferential equations which have no closed form solution.
4) In engineering systems where the radiative transfer mechanism is coupled with conduction or convection, the geometric configuration may become complex.

As noted in the Introduction, these problems can be simplified by utilizing the finite-element method of analysis. Although the underlying ideas of the FEM were first discussed in 1943 by Courant,[2] the method is generally attributed to aircraft structural engineers[3] of the 1950s, who used it in the analysis of large systems of structural elements in aircraft. Application of the FEM to nonstructural problems, such as fluid flows and electromagnetism was initiated by Zienkiewiez, Mayer, and Cheung[4] in 1965 and to various problems of nonlinear mechanisms by Oden and others in 1972. Since these initial studies, this analytical tool has found applications in diverse areas of continuum analysis to include heat transfer,[5] rarefied gasdynamics,[6] fluid mechanics,[7] and magnetohydrodynamics.[8]

Since 1970, there have been a few studies involving a FEM analysis of radiation heat-transfer processes. To the best of the authors' knowledge, none of these studies have been concerned with a discussion of absorbing, emitting, and scattering medium. One of the earlier studies was carried out by Lee,[9] who was concerned with the treatment of general radiative effects on a diffuse gray surface. Lee's work, which utilized both triangular (two-dimensional) and multifaceted (one-dimensional) bar elements, was connected with the development of a general purpose heat-transfer computer program, the NASTRAN Thermal Analyzer, which permits a unified finite-element representation for both thermal and structural analyses.

Wang and Aguirre-Ramirez[10,11] undertook two studies which utilized the FEM in solving integral equations, namely the Fredholm equation of the second kind, which arises in radiative heat transfer. In their first paper,[10] they used a weighted residual FEM to study a system composed of two plates which had surfaces that radiate and reflect in a diffuse manner and which possessed identical gray body emissivities. Both plate surfaces were assumed to be maintained at the same temperature. In their second study,[11] they extended their first analysis to include plates having different temperatures. In both studies, nonparticipating media were considered and it was found that the finite-element solutions and the exact solutions were in good agreement.

In 1975, Wassel, Edwards, and Catton[12] studied simultaneous nongray radiation and thermal diffusion in a thermally and hydrodynamically established laminar and turbulent pipe flow with uniform internal heat generation. The governing integrodifferential energy equation was solved using the Galerkin technique for the laminar case and finite difference and numerical iteration techniques for turbulent flow. Simpson's rule was used to perform the integrations involved in the Galerkin method.

Reddy and Murty[13] in 1978 used the FEM to solve one-dimensional linear and nonlinear integral equations that arise in radiative heat-transfer and laminar boundary-layer theory. The Galerkin method was used to construct an approximation of the differential equations. Three radiation problems were considered: 1) radiation heat transfer between two nonblack, finite, parallel plates, 2) the Bratu problem which arises in connection with diffusion of heat generated by positive temperature-dependent sources and in magnetohydrodynamics, and 3) the H equation of Chandrasekhar which governs the variation of intensity in a medium characterized by an absorption coefficient and an emission coefficient. The FEM results, when compared to the exact values, proved to be good.

Three other studies involving the FEM and radiative heat transfer were conducted as summarized below: 1) Lee and Jackson[14] in 1975 used the FEM (NASTRAN computer program) to solve a combined radiative-conductive heat-transfer problem with mixed diffuse specular surface characteristics for a large space telescope, 2) Eslinger and Chung[15] in 1978 presented a finite-element solution for the heat transfer from a radiating and convecting fin and fin arrays, where it was assumed that the fin surfaces were gray, the environment black, and the fluid transparent, and 3) Higashihara[16] in 1979 used the FEM in the digital simulation of convecting flow generated by the radiative cooling of the ground surface.

In this study, while the governing equations for a multi-dimensional system are derived, the problem chosen to test the validity of the radiation model to be developed below, is that of a gray one-dimensional absorbing, emitting, and scattering medium between infinite parallel plates. This particular problem was chosen because it offers a valid evaluation of the analytical approach, since ample data are available for this particular system and because it avoids unnecessary geometrical and computational complexities.

Two cases are considered. The first is that of a system in radiative equilibrium, which is the most sensitive to the accuracy of the radiation model and thus can provide accuracy guidelines on which to base subsequent studies.

In the second case, the radiation model developed for the first case is linked with a second finite-element model to study the energy processes in a participating medium with conduction considered. Since ample attention has been given to this problem by previous investigators (e.g., Ozisik[17]), the primary purpose of this study is to uncover problem areas encountered in implementing the link between the radiation model and the FEM.

The Finite-Element Methodology

Before proceeding with the derivation of the radiation/FEM model, a brief overview of the FEM will be presented. The following discussion is based on more complete presentations provided by Strang and Fix[18] and Heubner.[19]

Suppose that the problem to be solved is in variational form; that is, it may be required to find the function u which minimizes a given expression. This minimizing property leads to a differential equation for u for which an exact solution is normally impossible and some approximation is necessary. According to the methods of Rayleigh-Ritz or Galerkin (a method of weighted residuals), a finite number of trial functions $\phi_1, \phi_2, \ldots, \phi_n$ are chosen. Then from all their possible linear combinations $\Sigma q_j \phi_j$, one which is minimizing is found. The unknown weights q_j are determined, not by a differential equation, but by a system of N discrete algebraic equations, which can be solved using computer techniques. The minimizing process automatically seeks out the combination which is closest to u. Therefore, the trial functions ϕ_j must be chosen in such a way that they are convenient enough for u to be computed and minimized, but yet at the same time be general enough to approximate closely the unknown solution u.

As might be expected, achieving convenience and computability is the real difficulty. This is where finite elements find their applicability. It starts by dividing the structure or region of physical interest into smaller pieces called "finite elements." These pieces must be easy for the computer to record and identify, such as, triangles, squares, etc. Within each piece, interpolation functions (which are also the weighting functions, ϕ_j, used in the Galerkin method[9]), which are given an extremely simple form, normally a polynomial of at most fifth degree, are defined over the element. Boundary conditions are then imposed locally along the edge of the element, and then globally along a more complicated boundary. Finally, the elemental solutions are compiled to give a global solution. The accuracy of this approximation method can be increased either by using more complex trial functions or by using the same polynomials with smaller subdivisions.

Since it is more accurate to represent bodies with curved boundaries by elements with curved sides, a method for transferring the simple, straight-sided triangular, square, etc., elements used in the FEM into ones that match the actual physical body must be developed. To do this, the x,y coordinates will be expressed in terms of curvilinear coordinates, so that

$$x = x(\xi, \eta) \text{ and } y = y(\xi, \eta) \qquad (1)$$

where the functions used to relate the x,y coordinate system to the ξ, η coordinate system are referred to as "shape functions" and the new coordinate system as a "natural coordinate system."

When the number of nodes used to define the shape functions is equal to the number of nodes that are used to define the interpolation function (where both sets of nodes are independent of each other), that is, when the shape functions are identical to the interpolation functions defined over the elements, the elements are said to be isoparametric. The isoparametric transformations need not decrease the order of accuracy, provided these transformations are uniformly smooth. In this sense, the isoparametric technique is the best to use for second-order equations and curved boundaries.[18] A series of eleven papers covering several aspects of curved finite elements is presented in Ref. 19.

Basic Continuum Radiation Equations of Transfer

In an absorbing, emitting, and scattering medium, the interactions of the radiation flux with the medium may be described locally with little loss of generality. Considering a monochromatic beam of radiation, as shown in Fig. 1, its intensity is designated by I_λ and defined as the local net transfer of radiant energy per unit area normal to the beam per unit solid angle per unit time. As the beam traverses a path of length ds, it will undergo attenuation as a result of local absorption and scattering within the medium. The amount absorbed and scattered within a differential volume is proportional to I_λ and the path length ds. Thus the energy absorbed and scattered per unit time and solid angle is

$$\kappa_\lambda I_\lambda ds + \gamma_\lambda I_\lambda ds = \beta_\lambda I_\lambda \, ds \qquad (2)$$

where κ_λ is the monochromatic absorption coefficient, γ_λ is the monochromatic scattering coefficient, and β_λ is the monochromatic extinction coefficient.

Assuming that absorption and scattering processes are isotropic, the energy attenuated per unit time and volume is

$$\dot{Q} = \dot{Q}_a + \dot{Q}_s = \kappa_\lambda \int_{4\pi} I_\lambda \, d\omega + \gamma_\lambda \int_{4\pi} I_\lambda \, d\omega \qquad (3)$$

where $\dot{Q}_a$ is the energy absorbed per unit time and volume and $\dot{Q}_s$ is the energy scattered per unit time and volume.

Allowing dI_λ to denote the differential change in radiation intensity in traversing the path length ds, Eq. (2) becomes

$$-dI_\lambda = \beta_\lambda I_\lambda \, ds \qquad (4)$$

or solving for I_λ gives

$$I_\lambda = I_{o\lambda} \, e^{-\beta_\lambda s} \qquad (5)$$

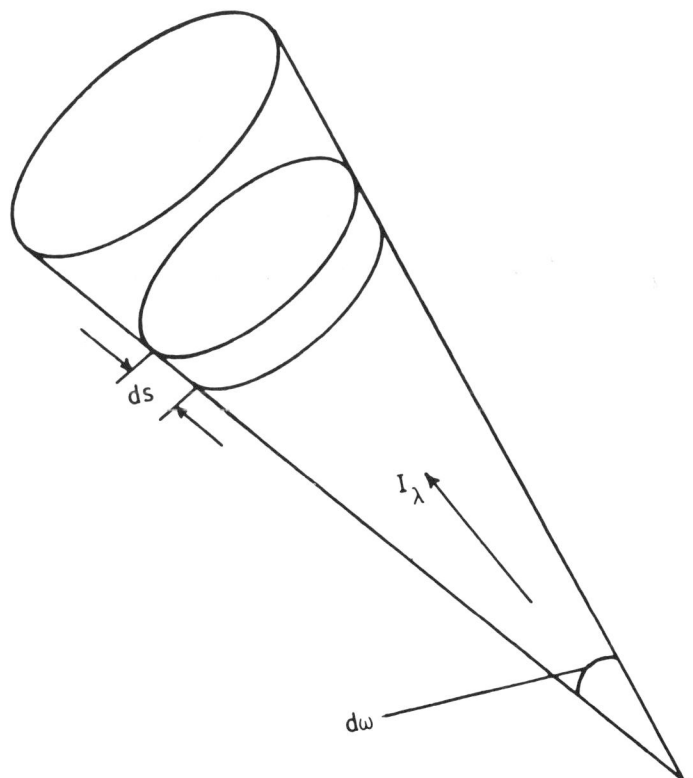

Fig. 1 General beam of radiant energy defining the absorption and scattering of energy.

where $I_{o\lambda}$ is the intensity of the monochromatic beam at $s = 0$. This assumption allows the intensity of the beam to radiate energy within an absorbing, scattering medium to follow an exponential decay with path length, so that

$$\lambda_p = 1/\beta_\lambda \qquad (6)$$

is the photon mean free path and

$$\tau_o = x_o/\lambda_p \qquad (7)$$

is the optical depth of the medium.

Using the extinction concept, a monochromatic irradiation, H_λ, may be defined as

$$H_\lambda = \beta_\lambda \int_{4\pi} I_\lambda \, d\omega \qquad (8)$$

so that Eq. (3) may be rewritten as

$$\dot{Q}_a + \dot{Q}_s = \left[(\kappa_\lambda + \gamma_\lambda)/\beta_\lambda\right] H_\lambda \qquad (9)$$

The local emission of monochromatic radiation per unit volume may be written as

$$\dot{Q}_e = 4\kappa_\lambda E_{b\lambda}(T) \qquad (10)$$

where $E_{b\lambda}(T)$ is Planck's blackbody emission function. Under the assumption of a gray medium (optical properties independent of wavelength), the blackbody emission of radiation $E_{b\lambda}(T)$ may be expressed by the Stefan-Boltzmann equation as

$$E_b(T) = \sigma T^4 \qquad (11)$$

where σ is the Stefan-Boltzmann constant. Thus, Eq. (10) may be rewritten as

$$\dot{Q}_e = 4\kappa \sigma T^4 . \qquad (12)$$

For a more complete discussion of this material refer to Sparrow and Cess.[20]

Radiation/Finite-Element Composite Model

Overview

In applying the FEM to continuum radiation processes, a segmented approach is employed. First, the radiation transfer

CONDUCTION AND RADIATION IN A PARTICIPATING MEDIUM

processes within the participating media will be described by a model specifically formulated for compatibility with the finite-element model. Then a second finite-element model will be developed that will accept the radiation transfer data from the radiation model and extend the analysis to include conduction effects. To facilitate calculations, similar computational algorithms and the same information describing the physical characteristics of the system in question will be used in both models.

To develop the radiation model, the following approach is taken. The system domain in question is first divided into small subunits or elements as shown in Fig. 2. Over each element, it is assumed that the properties of the continuum are closely approximated by expressing them as a function of the property values at the corner of nodal points of the element and a suitable interpolation function spanning the element. The properties within the two-dimensional elements, shown in Fig. 2, will be expressed with a linear interpolation function based on the four nodal points of the element. To attain a higher accuracy, a higher order interpolation scheme would be used, which would require nodal points within the element in addition to those at the corners of the element.

Using this discretization technique, the governing differential equation for the domain in question would be written in terms of the nodal values. The Galerkin method is then used to solve the governing differential equation. To reduce the computational complexities involved in performing the integra-

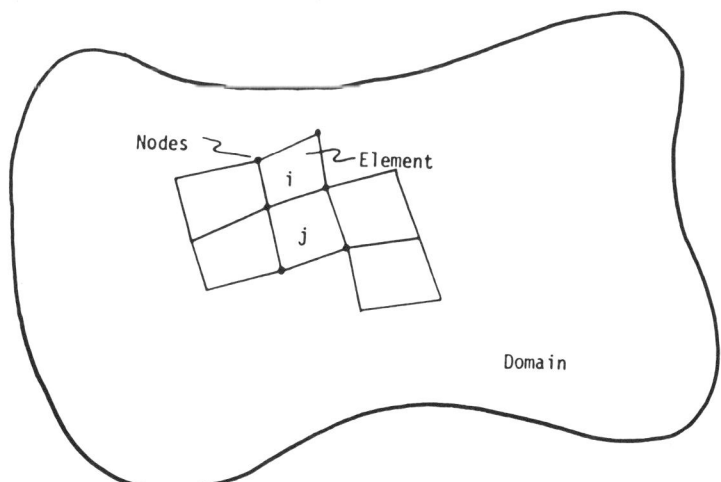

Fig. 2 Finite-element discretization of a generalized system.

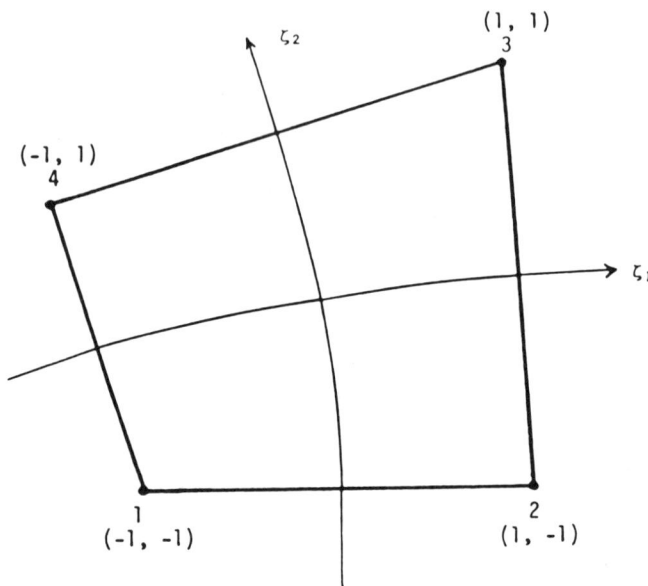

Fig. 3 The isoparametric element.

tion encountered in the Galerkin method, an isoparametric element shown in Fig. 3 will be utilized. This element will be considered to contain an orthogonal coordinate system with the origin at the centroid of the element. The nodal points lie exactly one unit in each coordinate direction from the origin. The integration required to satisfy the Galerkin condition is then performed piecewise, element by element using the Gaussian quadrature.

General Conservation Equations

In approaching the problem of interfacing continuum radiation effects with a finite-element model, it is necessary to define the information which is to be provided by the radiation model. To do this, the basic conservation laws for a moving fluid will now be considered.

Neglecting the radiation pressure tensor, a valid assumption at conditions generally encountered in engineering systems, the equations of conservation of momentum will remain unchanged in a radiative participating medium. The conservation of energy equation is given by

$$\rho C_v \frac{DT}{Dt} = - \text{div } q - p \text{ div } u + \mu \phi \tag{13}$$

where u = fluid velocity vector; q = heat flux vector; ϕ = Rayleigh dissipation function; p = fluid static pressure; div q = net energy addition to the differential volume; p div q = irreversible compression work done by the differential volume; $\mu\phi$ = energy dissipated by viscous effects.

For a nonparticipating medium, q represents thermal conduction within the medium and may be expressed as

$$q = -k \text{ grad } T \tag{14}$$

where k is the thermal conductivity of the fluid.

For a participating medium, which absorbs, emits, and scatters radiant energy, an additional radiation flux q_r must be added to Eq. (14), so that

$$q = -k \text{ grad } T + q_r \tag{15}$$

Incorporating Eq. (15) into the energy equation, Eq. (13), an augmented energy equation for a participating medium is obtained; that is,

$$\rho C_v \frac{DT}{DT} = \text{div} (k \text{ grad } T) - p \text{ div } u + \mu\phi - \text{div } q_r \tag{16}$$

Radiation Model Derivation

The last term in Eq. (16), div q_r, may be interpreted as a radiative source term, representing the net addition of energy to the differential volume by radiative mechanisms. In interfacing with a finite-element representation of the medium, the radiation model will be required to supply the radiative source term as a function of position. To provide this data, a thermal radiation energy balance is made on a differential volume. Consider a general system shown in Fig. 4 containing an absorbing, emitting, and scattering medium bounded by opaque surfaces. The surfaces will be assumed to be diffuse gray emitters and reflectors and the medium will be assumed to be gray and have isotropic emission, absorption, and scattering characteristics. In performing the radiant energy balance on the differential balance, the radiosity, w_i, of the element is introduced and defined as the sum of the emitted and scattered energy leaving the differential volume. That is,

$$w_i = \dot{Q}_{ei} + \dot{Q}_{si} \tag{17}$$

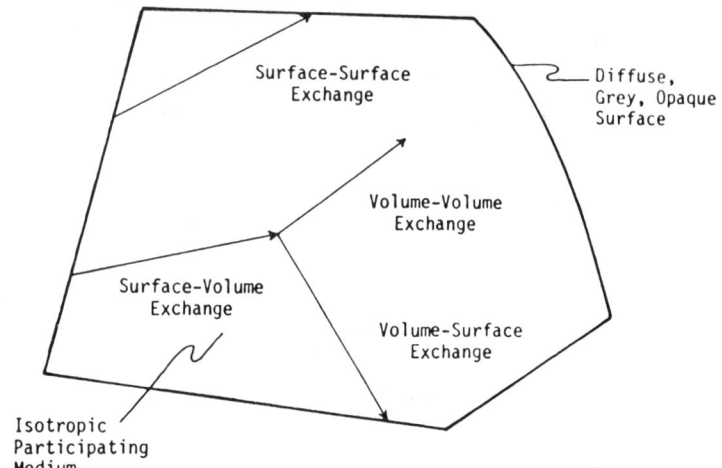

Fig. 4 The generalized participating media system.

where $\dot{Q}_{ei}$ is the emitted energy at differential volume i defined in Eq. (12), and $\dot{Q}_{si}$ is the scattered energy at differential volume i defined by the second term of Eq. (9).

Expanding Eq. (17) in terms of Eqs. (9) and (12), the radiosity for a differential element located within the participating medium is

$$w_i = 4 \kappa \sigma T_i^4 + (\gamma/\beta) H_i \qquad (18)$$

and for a differential element located on an opaque surface by

$$w_i = \varepsilon \sigma T_i^4 + (1-\varepsilon) H_i \qquad (19)$$

where ε is the surface emissivity.

The irradiation terms in Eqs. (18) and (19) may be expressed in terms of the radiosity of all the differential volumes of the system and an exchange or view factor, which gives the fraction of the radiosity at volume position v, which is available for scatter or reflection at position i. Thus

$$H_i = \int_V F_{vi} \, w(v) \, dv \qquad (20)$$

where F_{vi} = exchange factor from v to i; and $w(v)$ = system

CONDUCTION AND RADIATION IN A PARTICIPATING MEDIUM

radiosity which is a function of position. The exchange factors for the four possible types of radiation exchange among volume and surface elements are defined in Table 1. The one-dimensional exchange factors used in the calculations are given in Table 2.

Incorporating this irradiation function into Eqs. (18) and (19), they can be rewritten as

$$W_i = 4 \kappa \sigma T_i^4 + \frac{\gamma}{\beta} \int_V F_{vi} w(v) \, dv \tag{21}$$

and

$$W_i = \epsilon \sigma T_i^4 + (1-\epsilon) \int_V F_{vi} w(v) \, dv \tag{22}$$

Dividing Eq. (21) by 4κ and Eq. (22) by ϵ and solving the resulting expressions for σT_i^4, it is found that

$$\sigma T_i^4 = \frac{W_i}{4\kappa} - \frac{\gamma}{4\kappa\beta} \int_V F_{vi} w(v) \, dv \tag{23}$$

and

$$\sigma T_i^4 = \frac{W_i}{\epsilon} - \frac{1-\epsilon}{\epsilon} \int_V F_{vi} w(v) \, dv \tag{24}$$

Table 1 Four Classes of Exchange Factors, $_v^a F_{vi}$

v-i pairs	Surface		Volume	
Surface	$F_{vi} = \dfrac{\cos\phi_v \cos\phi_i dA_i e^{-\beta s}}{\pi s^2}$		$F_{vi} = \dfrac{\beta \cos\phi \, dv_i e^{-\beta s}}{\pi s^2}$	
Volume	$F_{vi} = \dfrac{dA_i \cos\phi \, e^{-\beta s}}{4\pi s^2}$		$F_{vi} = \dfrac{\beta dv_i e^{-\beta s}}{4\pi s^2}$	

^a_v For the system under consideration (see Fig. 5):

$$dA_i = 2\pi s^2 \tan\phi \, d\phi/\cos^2\phi$$

and

$$dv_i = 2\pi s^2 \tan\phi \, d\phi/\cos^2\phi$$

Table 2 One-Dimensional Exchange Factors[a]

v-i pairs	Surface	Volume
Surface	$F_{vi} = \int_1^\infty \dfrac{2e^{-\beta st}}{t^3}\,dt$ $= 2E_3(\beta s)$	$F_{vi} = \int_1^\infty \dfrac{\beta e^{-\beta st}}{t^2}\,dt\,dx$ $= \beta E_2(\beta s)$
Volume	$F_{vi} = \int_1^\infty \tfrac{1}{2}\dfrac{e^{-\beta st}}{t^2}\,dt$ $= \tfrac{1}{2}E_2(\beta s)$	$F_{vi} = \int_1^\infty \dfrac{2\beta e^{-\beta s/t}}{t}\,dt\,dx$ $= 3\beta E_1(\beta s)\,dx$

[a]
s = Distance between two infinite plane parallel surfaces
dA_v = Point source on surface v
dA_i = Annular ring swept by a beam of angular extent $d\phi$
dv_i = Volume of annular element at a constant angle ϕ
$s/\cos\phi$ = Path length through participating media from the point source to the annular differential area dA_i
t = sec ϕ
$E_1(\beta s)$ = First exponential integral
$E_2(\beta s)$ = Second exponential integral
$E_3(\beta s)$ = Third exponential integral

These definitions were incorporated into the various exchange factor expressions of Table 1, which were integrated from $\phi = 0$ to $\phi = \pi/2$ to give the one-dimensional exchange factors in this table.

To evaluate these integrals in the last two equations, the domain will be discretized following the procedure outlined for the FEM. For a domain divided into b elements, the integration over the domain may be carried out in a piecewise manner, element by element, so that Eqs. (23) and (24) become

$$\sigma T_i^4 = \frac{w_i}{4\kappa} - \frac{\gamma}{4\kappa\beta} \sum_{j=1}^{b} \int_{v_j} F_{v_{ji}}\, w(v_j)\, dv_j \qquad (25)$$

CONDUCTION AND RADIATION IN A PARTICIPATING MEDIUM

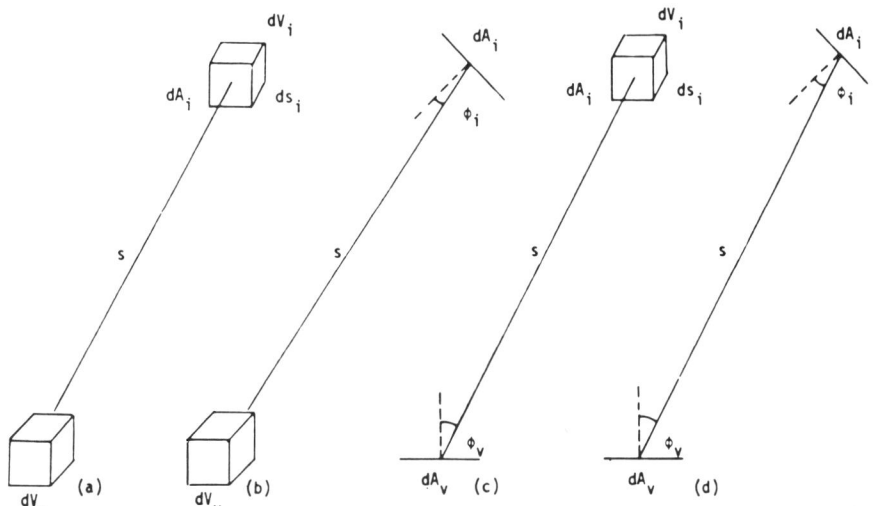

Fig. 5 Geometrical configuration for derivation of differential exchange factors.

and

$$\sigma T_i^4 = \frac{W_i}{\varepsilon} - \frac{1-\varepsilon}{\varepsilon} \sum_{j=1}^{b} \int_{v_j} F_{v_{ji}} w(v_j) \, dv_j \qquad (26)$$

respectively. Choosing the elements used in the discretization process to be isoparametric, the integrals over each element can be closely approximated by a Gaussian quadrature. For an mth order quadrature scheme applied to a dth dimensional system, Eqs. (25) and (26) are rewritten as

$$\sigma T_i^4 = \frac{W_i}{4\kappa} - \frac{\gamma}{4\kappa\beta} \sum_{j=1}^{b} \sum_{h=1}^{d} \sum_{p=1}^{m} \hat{C}_p F_{v_{ji}} w_j(\hat{a}_p) |J| \qquad (27)$$

and

$$\sigma T_i^4 = \frac{W_i}{\varepsilon} - \frac{1-\varepsilon}{\varepsilon} \sum_{j=1}^{b} \sum_{h=1}^{d} \sum_{p=1}^{m} \hat{C}_p F_{v_{ji}} w_j(\hat{a}_p) |J|, \qquad (28)$$

respectively, where C_p = quadrature weights; a_p = Gaussian points; and $|J|$ = determinant of the transformation matrix from the isoparametric system to Cartesian system.

Following the development of the finite-element approach, the radiosity within each element may be expressed as a function of the element and the interpolation function spanning the element. At an arbitrary point within an element containing ℓ nodes, the radiosity may be expressed as

$$w(v) = \sum_{e=1}^{\ell} \Omega_e(v) w_e$$

where $\Omega_e(v)$ = the eth nodal component of the interpolation function and w_e = the radiosity at node e.

Incorporating this interpolation scheme into the radiosity equations, Eqs. (27) and (28), one obtains

$$\sigma T_i^4 = \frac{w_i}{4\kappa} - \frac{\gamma}{4\kappa\beta} \sum_{j=1}^{b} \sum_{h=1}^{d} \sum_{p=1}^{m} \sum_{e=1}^{\ell} \hat{C}_p \times F_{v_{ji}} \Omega_e(\hat{a}_p) w_e \, J \quad (29)$$

and

$$\sigma T_i^4 = \frac{w_i}{\varepsilon} - \frac{1-\varepsilon}{\varepsilon} \sum_{j=1}^{b} \sum_{h=1}^{d} \sum_{p=1}^{m} \sum_{e=1}^{\ell} \hat{C}_p$$
$$\times F_{v_{ji}} \Omega_e(\hat{a}_p) w_e |J|. \quad (30)$$

If the differential volume location i is limited to nodal points only, the above equations may be employed over a n node system to yield n simultaneous linear algebraic equations in n unknown radiosities. In matrix form, this system of equations may be expressed as

$$[\psi_{ij}] [w_j] = [\Delta_i] \quad (31)$$

where $[\psi_{ij}]$ = coefficient matrix; $[w_j]$ = unknown nodal radiosities; and $[\Delta_i] = \sigma T_i^4$. Solving Eq. (31) yields

$$[w_i] = [\psi_{ij}]^{-1} [\Delta_i] \quad (32)$$

or

$$w_i = \sum_{n=1}^{n} \psi_{in}^{-1} \sigma T_n^4 \quad (33)$$

In the augmented energy equation, the div q_r, representing the radiative source term, is defined as the net radiant energy addition to the differential volume or surface element. For a <u>surface element</u>, the radiative source term consists of

$$-\text{div } q_r = \dot{Q}_a - \dot{Q}_e = \varepsilon H_i - \varepsilon \sigma T_i^4 \qquad (34)$$

The radiosity equation for a surface element is expressed as

$$W_i = \dot{Q}_e + \dot{Q}_s = \varepsilon \sigma T_i^4 + (1-\varepsilon) H_i \qquad (35)$$

Combining Eqs. (34) and (35) to eliminate H_i yields upon simplification

$$-\text{div } q_r = \frac{\varepsilon}{1-\varepsilon} (W_i - \sigma T_i^4) \qquad (36)$$

Expressing the radiosity in terms of the n nodal temperatures as defined in Eq. (33), Eq. (36) becomes

$$-\text{div } q_r = \frac{\varepsilon}{1-\varepsilon} \left[\left(\sum_{\eta=1}^{n} \psi_{i\eta}^{-1} \sigma T_\eta^4 \right) - \sigma T_i^4 \right] \qquad (37)$$

Employing a similar development to <u>volume differential elements</u>, the radiative source term may be defined as

$$-\text{div } q_r = \dot{Q}_a - \dot{Q}_e = (\kappa/\beta) H_i - 4 \kappa \sigma T_i^4 \qquad (38)$$

The radiosity for the volume element is given by

$$W_i = \dot{Q}_e + \dot{Q}_s = 4 \kappa \sigma T_i^4 + (\gamma/\beta) H_i \qquad (39)$$

Combining Eqs. (38) and (39) to eliminate H_i yields upon simplification

$$-\text{div } q_r = (\kappa/\beta) (W_i - 4 \beta \sigma T_i^4) \qquad (40)$$

Expressing the radiosity term in Eq. (40) in terms of the n nodal temperature gives

$$-\text{div } q_r = \frac{\kappa}{\gamma} \left[\left(\sum_{\eta=1}^{n} \psi_{i\eta}^{-1} \sigma T_\eta^4 \right) - 4 \beta \sigma T_i^4 \right] \qquad (41)$$

Equations (37) and (41) when used in Eq. (16), comprise the fundamental equations to be used in the solution of the problem under consideration. The goal now is to develop the working equations and to verify them.

Composite Model Verification

Radiative Equilibrium

A condition of radiative equilibrium exists when the radiation processes are the principal energy transport mechanism. In examining the general energy equation, Eq. (16), all terms are zero with the exception of div q_r, the radiative source term, so that the energy equation reduces to

$$\text{div } q_r = 0. \tag{42}$$

From Eqs. (37) and (41), the radiative source may be expressed in terms of the radiation properties and the nodal temperatures of the walls and within the medium as follows:

$$\frac{\varepsilon}{1-\varepsilon}\left[\left(\sum_{\eta=1}^{n} \psi_{i\eta}^{-1}\ \sigma T_\eta^4\right)\right]\sigma T_i^4 = 0 \tag{43}$$

and

$$\frac{K}{\gamma}\left[\left(\sum_{\eta=1}^{n} \psi_{i\eta}^{-1}\ \sigma\ T_\eta^4\right) - 4\ \beta\sigma\ T_i^4\right] = 0 \tag{44}$$

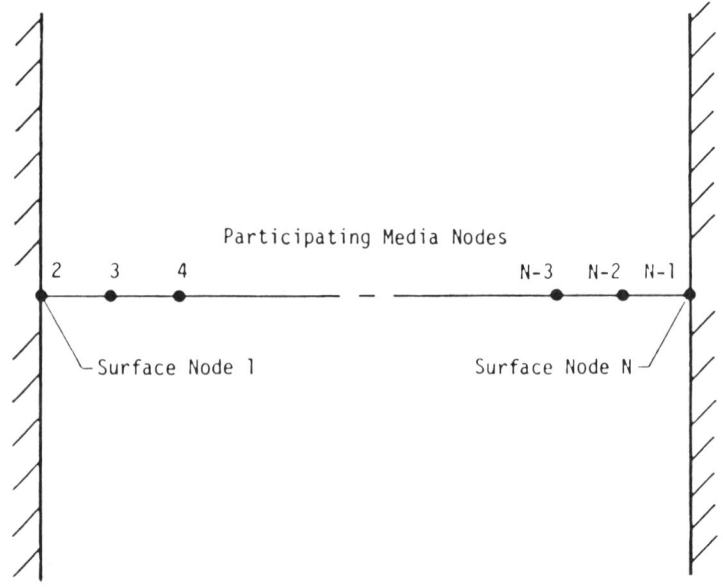

Fig. 6 General element layout of verification model.

Requiring the equilibrium equation to be satisfied at each nodal point, Eqs. (43) and (44) may be applied to each surface and volume node as appropriate to yield n simultaneous algebraic equations in n unknown temperatures linear with respect to T_i^4. The elemental discretization of the one-dimensional medium is performed as shown in Fig. 6. As a result of discontinuities in physical properties of the walls of the enclosure, different surface and volume nodes are required to occupy the same position in space.

Combined Conduction and Radiation

The radiation model used in studying the radiation processes in radiative equilibrium is linked with a simple one-dimensional finite-element model to access its utility in studies involving other energy transport mechanisms. The energy equation for a one-dimensional conducting-radiating system may be reduced to

$$k \frac{\partial^2 T}{\partial x^2} - \text{div } q_r = 0 \tag{45}$$

As noted earlier, the Galerkin method was selected to solve Eq. (45) due to its superior generality over other methods. Thus, the Galerkin method requires the following equality to be satisfied over the system domain

$$\int_x \left[k \frac{\partial^2 T}{\partial x^2} - \text{div } q_r \right] \Omega_i \, dx = 0 \tag{46}$$

where Ω_i is the ith component of the interpolation function chosen to span the element. The integral expression may be rearranged to yield

$$\int_x k \frac{\partial^2 T}{\partial x^2} \Omega_i \, dx = \int_x \text{div } q_r \Omega_i \, dx \tag{47}$$

Employing the Green-Gauss theorem, the order of the differential under the integral on the left may be reduced as follows:

$$\int_x k \frac{\partial T}{\partial x} \frac{\partial \Omega_i}{\partial x} \, dx - k \frac{\partial T}{\partial x} = \int_x \text{div } q_r \Omega_i \, dx \tag{48}$$

The second term on the left side describes the heat flux at the system boundaries. Since there will be no imposed flux in this sample problem, this term is zero. To model the radiation-conduction system, a one-dimensional linear isoparametric element is chosen. Thus the temperature within each element may be expressed as

$$T(\zeta) = \Omega_1(\zeta) \, T_1 + \Omega_2(\zeta) \, T_2 \tag{49}$$

where $T(\zeta)$ = temperature at a point corresponding to the isoparametric coordinate ζ; T_1 = temperature at nodal point 1 located at $\zeta = -1.0$; T_2 = temperature at nodal point 2 located at $\zeta = 1.0$; $\Omega_1(\zeta)$ = first nodal component of the interpolation function; and $\Omega_2(\zeta)$ = second nodal component of the interpolation function.

The interpolation functions are defined to be

$$\Omega_1(\zeta) = (1-\zeta)/2$$

and

$$\Omega_2(\zeta) = (1+\zeta)/2 \tag{50}$$

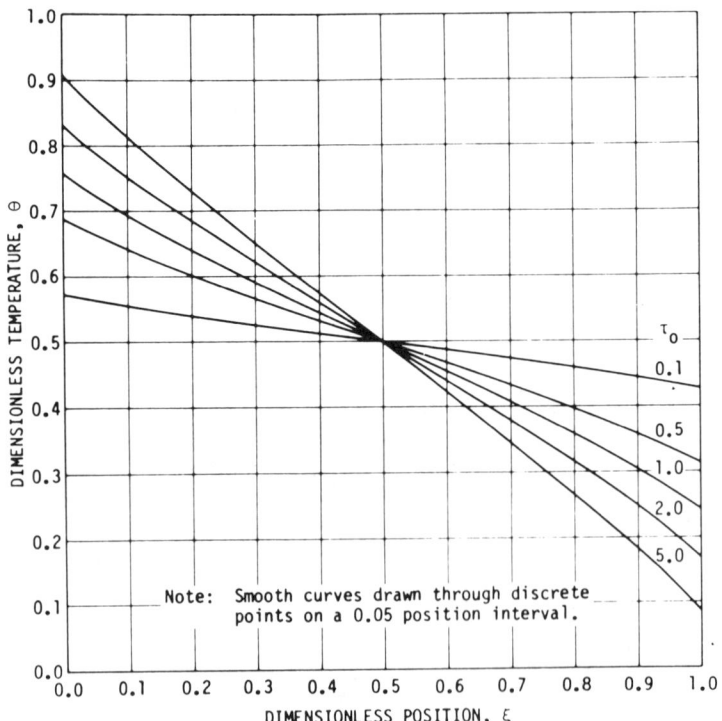

Fig. 7 Radiative equilibrium results.

Introducing Eq. (49) into Eq. (48) yields

$$\int_X kT_1 \frac{\partial \Omega_1}{\partial x} \frac{\partial \Omega_i}{\partial x} + \int_X kT_2 \frac{\partial \Omega_2}{\partial x} \frac{\partial \Omega_i}{\partial x} = \int_X \text{div } q_r \Omega_i \, dx \quad (51)$$

where the Gaussian quadrature integration scheme is used to evaluate the above integral expression piecewise over the system domain. The individual element equations are assembled into a global expression in which the nodal temperatures are the only unknowns. This global system of equations may be written in the following matrix form:

$$[S_{ij}] [T_i] = [B_i] \quad (52)$$

The $[S_{ij}]$ matrix is commonly referred to as the "stiffness" matrix and is derived from the assembly of terms on the left side of Eq. (51); that is,

$$[S_{ij}] = \int_X k \frac{\partial \Omega_i}{\partial x} \frac{\partial \Omega_j}{\partial x} \, dx \quad (53)$$

The $[B_i]$ matrix will be called the "boundary" vector and is composed of an assembly of the terms on the right side of Eq. (51); that is,

$$[B_i] = \int_X \text{div } q_r \, \Omega_i \, dx \quad (54)$$

The $[T_i]$ matrix is composed of the unknown nodal temperatures.

Since Eqs. (37) and (41) were used to evaluate the terms in the boundary vector, and since the dependent variable (temperature) occurs on both sides of Eq. (52), linear on the left side and raised to the fourth power on the right, this system of equations [Eq. (52)] must be solved using an iterative approach. The stiffness matrix $[S_{ij}]$ is inverted using the standard Gauss-Jordan inversion method. The boundary vector $[B_i]$ is calculated using a series of initial temperatures. A new set of nodal temperatures are calculated using the boundary vector and inverted stiffness matrix. This new set of temperatures can then be used to calculate a new boundary vector. This process is continued until the desired convergence is obtained. This iterative procedure imposes no severe penalties with respect to computer time usage since the stiffness matrix is inverted only once and adequate convergence is usually obtained in 20 to 30 iterations.

Some Numerical Results

Radiative Equilibrium

A preliminary study was performed to define accuracy guidelines for subsequent cases. A system consisting of an isotropic emitting and absorbing medium between black isothermal walls was chosen. The data of Heaslett and Warming[21] were used as a source of comparison. Since Heaslett and Warming presented their results in dimensionless form in terms of a dimensionless position parameter given by

$$\xi = x/L \quad (55)$$

where ξ = dimensionless position, x = position between the infinite parallel walls, L = distance between the infinite parallel walls, and a dimensionless temperature given by

$$\theta = (T_x^4 - T_1^4)/(T_1^4 - T_2^4) \quad (56)$$

where θ = dimensionless temperature, T_x = local temperature at position x, T_1 = temperature at wall 1, and T_2 = temperature at wall 2. These parameters were adopted for this portion of the present study.

Finally, optical depths of 2 and 5 were chosen. For each of these two optical depths, the participating medium was divided into elements of optical depth 0.500, 0.250, and 0.125. The results of the radiation model were compared with the results of Heaslett and Warming at $\xi = 0.0$ and $\xi = 0.5$ where tabulated data are available. The radiation model errors of the cases considered are presented as a percent of full scale (1.0) and are tabulated in Table 3 with the predicted and the "exact" values. From these data, the following three observations can be made:

1) The model errors for the cases considered are consistently less than 1%.

2) Except for one case ($\xi = 0.5$, $\tau_0 = 5$), as the element length decreases, the percent error decreases, which is to be expected.

3) For the exception noted in 2) above, it can also be seen that the error for all three $\tau_0^{(e)}$ is considerably higher than for all the other cases considered.

CONDUCTION AND RADIATION IN A PARTICIPATING MEDIUM

Since the differences in the Heaslett and Warming data and the model data were less than 1%, which is acceptable for most engineering analysis problems, the element size to be selected for the studies to follow will be chosen to enable the construction of smooth graphical results rather than for any other reasons.

Using a radiation model which divides the radiating medium into 20 equal elements of emitting-absorbing medium with optical depths of 0.1, 0.5, 1.0, 2.0, and 5.0, the above study was extended, where the results presented in Fig. 7 show no graphically observable difference from the data of Heaslett and Warming.

Combined Conduction and Radiation

The conduction-radiation model was used to repeat portions of a study by Viskanta and Grosh[22] of an absorbing-emitting medium between finite isothermal black walls with conduction in the medium considered. For comparison purposes, the dimensionless groups used by Viskanta and Grosh were adopted. The dimensionless temperature within the medium was expressed as $\theta = T/T^*$, where θ = dimensionless temperature; T = local temperature within the medium; and T^* = a reference temperature chosen as the temperature of one boundary. A dimensionless conduction-radiation parameter, N, was defined as

$$N = k\beta/4 \ \sigma T^{*3} \qquad (57)$$

where k = thermal conductivity; β = extinction coefficient; and σ = Stefan-Boltzmann constant. This conduction-radiation parameter provides a means for measuring the relative importance of conduction and radiation as energy transport mechanisms. That is, for large values of N conduction predominates, while for low values of N radiation is the dominating factor.

To facilitate the compatibility between the radiation-conduction model and the radiation equilibrium model, the dimensionless position and optical depth parameters remain unchanged from the radiative equilibrium study. An optical depth of 1.0 was chosen with the medium divided into 20 equal elements. Dimensionless boundary temperatures of $\theta(0) = 0.5$ and $\theta(0) = 0.0$ were considered and for each temperature given on the boundary two conduction-radiation parameters (i.e., N = 0.1 and 0.01) were used. The results are presented graphically in Figs. 8 and 9 and agree well with the results of Viskanta and Grosh, showing no graphically observable differences. Note that as N increases in magnitude, the curves approach the

Table 3 Comparison of Radiation Model
Results with Published Data [a]

Position ξ	Optical depth		Dimensionless temperature θ		Percent error
	System	Element	Model results	Published results	
0.0	5.0	0.500	0.90932	0.91008	0.076
0.0	5.0	0.250	0.90962	0.91008	0.046
0.0	5.0	0.125	0.91022	0.91008	0.014
0.0	2.0	0.500	0.83213	0.83075	0.138
0.0	2.0	0.250	0.83129	0.83075	0.054
0.0	2.0	0.125	0.83118	0.83075	0.043
0.5	5.0	0.500	0.49236	0.500	0.764
0.5	5.0	0.250	0.49906	0.500	0.094
0.5	5.0	0.125	0.50286	0.500	0.286
0.5	2.0	0.500	0.49689	0.500	0.311
0.5	2.0	0.250	0.49840	0.500	0.160
0.5	2.0	0.125	0.50031	0.500	0.031

[a] Percent error is defined as percent of full scale (1.0).

pure conduction case, which are depicted by the straight lines which connect the points (0.1, 0.0) and (1.0, 1.0) in Fig. 8 and the points (0.5, 0.0) and (1.0, 1.0) in Fig. 9.

In performing the iterative solution to the cases considered above, the method initially resulted in nonconvergence due to an extreme oscillating nature of the proposed algorithm. The covergence problem was solved by using a damping factor in the solution procedure. This damping factor was applied to each new nodal temperature to effectively tie the new calculated temperature to its old value by a certain fraction. This damping process was expressed as

$$T_d = D T_n + (1-D) T_o \qquad (58)$$

where T_d = the damped temperature; T_n = the calculated new temperature; T_o = the old temperature; and D = the damping factor. For the cases considered in this study, damping factors between 0.1 and 0.05 provided satisfactory results, and convergence was obtained in 30 to 40 iterations.

CONDUCTION AND RADIATION IN A PARTICIPATING MEDIUM

The radiation-conduction model developed above was also used to repeat portions of a study by Ozisik[17] of an absorbing, emitting, and scattering medium with conduction considered. All dimensionless groups used in the absorbing-emitting medium study were employed with the addition of the albedo of scatter, ω_s, defined as

$$\omega_s = \gamma/\beta \tag{59}$$

where γ = scattering coefficient and β = extinction coefficient. The two limiting cases, $\omega_s = 0$ and $\omega_s = 1$, characterize a purely absorbing and emitting medium and a purely scattering medium, respectively. As pointed out by Ozisik,[17] 1) the interaction of radiation with conduction is greatest for $\omega_s = 0$, 2) the radiation has no effect on the temperature distribution, hence the temperature distribution is the same as that for pure conduction, when $\omega_s = 1.0$.

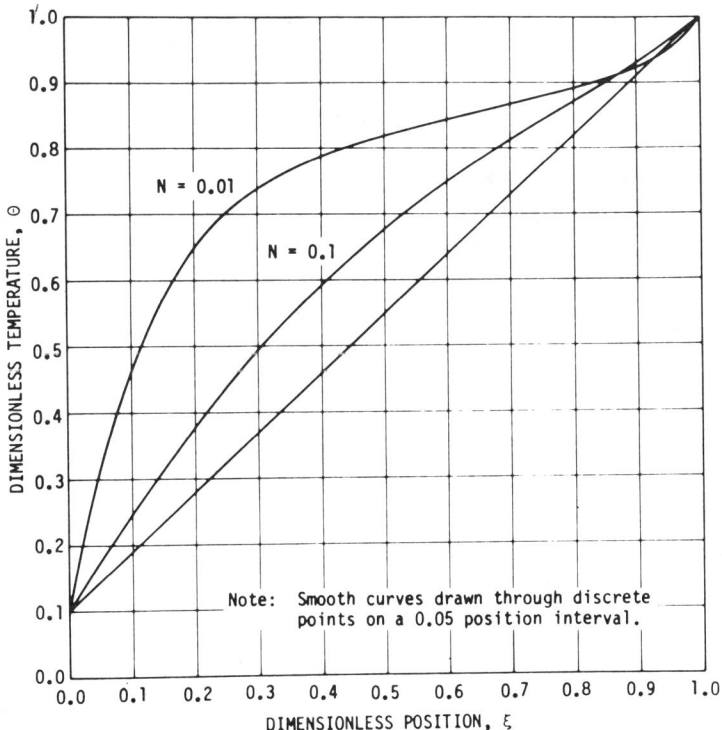

Fig. 8 Radiation and conduction in an absorbing-emitting media: $\Theta(0) = 0.1$, $\tau_0 = 1.0$.

For this study the walls of the plane parallel enclosure were assumed to be black emitters of radiation. An optical depth of 1.0 and a conduction-radiation parameter of 0.05 were considered. Dimensionless boundary temperatures of 0.5 and 0.0 were studied with albedo of scatter parameters of 0.0, 0.5, 0.9, and 1.0 applied to each. The results are presented in Figs. 10 and 11 and are found to agree well with the results of Ozisik.

Conclusion

For the problems considered, it has been shown that the FEM offers a viable means of solving continuum thermal radiation, as well as other energy transport problems. The radiation model developed to solve these problems was limited by the following assumptions:

1) The radiating medium is gray.

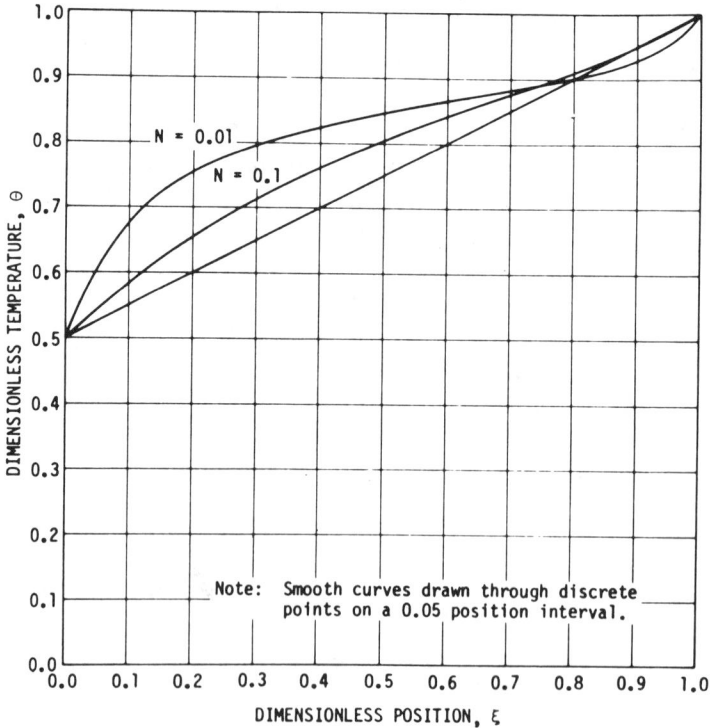

Fig. 9 Radiation and conduction in an absorbing-emitting media: $\Theta(0) = 0.5$, $\tau_0 = 1.0$.

2) The absorption, emission, and scattering characteristics of the medium are isotropic.

3) The optical properties of the medium are independent of temperature.

4) The boundary surfaces are opaque and gray with diffuse emission and reflection characteristics.

However, two of the assumptions outlined above may be generalized somewhat at the expense of computer resource usage. The gray medium assumption may be generalized to account for wavelength dependent properties through the use of a band approach. A two- or three-band model would impose only moderate computer usage penalties. The assumption of temperature-independent properties could be handled by employing an iterative approach involving the solution of the system with fixed properties and subsequent re-evaluation of the properties and radiosity network. Although certainly feasible, the approach

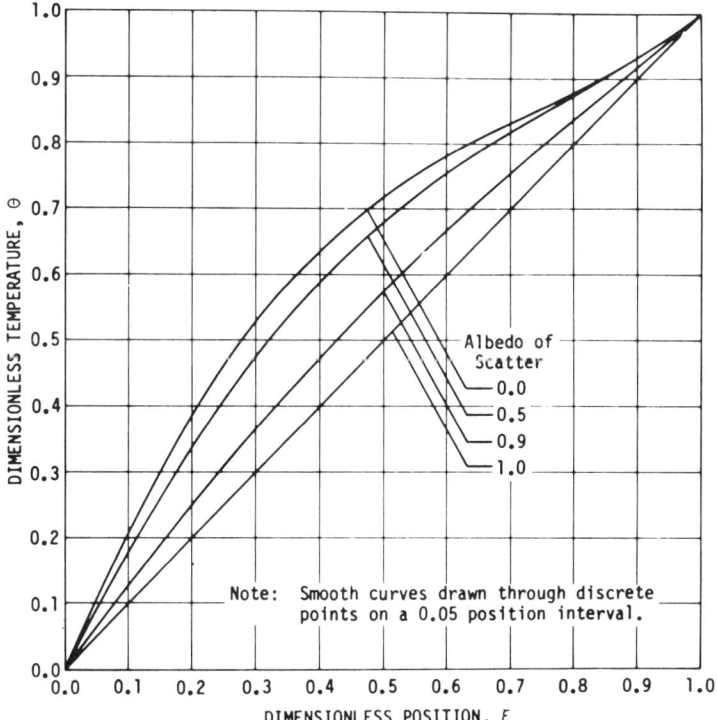

Fig. 10 Radiation and conduction in an absorbing-emitting and scattering media: $\theta(0) = 0.0$, $\tau_0 = 1.0$, $N = 0.05$.

would be quite time consuming, limiting its applicability to small systems. The assumption regarding the isotropic characteristics of the radiation processes within the medium is considered fundamental to the development of the radiation model. Systems requiring directional properties are better analyzed using techniques such as the Monte Carlo method.

For most digital computers a maximum practical problem size of approximately 200 nodes is indicated. This limitation arises mainly from the size of the coefficient matrix of the radiosity equation. A 200 node problem would require 40k words of storage to define the radiosity equation. In the general case, no zero elements will occur in the radiosity matrix since all nodes of a given system can interact with all other nodes. This limits extensive code savings through storing only non-zero terms. The use of peripheral mass storage and "out of core" matrix inversion algorithms would extend the problem size capabilities to limits dictated by economics.

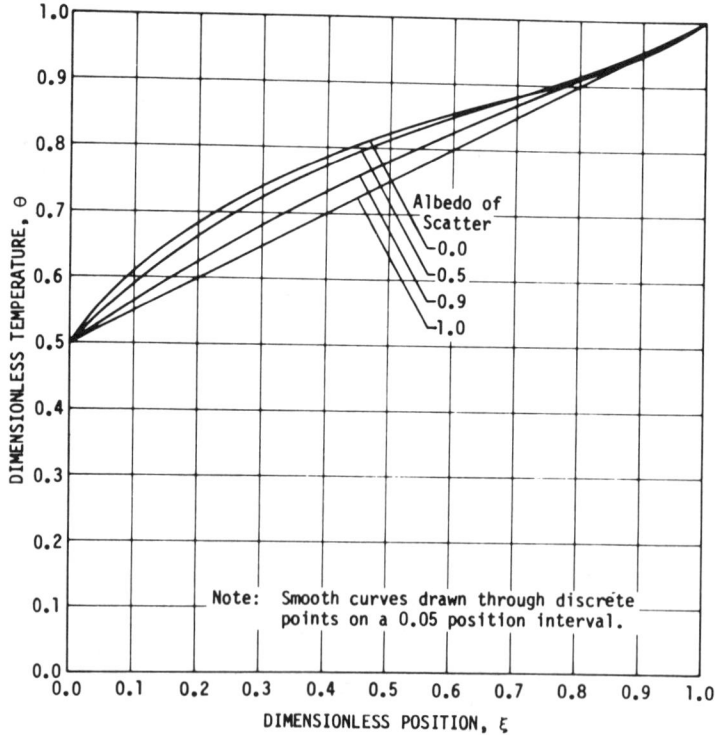

Fig. 11 Radiation and conduction in an absorbing-emitting and scattering media: $\Theta(0) = 0.5$, $\tau_0 = 1.0$, $N = 0.05$.

In summary, the thermal radiation finite-element modeling approach offers the following advantageous features in engineering problems;

1) Geometric generality and computational algorithm which is readily implemented on the digital computer.

2) Minimization of the quantity of input data necessary to describe the system in question because of the requirement for a common data base by both the radiation model and the FEM.

3) Provision by the discretization theory of the FEM of a functional basis for extension of the radiation model to higher order approximations.

As a final note, the authors of this paper would like to recognize the paper by Fernandes, Francis, and Reddy.[23] Their paper also considers utilization of the Galerkin finite-element method in solving the problem of combined conductive and radiative heat transfer in a participating medium. Examination of their paper will show that their results complement and support the results presented in this paper.

Acknowledgment

Work done by STW and REF was supported by Marshall Space Flight Center/NASA under Contract NAS8-28097.

References

[1] Chung, T. J., Finite Element Analysis in Fluid Dynamics, McGraw-Hill, Inc., New York, 1978.

[2] Turner, M. J., Clough, R. W., Martin, H. C., and Topp, L. P., "Stiffness and Deflection Analysis of Complex Structures," Journal of Aeronautical Science, Vol. 23, No. 9, 1956, pp. 805-823.

[3] Courant, R., "Variational Methods for the Solution of Problems of Equilibrium and Vibrations," Bulletin of American Mathematical Science, Vol. 49, 1943, pp. 1-23.

[4] Zienkiewicz, O. C., Mayer, P., and Cheung, Y. K., "Solution of Anisotropic Seepage Problems by Finite Elements," Proceedings of the ASCE, Vol. 92, EMI, 1966, pp. 111-120.

[5] Gallagher, R. H. and Mallett, R. H., "Efficient Solution Processes for Finite Element Analysis of Transient Heat Conduction," *Journal of Heat Transfer*, ASME, Vol. 93, No. 3, 1971, pp. 257-263.

[6] Oden, J. T., Wu, S. T., and Chung, T. J., "Analysis of Rarefied Gas Flow Through an Arbitrary Cross Section by the Finite Element Method," *Developments in Mechanics*, Vol. 7, 1973, pp. 289-299.

[7] Oden, T. J. and Wellford, L. C., "Analysis of Flow of Viscous Fluids by the Finite Element Method," *AIAA Journal*, Vol. 10, 1972, pp. 1590-1599.

[8] Wu, S. T., "Unsteady MHD Duct Flow by the Finite Element Method," *International Journal for Numerical Methods in Engineering*, Vol. 6, 1973, pp. 3-10.

[9] Lee, Hwa-Ping, "Application of Finite-Element-Method in the Computation of Temperature with Emphasis on Radiative Exchanges," AIAA 7th Thermophysics Conference, AIAA Paper No. 72-274, 1972.

[10] Wong, J. P. and Aguirre-Ramirez, G., "Numerical Solution of Integral Equations by Finite Element Method," Paper presented at 12th Annual Meeting of Society of Engineering Science, 1975.

[11] Wong, J. P. and Aguirre-Ramirez, G., "An Application of Finite Elements to a Radiation Problem," Paper presented at 8th Southeastern Conference on Theoretical and Applied Mechanics, 1977.

[12] Wassel, A. T., Edwards, P. K., and Catten, I., "Molecular Gas Radiation and Laminar or Turbulent Heat Diffusion in a Cylinder with Internal Heat Generation," *International Journal of Heat and Mass Transfer*, Vol. 18, 1975, pp. 1267-1276.

[13] Reddy, J. N. and Murty, V. D., "Finite-Element Solution of Integral Equations Arising in Radiative Heat Transfer and Laminar Boundary-Layer Theory," *Numerical Heat Transfer*, Vol. 1, 1978, pp. 389-401.

[14] Lee, Haw-Ping and Jackson, C. E., Jr., "Finite-Element Solution for a Combined Radiative-Conductive Analysis with Mixed

Diffuse-Specular Surface Characteristics," AIAA 10th Thermodynamics Conference, AIAA Paper No. 75-682, 1975.

[15] Eslinger, R. G. and Chung, B. T. F., "Periodic Heat Transfer in Radiating and Convecting Fins or Fin Arrays," AIAA, 1978.

[16] Higashihara, Hiromichi, "Micrometeorological Flow Induced by Heat Radiation of Ground Surface," 1979 (Other bibliographical data not available.)

[17] Ozisik, M. N., Radiative Transfer and Interactions with Conduction and Convection, John Wiley & Sons, New York, 1973.

[18] Strang, G., and Fix, G. J., An Analysis of the Finite Element Method, Prentice-Hall, Inc., Englewood Cliffs, N. J.

[19] Radin, Ervin Y., Editor, Computers and Mathematics with Applications, An International Journal, Vol. 5, No. 4, Dec. 1979.

[20] Sparrow, E. M., and Cess, R. D., Radiative Transfer, McGraw-Hill, Inc., New York, 1967.

[21] Heaslett, M. A. and Warming, R. F., "Radiative Transport and Wall Temperature Slip in an Absorbing Planar Medium," International Journal of Heat and Mass Transfer, Vol. 8, 1965, pp. 979-994.

[22] Viskanta, R. and R. J. Grosh, "Heat Transfer by Simultaneous Conduction and Radiation in an Absorbing Medium," Journal of Heat Transfer, Vol. 84C, 1962, pp. 63-72.

[23] Fernandes, R., Francis, J., and Reddy, J. N., "A Finite Element Approach to Combined Conductive and Radiative Heat Transfer in a Planar Medium," AIAA 15th Thermophysics Conference, July 1980.

[24] Heuber, K. H., The Finite Element Method for Engineers, John Wiley & Sons, New York, 1975.

A FINITE-ELEMENT APPROACH TO COMBINED CONDUCTIVE AND RADIATIVE HEAT TRANSFER IN A PLANAR MEDIUM

R. Fernandes,[*] J. Francis,[+] and J. N. Reddy[≠]
University of Oklahoma, Norman, Okla.

Abstract

Combined conductive and radiative heat transfer has been the subject of numerous investigations in recent years. The exact solution to the nonlinear integrodifferential equations involved is seldom possible, and typically one must resort to numerical approximations. The problem of heat transfer by simultaneous conduction and radiation in an absorbing, emitting, and scattering medium of plane parallel geometry has been solved by various numerical methods. The present study was undertaken to solve this problem using the Galerkin finite-element method. Results are presented for temperature profiles and heat flux at one boundary for both steady-state and transient cases.

Nomenclature

$D = 1 - 4(1-\varepsilon_1)(1-\varepsilon_2)E_3^2(\tau_o)$

$E_n(\tau) = \int_0^1 \mu^{n-2} \exp(-\tau/\mu) \, d\mu$

G = incident radiation
$I^+(\tau,\mu)$ = radiant intensity in the positive μ direction
$I^-(\tau,\mu)$ = radiant intensity in the negative μ direction
$I_b(T)$ = blackbody intensity
$I_o, \tilde{I}, \hat{I}, I_M, I_T$ = integral definitions
J = radiosity

Presented as Paper 80-1487 at the AIAA 15th Thermophysics Conference, Snowmass, Colo., July 14-16, 1980. Copyright © 1980, by J. E. Francis. Published by the American Institute of Aeronautics and Astronautics with permission.
    *Graduate Assistant.
    +Professor.
    ≠Professor (presently with Department of Engineering Science and Mechanics, Virginia Polytechnic Institute and State University, Blacksburg, Va.).

k	= thermal conductivity
M	= number of nodes used in the model
N	= dimensionless parameter defined as $k\beta/4n^2\sigma T_1^3$
n	= index of refraction
Q	= radiative heat flux in the x direction
q"	= total (conductive and radiative) heat flux
T	= absolute temperature
β	= extinction coefficient, $\beta = k + \sigma$
ε	= emissivity of a surface
η	= dimensionless incident radiation, $\eta = G/n^2\sigma T_1^4$
θ	= dimensionless temperature, $\theta = T/T_1$
κ	= absorption coefficient
μ	= cosine of polar angle
σ	= scattering coefficient, also Stefan-Boltzmann constant
τ	= optical depth of radiating material defined as $\tau = \int_0^x \beta dx$
τ'	= dummy integration variable
τ_o	= optical thickness of radiating material
Φ	= dimensionless radiative heat flux defined as $\Phi = Q/n^2\sigma T_1^4$
ϕ	= azimuthal angle
χ	= dimensionless radiosity, $\chi = J/n^2\sigma T_1^4$
Ψ	= dimensionless heat flux, $\Psi = q''/k\beta T_1$
ω_o	= albedo for single scattering, $\omega_o = \sigma/\beta$

Subscripts

1,2 refers to walls 1 and 2, respectively

Introduction

The study of heat transfer in insulating materials often involves simultaneous conductive and radiative heat transfer. This is also true in many other problems of practical importance to engineers including a myriad of problems with porous materials. Numerous investigators have studied combined conductive and radiative heat transfer in a planar medium [1-8].

Viskanta [1] studied combined conductive and radiative heat transfer in a planar medium with isotropic scattering and diffuse boundaries using a direct iterative scheme. Lii and Ozisik [8] studied the same problem including specular boundary conditions using the case normal mode expansion scheme, while Crosbie and Viskanta [4] studied the planar problem using the method of successive approximations. In this study we have addressed the planar problem with the Galerkin finite-element approach to a planar absorbing and emitting medium with isotropic scattering and diffuse boundaries. Important features of this method are its easy adaption to variable element sizes, that for the steady-state problems the boundary conductive heat flux may be directly determined, and that the method may be readily adopted to other geometries.

Analysis

For the plane parallel geometry shown in Fig. 1, the problem is axisymmetric. The physical model used is a plane gray layer of an isotropic scattering, absorbing, and emitting medium bounded by two diffuse surfaces at known temperatures. The properties are considered to be temperature independent.

The energy equation is represented by:

$$\rho c_p \frac{\partial T}{\partial t} + \frac{\partial}{\partial x}\left[-k \frac{\partial T}{\partial x} + Q\right] = 0 \tag{1}$$

The radiative flux term is given in terms of intensities in the positive and negative directions as

$$Q(\tau) = 2\pi \int_0^1 [I^+(\tau,\mu) - I^-(\tau,\mu)]\,\mu d\mu$$

The intensities I^+ and I^- are determined by the solution of the transport equation written in the positive and negative directions:

$$\frac{dI^+}{d\tau}(\tau,\mu) + \frac{1}{\mu} I^+(\tau,\mu) = \frac{\kappa n^2}{\mu \beta} I_b(T) + \frac{\sigma}{4\pi\beta\mu} G(\tau)$$

$$-\frac{dI^-}{d\tau}(\tau,\mu) + \frac{1}{\mu} I^-(\tau,\mu) = \frac{\kappa n^2}{\mu \beta} I_b(T) + \frac{\sigma}{4\pi\beta\mu} G(\tau)$$

where the irradiation $G(\tau)$ is given by:

$$G(\tau) = 2\pi \int_0^1 [I^-(\tau,\mu') + I^+(\tau,\mu')]\,d\mu'$$

The radiative boundary conditions used for diffuse surfaces are:

$$J_1 = \pi n^2 \varepsilon_1 I_b(T_1) + \rho_1 \int_0^{2\pi}\int_0^1 I^-(0,\mu)\,\mu d\mu d\phi$$

$$J_2 = \pi n^2 \varepsilon_2 I_b(T_2) + \rho_2 \int_0^{2\pi}\int_0^1 I^+(\tau_o,\mu)\,\mu d\mu d\phi$$

where $J_1 = \pi I^+(0)$ and $J_2 = \pi I^-(\tau_o)$.

A FINITE-ELEMENT APPROACH TO HEAT TRANSFER 95

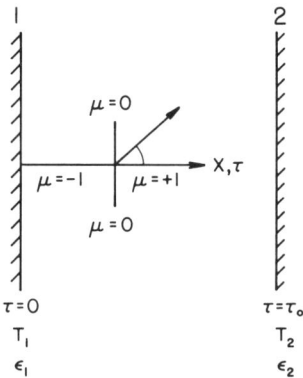

Fig. 1 Coordinate geometry.

Steady-State Solution

For the steady-state problem, nondimensionalizing the energy equation and boundary conditions yields:

$$\frac{d^2\theta}{d\tau^2} = \frac{(1-\omega_o)}{N}[\theta^4(\tau) - \frac{1}{4}\eta(\tau)] \quad (2)$$

where

$$\eta(\tau) = 2\{\chi_1 E_2(\tau) + \chi_2 E_2(\tau_o-\tau) + I_o(E_1(|\tau-\tau'|)\} \quad (3)$$

$$\chi_1 = [\varepsilon_1 + 2(1-\varepsilon_1)\{I_o(E_2(\tau'))+E_3(\tau_o)[\varepsilon_2\theta_2^4+2(1-\varepsilon_2)$$
$$I_o(E_2(\tau_o-\tau'))]\}]/D \quad (4)$$

$$\chi_2 = [\varepsilon_2\theta_2^4+2(1-\varepsilon_2)\{I_o(E_2(\tau_o-\tau'))+E_3(\tau_o)[\varepsilon_1$$
$$+2(1-\varepsilon_1) I_o(E_2(\tau'))]\}]/D \quad (5)$$

$$\Phi(\tau) = 2[\chi_1 E_3(\tau)-\chi_2 E_3(\tau_o-\tau)+\tilde{I}(E_2(\tau-\tau'))$$
$$- \hat{I} (E_2(\tau'-\tau))] \quad (6)$$

$$I_o(E_i(\xi)) = \int_0^{\tau_o}[(1-\omega_o)\theta^4(\tau') + \frac{\omega_o}{4} \eta(\tau')]E_i(\xi)d\tau'$$

$$\tilde{I}(E_i(\xi)) = \int_0^{\tau}[(1-\omega_o)\theta^4(\tau') + \frac{\omega_o}{4} \eta(\tau')]E_i(\xi)d\tau'$$

$$\hat{I}(E_i(\xi)) = \int_{\tau}^{\tau_o}[(1-\omega_o)\theta^4(\tau') + \frac{\omega_o}{4} \eta(\tau')]E_i(\xi)d\tau'$$

The total heat transfer is given by $q'' = -k(dT/dx) + Q$ which is nondimensionalized as

$$\Psi = -\frac{d\theta(\tau)}{d\tau} + \frac{1}{4N}\Phi(\tau) \tag{7}$$

The finite-element procedure is to divide the domain 0 to τ_o into a finite number (M-1) of linear elements. See Fig. 2. The unknown functions $\theta(\tau)$, $n(\tau)$, $\theta^4(\tau)$ are approximated over each element in the form

$$F_e(\tau) = \sum_{i=1}^{r} \psi_e^{(i)}(\tau) F_e^{(i)} \tag{8}$$

where $F_e(\tau)$ stands for any one of $\theta(\tau)$, $n(\tau)$, or $\theta^4(\tau)$ and $\psi_e^{(i)}(\tau)$ are the local interpolating functions. For linear interpolation $r = 2$. The interpolating functions for θ and θ^4 must be consistent such that the form chosen for one dictates the form of the other.

The integral terms are represented by a summation of integrals over the individual elements in the domain. Substitution of the assumed form of the unknown functions into the governing equations, multiplication by the shape function, and integating over the individual elements constitutes the Galerkin formulation. This procedure involves integration by parts of the terms involving second derivatives and results in equations for each element of the form:

$$-\int_{\tau_e^{(1)}}^{\tau_e^{(2)}} \frac{d}{d\tau}\psi_e^{(j)}(\tau) \frac{d}{d\tau} \sum_{i=1}^{r} \theta_e^{(i)} \psi_e^{(i)}(\tau) \, d\tau$$

$$+ \psi_e^{(j)} \frac{d\theta}{d\tau} \bigg|_{\tau_e^{(1)}}^{\tau_e^{(2)}}$$

$$= \frac{(1-\omega_o)}{N} \int_{\tau_e^{(1)}}^{\tau_e^{(2)}} \sum_{i=1}^{r} \theta 4_e^{(i)} \psi_e^{(i)}(\tau) \psi_e^{(j)}(\tau) d\tau$$

$$- \frac{(1-\omega_o)}{4N} \int_{\tau_e^{(1)}}^{\tau_e^{(2)}} \sum_{i=1}^{r} n_e^{(i)} \psi_e^{(i)}(\tau) \psi_e^{(j)}(\tau) d\tau \tag{9}$$

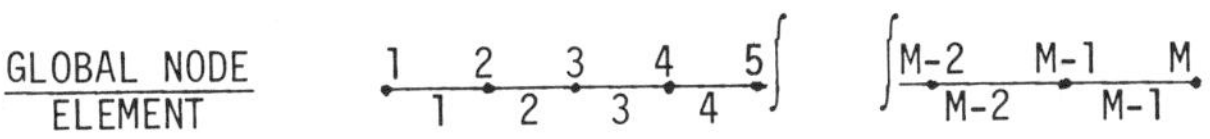

GLOBAL NODE	1 2 3 4 5	M-2 M-1 M
ELEMENT	1 2 3 4	M-2 M-1

Fig. 2. Relationship between global nodes and elements.

and

$$\int_{\tau_e^{(1)}}^{\tau_e^{(2)}} \sum_{i=1}^{r} \psi_e^{(i)}(\tau) \eta_e^{(i)} \psi_e^{(j)}(\tau) d\tau$$

$$= \frac{2\varepsilon_1 + 4(1-\varepsilon_1)E_3(\tau_o)\varepsilon_2\theta_2^4}{D} \int_{\tau_e^{(1)}}^{\tau_e^{(2)}} E_2(\tau)\psi_e^{(j)}(\tau) d\tau$$

$$+ \frac{(1-\varepsilon_1)\omega_o}{D} \int_{\tau_e^{(1)}}^{\tau_e^{(2)}} E_2(\tau)\psi_e^{(j)}(\tau) d\tau \quad I_M(E_2(\tau'))$$

$$+ \frac{2(1-\varepsilon_1)E_3(\tau_o)(1-\varepsilon_2)\omega_o}{D} \int_{\tau_e^{(1)}}^{\tau_e^{(2)}} E_2(\tau)\psi_e^{(j)}(\tau) d\tau$$

$$\times I_M(E_2(\tau_o-\tau'))$$

$$+ \frac{2\varepsilon_2\theta_2^4 + 4(1-\varepsilon_2)E_3(\tau_o)\varepsilon_1}{D} \int_{\tau_e^{(1)}}^{\tau_e^{(2)}} E_2(\tau_o-\tau)\psi_e^{(j)}(\tau) d\tau$$

$$+ \frac{(1-\varepsilon_2)\omega_o}{D} \int_{\tau_e^{(1)}}^{\tau_e^{(2)}} E_2(\tau_o-\tau)\psi_e^{(j)}(\tau) d\tau \quad I_M(E_2(\tau_o-\tau'))$$

$$+ \frac{2(1-\varepsilon_2)E_3(\tau_o)(1-\varepsilon_1)\omega_o}{D} \int_{\tau_e^{(1)}}^{\tau_e^{(2)}} E_2(\tau_o-\tau)\psi_e^{(j)}(\tau) d\tau$$

$$\times I_M(E_2(\tau'))$$

$$+ \frac{4(1-\varepsilon_1)(1-\omega_o)}{D} \int_{\tau_e^{(1)}}^{\tau_e^{(2)}} E_2(\tau)\psi_e^{(j)}(\tau) d\tau \quad I_T E_2(\tau'))$$

(Equation continued on next page.)

$$+ \frac{8(1-\varepsilon_1)E_3(\tau_o)(1-\varepsilon_2)(1-\omega_o)}{D} \int_{\tau_e^{(1)}}^{\tau_e^{(2)}} E_2(\tau)\psi_e^{(j)}(\tau)d\tau$$

$$\times \quad I_T(E_2(\tau_o-\tau'))$$

$$+ \frac{4(1-\varepsilon_2)(1-\omega_o)}{D} \int_{\tau_e^{(1)}}^{\tau_e^{(2)}} E_2(\tau_o-\tau)\psi_e^{(j)}(\tau)d\tau \quad I_T(E_2(\tau_o-\tau'))$$

$$+ \frac{8(1-\varepsilon_2)E_3(\tau_o)(1-\varepsilon_1)(1-\omega_o)}{D} \int_{\tau_e^{(1)}}^{\tau_e^{(2)}} E_2(\tau_o-\tau)\psi_e^{(j)}(\tau)d\tau$$

$$\times \quad I_T(E_2(\tau'))$$

$$+ 2(1-\omega_o)\int_{\tau_e^{(1)}}^{\tau_e^{(2)}} I_T(E_1(|\tau-\tau'|))\psi_e^{(j)}(\tau)d\tau$$

$$+ \frac{\omega_o}{2}\int_{\tau_e^{(1)}}^{\tau_e^{(2)}} I_M(E_1(|\tau-\tau'|))\psi_e^{(j)}(\tau)d\tau \qquad (10)$$

where

$$I_M(E_i(\xi)) = \sum_{f=1}^{M-1} \int_{\tau'_f{}^{(1)}}^{\tau'_f{}^{(2)}} \sum_{i=1}^{r} \psi_f^{(i)}(\tau')n_f^{(i)} E_i(\xi)d\tau'$$

and

$$I_T(E_i(\xi)) = \sum_{f=1}^{M-1} \int_{\tau'_f{}^{(1)}}^{\tau'_f{}^{(2)}} \sum_{i=1}^{r} \theta 4_f^{(i)} \psi_f^{(i)}(\tau') E_i(\xi)d\tau'$$

The simplest interpolating functions which satisfy the conditions for completeness and compatibility in this problem are the linear interpolating functions.

$$\psi_e^{(i)}(\tau) = (\tau_e^{(i+1)} - \tau)/(\tau_e^{(i+1)} - \tau_e^{(i)}) \qquad (11)$$

and

$$\psi_e^{(i+1)}(\tau) = (\tau - \tau_e^{(i)})/(\tau_e^{(i+1)} - \tau_e^{(i)})$$

where the superscripts denote the node index and the subscripts denote the element index.

A FINITE-ELEMENT APPROACH TO HEAT TRANSFER

The chosen shape functions are substituted into the equations and the integrals are evaluated.

Since we have more equations than unknowns, the equations must be assembled in the normal manner.[10,11] Setting $j = 1,2$ the flux term $\psi_e^{(j)} \frac{d\theta}{d\tau} \Big|_{\tau_e(1)}^{\tau_e(2)}$ in Eq. (9) reduces to

$-\frac{d\theta}{d\tau}\Big|_{\tau_e(1)}$ for $j = 1$, and $+\frac{d\theta}{d\tau}\Big|_{\tau_e(2)}$ for $j=2$. Thus from Eqs. (9) and (10) we see that for each element "e" we have 4 equations and 6 unknowns, i.e., θ_e, η_e, $d\theta/d\tau$ which are unknown at both nodes 1 and 2 of each element.

Thus for M-1 elements we have 4(M-1) equations and 6(M-1) unknowns. But the elements are connected such that the second node of an element is shared by the first node of the right side element. (See Fig. 2.) Thus $\frac{d\theta}{d\tau}\Big|_{\tau_e(2)} = \frac{d\theta}{d\tau}\Big|_{\tau_{e+1}(1)}$ since $\tau_e(2)$ and $\tau_{e+1}(1)$ are the same location. In order to reduce the number of unknowns, the equation corresponding to the second node i.e., $j = 2$ of element e is added to the equation corresponding to the first node i.e., $j = 1$ of element e+1. The only equations which remain unaffected are the equations for the first node of element 1 and the equation for the second node of element M-1. Adding the equations in this manner eliminates the conductive flux unknowns at the interior nodes. Thus we have 2M equations with 2M+2 unknowns, the unknowns being θ and η at each node $1, 2, 3 \ldots$, M plus the conductive heat flux $d\theta/d\tau$ at global node 1 and at node M. These 2M equations together with the two prescribed boundary conditions at $\tau = 0$, $\theta^{(1)} = 1$ at $\tau = \tau_o$, $\theta^{(M)} = \theta_2$ can now be solved by any suitable method.

The results obtained for the steady-state problem compare quite favorably with those of previous authors and are presented in Tables 1-4 and Figs. 3-6. These results are discussed in the results section of this paper.

Transient Solution

Nondimensionalizing Eq. (1) yields:

$$-\frac{\partial \theta}{\partial t^*} + \frac{\partial^2 \theta}{\partial \tau^2} = \frac{(1-\omega_o)}{N} [\theta^4(\tau, t^*) - \frac{n(\tau, t^*)}{4}] \quad (12)$$

where $t^* = (k\beta^2 t)/(\rho C_p)$. All other terms including the boundary conditions are as discussed previously. The initial condition is that at $t^* = 0$, $\theta = \theta_2$.

Equation (12) along with Eq. (10) and the known initial condition are the governing equations for the transient case.

Denoting the present time step as n and the next time step as n+1 we have, using the Crank-Nicolson scheme,

$$\{\theta\}_{n+1} = \{\theta\}_n + \frac{\Delta t}{2} \{\dot{\theta}\}_{n+1} + \frac{\Delta t}{2} \{\dot{\theta}\}_n \qquad (13)$$

where Δt is the time step. Applying Galerkin's formulation as before to Eq. (12) we have

$$-\int_{\tau_e^{(1)}}^{\tau_e^{(2)}} \sum_{i=1}^{2} \psi_e^{(i)}(\tau) \frac{\partial \theta_e^{(i)}}{\partial t^*} \psi_e^{(j)}(\tau) d\tau - \int_{\tau_e^{(1)}}^{\tau_e^{(2)}} \frac{d}{d\tau} \psi_e^{(j)}(\tau)$$

$$\frac{d}{d\tau} \sum_{i=1}^{2} \theta_e^{(i)} \psi_e^{(i)}(\tau) d\tau + \psi_e^{(j)} \frac{d\theta}{d\tau} \Big|_{\tau_e^{(1)}}^{\tau_e^{(2)}}$$

$$= \frac{(1-\omega_o)}{N} \int_{\tau_e^{(1)}}^{\tau_e^{(2)}} \sum_{i=1}^{2} \theta_e^{4(i)} \psi_e^{(i)}(\tau) \psi_e^{(j)}(\tau) d\tau$$

$$- \frac{(1-\omega_o)}{4N} \int_{\tau_e^{(1)}}^{\tau_e^{(2)}} \sum_{i=1}^{2} n_e \psi_e^{(i)}(\tau) \psi_e^{(j)}(\tau) d\tau \qquad (14)$$

On adding the appropriate equations in the manner described in the previous section we obtain from Eq. (14) the following matrix form for the M nodes

$$[G]_{M \times M} \{\dot{\theta}\}_{M \times 1} + [A1]_{M \times M} \{\theta\}_{M \times 1} + [B1]_{M \times M} \{n\}_{M \times 1} + [C1]_{M \times M} \{\theta^4\}_{M \times 1}$$

$$+ [D1]_{M \times M} \{\partial \theta / \partial \tau\}_{M \times 1} = 0 \qquad (15)$$

where [A1], [B1], [C1], [D1], and [G] are the coefficient matrices. For example [A1] is determined from assembly of the term

$$-\int_{\tau_e^{(1)}}^{\tau_e^{(2)}} \frac{d\psi_e^{(j)}}{d\tau}(\tau) \frac{d\psi_e^{(i)}}{d\tau}(\tau) d\tau$$

Table 1 Comparison of results with Viskanta[1a]

ω_0	$d\theta/d\tau$ $\tau=0$		$\Phi(0)$		Ψ	
	Viskanta[1]	This Paper	Viskanta[1]	This paper	Viskanta[1]	This paper
			$N = \infty$			
0	-0.5	-0.5	0.5717	0.571	0.5	0.5
0.25	-0.5	-0.5	0.5708	0.570	0.5	0.5
0.5	-0.5	-0.5	0.5662	0.566	0.5	0.5
0.75	-0.5	-.05	0.5661	0.552	0.5	0.5
1.0	-0.5	-0.5	0.5191	0.518	0.5	0.5
			$N = 0.10$			
0	-0.6415	-0.597	0.5327	0.539	1.9733	1.944
0.25	-0.5495	-0.565	0.5393	0.542	1.8978	1.920
0.50	-0.5175	-0.506	0.5428	0.542	1.8745	1.860
0.75	-0.4863	-0.482	0.5404	0.541	1.8373	1.834
1.0	-0.500	-0.500	0.5290	0.518	1.8225	1.795

[a] $\theta_1 = 1$; $\theta_2 = 0.5$; $\varepsilon_1 = \varepsilon_2 = 1.0$; $\tau_0 = 1$.

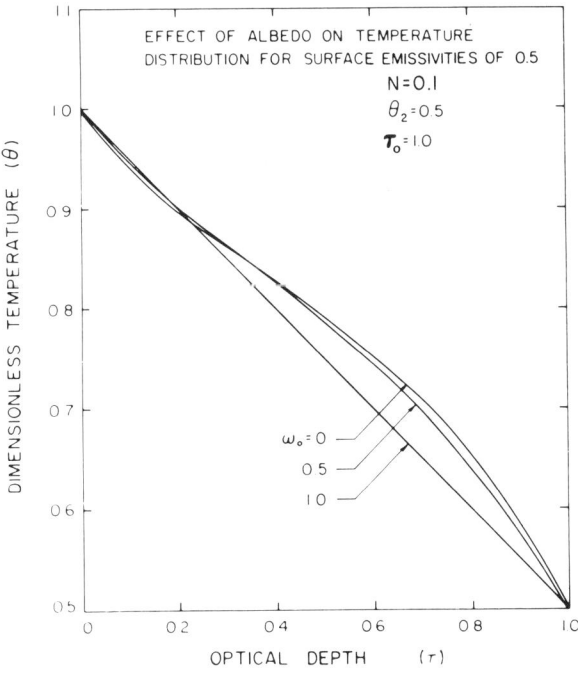

Fig. 3 Effect of albedo on temperature distribution.

Table 2[a]

$\varepsilon_1=\varepsilon_2$	$\frac{d\theta}{d\tau}\big|_{\tau=0}$		$\Phi(0)$		Ψ	
	Viskanta[1]	This paper	Viskanta[1]	This paper	Viskanta[1]	This paper
			(a) $\omega_0 = 0$			
1.0	-0.5019	-0.502	0.5656	0.565	0.6433	0.643
0.75	-0.5126	-0.512	0.4114	0.411	0.6154	0.614
0.5	-0.524	-0.524	0.2676	0.267	0.5909	0.590
0.25	-0.5376	-0.536	0.1332	0.131	0.5709	0.568
0.1	-0.5410	-0.543	0.0514	0.052	0.5538	0.556
0	-0.5469	-0.548	0.0	0.0	0.5469	0.548
			(b) $\omega_0 = 0.5$			
1.0	-0.4967	-0.496	0.5626	0.563	0.6374	0.637
0.75	-0.5034	-0.503	0.4142	0.415	0.6069	0.607
0.5	-0.5109	-0.510	0.2733	0.273	0.5792	0.579
0.25	-0.5187	-0.519	0.1366	0.137	0.5529	0.553
0.1	-0.5425	-0.525	0.0549	0.055	0.5382	0.539
0	-0.5288	-0.529	0.0	0.0	0.5288	0.529
			(c) $\omega_0 = 1.0$			
1.0	-0.5	-0.5	0.5251	0.518	0.6313	0.629
0.75	-0.5	-0.5	0.3810	0.378	0.5952	0.595
0.5	-0.5	-0.5	0.2505	0.245	0.5626	0.561
0.25	-0.5	-0.5	0.1268	0.119	0.5317	0.530
0.1	-0.5	-0.5	0.0525	0.047	0.5131	0.512
0	-0.5	-0.5	0.0	0.0	0.5	0.5

[a] $N = 1.0$, $\theta_1 = 1.0$, $\theta_2 = 0.5$, $\tau_0 = 1.0$.

with i = 1,2, j = 1,2 and e = element index terms. The other coefficient matrices are determined in a similar manner.

Multiplying Eq. (13) by [G] gives

$$[G]\{\dot{\theta}\}_{n+1} = \frac{2}{\Delta t} [G]\{\theta\}_{n+1} - \frac{2}{\Delta t} [G]\{\theta\}_n - [G]\{\dot{\theta}\}_n \qquad (16)$$

Evaluating Eq. (15) at time n+1 and using Eq. (16) we have on simplifying

$$[(2)/(\Delta t) [G] + [A1]]\{\theta\}_{n+1} + [C1]\{\theta^4\}_{n+1} + [B1]\{n\}_{n+1}$$
$$+ [D1]\{\partial\theta/\partial\tau\}_{n+1} = [G]\{\dot{\theta}\}_n + (2)/(\Delta t) [G]\{\theta\}_n \qquad (17)$$

Solution to Eqs. (16) and (17) together with the equations resulting from Eq. (10) give the transient solution.

Results

Steady-State Results

The results obtained were compared with those of other investigators.[1,4,8] Nondimensional temperature values are always in good agreement with published results. However, the initial slope comparisons for temperature are not always good. For cases where radiation dominates the mode of heat transfer elaborate schemes must be used for updating θ^4 in order to

Table 3 Comparison of results with Lii and Ozisik[8a]

N	ω_o	$-\frac{d\theta}{d\tau}\Big\|_{\tau=0}$	Lii and Ozisik[8]	Φ	Ψ	Lii and Ozisik[8]
0.5	0	−0.940	−0.9396	0.717	1.299	1.2981
	0.5	−0.947	−0.9491	0.684	1.289	1.2884
	1.0	−1.0	−1.0	0.552	1.276	1.2767
0.1	0	−0.854	−0.8520	0.666	2.519	2.5171
	0.5	−0.821	−0.8513	0.655	2.458	2.4437
	1.0	−1.0	−1.00	0.552	2.381	2.3835
0.05	0	−0.894	−0.8986	0.632	4.054	4.053
	0.5	−0.786	−0.7908	0.630	3.936	3.8845
	1.0	−1.0	−1.0	0.552	3.762	3.7670

[a] $\theta_1 = 1$, $\theta_2 = 0$, $\tau_o = 1$, $\varepsilon_1 = \varepsilon_2 = 1$

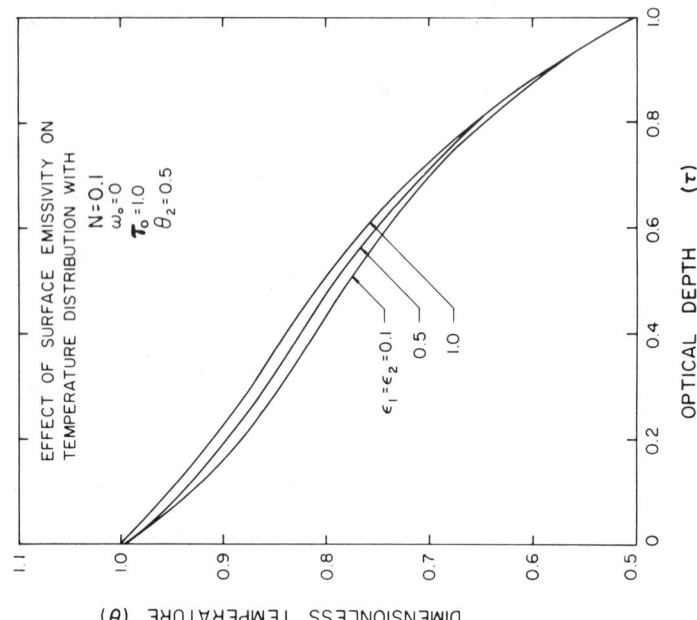

Fig. 5 Effect of surface emissivity on temperature distribution.

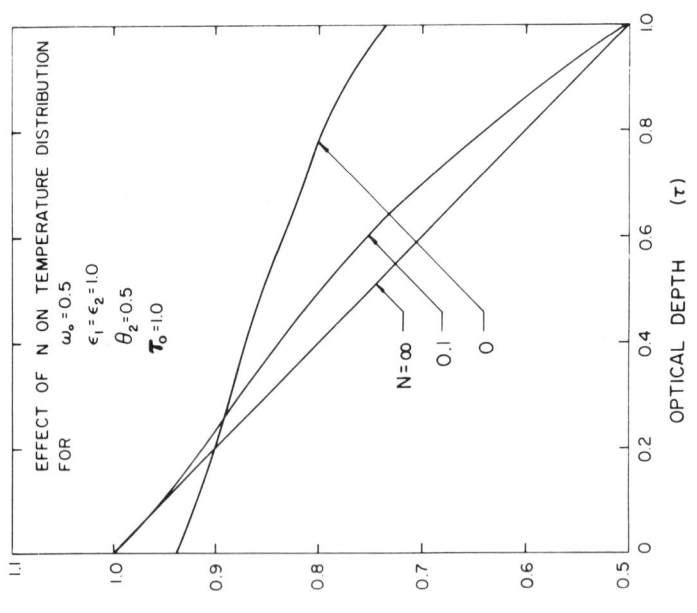

Fig. 4 Effect of dimensionless parameter N on temperature distribution.

A FINITE-ELEMENT APPROACH TO HEAT TRANSFER 105

Table 4 Comparison of results with Crosbie[4][a]

τ	$\theta = T/T_2$	
	Present study	Crosbie[1]
0	0.5	0.5
0.1	0.606	0.605
0.2	0.685	0.683
0.3	0.745	0.742
0.4	0.79	0.786
0.5	0.826	0.822
0.6	0.856	0.853
0.7	0.884	0.881
0.8	0.913	0.911
0.9	0.947	0.947
1.0	1.0	1.0

	$\left.\dfrac{d\theta}{d\tau}\right|_{\tau=0}$	$\Phi(0)$	ψ
Present Study	1.215	-0.427	-0.670
Crosbie	1.206	-0.42346	-0.6647

$\Phi(\tau) = Q(\tau)/n^2\sigma T_2^4$
$\psi(\tau) = q''/n^2\sigma T_2^4 = -4N\, d\theta(\tau)/d\tau + \Phi(\tau)$

[a] $N = 0.05$, $\omega_0 = 0$, $\varepsilon_1 = \varepsilon_2 = 1$, $\theta_1 = 0.5$, $\theta_2 = 1$, $\tau_0 = 1$ where $\theta = T/T_2$ and $N = k\beta/4n^2\sigma T_2^3$
For this case the nondimensionalization is done with respect to T_2.

obtain convergence. This results in more computer time required to solve a set of equations. The steady-state results for all but very small values of $(1-\omega_0)/N$ required the approximating functions for temperature to be modified from the conventional linear form to improve accuracy. Since in the steady state the solutions for η and θ^4 are approximately linear, we have chosen a linear approximation for them across each element.

A comparison of results obtained for the steady-state case is presented in Tables 1-4 and Figs. 3-6. When comparing with Viskanta[1] (Tables 1 and 2) good initial slope results are obtained for low $(1-\omega_0)/N$ i.e., conduction dominating. For large $(1-\omega_0)/N$, i.e., radiation dominating, the results are not as good even though the initial slope is still within 10%. When comparing to Crosbie[4,12] and Lii,[8] i.e., Tables 3 and 4, good results are obtained even for radiation dominated cases. In the former case the nondimensionalization was done in a different manner so that the conduction radiation parameter N

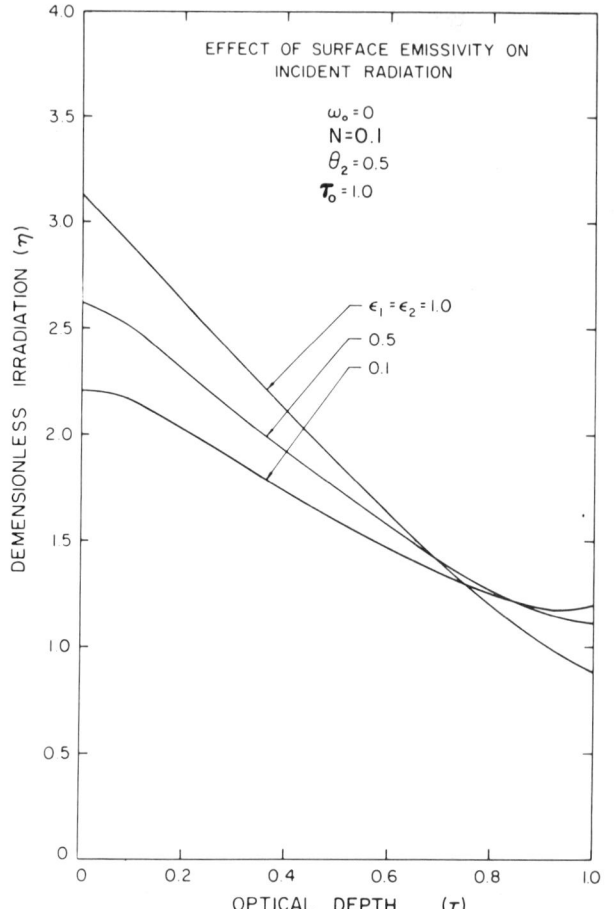

Fig. 6 Effect of surface emissivity on incident radiation.

is replaced with a new parameter. The appropriate parameters are defined with the tabular results.

Figure 3 shows the effect of scattering on the temperature distribution. An albedo of 1 implies that the media does not absorb but only scatters radiation and thus we would expect a straight line. But for $\omega_o \neq 1.0$ radiation in the medium is absorbed and we would expect the temperature of the medium to be nonlinear.

Figure 4 shows the effect of N on the temperature distribution. Given an ω_o other than 1, decreasing N increases the radiation in the media, thereby increasing the average temperature.

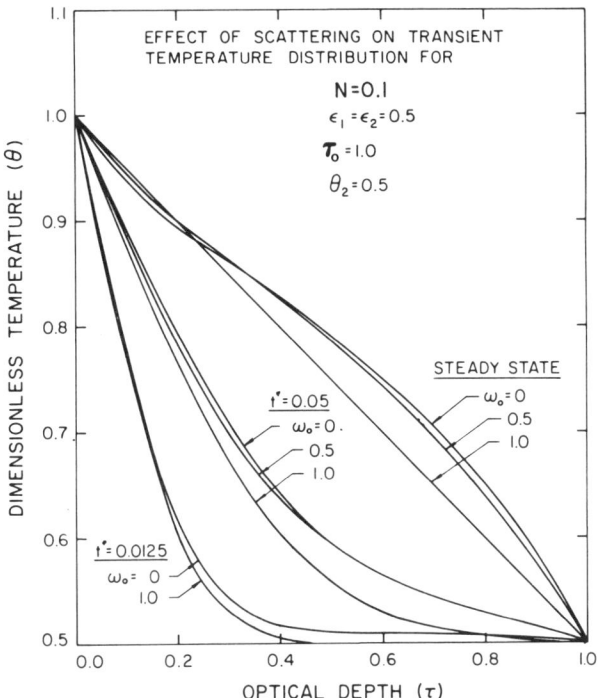

Fig. 7 Effect of scattering on transient temperature distribution.

Figure 5 shows the effect of surface emissivities on the temperature distribution. It is seen that the initial slope increases with a decrease in emissivity.

Figure 6 shows the effect of surface emissivities on the irradiation. At the hot surface, i.e., $\tau = 0$, the irradiation decreases as the emissivity decreases since less radiation is emitted. At the cold surface, i.e., $\tau = \tau_o$, a larger fraction of the incident radiation is reflected as emissivity is decreased and thus the irradiation increases.

Transient Results

The formulation for the transient case is similar to that of the steady-state case except that there is an added transient term in the conservation of energy equation. In the transient case θ^4 is not linear, so the assumption used in the steady-state formulation is not valid. A linear approximation for θ in the transient term, a linear profile for θ in the conduction term, and a linear approximation for η in the radiation term are good approximations for the transient problem.

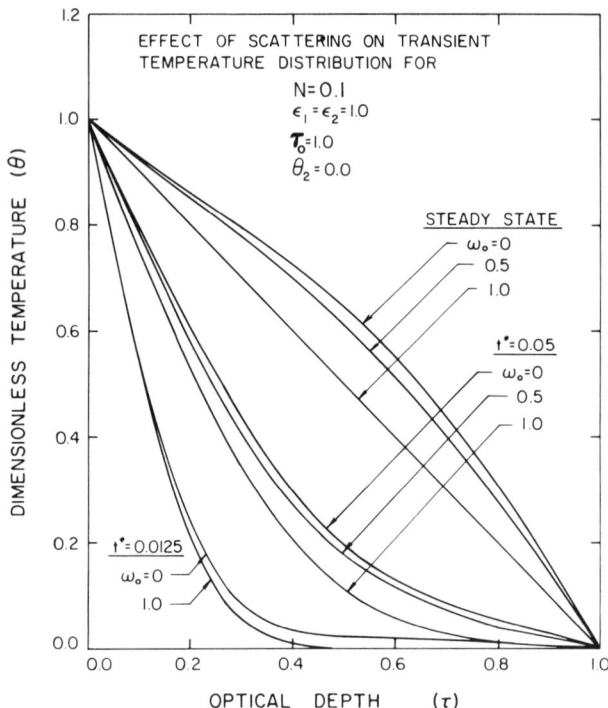

Fig. 8 Effect of scattering on transient temperature distribution.

For the transient case the initial slope was calculated from temperature results while for the steady-state case it was solved as a part of the system of equations.

The results obtained are in good agreement with those obtained by Lii and Ozisik.[8] Figures 7 and 8 show the effect of scattering on the temperature distribution for different cases.

This paper thus shows that the Galerkin finite-element technique is a viable means for solving the planar problem for heat transfer by simultaneous conduction and radiation through an absorbing, emitting, and scattering medium between two reflecting parallel plates.

References

[1] Viskanta, R., "Heat Transfer by Conduction and Radiation in Absorbing and Scattering Materials," *Journal of Heat*

Transfer, _Transactions of ASME_, Series C, Vol. 87, No. 1, Feb. 1965, pp. 143-150.

[2] Amlin, D. W. and Korpela, S. A., "Influence of Thermal Radiation on the Temperature Distribution in a Semi-Transparent Solid," _Journal of Heat Transfer, Transactions of ASME_, Vol. 101, Feb. 1979, pp. 76-80.

[3] Viskanta, R. and Grosh, R. J., "Heat Transfer by Simultaneous Conduction and Radiation in an Absorbing Medium," _ASME Journal of Heat Transfer_, Series C, Vol. 88, 1962, pp. 63-72.

[4] Crosbie, A. L. and Viskanta, R., "Interaction of Heat Transfer by Conduction and Radiation in a Nongray Planar Medium," _Warme-und Stoffubertragung_, Bd. 4, 1971, pp. 205-212.

[5] Weston, K. C. and Hauth, J. L., "Unsteady, Combined Radiation and Conduction in an Absorbing, Scattering, and Emitting Medium," _ASME Journal of Heat Transfer_, Vol. 95, Aug. 1973, pp. 357-364.

[6] Viskanta, R. and Grosh, R. J., "Heat Transfer by Simultaneous Conduction and Radiation," _International Journal of Heat Mass Transfer_, Vol. 5, 1962, pp. 729-734.

[7] Berganam, J. B. and Seban, R. A., "Heat Transfer by Conduction and Radiation in Absorbing and Scattering Materials," _ASME Journal of Heat Transfer_, Vol. 93, May 1971, pp. 236-238.

[8] Lii, C. C. and Ozisik, M. N., "Transient Radiation and Conduction in an Absorbing, Emitting, Scattering Slab with Reflective Boundaries," _International Journal of Heat Mass Transfer_, 13, 1970, pp. 517, 722.

[9] Viskanta, R. and Grosh, R. J., "Heat Transfer in a Thermal Radiation Absorbing and Scattering Medium," _International Developments in Heat Transfer_, Part IV, ASME, New York, 1961, pp. 820-828.

[10] Wong, J. P. and Aguirre-Ramirez, G., "Numerical Solution of Integral Equations by Finite Element Method," _Proceedings of the 12th Annual Meeting of the Society of Engineering Science_, University of Texas at Austin, 1975.

[11] Reddy, J. N. and Murty, V. D., "Finite-Element Solution to Integral Equations Arising in Radiative Heat Transfer and Laminar Boundary-Layer Theory," _Numerical Heat Transfer_, Vol. 1, 1978, pp. 389-401.

[12] Crosbie, A., Personal correspondence.

HEAT TRANSFER IN IRRADIATED SHALLOW LAYERS OF WATER

M. Behnia* and R. Viskanta[†]

Purdue University, West Lafayette, Ind.

Abstract

The paper reports on an experimental and analytical investigation of temperature structure in a shallow layer of water heated by an external radiation shource. Experiments have been performed in a test cell of known bottom reflection characteristics and suitable for optical observations. A shadowgraph and a Mach-Zehnder interferometer were used to visualize and measure the unsteady temperature profile in the water during irradiation. Experimental results obtained show that absorption of radiant energy by the substrate and/or by the water near the bottom results in an unstable situation and that natural convection flow develops. However, with continued heating the flow is suppressed. Analytical results, supported by the data, indicate that the spectral radiation flux incident on the water surface, the spectral radiation characteristics of the bottom, and the depth of the water have a decisive influence on the temperature distribution.

Nomenclature

c = specific heat
D = layer depth
$E_n(x)$ = exponential integral function,
$$\int_0^1 e^{-x/\mu} \mu^{n-2} d\mu$$

Presented as Paper 80-1519 at the AIAA 15th Thermophysics Conference, Snowmass, Colo., July 14-16, 1980. Copyright © American Institute of Aeronautics and Astronautics, Inc., 1980. All rights reserved.

*Heat Transfer Laboratory, School of Mechanical Engineering (presently with Department of Mechanical Engineering Teheran University, Teheran, Iran).

†Professor, Heat Transfer Laboratory, School of Mechanical Engineering.

HEAT TRANSFER IN IRRADIATED SHALLOW WATER LAYERS

F	= local radiative flux given by Eq. (5)
F_b^-	= radiative flux leaving (reflected) from the bottom given by Eq. (6)
F_{inc}^o	= external radiation flux incident on surface
g	= gravitational constant
k	= thermal conductivity of water
q	= heat flux
q^*	= dimensionless heat flux, q/F_{inc}^o
T	= temperature
$T_3(x)$	= exponential integral transmission function, $\int_0^1 \tau(\mu') \exp(-x/\mu) \gamma \mu' d\mu'$
t	= time
z	= distance measured from the surface
α	= thermal diffusivity or absorptance
β	= thermal expansion coefficient
γ_λ	= interreflection function defined as $[1 - \rho_{b\lambda}\rho_\lambda(\mu)\exp(-2\kappa_\lambda D/|\mu|)]^{-1}$
Θ	= dimensionless temperature, $\Theta = k(T - T_i)/F_{inc}^o D$
κ	= absorption coefficient or von Kármán constant
λ	= wavelength
μ	= direction cosine
ξ	= dimensionless distance, z/D
ρ	= density or reflectance
τ	= dimensionless time (Fourier number), $\alpha t/D^2$
Φ	= dimensionless radiative flux, F/F_{inc}^o
ψ_0	= fraction of incident flux absorbed at surface

Subscripts

b	= bottom or beam component
con	= convection
d	= diffuse component
i	= initial (uniform)
lat	= latent heat flux
o	= surface
rad	= radiation heat flux

Superscripts

o	= the air side

\+   = positive (downward) z direction
\-   = negative (upward) z direction

Introduction

The maintenance of thermal stratification in shallow[1] and salt gradient stratified[2-4] solar ponds is a problem of current engineering concern. Fluid motion in thermally stratified liquids, induced by buoyancy forces due to heating from below, is also of special interest currently in problems concerned with impounded bodies of water (i.e., lakes, reservoirs, ponds) and seasonal energy storage in artificial lakes. Natural convection plays an important role in establishing the vertical transport of energy and species in a thermally stratified water and in determining the temperature (density) distribution. For example, the buoyancy generated mixing processes in natural water bodies are important in the dispersal of material pollutants, in the transport of nutrients and biota, and in the dispersal of thermal effluents in cooling ponds and lakes.[5]

In a deep layer of water, the effect of the bottom on radiation transfer and on the temperature distribution is not expected to be significant. This is due to the penetration of the shortwave solar radiation only to a finite depth. For example, the percentage of surface irradiance reaching different depths of various types of ocean waters has been reported by Jerlov.[6] For clear oceanic water, less than 90% of the incident solar energy is absorbed within the first 30 m, while for very turbid coastal water more than 99% is absorbed within the first 6 m. It is therefore expected that only in shallow waters will the bottom and its radiation characteristics affect radiation transfer and volumetric heating rate. The penetration of a large fraction of the shortwave solar radiation in water will result in the heating of the bottom. Radiation transfer in layers of water have been studied and a comprehensive list of references can be found elsewhere.[6-8] Temperature distribution in a quiescent layer of water being stratified by radiant heating in a laboratory tank has been measured, and the good agreement between data and predictions[9,10] provides an indirect confirmation of the validity of the phenomenological radiation transfer models and gives some confidence in the approach used. However, it should be emphasized that neither the interest nor experimental conditions were such that the effect of the bottom on radiation transfer and temperature distribution could be studied.[9,10] There is some qualitative evidence[2,3,11,12] that natural convection motion is established near the bottom in a shallow layer of water heated by a radiation source from above. Heat transfer in shallow, irradiated layers of water does not appear to have been studied quantitatively.

It is the purpose of this paper to report on a combined experimental and analytical investigation of temperature structure in a shallow layer of water heated by an external radiation source. Carefully controlled laboratory experiments were performed using a test cell which was built to model a shallow layer of water with known bottom radiation characteristics. A Mach-Zehnder interferometer was used to visualize and measure the unsteady temperature profile in the water during heating. A shadowgraph and a thermistor located near the bottom were used to detect instabilities produced by heating of the bottom of the test cell due to absorption of radiation by the bottom or the water in its vicinity.

Experiments

Test Apparatus

A Mach-Zehnder interferometer was selected as a diagnostic tool for measuring the unsteady temperature distribution and inferring the fluid motion. The interferometer is an ideal instrument to study two-dimensional transport phenomena.[13] Since the instrument senses differences in the refractive index of fluid, it requires no physical contact of a foreign object with the fluid and does not disturb or distort the temperature field. The interferometer used in the study was of typical rectangular design with 25 cm diam optics. A He-Ne laser served as a light source, and a system of lenses with 25 cm diam parabolic mirrors produced a collimated beam.

The fluid motion was observed with a shadowgraph technique. This was simply achieved by blocking the collimated beam in the reference leg of the interferometer. The beam in the test leg of the interferometer (which passed through the test cell) created a shadowgraph image which was photographed. A calibrated thermistor (YSI44201 with 0.2 cm bead diam) was installed approximately 0.6 cm above the inside bottom of the test cell to detect local temperature fluctuations caused by fluid parcels departing from the vicinity of the plate.

A rectangular test cell (Fig. 1) with the inside dimensions of 10 cm along the optical path, 25 cm wide, and 20 cm high, was placed in the test leg of the interferometer. Optical quality glass windows, 15 cm wide and 1.5 cm thick, were

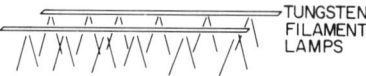
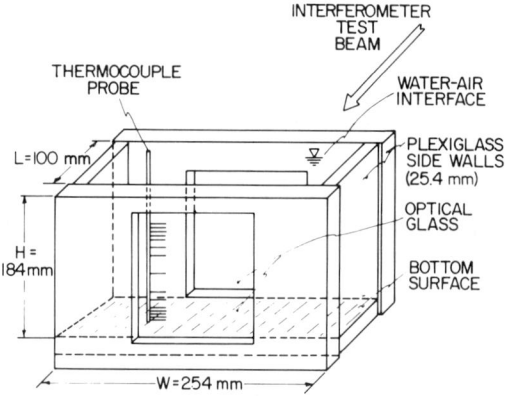

Fig. 1 Schematic diagram of the test cell.

installed into the plexiglass walls. A larger test cell would have been highly desirable. Unfortunately, the choice of diagnostics and the size of the interferometer optics did not permit this. The bottom of the cell was coated with a black paint or covered with a thin aluminum foil to simulate a highly absorbing and a highly reflecting surface. Two calibrated type T thermocouples were epoxied into the inlet and outlet tubes of the heat source/sink. A type T thermocouple was installed in the center of the bottom plate. Styrofoam insulation, 3 cm thick, was placed along the side walls and covered the optical windows. The bottom of the heat source/sink was insulated with 2.5 cm thick glass wool. The insulation from the windows could be easily removed in order to take photographs of the interference fringe pattern.

Procedure and Data Reduction

The test cell was cleaned and then filled with distilled water, covered, and left undisturbed for some time. The reason for this was to eliminate all convective currents which are normally present and also to attain a uniform ambient room temperature. After it had been determined that convective currents in the test cell are absent, all motors and compressors in the laboratory were turned off and the interferometer adjusted to an infinite fringe (a single all light or all dark fringe). A thin sheet metal shield, covered with aluminum foil, was placed approximately 30 cm above the test cell and covered the entire enterferometer leg in which the test cell rested. A hole in the shield protected other components from radiant heating and reduced free convection currents within the interferometer leg.

The water was heated from above by means of two high-temperature tungsten filament lamps in parabolic reflectors which known spectral radiation characteristics. The reflectors were designed to provide a radiation beam within a 5 cm wide, 25 cm long rectangular region. After the interferometer was aligned to the infinite fringe, the optical windows of the test cell were covered with styrofoam insulation to prevent cooling from the sides during heating. A few sheets of glass were placed directly above the test cell surface to attenuate the incident infrared radiation. The heating lamps were set at a predetermined voltage to give a desired filament temperature. The heating of the water with the radiant heaters simulating the solar irradiation was then initiated.

At prescribed time intervals, the optical window covers were removed, the interference fringe patterns photographed, and the thermocouple emf readings simultaneously recorded. The photographs were taken with a 35 mm Nikon F2-AS camera on Kodak Plus-X 125 ASA film. The distances between the water surface and the reference thermocouple needed for data reduction were measured before and after each experiment using a cathetometer. The incident radiative flux was measured at the start and at the end of each experiment using an Eppley radiometer.

The position of the interference fringes was measured using a vernier microscope accurate to $\pm$ 0.01 mm (corresponding to an actual distance of approximately $\pm$ 0.025 mm). Subsequently, the interferograms were interpreted using the known relation between index of refraction and temperature data[14] to obtain the temperature profiles. A single reference temperature needed to interpret the interferograms was measured using the previously mentioned type T thermocouple. The estimated accuracy of the temperature measurement is about $\pm$ 0.1 °C.

Analysis

Physical Model and Basic Equations

The purpose of the analysis is to model a shallow layer of water which is heated by solar radiation and to determine the effect of the depth and bottom characteristics on radiation transfer and the temperature distribution in the water. To make the problem mathematically and computationally tractable it is assumed that the layer is one-dimensional and is bounded by an opaque, adiabatic bottom. The free surface is exposed to a uniform radiation flux and is cooled by convection, latent energy transport, and longwave thermal radiation. The incident radiation flux is separated into "shortwave" ($0.1 < \lambda < 3$ μm) radiation of solar origin and "longwave" ($3 < \lambda < 150$ μm) radiation of terrestial origin. Because for $\lambda > 1$ μm the spectral absorption coefficient of water is of the order of 10^4 cm^{-1},[15] a large fraction of the incident flux will be absorbed in the immediate vicinity of the surface and thereby thermally stratify the upper layers.

For a one-dimensional layer of constant depth and infinite lateral extent, i.e., the edge effects are neglected, the energy equation for a quiescent, constant property incompressible fluid under the hydrostatic equilibrium condition is

$$\rho c \frac{\partial T}{\partial t} = k \frac{\partial^2 T}{\partial z^2} - \frac{\partial F}{\partial z} \qquad (1)$$

At the air-water interface ($z = 0$) the boundary condition,

$$-k \frac{\partial T}{\partial z}\bigg|_{z=0} = q_o(t) = q_{con} + q_{lat} + q_{rad} \qquad (2)$$

assumes that heat transfer occurs by convection of sensible and latent energy and longwave radiation of both solar and terrestial origin. At the bottom ($z = D$) the boundary condition is

$$k \frac{\partial T}{\partial z}\bigg|_{z=D} = q_b(t) = \int_0^\infty \alpha_{b\lambda} F_\lambda^+(z = D) d\lambda \qquad (3)$$

Equations (2) and (3) represent instantaneous energy balances at the two boundaries. The initial condition for the temperature distribution is taken as

$$T(z,0) = T_i \qquad (4)$$

Radiative Transfer

Radiative transfer in shallow layers of water bounded by a smooth free surface and diffusely reflecting bottom has been analyzed.[7,8,16] In all the models, the volumetric rate of shortwave radiant energy emission was assumed to be negligible in comparison with the rate of absorption of external radiation. Emission of longwave radiation was considered to be a surface phenomena because for $\lambda > 1$ μm water is effectively opaque. Scattering of shortwave radiation, including nonisotropic effects, has been accounted for. However, for relatively clean, shallow layers of water scattering effects are not expected to be significant and can therefore be neglected. In view of this, the analysis of Viskanta and Toor[7] is considered to be appropriate and will be used in predicting radiative transfer in water.

If as an approximation the radiative flux incident on the water surface $F^o_{inc,\lambda}$ is separated into a beam ($F^o_{b\lambda}$) and a diffuse ($F^o_{d\lambda}$) components, the local radiative flux can be expressed as[7]

$$F(z) = F^+(z) - F^-(z)$$

$$= 2 \int_0^\infty \{\frac{1}{2} \mu^o \tau_\lambda(\mu^o) F^o_{b\lambda} \gamma_\lambda(\kappa_\lambda D, |\mu|) e^{-\kappa_\lambda z/\mu}$$

$$+ F^o_{d\lambda} T_3[\kappa_\lambda(D-z)] - F^-_{b\lambda} E_3[\kappa_\lambda(D-z)]\} d\lambda \tag{5}$$

where the upward radiative flux leaving the bottom is given by

$$F^-_{b\lambda} = 2\rho_{b\lambda} \{(1/2)\mu^o \tau_\lambda(\mu^o) F^o_{b\lambda} \gamma_\lambda(\kappa_\lambda D, |\mu|) e^{-\kappa_\lambda D/\mu}$$

$$+ F^o_{d\lambda} T_3(\kappa_\lambda D) \tag{6}$$

The direction cosine μ inside th water is related to the direction cosine μ^o in air by Snell's law of refraction. The transmissivity $\tau_\lambda(\mu^o)$ of the air-water interface is predicted by Fresnel equations. Since the radiative transfer was considered to be quasi-steady, Eq. (6) for the local radiative flux is valid where the radiation flux incident on the water surface $F^o_{inc,\lambda}(=\mu^o F^o_{b\lambda} + F^o_{d\lambda})$ is a function of time.

Method of Solution

The model equations were solved using an implicit finite difference scheme. The equations were first nondimensionalized such that the energy Eq. (1) become

$$\frac{\partial \Theta}{\partial \tau} = \frac{\partial^2 \Theta}{\partial \xi^2} - \frac{\partial \Phi}{\partial \xi} \tag{7}$$

and the boundary and initial conditions, Eqs. (2)-(4), became

$$-\left.\frac{\partial \Theta}{\partial \xi}\right|_{\xi=0} = \psi_0 + q_0^* \quad \text{at } \xi = 0 \tag{8}$$

$$\left.\frac{\partial \Theta}{\partial \xi}\right|_{\xi=1} = q_b^* \quad \text{at } \xi = 1 \tag{9}$$

and

$$\Theta(\xi,0) = 0 \tag{10}$$

respectively. In Eq. (8) ψ_0 represents the fraction of the incident radiation flux absorbed at the surface in the long-wave part of the spectrum where water is opaque to radiation. This fraction is determined from the solution of the radiative transfer problem, Eqs. (5) and (6). It has been found[9] that such a separation is realistic physically and improves the accuracy of the results.

Details of the numerical calculations are given elsewhere[17]. It should suffice to note that the total radiative flux $F(z)$ was calculated using spectral absorption coefficient data for water[15] by first specifying the spectral incident flux $F^Q_{inc,\lambda}$. A nonuniform grid spacing was used, with a much finer spacing near the free surface. The heat flux at the free surface is a function of the surface temperature T_0, the radiation characteristics of water, and the ambient parameters, e.g., air temperature and humidity, mode of cooling (forced or free convection), etc. An analysis for modeling heat transfer at the air-water interface is described elsewhere.[9,17]

Results and Discussion

Radiative Transfer

There are a large number of independent parameters, and therefore it is possible to present only some sample results. The effect of the spectral distribution of the incident radiation on the total radiative flux and its divergence are given in Figs. 2 and 3, respectively. The results are given for solar radiation (air mass of unity) and radiation emitted by

HEAT TRANSFER IN IRRADIATED SHALLOW WATER LAYERS   119

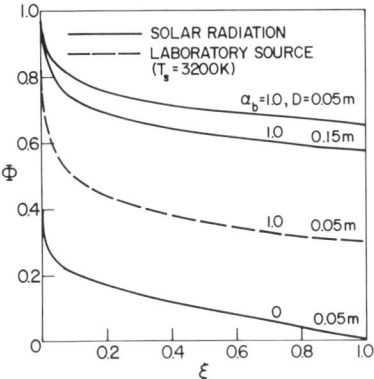

Fig. 2 Local radiative flux in water; incident solar radiation is for air mass of unity, $F^0_{d\lambda} = 0$.

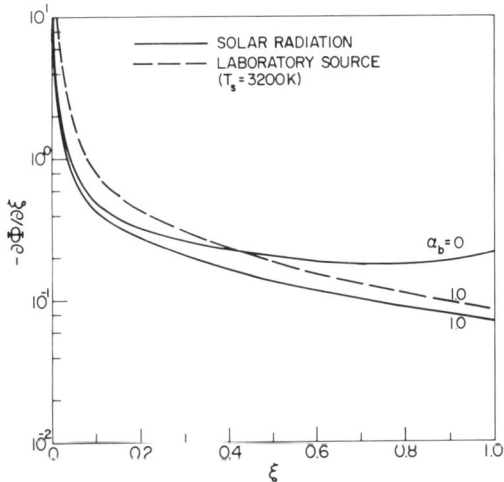

Fig. 3 Local radiative flux divergence in water; incident solar radiation is for air mass of unity, $F^0_{d\lambda} = 0$.

heaters used in the laboratory phase of the study and operating at a temperature of 3200 K. In the calculations, the optical properties of distilled water as given by Hale and Querry[15] were used. Since it has been shown elsewhere[7] that the effects of directional distribution of the incident radiation field on the radiative flux divergence are relatively small, the results reported are for the case of $F^0_{d\lambda} = 0$.

Figure 2 shows that for the same water layer depth and bottom characteristics there is greater attenuation of radiation emitted by the source than by the Sun. This is due to the fact that a much larger fraction of the radiation emitted by the laboratory source is in the longer wavelength part of the spectrum where water is a relatively strong absorber of radiation. As expected, the results also show that the radiation flux at the perfectly reflecting bottom vanishes and that it is greatly reduced at the air-water interface.

The results for the radiation flux divergence ($\partial\Phi/\partial\xi$), presented in Fig. 3, show that the divergence decreases very sharply near the surface because of the strong attenuation of longwave ($\lambda > 1$ μm) radiation by the water. The radiation characteristics of the bottom influence the absorption of radiation primarily near the bottom. The difference between results near the surface ($\xi < 0.05$) for the perfectly abosorbing and reflecting bottoms is negligible. For a water layer 15 cm deep and $\xi > 0.05$ the divergence differs by less than 1% from that for a 5 cm deep layer. Separate curves could not be clearly drawn and are therefore not included in the figure. A comparison of the results reveals that even though the trends are the same, the divergences are significantly greater for the laboratory than for the solar radiation source. This greater absorption of radiation by water is due to the fact that a much larger fraction of the incident radiation is in the longwave part of the spectrum.

The temperature distributions predicted from the model equations are compared in Fig. 4. The results show that the temperature of water layer with a black substrate increases much more rapidly than the one with a reflecting substrate. This is due primarily to the absorption of radiation by the substrate which is transferred to the water above it. Because a significant fraction of energy incident is lost from a water layer substrate system having a perfectly reflecting bottom the temperature is lower. Comparision of Figs. 4a and 4b show that the temperature profiles are similar; however, the temperature at the bottom of the layer is lower for the deeper layer of water (Fig. 4b) because the radiation flux reaching the substrate is smaller (see Fig. 2). The results of Fig. 4 clearly show that for the black substrate the situation is inherently unstable, and that natural convection flow can be expected. Depending on the local rate of radiation absorption and the heating rate at the surface, buoyancy induced mixing may penetrate the air-water interface. Thermal stability of fluid layers with nonuniform volumetric heat sources has been studied and the effects of the relevant parameters established.[18] Since the model equation, Eq. (1), does not account for bouyancy-induced mixing, it is expected to overpredict the temperature near the bottom.

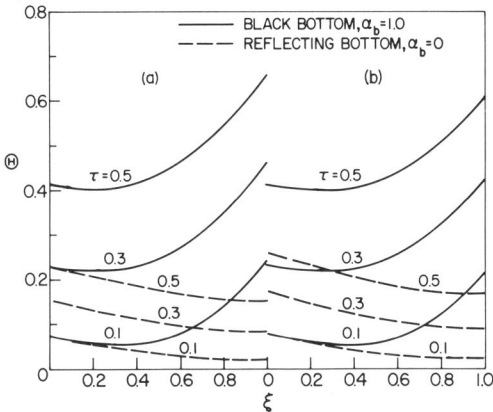

Fig. 4 Temperature profile during radiant heating with incident solar spectrum corresponding to air mass of unity ($F^0_{d\lambda} = 0$): a) D = 0.05 m and b) D = 0.15 m.

Experimental Observations

Some selected photographs of the interference fringe pattern illustrating the heating of an initially uniform layer of water are shown in Fig. 5 (the small dark image in the photographs is a turning mirror). In this particular experiment the bottom was coated with a "black" paint having a total absorptivity of $\alpha_b = 0.92$ and the layer of water was 10 cm deep. The reference thermocouple is seen in the photographs. The dark gray area at the top of the photographs is due to intense radiant heating of the water. The very large fringe density that has resulted makes it impossible to distinguish the fringes from one another in this particular region.

In Figs. 5a and 5b two clearly distinct regions can be demarcated. An intensely stratified region near the interface indicates that the surface is being heated, and the temperature decreases with depth in this region. Near the bottom of the test cell fringes are some distance apart indicating a less intense heating of this layer. The fringes are more closely spaced near the bottom than the overlying region. The overall interference fringe pattern can be explained by the fact that a large fraction of the incident radiation transmitted into the water is absorbed near the surface which results in decrease of temperature with depth. As the radiation propagates through the water it is partially attenuated. Upon reaching the bottom of the test cell the radiation is almost totally absorbed by the nearly black-coated bottom. Therefore, the bottom is heated and its temperature raised, which in turn causes an in-

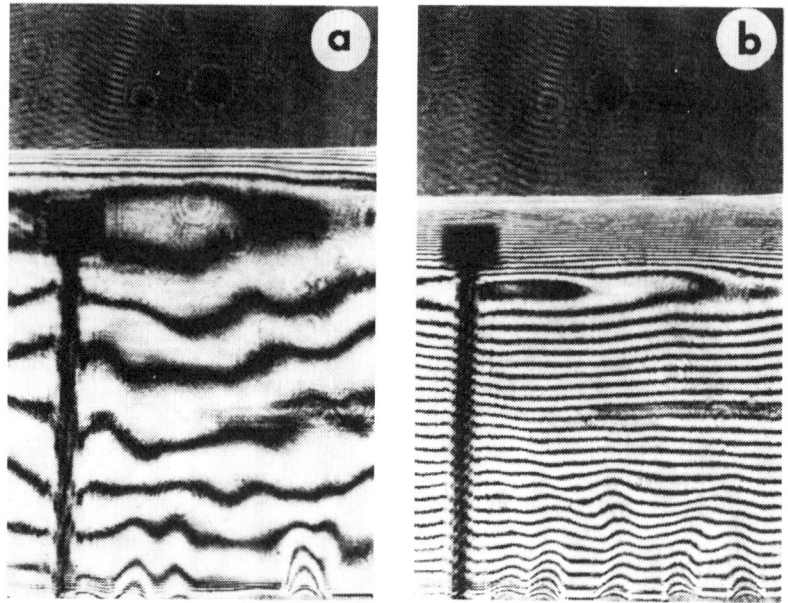

Fig. 5 Selected interferograms during radiant heating of an initially uniform layer of water, D = 0.1 m, $\alpha_b \simeq 0.92$, T_i = 20°C: a) t = 4 min and b) t = 16 min.

crease in the local temperature above the bottom. The particular shape of temperature profile near the bottom is also partially due to the reflected portion of the radiation reaching the bottom which is absorbed as it propagates upward (i.e., in the opposite direction of the incident beam). The heating of the bottom of the test cell was clearly observed by a continuous increase in the emf output of the thermocouple installed into the bottom plate.

Figure 5b shows a small region of relatively uniform temperature water about 6 cm above the bottom plate. In this region the fringes are reversed indicating a reversal in the temperature gradient. This particular profile shows that temperature decreases with depth (negative gradient) above the reversal and temperature increases with depth (positive gradient) below it. Interferograms recorded at different times suggested that the location of this temperature reversal was nearly fixed once it was established.

Absorption of radiation directly by the bottom of the test cell and the water in its vicinity results in establishment of natural convection motion if critical conditions are reached and exceeded. Figure 5b provides evidence that thermal plumes

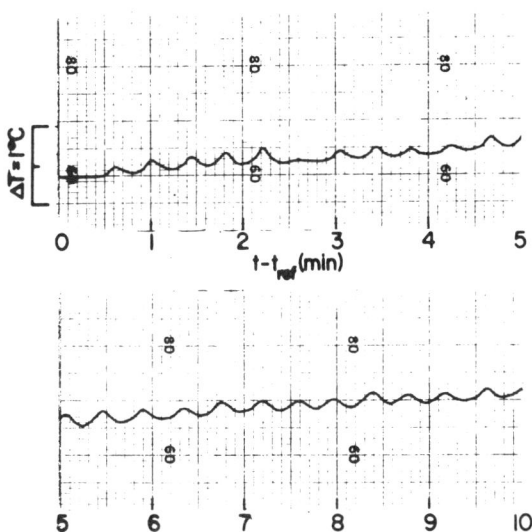

Fig. 6 Thermistor output for experiment described in Fig. 5.

are formed in the water right above the bottom of the test cell. The temperature fluctuations near the bottom were detected by the thermistor and recorded on a strip chart. Figure 6 presents a sample of strip chart output of the thermistor for this experiment. The chart indicates that the average temperature of the water at the location of the thermistor (about 6 mm above the bottom) increases with time. Also, fluctuations of about ± 0.15°C are detected. The temperature of a plume was approximately 0.15°C higher than the bulk temperature of the surrounding fluid. The temporal frequency of the plume generation is not fixed. However, for this particular experiment in a time period of 1 min output voltage peaks 2.6 times. This corresponds to an average plume departure frequency of about 2.6/min. Shadowgraph observations of instabilities and plume activity could not be recorded photographically even though they were clearly visible on a screen. This was due to small density differences and the averaging effect along the beam as it traversed the test cell.

Two typical photographs of the interference fringe patterns for a different experiment are shown in Fig. 7. In this experiment the bottom of the test cell was covered with a thin aluminum foil. The interferograms do not cover the entire water layer depth and show only the region of primary interest near the bottom. In the photographs the squares in the background are 1 cm × 1 cm. The greater fringe density near the bottom (rather than some distance above) indicates a steeper

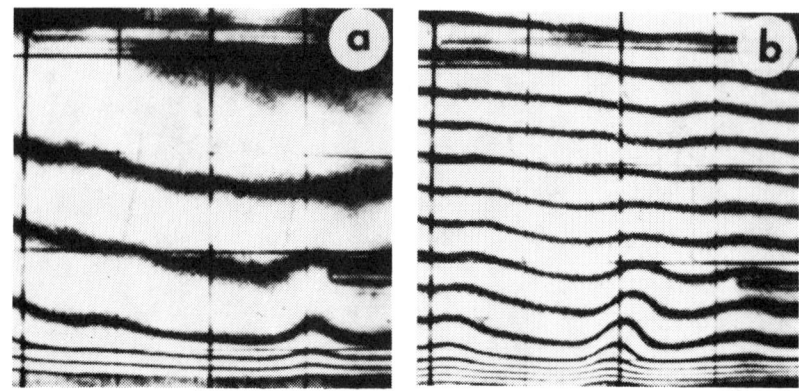

Fig. 7 Photographs of interference fringes during radiant heating of an initially uniform layer of water, $D = 0.1$ m, $\alpha_b \simeq 0.06$, $T_i = 22.5°C$: a) $t = 3$ min and b) $t = 6$ min.

temperature gradient, and the increase in fringe density with time (e.g., compare Figs. 7a and 7b) provides evidence that the water continues to be stratified. Comparison of Figs. 5 and 7 shows that the stratification near the bottom is more intense for the highly absorbing than for the highly refelcting test cell bottoms.

Comparison of Temperature Predictions with Data

A comparison of the measured and predicted temperature distributions shown in Fig. 8 indicates good agreement between the results for $\xi < 0.5$. There is some discrepancy between data and predictions in the vicinity of the air-water surface which is attributed to an inadequate modeling of the instantaneous heat flux at the water surface $q_o^*(\tau)$. In the experiment discussed a glass sheet was placed a few centimeters above the water surface. As a result of a rather complex physical situation, modeling of longwave radiation and latent energy transport at the surface may not have been completely satisfactory. The analytical model is inadequate for predicting the temperature structure of water near the bottom of the test cell because it does not take into account the natural convection which was experimentally observed. Experimental results show that because of the effective mixing the temperature of water for $\xi > 0.5$ is practically uniform. Because the tungsten filament lamps operating at high temperature aged (degraded) during the experiment as a result of opaque deposits on the quartz envelope, the experiments were of a relatively short

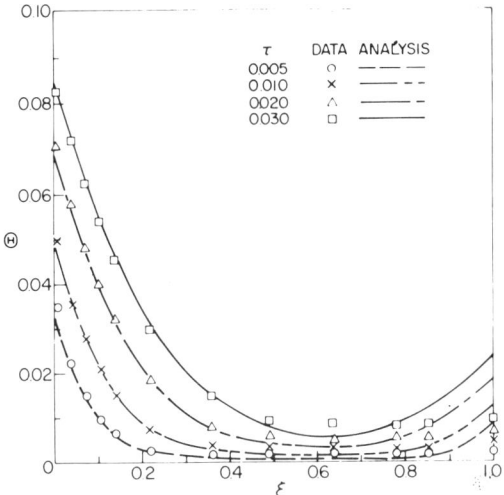

Fig. 8 Comparison of measured and predicted temperature distributions during radiant heating of an initially uniform layer of water: $D = 0.14$ m, $T_i = 21.75°C$, $F^0_{inc} = 2170$ W/m^2, $\alpha_b \simeq 0.92$.

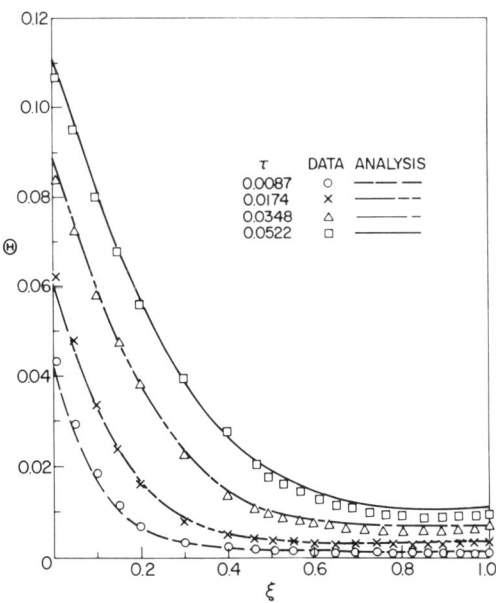

Fig. 9 Comparison of measured and predicted temperature distributions during heating of an initially uniform layer of water: $D = 0.1$ m, $T_i = 22.3°C$, $F^0_{inc} = 2060$ W/m^2, $\alpha_b \simeq 0.06$.

duration (typically not lasting more than 60 min). The radiation flux incident on the water surface was measured before and after the experiment and an average value used in the data reduction and analysis. Because of the intense heating, the interference fringe density in the upper layers of water was very high, making the interpretation of fringes impossible. Therefore, the temperature distributions for this experiment were measured with thermocouples.

Attempts were made to account for turbulent mixing of water near the bottom of the cell by using the turbulent conductivity derived by assuming that production of turbulent kinetic energy by buoyancy is in quasi-steady equilibrium with the dissipation of turbulent kinetic energy.[19] The effective conductivity of water in the unstable region near the bottom was approximated by

$$k_{eff} = k + \rho c [\kappa (D-z)]^2 (g\beta \frac{\partial T}{\partial z})^{1/2} \qquad (11)$$

Unfortunately, use of this expression in unsteady one-dimensional energy equation did not yield meaningful results because it greatly overpredicted the turbulent conductivity.

A comparison of measured and predicted temperature distributions for an experiment in a test cell with a highly reflecting bottom (covered with aluminum foil) is shown in Fig. 9. Because the fringe density near the water surface was very large, only the temperatures in the bottom region were determined interferometrically. Inspection of the figure reveals relatively good agreement between the data and predictions (even near the bottom). This is primarily due to the fact that only a small amount of radiant energy was directly absorbed by the bottom. Even though the thermistor and the interference fringes indicated plume activity during early times, the buoyance-driven mixing was much weaker than for the case of a highly absorbing bottom and was eventually completely suppressed as the heating continued.

Conclusions

The experimental and analytical results clearly demonstrate that absorption of radiation by the water and bottom can produce an unstable situation that may result in natural convection mixing and energy transport in the layer immediately adjacent to the bottom. Even though the spectral characteristics of incident solar radiation could not be simulated in the laboratory, the fact that a larger fraction of the solar energy is in the shortwave part of the spectrum suggests that the phenomena observed in the experiments is expected to occur in relatively shallow quiescent waterbodies and solar ponds. However, no detailed observations giving evidence of this particular phenomenon appear to have been reported in the literature.

There is a need for models to predict buoyancy-generated turbulence and mixing in thermally stratified fluid layers which are being heated by absorption of external radiation and simultaneously cooled from the surface by sensible and latent energy transport and longwave radiation and/or heated from below by absorption of radiation by the bottom. Such models could then be used in thermal hydraulics equations to predict flow and thermal structure in solar ponds, natural waters, and artificial lakes for seasonal energy storage.

Acknowledgments

This work was supported in part by the United States Department of Interior, Office of Water Research and Technology, under Matching Grant Project OWRT B-077-IND. The authors wish to ackowledge the assistance of Mr. C.-J. Ho with data reduction and calculations.

References

[1] Savage, S. B., "Solar Pond," Solar Energy Engineering, edited by A. M. Sayigh, Academic Press, New York, 1977, pp. 217-232.

[2] Rabl, A. and Nielson, C., "Solar Ponds for Spacing Heating," Solar Energy, Vol. 17, No. 1, 1975, pp. 1-12.

[3] Styris, D. L., Harling, O. K., Zaworski, R. J., and Leshuk, J., "The Non-Convective Solar Pond Applied to Building and Space Heating," Solar Energy, Vol. 18, No. 3, 1976, pp. 245-252.

[4] Jaydev, T. S. and Henderson, J., "Salt Concentration Gradient Solar Ponds - Modeling and Optimization," Solar Energy Research Institute Report SERI/TP-34-277; also paper presented at the Annual Meeting of the International Solar Energy Society, Atlanta, Ga., May 28-June 1, 1979.

[5] Zaric, Z., Ed., Thermal Effluent Dispersal From Power Generation, Hemisphere Publishing Corporation, Washington, D.C., 1977.

[6] Jerlov, N. G., Marine Optics, Elsevier Publishing Co., Amsterdam, 1976.

[7] Viskanta, R. and Toor, J. S., "Absorption of Solar Radiation in Ponds," Solar Energy, Vol. 21, No. 1, 1978, pp. 17-25.

[8] Daniel, K. J., Laurendeau, N. M., and Incropera, F. P., "Prediction of Radiation Absorption and Scattering in Turbid Water Bodies," ASME Journal of Heat Transfer, Vol. 101, No. 1, 1979, pp. 63-67.

[9] Snider, D. M. and Viskanta, R., "Radiation Induced Stratification in Surface Layers of Stagnant Water," ASME Journal of Heat Transfer, Vol. 97, No. 1, 1975, pp. 35-40.

[10] Bloss, S. and Grigull, U., "Temperaturverteilung in tiefen und flachen Seen," Wärme-und Stoffubertragung, Vol. 11, No. 2, 1978, pp. 119-130.

[11] Nielsen, C. E., "Conditions for Absolute Stability of Salt Gradient Solar Ponds," Sun, edited by F. deWinter and M. Cox, Pergamon Press, New York, 1978, pp. 1176-1180.

[12] Leshuk, J. P., Zaworski, R. J., Styris, D. A., and Harling, O. K., "Solar Pond Stability Experiments," Solar Energy, Vol. 21, No. 3, 1978, pp. 237-244.

[13] Hauf, W. and Grigull, U., "Optical Methods in Heat Transfer," Advances in Heat Transfer, Vol. 6, edited by J. P. Hartnett and T. F. Irvine, Jr., Academic Press, New York, 1970, pp. 191-362.

[14] Tilton, L. S. and Taylor, J. K., "Refractive Index and Dispersion of Distilled Water for Visible Radiation at Temperatures 0 to 60°C," U. S. Bureau of Standards, Journal of Research, Vol. 20, No. 4, 1938, pp. 419-477.

[15] Hale, G. M. and Querry, M. R., "Optical Constants of Water in the 200 nm and 200 μm Wavelength Region," Applied Optics, Vol. 12, No. 2, 1973, pp. 555-563.

[16] Incropera, F. P. and Hauf, W. G., "A Three-Flux Method for Predicting Radiative Transfer in Aqueous Suspensions," ASME Journal of Heat Transfer, Vol. 101, No. 3, 1979, pp. 496-501.

[17] Behnia, M. and Viskanta, R., "Laboratory Study of Flow and Thermal Structures in Heated and/or Cooled Layers of Water,"

Purdue University, Water Resources Research Center Technical Report No. 123, Aug. 1979.

[18] Yücel, A. and Bayazitoglu, Y., "Onset of Convection in Fluid Layers with Non-uniform Volumetric Energy Sources," ASME Journal of Heat Transfer, Vol. 101, No. 4, 1979, pp. 666-671.

[19] Viskanta, R. and Parkin, J. R., "Laboratory Study of Unsteady Energy Transfer in Surface Layers of Stratified Water," Water Resources Research, Vol. 12, No. 6, 1976, pp. 1277-1285.

AN ANALYTICAL AND EXPERIMENTAL INVESTIGATION OF TEMPERATURE DISTRIBUTIONS IN LASER HEATED GASES

John P. Schuster,* Wei O. Li,† and William J. McLean‡
Cornell University, Ithaca, N.Y.

Abstract

High flux infrared lasers can be used to rapidly heat selectively absorbing gases for the purpose of studying gas phase reactions under strictly homogeneous conditions. In the present study we consider the temperature and velocity distributions produced in a laser heated reactor consisting of a closed cell containing Ar and the selective absorber SF_6, and irradiated with cw CO_2 laser light at 10.6 µm. A model of the transport processes in the cell is developed by solving the appropriate energy and momentum equations using a numerical finite difference procedure subject to the Boussinesq approximation. The absorption of infrared laser radiation is modeled as nonuniform internal heat generation which varies according to the local absorptivity in the sample cell. It is found that the temperature and velocity distributions in the cell are largely determined by the buoyant flow generated by the nonuniform internal heating. This flow sustains large radial temperature and velocity gradients. Fine wire thermocouples are used to measure gas temperatures in a practical cell and good agreement is found between model predictions and experimental results except where higher laser fluxes or high SF_6

Presented as Paper 80-1522 at the AIAA 15th Thermophysics Conference, Snowmass, Colo., July 14-16, 1980. Copyright © American Institute of Aeronautics and Astronautics, Inc., 1980. All rights reserved.

*Research Assistant, Sibley School of Mechanical and Aerospace Engineering; (presently with Scientific Research Laboratories, Ford Motor Company, Dearborn, Mich.).

+Research Assistant, Sibley School of Mechanical and Aerospace Engineering.

‡Associate Professor, Sibley School of Mechanical and Aerospace Engineering; (presently with Combustion Sciences Department, Sandia National Laboratories, Livermore, Calif.).

concentrations result in complex absorption processes which are not adequately approximated by the model.

Introduction

The availability of high flux molecular lasers operating in the infrared, coupled with the knowledge that several molecular species exhibit quite large absorption cross sections for such infrared radiation, makes it possible to generate high temperatures for studying gas phase chemical reactions under strictly homogeneous conditions. The lack of hot walls, as would be found in a flow reactor experiment, eliminates the occurrence of catalytic wall reactions, thus rendering this laser technique homogeneous in the chemical kinetic sense; i.e., all reactions take place in the gas phase. For example, Shaub and Bauer[1] have described a "laser powered homogeneous pyrolysis" technique in which a cw CO_2 laser of modest power is used to irradiate a sample cell containing argon (Ar) and sulphur hexafluoride For sufficiently high laser fluxes the SF_6 exhibits a very high absorption coefficient for 10.6 μm radiation, and a considerable fraction of the incident laser energy can be absorbed. At temperatures below about 1500 K the SF_6 is stable and the absorbed energy is converted to thermal energy quite rapidly (~ 1 ms) by V-T relaxation of the SF_6. In this way a high temperature bath of Ar containing a small fraction of SF_6 is produced, and chemical reactions in reagents of interest may be studied under the strictly homogeneous conditions made possible by the absence of hot walls.

The laser heating technique can be thought of as complementary to shock tube techniques where homogeneously heated environments are also generated, but at much higher temperatures and for much shorter test times. Although like the shock tube the laser heating process is a homogeneous one, it does not produce the uniform temperature which the shock tube does. Rather, a temperature distribution is produced in the sample cell which depends upon the cell geometry, the intensity distribution in the laser beam, the absorption processes of the SF_6, and the transport of heat by both thermal conduction and natural convection within the sample cell. Our own interest in laser heating of gases was stimulated by our desire to study the pyrolysis of synthetic fuels and their components at moderate temperatures for the purpose of examining the chemical dynamics associated with the formation of soot precursors.

The present study was undertaken with the objective of obtaining a better understanding of the thermal environment and gas flowfield, under both transient and steady-state conditions, produced as a result of laser irradiation of selectively absorbing gases. Because the absorption of the infrared laser energy and the subsequent V-T relaxation processes are fast relative to establishment of buoyancy driven flowfields, it is possible to analyze the system by solving the governing momentum and energy conservation equations with the laser heating modeled by a spatially and temporally nonuniform internal heat generation term. In addition to the analytical studies, fine wire thermocouple measurements of temperature distributions in confined laser heated gases were carried out to experimentally examine the reliability of the numerical calculations. In the following sections the analytical and experimental procedures and the major results are described. Comparisons of theory and experiment are also presented.

Analytical Studies

Governing Equations

The analysis of the temperature and velocity distributions produced in a cell filled with an absorbing gas is carried out by solving the governing energy and momentum equations by a numerical finite difference procedure. The geometry chosen is a vertically oriented circular cylinder with the laser light incident on the bottom window and exiting through a window on the top of the cell. The vertical cylinder geometry is chosen because it requires only a two-dimensional axisymmetric numerical solution, and also because earlier studies indicated that a vertical cell heated from below gives a higher degree of chemical reaction due to the presence of vigorous buoyant mixing processes.

The equations solved are the unsteady incompressible energy and momentum equations with the important buoyancy force term incorporated via the Boussinesq approximation. Several simplifying assumptions are made in solving the problem. Angular variations in heating due to laser intensity variations are assumed to average out, so that an axisymmetric calculation is possible. Axial and radial variations in the laser heating source term are accounted for. The cell walls and end windows are assumed to remain at the initial temperature, Θ_0, of 300 K, and fluid properties such as viscosity, ν, thermal conductivity, k, thermal diffusivity, a, and volume expansion coefficient, β, are assumed constant and equal to their values at 300 K. Density is assumed constant except for the generation of buoyancy forces (Boussinesq approximation).

Using the assumptions described above, dimensionless energy and momentum equations are written for internal heat generation in a vertical circular cylinder of radius A and height L. The dimensionless variables are written in terms of the dimensional variables as follows:

$$t = (a/A^2)\tau \quad R = r/A \quad Z = z/A$$

$$U = (A/a)u \quad V = (A/a)v \quad \psi = \Psi/Aa$$

$$\omega = (A^2/a)\Omega \quad \text{and} \quad T = (\theta - \theta_0)/(\dot{q}'''_{max}A^2/k)$$

Here r and z are the radial and axial coordinates, u and v are the radial and axial velocities, τ is time, θ is temperature, and ψ and Ω are the stream function and vorticity. The largest volumetric heat source value, $\dot{q}'''_{max}$, found at any individual finite difference grid volume is used to scale the temperature.

The energy equation becomes

$$\frac{\partial T}{\partial t} + \frac{1}{R}\frac{\partial}{\partial R}(RUT) + \frac{\partial}{\partial Z}(VT) = \frac{1}{R}\frac{\partial}{\partial R}(R\frac{\partial T}{\partial R}) + \frac{\partial^2 T}{\partial Z^2} + \frac{\dot{q}'''(R,Z,t)}{\dot{q}'''_{max}}$$

except at the centerline where it is

$$\frac{\partial T}{\partial t} + 2T\frac{\partial U}{\partial R} + \frac{\partial (VT)}{\partial Z} = 2\frac{\partial^2 T}{\partial R^2} + \frac{\partial^2 T}{\partial Z^2} + \frac{\dot{q}'''(R,Z,t)}{\dot{q}'''_{max}}$$

A Taylor series expansion was used to obtain the above form of the energy equation which was needed in order to avoid numerical difficulties caused by the 1/R term in the general equation.

Cross differentiating the R and Z momentum equations and subtracting them to eliminate pressure terms leads to the azimuthal vorticity equation:

$$\frac{\partial \omega}{\partial t} + \frac{\partial}{\partial R}(U\omega) + \frac{\partial}{\partial Z}(V\omega) = -RaPr\frac{\partial T}{\partial R} + Pr[\frac{\partial}{\partial R}(\frac{1}{R}\frac{\partial}{\partial R}(R\omega)) + \frac{\partial^2 \omega}{\partial Z^2}]$$

The Prandtl number and modified Rayleigh number are $Pr = \nu/a$ and $Ra = g\beta A^3 (\dot{q}'''_{max} A^2/k)/\nu a$, where g is the acceleration of gravity. The ratio of the two diffusivities, Pr, is a comparison of viscous and thermal effects. The Rayleigh number is a ratio of the buoyant and thermal heating forces. High values of the Rayleigh number are indicative of very vigorous convective flows. In this investigation the Rayleigh number ranged from 1000 to 10,000.

The continuity equations and the definitions of stream function and vorticity provide the following relations:

$$U = -\frac{1}{R}\frac{\partial \psi}{\partial Z}$$

$$V = \frac{1}{R}\frac{\partial \psi}{\partial R}$$

and

$$\frac{1}{R}\frac{\partial^2 \psi}{\partial Z^2} + \frac{\partial}{\partial R}\left(\frac{1}{R}\frac{\partial \psi}{\partial R}\right) = -\omega$$

For $t = 0$ the initial conditions are for all R and all Z: $T = \omega = \Omega = 0$, $\dot{q}'''(R,Z,t) = 0$.

Assuming symmetry at the centerline, and no slip conditions, constant temperature, and impermeability at the walls, the boundary conditions for $t \geq 0$ are:

$R = 0$, for all Z: $\frac{\partial T}{\partial R} = 0$, $\psi = \omega = 0$

$R = 1$, for all Z: $T = \psi = 0$

$Z = 0$, for all R: $T = \psi = 0$

$Z = L/A$, for all R: $T = \psi = 0$.

The input parameters to the problem are Ra, Pr, L, A, and $\dot{q}'''(R,Z,T)$. The heat source term is the major input parameter and is discussed further in the following section.

Heat Source Term

Perhaps the most difficult aspect of the analytical study is the determination of a satisfactory relationship between the heat source term and the laser intensity distribution. For a particular grid volume the heat generation rate is of course equal to the difference between the incoming and outgoing laser energy fluxes. It is expected from the Beer-Lambert law that the absorption within that volume is exponentially dependent upon the optical path length, the absorption cross section, and the number density of the absorbers. That is

$$I_o/I_i = \exp(-\sigma n Z)$$

where I_o/I_i is the ratio of outgoing to incoming laser intensity, σ is the absorption cross section, n is the number density of absorbers, and Z is the axial length. For the relatively high laser fluxes considered in the present analysis, the absorption cross section is related to very complex chemical and physical processes involving excited state absorption, and analytical expressions for this cross section are not available.[2-5] Instead, it is assumed that the absorption over the entire length of the sample cell can be expressed by a Beer-Lambert law, and an effective absorption coefficient is derived by matching the total computed absorbed energy to the difference between the measured input and output laser energies. Thus an overall absorption coefficient α_0 is determined from

$$\alpha_0 = -\frac{1}{L} \ln \left(\frac{E_{out}}{E_{in}}\right)$$

where the measured laser energies are E_{out} and E_{in}.

Since the effect of the local heat generation due to the absorption process is to locally raise the temperature and lower the density, the number density of the absorbers decreases during the heating process. This effect is accounted for in the calculations by adjusting the local heat generation term in inverse proportion to the local temperature, while assuring that the total absorption in the cell is constant. For example for a given grid point (I,J), the local absorption coefficient $\alpha(I,J)$ is given by

$$\alpha(I,J) = \alpha_0 \frac{T_m}{T(I,J)}$$

where T_m is the mass mean temperature for the entire cell and $T(I,J)$ is the local temperature at grid point (I,J). The heat generation term in a given grid volume is then determined by using an average value of α for that volume to determine the amount of laser energy absorbed per unit volume. The radial variation in the heat generation term is a direct function of the radial intensity distribution in the incoming laser light. Calculations are carried out for various intensity distributions, both Gaussian and non-Gaussian, and in the experimental program an attempt is made to determine this distribution.

Numerical Procedures

The governing vorticity and energy equations are written in finite difference form and solved for temperature and vorticity over one-half of the symmetrical cell using a conservative upwind differencing technique.[6] The use of this numerical method allowed for the calculation of heat transfer rates to the enclosure walls, and also enhanced the stability of the numerical solution. With the vorticity known, the equation relating the stream function to vorticity is solved by an iterative, successive-over-relaxation technique. The boundary conditions are obtained as described through the no-slip conditions and by assuming the walls and windows remain at constant temperature. At time zero an internal heat generation function determined as described above is imposed on the initially isothermal gas mixture. The upwind differencing method was chosen over several others[7] because it requires no mesh size restriction, and because it conserves energy and vorticity both locally and overall. The former characteristic of this method alleviates the need for large amounts of computer storage and calculation time, while the latter characteristic permits calculation of the heat transfer rates to and from the enclosure. Furthermore, this method has good stability properties at high Rayleigh numbers. Although this differencing scheme formally has only first order accuracy, it can be shown that in practice the method is nearly second order accurate.[8] The truncation error, which introduces false (i.e., numerical) viscosity and heat diffusivity into the numerical solutions of the vorticity and temperature equations, becomes significant only at high Rayleigh numbers where it actually enhances fluid stability.

Further details of the differencing and numerical integration procedures, including stability considerations, are given elsewhere.[9] Lax and Richtmeyer[10] point out that if their stability condition is fulfilled, and if the finite difference approximations are consistent with the differential equations, and if the initial conditions meet certain requirements, then the solution of the finite difference equations will coverge to the solution of the differential equations as ΔR, ΔZ, and Δt approach zero. Solution of a baseline case using 91, 231, 299, and 435 mesh points was done in order to check for such convergence. Figure 1 indicates how the steady-state temperature at the midheight of the centerline approaches an asymptotic value as the number of mesh points, N, is increased. The higher accuracy calculations also predict lower temperatures than do the less accurate lower N cases. These two trends are similar to previously reported results[7] of calculations with several grid sizes using the conservative upwind differencing scheme.

Most of the calculations in this investigation are carried out with a mesh having 11 radial and 21 axial grid points. More points are used only when the additional expense in increased computer time seems warranted, such as in high pressure (high Rayleigh number) conditions. It may be argued that an 11 x 21 mesh is somewhat coarse. Indeed, cell Reynolds numbers for the flows produced in this study are large enough to suggest that truncation errors may be significant. However, the prime motive behind this study is not to obtain high numerical accuracy, but rather to investigate different heating and flow configurations. The results in the preceding paragraph are cited as evidence that the calculations are of sufficient accuracy for the purposes of this work.

The numerical model was initially tested on a uniform heat generation case. Qualitatively the results compared well with the isotherms and streamlines obtained in calculations for a vertical cylinder with internal heat generation which were reported by Kee et al.[10]

Results of Parametric Study

The numerical model described above was used to determine the effect of various parameters on the velocity and temperature distributions inside a heated cell. The "baseline" conditions, selected as typical of our experiments, are as

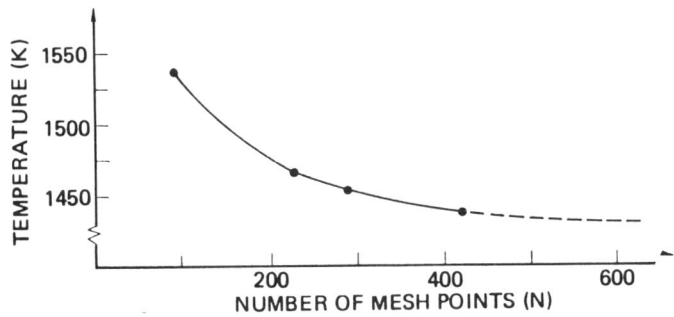

Fig. 1 Steady-state mid-centerline temperature as a function of the total number of mesh points.

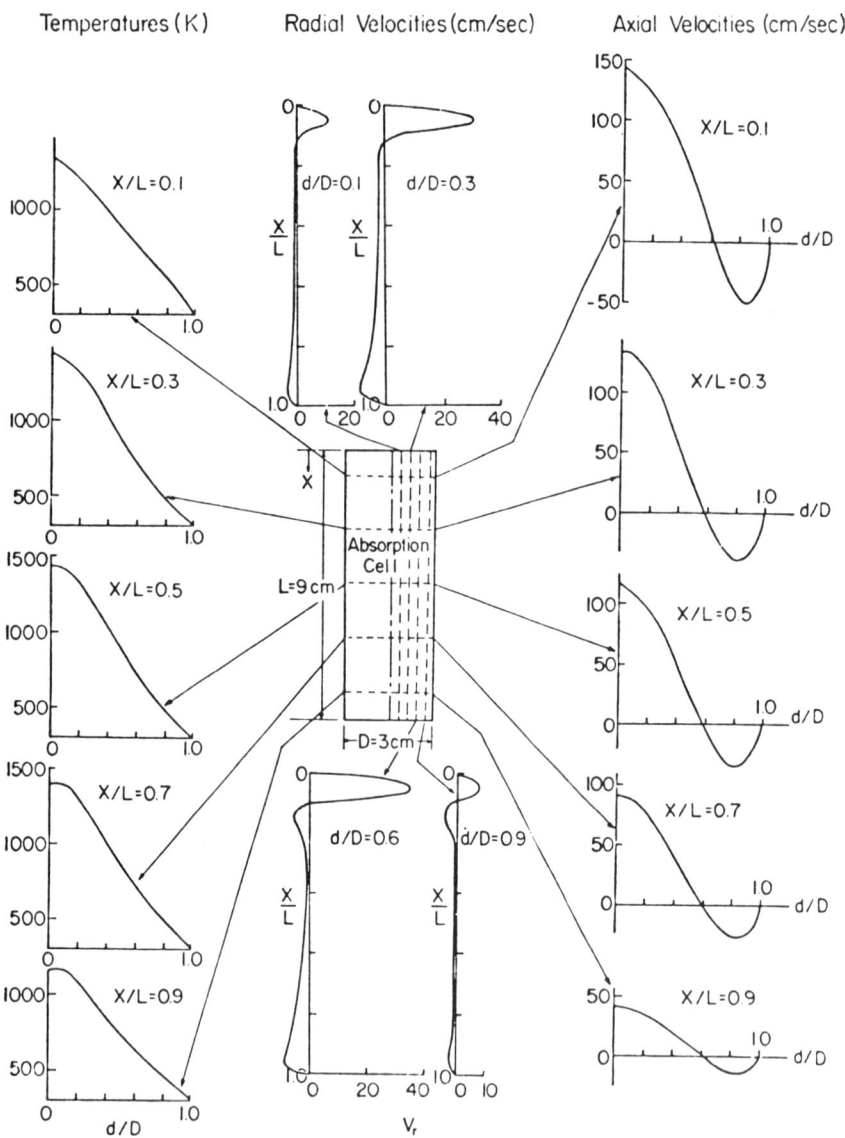

Fig. 2 Steady-state temperature and velocity distributions for the baseline case.

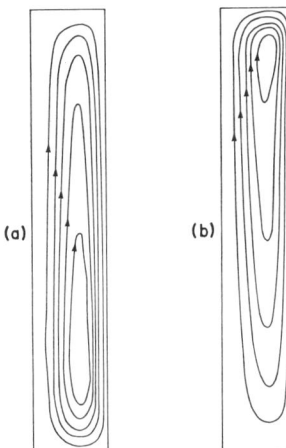
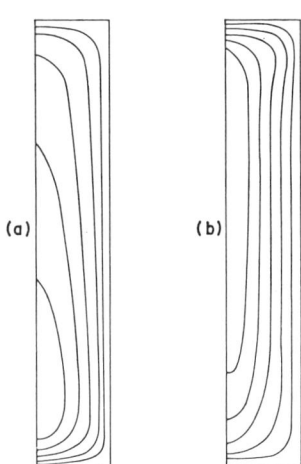

Fig. 3 Streamlines early in heat-up transient (a), and at steady state (b).

Fig. 4 Isotherms early in heat-up transient (a), and at steady state (b).

follows. The cell is cylindrical, 3 cm in diameter and 9 cm high. It contains 99 Torr Ar and 1 Torr SF_6, initially at room temperature. The cw CO_2 laser provides 30 W input power of which 20 W are absorbed by the gas in the cell. The laser intensity is assumed to be distributed radially as a Gaussian function, with the intensity at a radius of 1.05 cm equal to 59% of the centerline intensity.

A montage of steady-state temperature and velocity profiles for the baseline case is shown in Fig. 2. Since all heat losses are to the cool walls and because the heat source is more concentrated in the center of the cell, a pronounced radial temperature gradient is indicated in Fig. 2, with the temperature changing from 300 K at the wall to as high as 1400 K along the centerline. The velocity profiles indicate the expected buoyancy driven convective flow patterns, where a strong buoyant upward flow exists in the hot central core region, with cooler regions along the outer walls exhibiting downward flow. Radial velocities are small except near the windows at the top and bottom of the cell, where the convection currents move gas from the central core to the outer region at the top and back inward towards the center at the bottom.

The temporal development of the streamlines and isotherms for a case similar to that shown in Fig. 2 can be seen in Figs. 3 and 4 where the streamlines and isotherms in the right half of an azimuthal plane are shown early in the heat-up transient and also at steady state. At early times the maximum temperatures and velocities are found near the lower entrance window where most of the laser energy is absorbed. At steady state the buoyancy driven convective flow dominates the motion in the cell, and the flow is driven rapidly upward along the centerline, jets out to the sides near the top of the cell, and then falls slowly along the cool outer walls to return to the bottom of the cell. Relatively uniform temperatures are found at least along the centerline at steady state, with the maximum temperature occurring above midplane of the cell due to the convective transport.

Four cases besides the baseline case (case 1) have been analyzed. In each case one parameter is varied from the baseline condition. In case 2 the power absorbed is halved for the same input power. This corresponds to a lower SF_6 pressure, and leads to a more uniform axial distribution of internal heat generation, due to the exponential term in the laser absorption equation. Case 3 has one-half the total pressure, but the same SF_6 pressure, as the baseline case. In case 4 the laser energy is concentrated closer to the centerline by using a modified Gaussian intensity distribution. In case 5 the cell length is halved, and the resulting reduction in the optical path reduces the total energy absorbed by 56%.

The velocity and temperature fields for the five cases described above are shown in Figs. 5-7. In Fig. 5 it is seen that changing the radial distribution of the heat source term has a small effect on the vertical velocity, while reducing the power absorbed (cases 2 and 5) has the effect of reducing the buoyant upward flow. By reducing the total pressure by a factor of 10 (case 3), the buoyant motion is nearly completely eliminated and heat transfer is almost entirely by conduction. It is found that the general behavior of the velocity profile correlates with the Rayleigh number. Cases 1 and 4 and cases 2 and 5 have similar Rayleigh numbers and hence similar velocity fields, with the later two cases having a lower Rayliegh number and reduced buoyant motion. For the low pressure case (case 3) the Rayleigh number is about a factor of 10 lower than the baseline case and buoyant convective motion is nearly absent. For a given cell, lower laser powers or lower pressures produce lower Rayleigh numbers. For Rayleigh numbers less than about 1000 the temperature distribution is determined principally by thermal conduction. In contrast,

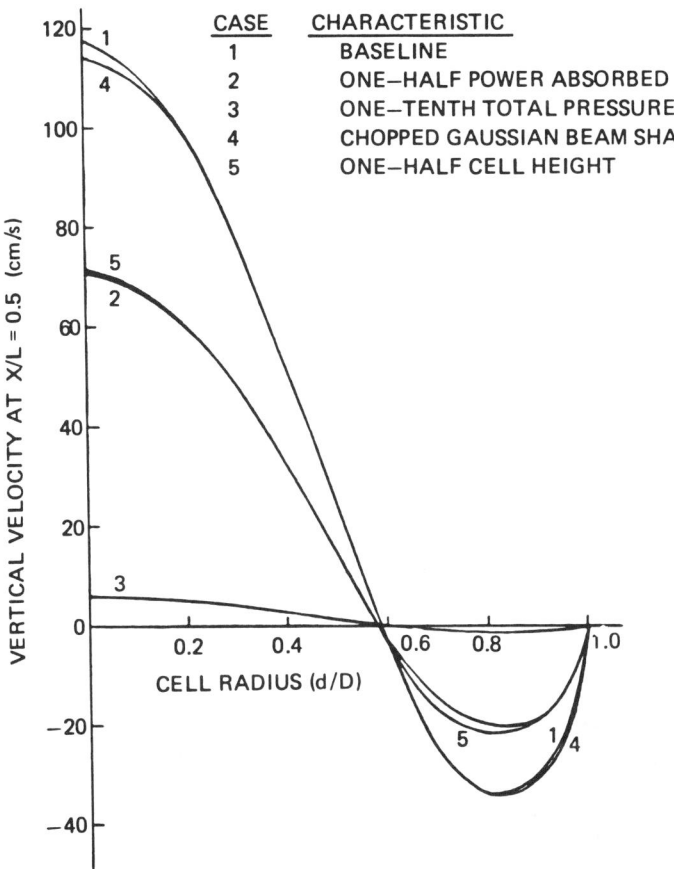

Fig. 5 Radial distribution of vertical velocity at midheight for various parametric cases.

for Rayleigh numbers exceeding 10,000, the flow is completely dominated by vigorous buoyancy driven circulation.

The steady-state radial temperature profiles shown in Fig. 6 also show the expected trends with the parameter variations. Cases 1 and 4 are again similar as are cases 2 and 5, which show the lower temperatures due to the reduced absorbed power. For low total pressure (case 3) the model predicts a very high centerline temperature primarily as a result of the reduced total heat capacity. Temperatures as high as predicted could not be reached in an actual sample cell because the rapidly increasing SF_6 heat capacity would limit temperature rise. Also, thermal decomposition of SF_6 would eventually limit absorption. The axial

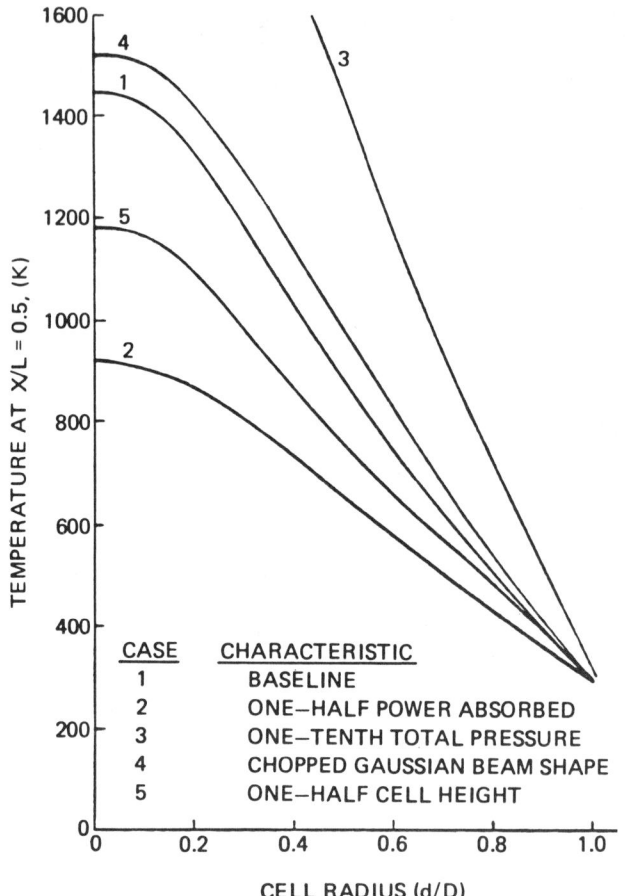

Fig. 6 Radial distribution of temperature at midheight for various parametric cases.

temperature profiles shown in Fig. 7 follow trends similar to the radial profiles. The temperatures are reasonably uniform along the axis except near the top and bottom windows where the gas temperature is reduced by heat transfer to the windows.

Since the numerical analysis integrates the unsteady equations until steady state is reached, the transient flowfield and temperature profiles are also available. The mass averaged gas temperature is shown as a function of time in Fig. 8 for the various cases. Except for the low pressure case, each system reaches steady state when the buoyant flow

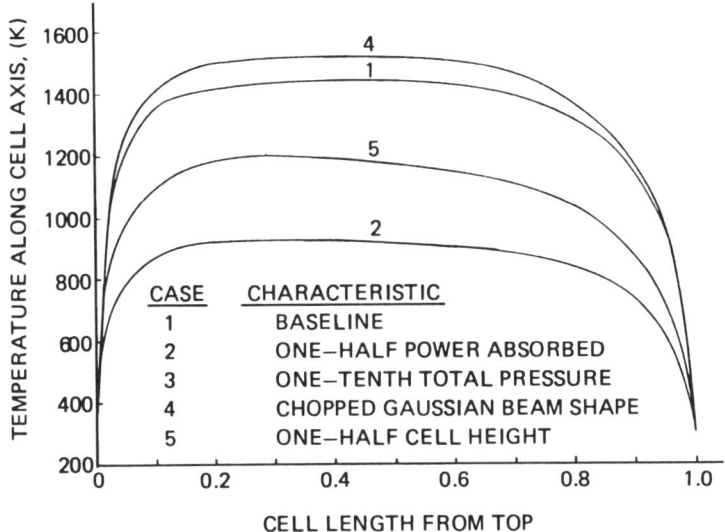

Fig. 7 Axial distribution of temperature along centerline for various parametric cases.

pattern is fully established in about 0.2 s. The conduction limited profile is much more quickly established as indicated by the shorter transient for case 3.

Experimental Studies

Apparatus and Procedure

A laboratory constructed cw CO_2 laser is used to provide 10.6 μm radiation for the experimental measurements of temperature distributions in laser heated gases. The 5 m long 2.5 cm bore Pyrex laser tube is water cooled with the high voltage electrodes placed at either end and at the center of the tube. A grating is used to tune the laser to the desired P and R branch transitions in the 10.6 μm band. The SF_6 absorption of interest occurs at wavelengths corresponding to the P(12) - P(32) lines in the 10.6 μm band. The laser power output is as high as 30 W when running on multiple lines and in multiple modes as in the present experiments. The sample cell is a stainless steel cylinder 3.5 cm in i.d. and 9.5 cm long with both ends o-ring sealed with infrared transparent, polished NaCl windows (95% transmission at 10.6 μm). The cell can be water cooled to maintain constant wall temperature, and provision is made to insert up to five thermocouples through

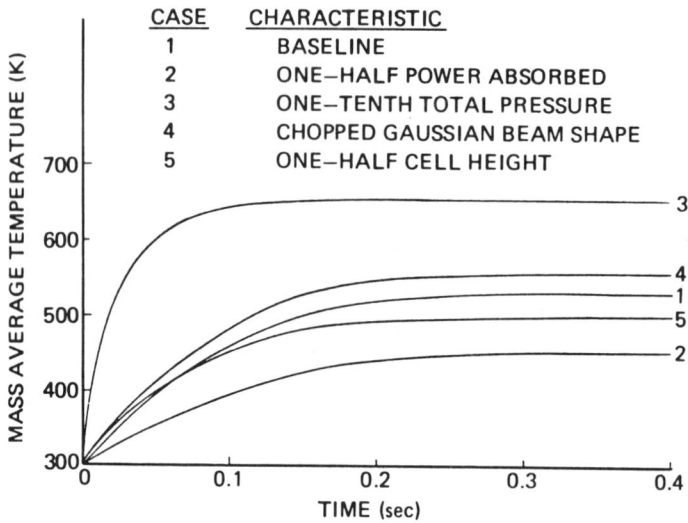

Fig. 8 Mass averaged gas temperature as a function of time for various parametric cases.

the cylinder wall at various locations along the length. These thermocouples may be slid radially inward or outward during the experiment without affecting the cell pressure. The locations of the thermocouple beads inside the cell can be determined to within 1 mm. Chromel-alumel thermocouples 0.12 mm in diameter are used for the measurements.

Because the thermocouples are used in a high temperature, high infrared radiative flux environment, it is necessary to consider the various heat fluxes, and determine the relationship between the indicated thermocouple temperature and the true gas temperature. The heat fluxes include direct infrared laser heating of the thermocouple bead, thermal radiation losses from the thermocouple, conduction losses along the thermocouple leads, and heat input from the surrounding gas by convective heat transfer.

A series of experiments was carried out to evaluate these various heat flux terms, and the details of this procedure are given elsewhere.[11] The results are summarized here. The emissivity of the thermocouple bead for the incoming 10.6 μm radiation is determined by measuring the thermocouple response in an evacuated cell under known irradiation conditions and by using the known total emissivity for chromel-alumel to compute the radiative losses. Conduction losses are evaluated by moving the thermocouple from regions

of small radial temperature gradients to regions of large
radial gradients and observing the resulting effect on the
measured temperature. The results of these investigations
show that for the highest power input, where the radiation
correction is largest, the thermocouple temperatures along
the cell centerline are no more than 25 K cooler than
the local gas temperature. Differences up to 50 K are found
in regions of steep radial temperature gradients due to the
additional conduction losses. This thermocouple accuracy is
of course well within the limits of the accuracy of the
numerical analysis given the numerous simplifying assumptions
used therein, and corrections are therefore not applied to
the thermocouple measurements.

A second set of supplementary experiments was carried
out to determine the approximate radial distribution of
laser intensity. The details of these experiments are also
reported elsewhere[11] and summarized here. The laser power
is recorded while an aperture is used to gradually reduce
the effective beam diameter to smaller and smaller openings.
By examining this measured power as a function of radius, it
is possible to compare the distribution to that produced by
either Gaussian or constant intensity distributions. It is
found that the actual radial intensity distribution is
neither Gaussian nor constant, but rather somewhere in
between. The intensity distribution is flatter and less
peaked than the Gaussian in the center of the cell, but
decreases somewhat more rapidly than a Gaussian distribution
at the outer radii of the cell.

Results and Discussion

Axial and radial temperature distributions in SF_6-Ar
mixtures in the 3 cm x 9 cm cell have been measured over a
range of laser powers from 6.4 W to 25 W, SF_6 partial
pressures from 0.4 to 2.3 Torr, and total pressures of 10
and 100 Torr. The measured distributions are compared with
those predicted using the numerical model. In general, it
is found that the best agreement between the calculations
and the measurements is obtained for relatively low laser
power, low SF_6 pressure, and 100 Torr total pressure.
This trend is thought to be caused by a number of factors.
First, low laser power, low SF_6 pressure and high Ar
pressure combine to produce the lowest temperature rise of
all experimental conditions. The low temperature rise in
turn means that the Boussinesq assumptions used in the model
more nearly approximate the actual flow conditions. Second,
it is to be expected that the lowest laser intensities and

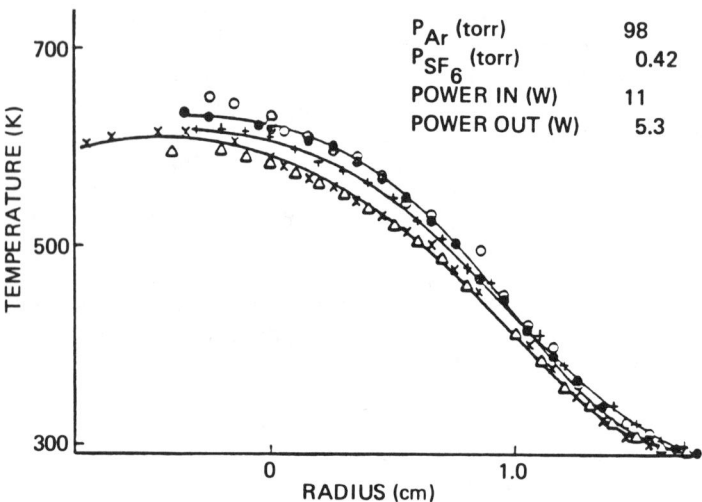

Fig. 9 Measured temperature distribution for low power input, low SF$_6$ pressure, and high total pressure. Distances above entrance window: x 1.2 cm, 0 3 cm, ● 4.75 cm, + 6.5 cm, △ 8.25 cm.

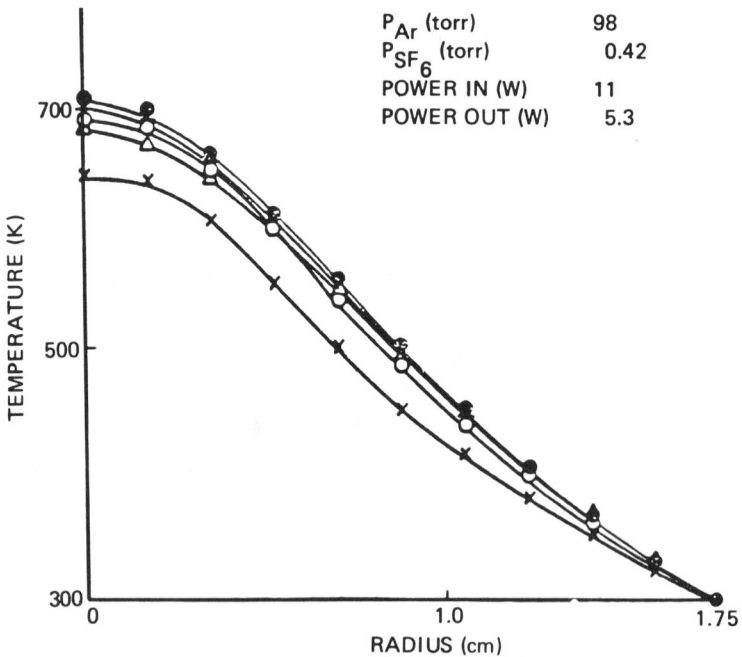

Fig. 10 Computed temperature distribution for same conditions as Fig. 9.

lowest gas temperatures lead to molecular absorption processes most nearly approximated by the Beer-Lambert law assumption. Furthermore, low total pressures always lead to higher temperatures under otherwise constant conditions, and we thus expect poorer agreement between model and experiment at lower total pressures.

Measured radial distributions of temperature for various axial locations for a case with modest laser power and low SF_6 pressure are shown in Fig. 9. Temperatures reach a maximum of about 650 K near the centerline, and some asymmetry is observed, probably due to asymmetry in the laser intensity distribution. Computed temperature profiles for the same conditions as Fig. 9 are shown in Fig. 10. The computations produce temperature levels similar to those measured, but with a somewhat more peaked profile. The calculations are carried out using a Gaussian laser beam intensity distribution, while the actual beam intensity is somewhat flatter than Gaussian in the center of the cell. This results in the flatter temperature profiles seen in Fig. 9. Computed and calculated axial temperature profiles for this same case are shown in Fig. 11. The computed and measured profiles are quite similar except for the somewhat higher predicted centerline temperature. We are encouraged by the agreement betweeen the measurements and predictions, and it is apparent that the vigorous bouyancy driven flow predicted by the model is in fact an important transport mechanism in the laser heated gases.

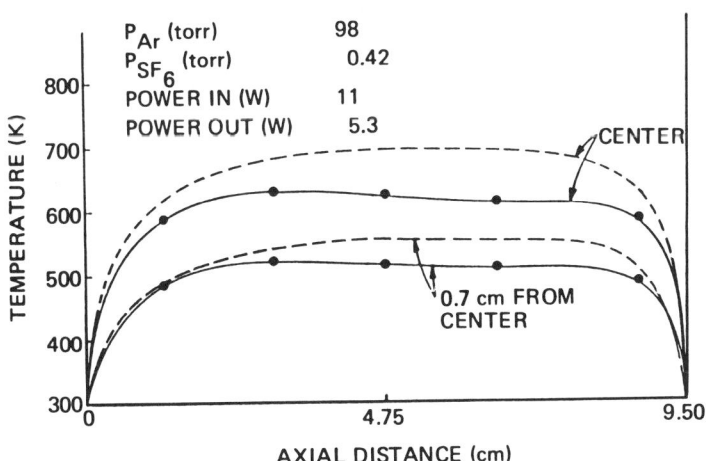

Fig. 11 Measured and computed axial distribution of temperature for same conditions as Fig. 9, —●— measured, --- computed.

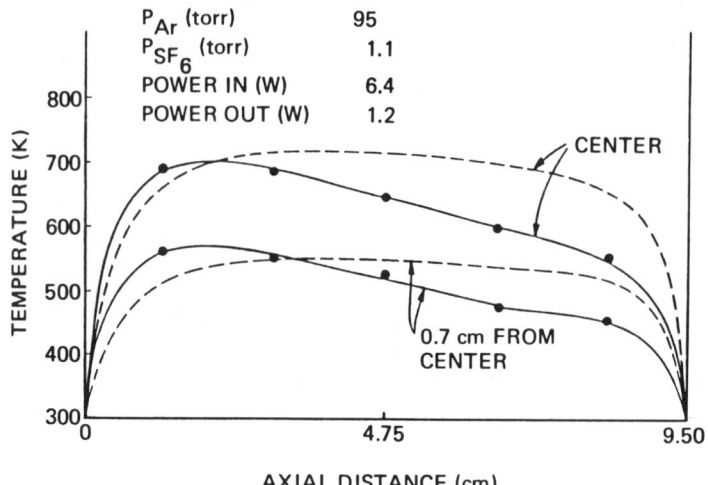

Fig. 12 Measured and computed axial distribution of temperature for higher SF_6 pressure.

As noted previously, increased SF_6 partial pressures and increased laser intensities result in more discrepancy between measured and calculated temperatures. For example, the axial temperature profiles for a case where the SF_6 pressure is 1.1 Torr and 80% of the incoming laser power is absorbed are shown in Fig. 12. Here the numerical model predicts a relatively uniform axial temperature profile except near the end windows, while the measurements indicate that the temperature is highest in the region near the bottom of the cell. This difference is evidently caused by higher than predicted heating rates in the lower portion of the cell.

SF_6 is known to exhibit "anti-bleaching" behavior,[12] where increased laser fluxes lead to higher than proportional absorption due to absorption by excited states. This behavior is also noted in the present experiments where increased laser fluxes at otherwise constant conditions produce much more than a proportional increase in absorbed energy. Thus while the simple Beer-Lambert absorption profile is adequate for relatively low laser energies and low SF_6 pressures, the approximation is not satisfactory under other conditions.

Further comparisons between measurements and predictions at high laser fluxes indicate that the heated region spreads radially outward more than predicted at the higher laser powers,

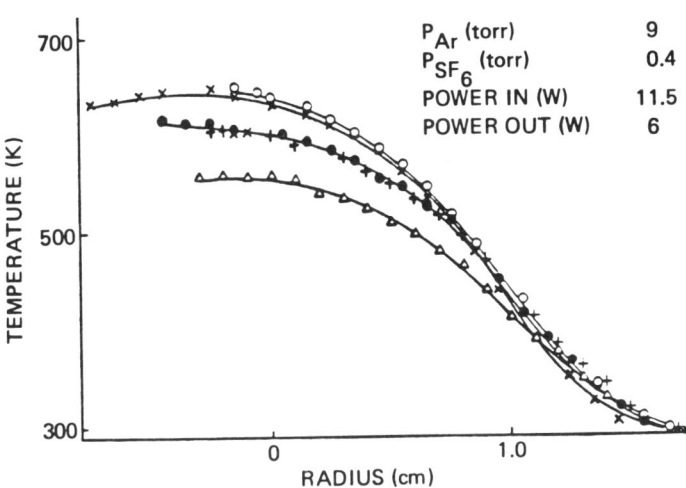

Fig. 13 Measured temperature distribution for low total pressure. Distance above entrance window: x 1.2 cm, O 3 cm, ● 4.75 cm, + 6.5 cm, △ 8.25 cm.

although the average temperature levels are in good agreement with predictions. We take this as further evidence of the complex absorption behavior of SF_6 under high laser fluxes.

For 10 Torr total pressure the agreement between prediction and model is not nearly so good. The measurements and predicitons for low SF_6 pressure and modest laser power input at 10 Torr are shown in Figs. 13 and 14. Here the predicted temperatures are in general several hundred degrees higher than measured, and the measured distribution is considerably flatter than predicted. These discrepancies are even more pronounced at higher laser powers and higher SF_6 pressures. With the lower total pressures the contribution to the total mixture heat capacity from SF_6 is ten times higher than in the previous higher total pressure experiments where the mixture was primarily argon, and thus had a heat capacity that remained nearly constant over a wide temperature range. The neglect in the model calculations of the rapid increase in SF_6 heat capacity with temperature is the probable cause for the predicted temperatures being higher than measured. The low total pressure cases produce the highest temperatures measured in these experiments. For 10 Torr total pressure and 1 Torr SF_6 pressure, and a laser input power of 23 W of which 16 W was absorbed, temperatures up to 1000 K in the center of the sample cell are recorded.

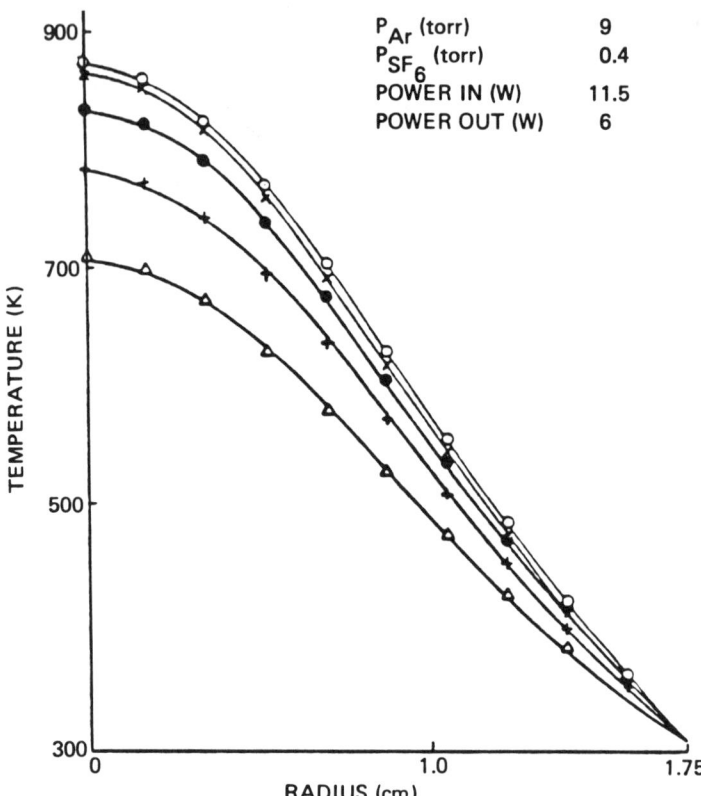

Fig. 14 Computed temperature distribution for same conditions as Fig. 13.

Conclusions

Temperature and velocity distributions in confined laser heated gases have been analytically and experimentally examined. Under conditions of interest for studying chemical reactions in such gases, it is found that the temperature distribution is highly nonuniform, and is largely determined by the intensity distribution in the incoming laser beam and by the buoyant flows which are induced in the cell. Although the analytical model employs the Boussinesq approximation under conditions of relatively large temperature change, reasonable agreement is still obtained between measured and calculated temperatures. The differences between the measured and computed temperature profiles are attributed to two major uncertainties in the model: first, the lack of detailed knowledge of the physical and chemical processes involved in absorption of 10.6 m radiation by SF_6, and second, the uncertainty as to the radial distribution of the laser intensity.

Acknowledgments

This work was supported in part by the Office of Naval Research, Project SQUID, through Contract Number 8960-35 to Cornell University. We gratefully acknowledge the helpful discussions and laboratory assistance provided by Professor S. H. Bauer and Dr. E. R. Lory of the Cornell University Chemistry Department. Professor K. E. Torrance of the Sibley School provided valuable advice on the numerical analysis.

References

[1] Shaub, W. M. and Bauer, S. H., International Journal of Chemical Kinetics, Vol. VII, 1975, p. 509.

[2] Burak, I., Steinfeld, J. I., and Sutton, D. G., Journal Quantum Spectroscopy Radiative Transfer, Vol. 9, 1969, p. 959.

[3] Wood, O.R., Gordon, P. L., and Schwarz, S. E., IEEE Journal of Quantum Electronics, Vol. QE-5, 1969, p. 502.

[4] Brunet, H., IEEE Journal of Quantum Electronics, Vol. QE-6, No. 11, 1970, p. 678.

[5] Nowak, A. V. and Lyman, J. L., Journal Quantitative Spectroscopy and Radiative Transfer, Vol. 15, 1975, p. 945.

[6] Torrance, K. E., and Rockett, J. A., Journal Fluid Mechanics Vol. 36, 1969, p. 33.

[7] Torrance, K. E., Journal of Research of the National Bureau of Standards, Section B: Mathematical Sciences, Vol. 72B, 1968, p. 281.

[8] Roache, P. J., Computational Fluid Dynamics, Hermosa Publishers, Albuquerque, 1976, p. 119.

[9] Shuster, J. P., Master of Science Thesis, Cornell University, 1979.

[10] Kee, R. J., Landram, C. S., and Miles, J. C., Journal Heat Transfer, Transaction ASME, Series C, Vol. 98 1976, p. 55.

[11] Li, Oi Wei, Master of Science Thesis, Cornell University, 1980.

[12] Oppenheim, U. P. and Melman, P., IEEE Journal of Quantum Electronics, Vol. QE-7, 1971, p. 426.

NUMERICAL METHODS FOR THE ANALYSIS OF LASER ANNEALING OF DOPED SEMICONDUCTOR WAFERS

J. R. Kirkpatrick,[*] G. E. Giles Jr.,[+] and R. F. Wood[†]
Union Carbide Corporation, Nuclear Division, Oak Ridge, Tenn.

Abstract

Certain classes of semiconductor devices can be manufactured by ion implantation of the dopant. Crystal damage resulting from the ion implantation can be annealed by melting the surface layer using a short, high-intensity laser pulse. The dopant atoms are redistributed during melting and resolidification. Analysis of this process requires studying two separate problems, each with its own difficulties. Heat-transfer calculations with spatially and temporally varying internal heat sources and change of phase give the depth of melting, duration of the melted layer, and resolidification velocity. Dopant transport calculations give the diffusion of the dopant in the liquid and include segregation at the phase boundary.

Nomenclature

C = heat capacity, J/C
c = dopant concentration, atoms/cm^3
D = diffusivity of dopant in liquid silicon, cm^2/s
$_iK_m$ = conductance from node i to m, W/C
k = segregation coefficient
p = material phase
Q = heat generation, W

Presented as Paper 80-1473 at the AIAA 15th Thermophysics Conference, Snowmass, Colo., July 14-16, 1980. This paper is declared a work of the U.S. Government and therefore is in the public domain.
[*]Computing Applications Specialist, Computer Sciences Division.
[+]Computer Applications Specialist, Computer Sciences Division.
[†]Senior Research Staff Member, Solid State Division, Oak Ridge National Laboratory.

$\dddot{q}$ = heat generation/unit volume, W/cm³
$\ddot{q}$ = incident laser energy, W/cm²
T = temperature, °C
t = time, s
x = depth, cm
α = absorption coefficient, cm⁻¹
Δt = time step size, s
Δx = space step size, cm
ρ = surface reflectivity
( ) = dependent variables

Subscript

i,m = node number

Superscript

n = time step number

Introduction

Certain classes of semiconductor devices have a very thin doped layer on the surface of a wafer of a semiconductor material. One way to dope the wafer is with ion implantation of the dopant. Unfortunately, this process causes massive damage to the crystal structure in the doped layers. This crystal damage must be annealed in order to produce a usable semiconductor device. One scheme for annealing the damage is thermal annealing which involves holding the wafer at high temperatures for a long time. An alternative is laser annealing.

Semiconductor wafers have been irradiated by short, high-intensity laser pulses which melt the surface layers. After resolidification, the surface layers were found to have regrown on the underlying crystal producing a monocrystal (liquid phase epitaxial regrowth). The dopant atoms were in substitutional sites within the regrown crystal. This process offers several advantages over thermal annealing. First, the laser annealed samples have superior electrical properties because the crystal damage is almost perfectly annealed and because more of the dopant atoms are in substitutional sites. The laser process uses less energy. The localized heating avoids the disadvantages of high temperature in the substrate. The process happens so rapidly that no significant amounts of impurities are picked up from the air and so the process need not be limited to vacuum. With a laser, the annealing may be restricted to a local area of the surface, if desired.

Calculations indicate that the surface of the silicon samples may stay molten for hundreds of nanoseconds. The diffusivities of most dopants in liquid silicon are large enough for the dopants to be significantly redistributed during this interval. Experimental measurements of dopant profiles before and after annealing show such redistribution. If the redistribution processes could be understood and controlled, it might be possible to tailor the dopant profiles for optimum electrical properties.

The analysis requires solution of two very different problems. First, a heat-transfer calculation must be done to determine the depth and duration of the molten phase. Then, using this melt information, a dopant transport calculation determines the motion of the dopant. Although diffusion is responsible for both heat and dopant transport, the diffusivities, microscopic space and time scales, and other controlling parameters are sufficiently dissimilar that the problems required different numerical techniques.

This paper is concerned with calculations to fit the results of experiments on silicon wafers. The experimental

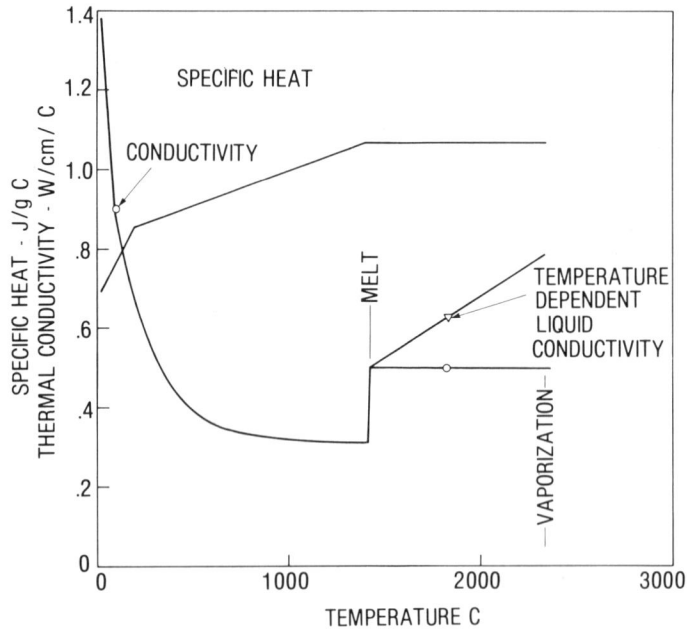

Fig. 1 Temperature dependent thermal properties for silicon.

data were taken from the same sources as reported in Refs. 1-3. The experimental methods are explained in these references.

Heat-Transfer Analysis Model

The single pulse, large beam diameter laser annealing process was modeled as one-dimensional heat diffusion with phase change and internal heat generation. The material can exist in several phases (amorphous, crystalline, polycrystalline, and liquid) which have different temperature-dependent thermal properties (Fig. 1), transition temperatures, and latent heats.[4,5] The wafer may initially be composed of layers of different phases.

The boundary conditions applied were a fixed temperature at an arbitrary depth to simulate the heat sink provided by the rest of the wafer and radiative heat transfer to ambient from the front (irradiated) surface. The depth of the fixed temperature rear surface was chosen to minimize computation resources required without significantly affecting the results within the region (< 2 μm) and time span of interest (< 1 μs).

The internal heating was produced by the interaction of the laser beam with the silicon wafer and is given by

$$\dddot{q}(x,t,p) = \dddot{q}(t)[1 - \rho(p)] e^{-\alpha(p)x} \qquad (1)$$

The absorption process depends on the material phase, laser wavelength, dopant concentration, and material temperature. Equation (1) is slightly more complicated if the absorption coefficient, α, is allowed to be stepwise constant with depth (for instance, in a model with an amorphous layer over a crystalline substrate). A more complicated equation is required if α is a function of dopant concentration or temperature. The more complicated forms for α have been utilized in a few calculations but gave no improvements for the present calculations. Most of the analyses presented here were performed with α constant or stepwise constant, simulating a one-, two-, or three-layer wafer.

The surface reflectivity, ρ, strongly influences the amount of absorbed energy and can be a function of the near surface material phase[6] and temperature and of the laser wavelength. Experimentally, the laser pulse has a roughly Gaussian variation with time, which has been modeled by a triangular function. Comparative studies using different pulse shapes have shown this approximation to be sufficiently accurate for our purposes, and it simplifies the analysis and reduces the computational times somewhat.

Table 1 contains a summary of the modeling parameters for the analyses presented.

Heat-Transfer Numerical Method

The complex nature of the physical process and the accuracy needed dictated the use of finite difference methods. A limited investigation showed that a modified explicit time integration technique[7] required the least computer resources for the desired accuracy. The Levy time integration technique, as implemented in HEATING5,[8] combines the classical explicit finite volume heat balance equation (2) with a three-time-level explicit heat balance equation (3)

$$T_i^{n+1} = T_i^n + \frac{\Delta t}{C_i}\left[Q_i^n + \sum_{m=i-1}^{i+1} {}_iK_m(T_m^n - T_i^n)\right] \quad (2)$$

$$T_i^{n+1} = T_i^n + \frac{1}{1+Z_i}\left\{\frac{\Delta t}{C_i}\left[\sum_{m=i-1}^{i+1} {}_iK_m(T_m^n - T_i^n) + Q_i^n\right] + Z_i(T_i^n - T_i^{n-1})\right\} \quad (3)$$

where

$$Z_i = \begin{cases} 0, & \Delta T \leq CFL_i \\ \frac{1}{2}\left(\frac{\Delta t}{CFL_i} - 1\right), & \Delta t > CFL_i \end{cases}$$

$$CFL_i = \frac{C_i}{\sum_{m=i-1}^{i+1} {}_iK_m} \quad (4)$$

The term CFL_i represents the maximum stable time step for node i which can be used by the classical explicit technique according to the Von Neumann criterion.[9]

Levy's modified explicit technique is stable for any time step, but is more expensive than a simple explicit technique. The calculations used a mixture of Levy and simple explicit technique. Whenever the stable time step for a cell was

Table 1   Model parameters

c Model
(Doped Crystalline Material)

Thermal Properties:
    Conductivity
        Solid   Ref. 4
        Liquid   0.5 W/cm C
    Specific heat
        Solid   Ref. 4
        Liquid   1.067 J/gm C
    Density
        Solid & Liquid   2.33 gm/cm^3
    Latent heat
        Fusion   1799.12 J/gm at 1410 C
        Vaporization   10613.5 J/gm at 2315 C

Optical Properties:
    Absorption coefficient 3×10^4 cm^{-1}
    Reflectivity
        Solid (ρ_S)   0.35
        Liquid (ρ_L)   0.6

c Model
(Bell)

Same as c Model except as follows:

Optical Properties:

$\alpha = 1 \times 10^5$ cm^{-1} $t_H = 33$ ns

a Model
(Amorphous Layer-Crystalline Substrate)

Thermal Properties:
    Latent heat of fusion   1079.47 J/g
    for amorphous to liquid
    transition

Optical Properties:
    Absorption coefficient:

        α_a - 5×10^4 cm^{-1} for $x < 1.5 \times 10^{-5}$ cm^{-1}

        α_c - 3×10^3 cm^{-1} for $x > 1.5 \times 10^{-5}$ cm^{-1}

greater than the specified calculational time step, the simple explicit was used for that cell; otherwise, the Levy technique was used.

In comparison with the explicit and two implicit techniques (HEATING5, TRUMP,[10] and HEATING5's implicit technique with specific heat modeling of latent heat), Levy's method produced accurate melt depth histories (± 2.5 nm) while requiring the least computational resources. The simple explicit required more resources than Levy's technique due to the stability limitations on the time step. The implicit techniques were more expensive because they required more computations per time step than the Levy technique while the implicit time step size required for accurate phase front location was not much larger than that for the Levy technique. In addition, the implicit techniques were less robust (failing to produce results for some models) than the explicit techniques due to the large latent heat and highly temperature dependent thermal properties of the wafer material.

The phase change was modeled by a straightforward energy accounting technique. The temperature of a cell is calculated from the finite difference heat balance Eq. (2) or (3) until the phase change (transition) temperature is reached. The temperature is then fixed until the net heat transferred is greater than the latent heat capacity of the cell. The temperature is then calculated by Eq. (2) or (3). This technique has proven to be more accurate than the incorporation of the latent heat into the temperature dependent heat capacity. Although the energy accounting system has proven to be robust, there are inherent errors in the calculated heat flux into a cell that is melting in any method that fixes the average temperature of the cell at the melt temperature. The calculational thermal gradients will be greater, less than, or equal to the actual gradient depending on the melt front location within the cell (Figs. 2a, 2b, and 2c). As shown in Fig. 2b, the cell calculated heat fluxes are correct only when the melt front is at the cell center (for uniform grid). Thus, using the melt temperature to calculate heat flux is only accurate when the front is at the cell center. The reduction of this error requires cell and time step sizes smaller than would be sufficient for thermal diffusion without phase change.

Several analytical problems were compared to determine the accuracy of the phase change technique. Figure 3 presents the melt front location as a function of time for a semi-infinite solid initially at melt temperature with a sudden change in surface temperature. The relative error in location

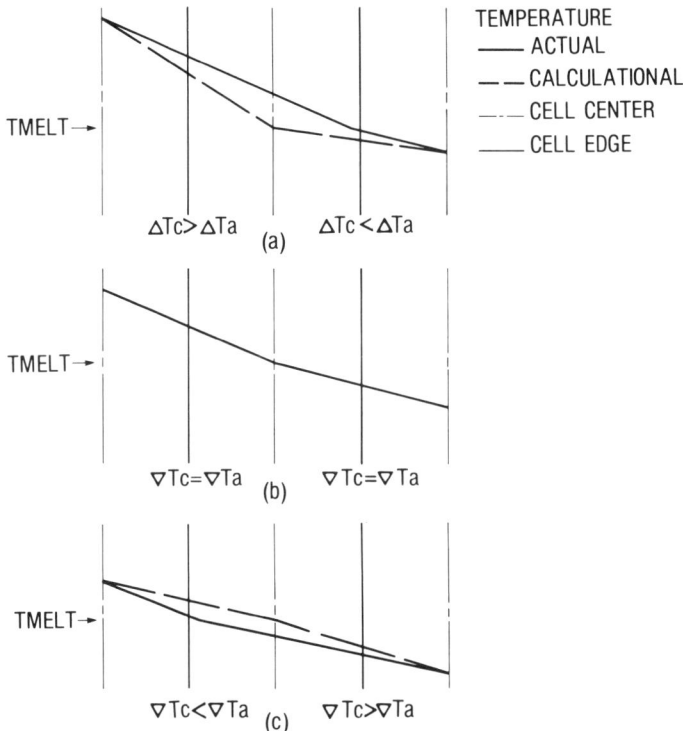

Fig. 2 Heat flux errors in the calculational technique.

is also presented in Fig. 3 and, after large errors at the start, is bounded to within a tenth of the cell width. An accurate location history can be calculated if the time step and cell size are chosen sufficiently small.

Heat-Transfer Results

The effects of the major modeling components are compared in Fig. 4 and justify the incorporation of the phase change technique and temperature dependent thermal properties into the model. Each curve represents the addition of the labeled modeling component progressing from top to bottom. All but the lower curve are temperature distributions at the end of a constant power pulse. The lower one is at the same time from pulse start but halfway through or at peak power for a triangular pulse. The triangular pulse model ultimately achieves nearly the same maximum temperature as the model immediately above in the figure and has a very similar melt depth history. A common approximation for preliminary analysis has been to

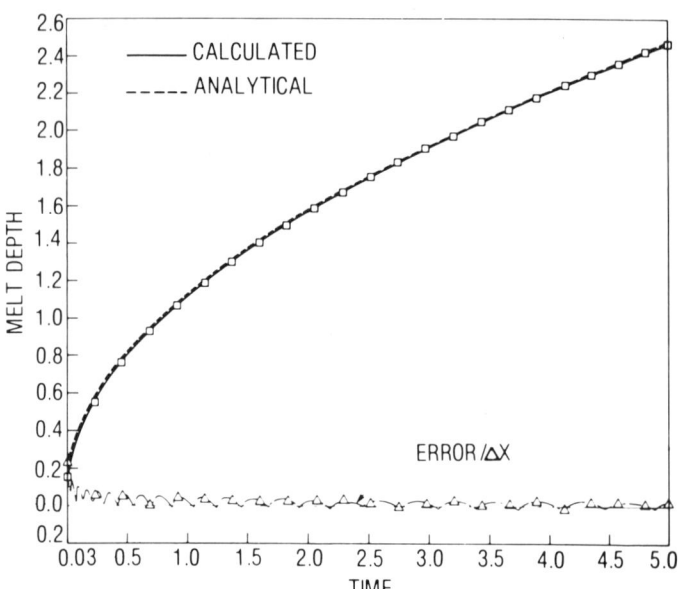

Fig. 3 Comparison of melt front locations from analytical solutions with finite difference calculations.

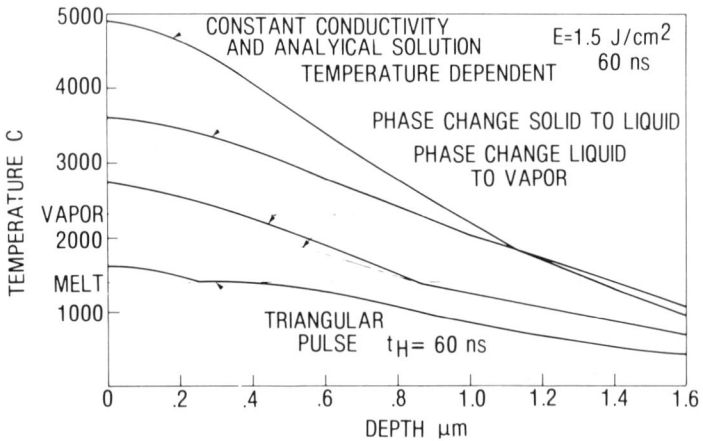

Fig. 4 Temperature distributions.

utilize constant average thermal properties and neglect phase change. The melt penetration is estimated from the maximum penetration of the melt temperature isotherm. This method can result in significant overestimation of the melt depth as shown in Fig. 4. This error is primarily due to the increased thermal gradients near the surface caused by the energy that should be stored in the latent heat of the material.

Transient temperature distributions for a "typical" laser pulse are shown in Fig. 5. The rapid decay of superheat after the pulse termination can be seen in the surface temperature history, Fig. 6. The location of the melt front is of prime

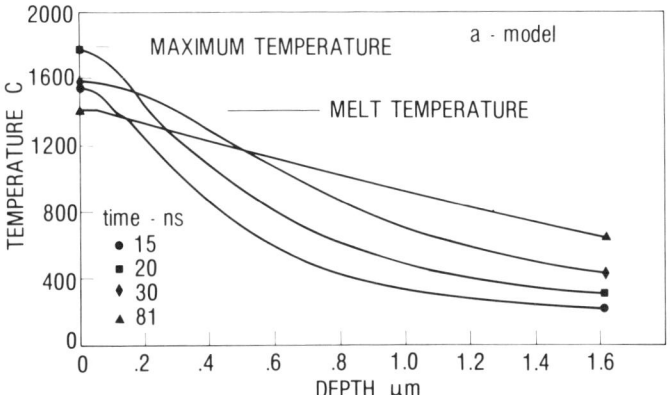

Fig. 5   Temperature distribution for a model, $E_L = 1.41$ J/cm^2.

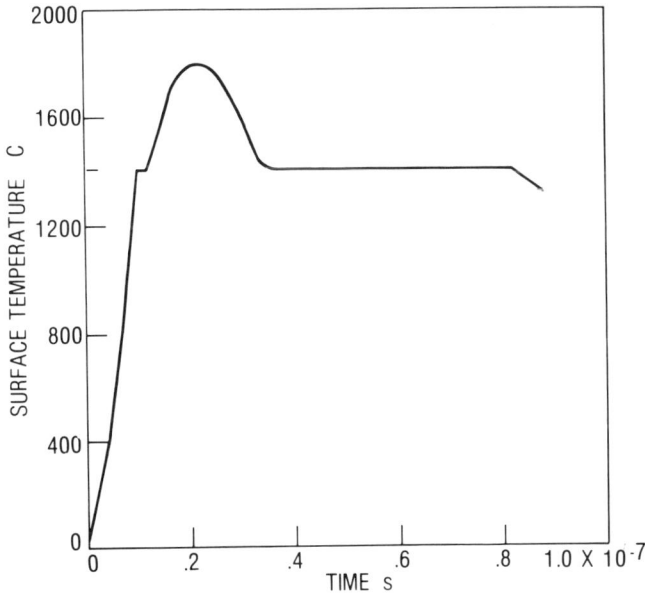

Fig. 6   Surface temperatures for $E_L = 1.41$ J/cm^2.

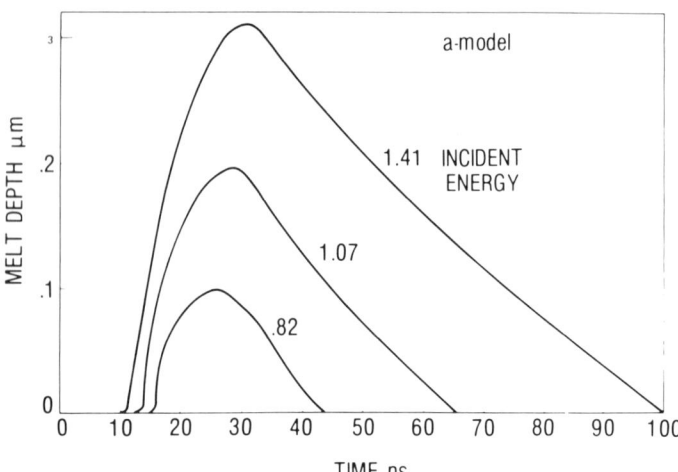

Fig. 7 Melt depth histories for arsenic-doped silicon laser annealing.

importance to the dopant diffusion calculation. Sample melt depth histories for several laser energies are shown in Fig. 7.

Figure 8 summarizes the maximum melt depth (penetration) for several models as a function of incident energy. As will be demonstrated in the following sections, the combination of calculated melt front histories and dopant transport calculations give good agreement with experimental dopant distributions. This process can provide an indirect verification of the calculated penetrations. The two experimental points with the highest energy on Fig. 8 (1.07 and 1.41 J/cm^2) were found from such a correlation. A more direct comparison with the experiments is possible for lower energy pulses which give penetrations that are less than the maximum depth of the dopant distribution before annealing. If the dopant is assumed to be mobile in the liquid phase only, then the before and after annealing distributions will exhibit redistribution for the melted region only. The region that has remained solid will have the same dopant distribution before and after annealing. Estimates of the penetration by examination of the distributions for the 0.82 and 1.07 J/cm^2 annealing experiments are presented in Fig. 8. The two methods of estimating the penetration agree for the 1.07 J/cm^2 and both estimates agree well with the calculated penetrations.

There are some data in the literature giving the results of measurements of surface melt duration.[11] Figure 9 shows

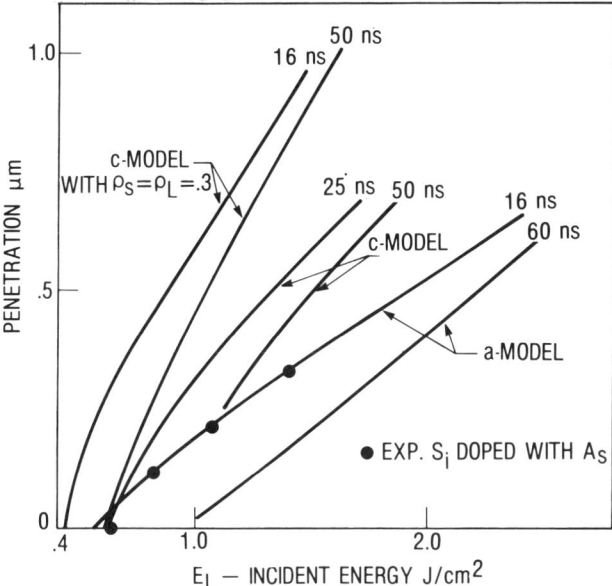

Fig. 8 Summary of penetrations for several models.

comparisons of this data with calculations. The model marked "Bell" in the figure was an adaptation of the c model that was more consistent with the experimental conditions in Ref. 11. Both the "Bell" and the c models show good agreement with the data.

Dopant Transport Calculations

Simple diffusion is assumed to be the mechanism for dopant transport in the molten region of the sample. The problem is complicated by the presence of the moving phase boundary (melt front) and the fact that the dopant may segregate at the melt front. This segregation phenomenon will be discussed in greater detail later. These complications have been effectively modeled as local perturbations on the general numerical algorithm for diffusion.

Diffusion is calculated by finite differences using a straightforward, fully implicit algorithm. In one dimension, the diffusion equation is

$$\frac{\partial c}{\partial t} = \frac{\partial}{\partial x} \left(D \frac{\partial c}{\partial x} \right) \qquad (5)$$

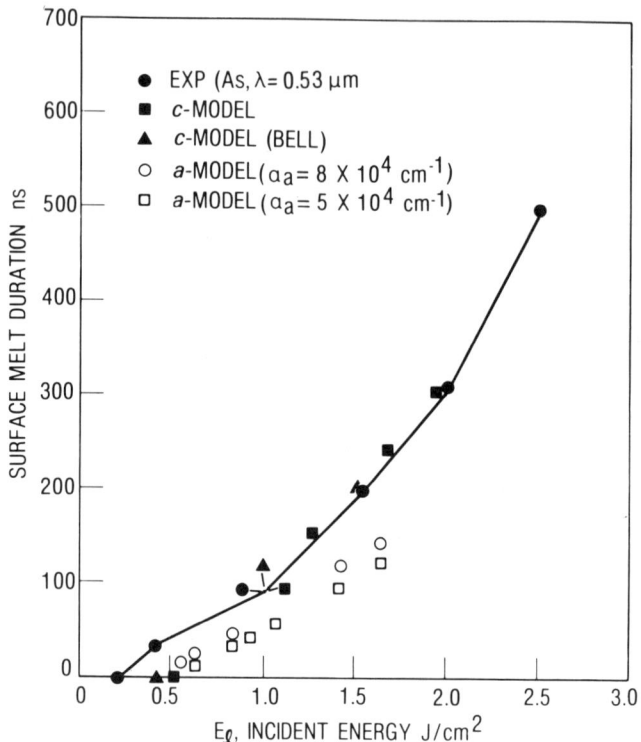

Fig. 9 Surface melt duration.

The dopant concentration for the finite difference algorithm $c_{i+1/2}$ is defined at the center of the computational cell $x_{i+1/2}$. The diffusive fluxes $D(\partial c/\partial x)$ are defined at the cell edges. They are

$$\left(D \frac{\partial c}{\partial x}\right)_i \approx D_i \frac{c_{i+1/2} - c_{i-1/2}}{x_{i+1/2} - x_{i-1/2}}$$

This flux approximation requires a diffusivity at x_i or, more exactly, an averaged diffusivity in the interval from $x_{i-1/2}$ to $x_{i+1/2}$. If the material is fully melted or fully solid in this interval, the value of D_i is obvious. Generally, it is not desirable to move the mesh points to place the melt front in the center of a cell. Therefore, we calculated an averaged diffusivity for the interval which contains the melt front based on adding the resistances in the subinterval which is melted and that which is solid. This approach has proven quite satisfactory in our calculations.

The boundary condition at the surface is that the flux is known. For most samples, the surface flux is zero. The boundary flux is added to the difference equation for the cell whose edge is the surface. The boundary condition used in the deep interior is one of zero flux and therefore the mesh should extend somewhat beyond the deepest that the melt front will penetrate. Thus, the diffusion at the interior boundary is negligible so that the zero flux condition is as accurate as the more nearly correct continuative condition.

The finite difference equation for an interior cell is

$$\frac{c_{i+1/2}^{n+1} - c_{i+1/2}^{n}}{\Delta t} = \frac{1}{x_{i+1} - x_i} \left(D_{i+1}^{n+1} \frac{c_{i+3/2}^{n+1} - c_{i+1/2}^{n+1}}{x_{i+3/2} - x_{i+1/2}} - D_{i}^{n+1} \frac{c_{i+1/2}^{n+1} - c_{i-1/2}^{n+1}}{x_{i+1/2} - x_{i-1/2}} \right) \quad (6)$$

where the superscript n+1 represents quantities at time $t = (n+1)\Delta t$. The set of equations for the total assembly of cells becomes a system of simultaneous linear equations for the c^{n+1}. This system is tridiagonal and may be solved by simple Gaussian elimination.

Segregation is a phenomenon which was mentioned briefly a few paragraphs earlier. This term denotes a tendency for some dopants to remain in the molten portions of a resolidifying sample rather than being incorporated into the newly formed solid layers. Segregation is well known in metallurgical processes. The strength of the segregation effect is given by the interface segregation coefficient k which is defined as the ratio of the dopant concentration on the solid side of the melt (resolidification) front to that on the liquid side precisely at the melt front. Values of this coefficient for various dopants in silicon are published in the literature. These have been measured under conditions in which the front is moving very slowly. When a resolidification front moves at high speed, the freezing material will not have time to reject its quota of dopant nor will the rejected dopant be able to diffuse fast enough to escape the oncoming front. Consequently, larger amounts of dopant will be frozen into the solid than would be expected with slower motion. The apparent segregation coefficient inferred by analyzing the dopant profiles left after the sample has refrozen will be closer to 1.0 (i.e., less apparent segregation) than the equilibrium value. The difference between this nonequilibrium segregation coefficient and the equilibrium value increases with increases in the resolidification rate. A sample of a material which is known to segregate under thermodynamic equilibrium conditions may show no signs of segregation after ultrarapid resolidification.

The modeling of segregation requires some significant alterations in the finite difference scheme. Fundamentally, the segregation process represents an internal "boundary" condition at the melt front. It acts like a pump transferring dopant from the layer which is refreezing to that which is remaining molten. Mathematically, the interface condition is

$$c_s/c_\ell = k \qquad (7)$$

where c_s represents the solid phase dopant concentration and c_ℓ is the liquid phase concentration, both measured at the melt front.

To deal with this internal condition, we chose to force the mesh to align with the melt front. Rather than move all the mesh points, some extra cells are created at the beginning of each time cycle and then removed at the end of the cycle. These cells are inserted into the cell which contained the melt front in such a way that the front was at the center of one of the new cells. The interface condition is applied to produce a two-sided dopant distribution in the cell which contains the melt front. The modified finite difference equation for this cell is

$$\frac{c_{i+1/2}^{n+1} - c_{i+1/2}^n}{\Delta t} = \frac{1}{x_{i+1} - x_i} \left[\frac{D_{i+1}}{x_{i+3/2} - x_{i+1/2}} \left(c_{i+3/2}^{n+1} - \frac{2k}{k+1} c_{i+1/2}^{n+1} \right) \right.$$
$$\left. - \frac{D_i}{x_{i+1/2} - x_{i+1/2}} \left(\frac{2}{k+1} c_{i+1/2}^{n+1} - c_{i-1/2}^{n+1} \right) \right] \qquad (8)$$

This assumes that the molten portion is at $x < x_{i+1/2}$. The two-sided nature of the dopant distribution at the interface is shown by the factors multiplying $c_{i+1/2}$.

The implicit finite difference algorithm is stable for any value of time step Δt. But stability does not guarantee accuracy. For problems with no segregation, the time step has to be limited so that the melt front takes at least two cycles to cross a cell when the cell is resolidifying. For problems with segregation, the time step limitations are more stringent. These limitations are dependent on the value of the segregation coefficient and will be discussed in a later section.

There is an artificial diffusivity which is inherent in the truncation errors of the finite difference algorithm. For very small but nonzero diffusivities, this synthetic diffusion can overpower the real diffusion. However, if the physical diffusivity is exactly zero, the artificial diffusivity is also zero. In view of the very small diffusivities in solid silicon, more accurate results are obtained by using a zero value of diffusivity in cells which are completely solid.

Comparison of Dopant Transport Calculations with Analytic Solutions

There is an analytic solution for the segregation problem under specified restrictions. Let a solidification front begin to move with a constant velocity V across a liquid of uniform initial dopant concentration c_0. After some time, the shapes of the curves of dopant concentration ahead of and behind the front will reach asymptotic forms. From Ref. 12, the asymptotic concentration behind the front will be the initial value c_0 while that ahead of the front is given by

$$c_\ell(x)/c_0 = 1 + [(1-k)/k] \, e^{-\frac{V}{D}x} \qquad (9)$$

where D is the liquid phase diffusivity and x is the distance measured from the moving front. Further, the exact concentration gradient in the already solidified material is given by

$$c_s(y)/c_0 = 1/2 \, \{1 + \mathrm{erf}(1/2 \sqrt{Vy/D}\,)$$
$$+ (2k-1) \, \exp[-k(1-k)(Vy/D)]$$
$$\mathrm{erfc}\,[(2k-1)/2 \sqrt{Vy/D}\,]\} \qquad (10)$$

where y is the distance from the point at which the front began moving. This function has a value of k for $y = 0$ and asymptotes to 1.0 for $y \to \infty$.

The accuracy of the computer code was first tested by comparison with the analyic solution. As might have been expected, numerical difficulties cropped up as the segregation coefficient was reduced from 1.0. Although the fully implicit diffusion algorithm is formally unconditionally stable, the combination of diffusion and segregation is not. In a calculation allowing 26.5 time steps for the front to cross a 12.5 Å cell, there was a sawtooth pattern to the concentration in the solid $k = 0.1$. This suggested an incipient instability. When the time step was cut in half, this sawtooth vanished, which confirmed the instability supposition. For a calculation with $k = 0.3$, the 26.5 cycles per cell transient time step did not show such a sawtooth.

In addition to stability, the time step also affects accuracy. For $k = 0.1$, the time step of 53 cycles per cell needed for stability gave a concentration curve which roughly matched that given by the analytic solution using a value of $k \approx 0.15$. Cutting the time step by another factor of 8.0 gave a numerical solution which followed the analytical curve for $k \approx 0.12$. The results for this case are plotted in Fig. 10. This small a time step is past the point of diminishing returns. The deviation of the calculated results from the analytic solution approaches zero as time step goes to zero, but calculations with the smaller time steps are uneconomical. Calculations using the same time steps were run for $k = 0.35$. Cutting the time step by another factor of 8.0 gave a curve which followed the analytic results for $k \approx 0.31$.

These problems are zoned with a space step of 12.5 Å. Initially it was thought that this value might be too large to follow the concentration curve accurately. A few calculations for $k = 0.1$ were run using half this value. It was found that the results were virtually identical to those from calculations using the larger space step, provided one compared problems with time steps using the same number of cycles to cross a cell. In other words, when both space step and time step were cut in half, the results were virtually identical. Thus, the 12.5 Å space step is quite sufficient. The accuracy is controlled by the time step only.

Dopant Transport Results

The techniques described in this paper have been applied to experiments using four different dopants: boron, arsenic, bismuth, and antimony. The results presented here are best matches produced after some iteration on the models and some of the constants. This latter point deserves some comment. Some of the parameters of the heat-transfer model, notably those which govern the deposition of the laser energy in the silicon, do not have values which are precisely known from other experiments. Also, the dopant segregation coefficients are not known very accurately. The diffusivities used were generally those from Kodera.[13] The dopant transport calculations used

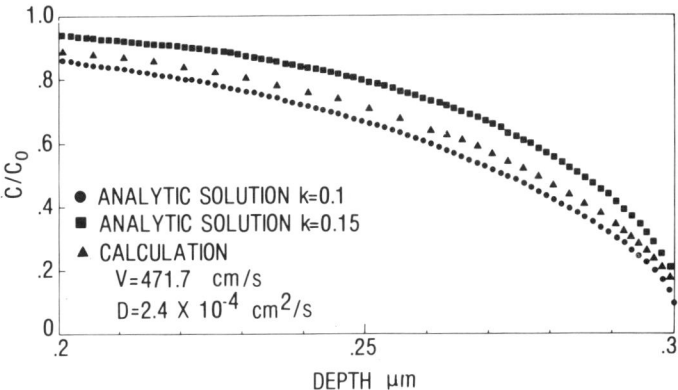

Fig. 10 Diffusion with segregation constant resolidification front velocity uniform initial condidtion.

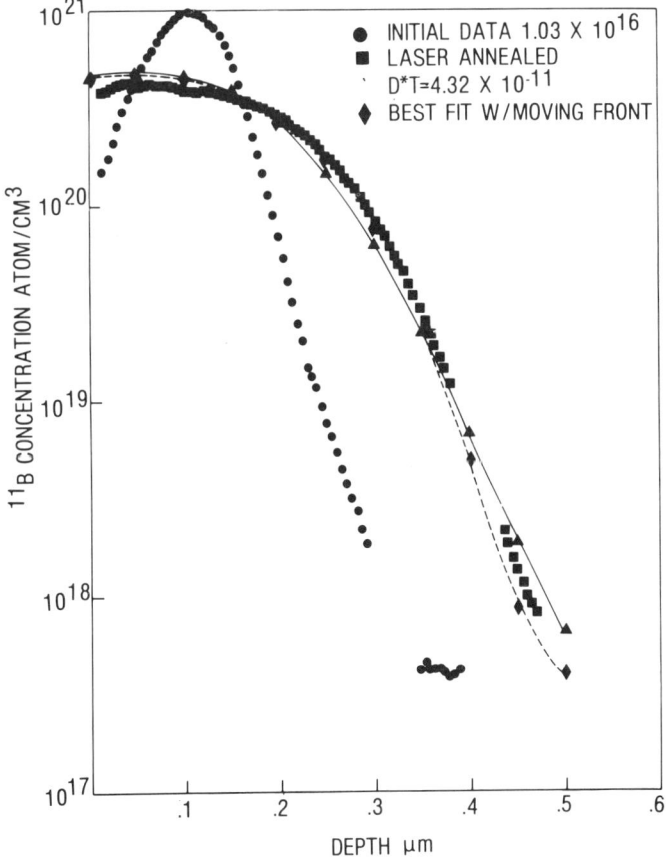

Fig. 11 Laser annealed silicon doped with boron.

melt front results from the heat-transfer calculations which, of course, have uncertainties associated with them because of the limited accuracy of the input data. The heat-transfer results had shown that melt depth histories produced by changes in many of the heat-transfer parameters tended to look somewhat alike when depth divided by maximum depth was plotted against time from first melt divided by maximum melt duration. Thus, a simulation of the effects of changing heat-transfer parameters could be provided for the dopant transport calculation by simply scaling the melt depths and times from a particular heat-transfer calculation. Dopant calculations varying both melt profile and segregation coefficient were done until a best match was found. Then, the resulting estimate of the maximum melt depth was used to refine the heat-transfer parameters by finding which values would give that depth. While this process was hardly exact, it did make useful contributions to calibrating the parameters.

The boron experiments were studied first and the results are shown in Fig. 11. Comparison of the experimental dopant profiles before and after laser annealing shows the degree to which the initial profile diffuses. There are two different calculational curves to compare with the data. The first was obtained by instantaneously melting the sample to "infinite" depth and letting the dopant diffuse with no melt front motion. This curve corresponds to the melt time duration which gave the best match to the data. The second was obtained using a melt front motion found from heat-transfer calculations. The instant melt results show a good match to the data, but the moving melt front calculation does slightly better. As it comes back toward the surface, the front presents a barrier to

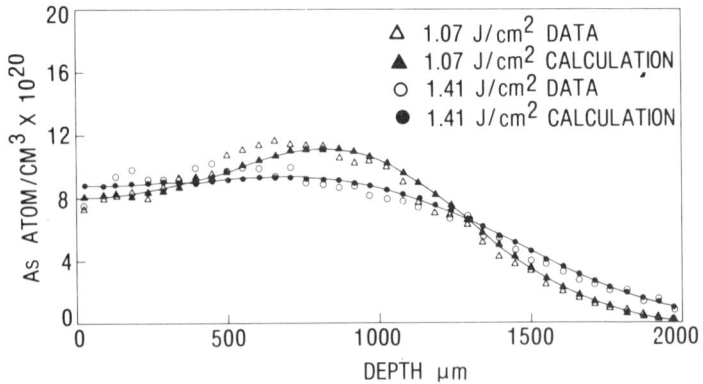

Fig. 12 Laser annealed silicon doped with arsenic.

further diffusion into the interior, thus steepening the annealed profile. These results were obtained without any segregation.

The next material to be considered is arsenic. The arsenic results are shown in Fig. 12. The as-deposited profile has been left off so that the figure will not be too busy. The figure shows results of annealing with two different laser energies. The experimental data were also fit with an instantaneous melt calculation. As was the case with boron, the moving melt front gave a better match than the instantaneous melt calculation. Arsenic shows a moderate tendency toward segregation (k = 0.3) under equilibrium conditions, but inclusion of segregation in the calculations with this value of k did not improve the fits to the data. Further, if segregation had been significant for these experiments, there should be a concentration spike at the surface which contains the dopant carried ahead of the moving front. There is no sign of such a spike in the data.

Bismuth is the only dopant considered here which shows clear signs of segregation. The surface spike in the data on Fig. 13 is unmistakable. Comparison of the calculations with and without segregation demonstrates the effect of segregation. The equilibrium segregation coefficient for bismuth is 0.0007 while the best agreement with experiment was obtained using a value of 0.3. This determination of the nonequilibrium k is

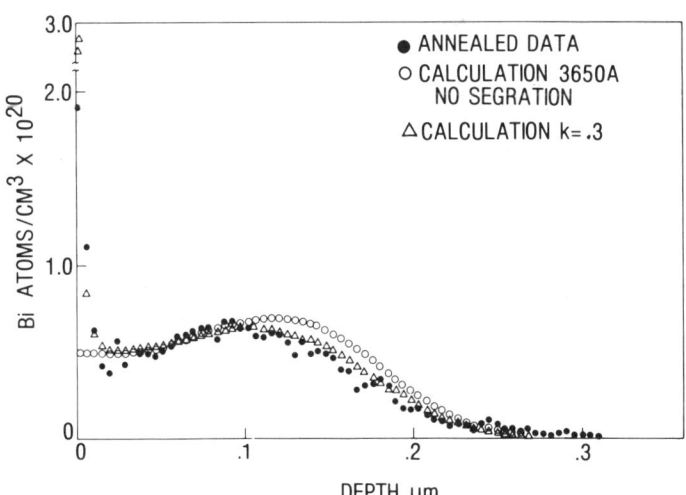

Fig. 13   Laser annealed silicon doped with bismuth.

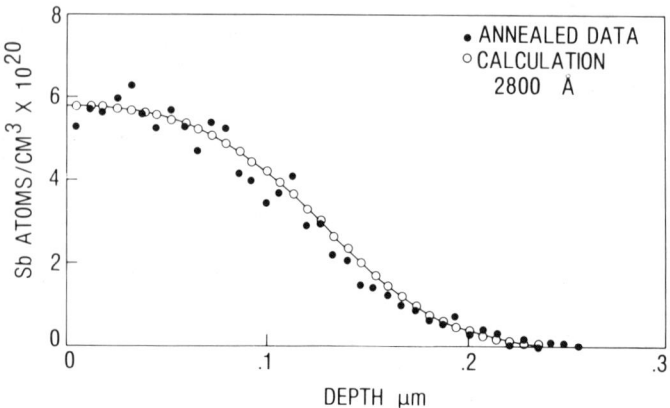

Fig. 14 Laser annealed silicon doped with antimony.

rather approximate because the quality of the match is little changed for values ranging from 0.2 - 0.4.

Antimony showed an unusual behavior. The integrals under the before and after annealing curves indicate a loss of ~ 11% of the dopant. There are at least three possible calculational models for this loss process: 1) a constant boundary flux during the entire time the surface is melted; 2) a boundary flux proportional to the concentration of dopant at the boundary; and 3) a high boundary flux early in the melting process when the surface temperatures are the highest, with a zero flux thereafter. The best match was obtained using the third option with the loss occurring in the first 10 ns of the approximately 170 ns total surface melt time. The antimony data does not show signs of segregation, although the material has an equilibrium segregation coefficient of 0.023. Calculations with segregation did not improve the match. The comparison of experimental and calculated profiles is shown in Fig. 14.

Conclusions

The combination of heat-transfer and dopant transport calculations can give good fits to experimentally measured dopant distributions. It is worth noting that the space scales for both calculations are at the lower edge of continuum. The success of the calculations suggests that statistical (noncontinuum) effects were not significant.

The finite difference dopant transport calculations with segregation can be made to match analytical solutions closely.

The expense associated with highly accurate fits rises rapidly for segregation coefficients below ~ 0.2 and can become prohibitive for very low values. We also note here in connection with computational requirements that the cost of the heat-transfer calculations may become great for short laser pulses and for high absorption coefficients.

The magnitude of the segregation effect in samples which recrystallize at high rates is much less than it is under equilibrium (very slow recrystallization) conditions. Arsenic and antimony, which segregate under equilibrium conditions, show no signs of segregation in the calculational fits to the results of the laser annealing experiments.

The theory of the physical phenomena occurring during pulsed laser annealing has been discussed recently in much greater detail in Refs. 14 and 15 and references therein.

Acknowledgments

The authors would like to express their appreciation to W. D. Turner, Computer Sciences Division, Union Carbide Corporation, Nuclear Division, Oak Ridge, Tennessee, and J. C. Wang, Solid State Division, Oak Ridge National Laboratory, Union Carbide Corporation, Nuclear Division, Oak Ridge, Tennessee for technical discussions and the authors of Refs. 1-3 who provided much of the experimental data before its publication. Research was performed at the Oak Ridge Gaseous Diffusion Plant and at the Oak Ridge National Laboratory, operated by Union Carbide Corporation, Nuclear Division, for the U. S. Department of Energy under U. S. Government Contract W-7405 Eng 26.

References

[1] White, C. W., Wilson, S. R., Appleton, B. R., and Young, F. W., Jr., "Supersaturated Substitutional Alloys Formed by Ion Implantation and Pulsed Laser Annealing of Group-III and Group-V Dopants in Silicon," Journal of Applied Physics, Vol. 51, No. 1, Jan. 1980, pp. 738-749.

[2] Wang, J. C., Wood, R. F., White, C. W., Appleton, B. R., Pronko, P. P., Wilson, S. R., and Christie, W. H., "Dopant Profile Changes Induced by Laser Irradiation of Silicon: Comparison of Theory and Experiment," Laser-Solid Interactions and Laser Processing, edited by H. J. Leamy and J. M. Poate, Owen Institute of Physics, New York, 1979.

[3] White, C. W., Pronko, P. P., Wilson, S. R., Appleton, B. R., Narayan, J., and Young, R. T., "Effects of Pulsed Ruby-Laser Annealing on As and Sb Implanted Silicon," Journal of Applied Physics, Vol. 50, No. 5, May 1979, pp. 3261-3273.

[4] Goldsmith, A. et al., Handbook of Thermodynamic Properties of Solid Materials-Elements, Vol. 1, Peragon Press, New York, 1961.

[5] Lyman, T., Metals Handbook, Vol. 1, American Society for Metals, Metals Park, Ohio, 1961.

[6] Auston, D. H., Surko, C. M., Venkatesan, T. N. C., Slusher, R. E., and Golovchenko, J. A., "Time Resolved Reflectivity of Ion-Implanted Silicon During Laser Annealing," Applied Physics Letters, Vol. 33, No. 5, Sept. 1978, pp. 437-440.

[7] Levy, S., "Use of 'Explicit Method' in Heat Transfer Calculations with Arbitrary Time Step," GE Report No. 68-C-282, General Electric, Schenectady, N. Y., Aug. 1968.

[8] Turner, W. D., Elrod, D. C., and Siman-Tov, I. I, "HEATING5 - An IBM 360 Heat Conduction Program," ORNL/CSD/TM-15, Oak Ridge, Tenn., March 1977.

[9] Richtmyer, R. D., and Morton, K. W., Difference Methods for Initial-Value Problems, 2nd Ed., Interscience Publishers, New York, 1967.

[10] Edwards, Arthur L., "TRUMP: A Computer Program for Transient and Steady State Temperature Distributions in Multidimensional Systems," UCRL-14754 rev. 3, Lawrence Livermore Laboratory, Sept. 1972.

[11] Auston, D. H., Golovchenko, J. A., Simons, A. L., Slusher, R. E., Smith, P. R., Surko, C. M., and Venkatesan, T. N. C., "Dynamics of Laser Annealing," Proceedings of Laser-Solid Interactions and Laser Processing Symposium, Boston, Mass., Nov. 28 - Dec. 1, 1978, pp. 11-16.

[12] Smith, V. G., Tiller, W. A., and Rutter, J. W., "A Mathematical Analysis of Solute Redistribution During Solidification," Canadian Journal of Physics, Vol. 33, No. 12, Dec. 1955, pp. 723-745.

[13] Kodera, H., "Diffusion Coefficients of Impurities in Silicon Melt," Japanese Journal of Physics, Vol. 2., No. 4, April, 1963, pp. 212-219.

[14] Wood, R. F., Giles, G. E., "Macroscopic Theory of Pulsed-Laser Annealing: I. Model for Thermal Transport and Melting," to be published in Physical Review B, 1981.

[15] Wood, R. F., Kirkpatrick, J. R., and Giles, G. E., "Macroscopic Theory of Pulsed-Laser Annealing: II. Dopant Diffusion and Segregation," to be published in Physical Review B, 1981.

Chapter II. Conduction Heat Transfer

THERMAL RESISTANCE OF TWO-DIMENSIONAL GEOMETRIES HAVING ISOTHERMAL AND ADIABATIC BOUNDARIES

G. E. Schneider[*] and M. M. Yovanovich[+]

University of Waterloo, Waterloo, Ontario, Canada

Abstract

An important class of two-dimensional, steady-state heat conduction problems with no internal heat generation is examined. The domain is bounded by coordinate surfaces of any two-dimensional planar coordinate system and the heat enters and leaves only through two isothermal contacts on the boundary. The location of the two contacts is arbitrary. It is demonstrated that the thermal resistance for these problems is invariant when the domain is subjected to a conformal transformation. A general procedure is presented for solving the above class of problems. The procedure consists of two fundamental steps, with each step consisting of four conformal transformations for the general case. Two worked examples are provided to demonstrate the application of this procedure.

Nomenclature

A = area, or location
B = location
C = location
C_1 = constant in Schwartz-Christoffel transformation
cn = Jacobian elliptic cosine amplitude function
D = location
E = location

Presented as paper 80-1471 at the AIAA 15th Thermophysics Conference, Snowmass, Colo., July 14-16, 1980. This paper is declared a work of the U.S. Governement and therefore is in the public domain.

[*]Assistant Professor, Thermal Engineering Group, Department of Mechanical Engineering.

[+]Professor, Thermal Engineering Group, Department of Mechanical Engineering.

F = location or incomplete elliptic integral of the first kind
f = transformation designation
G = location
g = transformation designation or metric coefficient
i = $\sqrt{-1}$
K = complete elliptic integral of the first kind
k = thermal conductivity, or modulus of elliptic functions
L = thickness
n = normal
p = plane designation
Q = heat flow rate
q = heat flux, or plane designation
r = plane designation, or radius
R = thermal resistance
S = surface
s = plane designation
sn = Jacobian elliptic sine amplitude function
T = temperature
t = plane designation
u = plane designation, or coordinate of curvilinear system
x = Cartesian coordinate
y = Cartesian coordinate
z = Cartesian coordinate
Z = plane designation
θ = angle
λ = modulus of elliptic functions
ρ = coordinate representation

Introduction

In the analysis of thermal systems, there are many applications in which the transfer of energy as heat is two-dimensional, is steady-state, and occurs in the absence of internal sources or sinks of thermal energy. For many of these applications, which may be single components of a larger system, the heat transfer occurs through the component from one portion of the boundary to another, with the remaining surface being adiabatic. Such problems occur in the modelling of solar collector plates,[1] the analysis of the hulls of ships,[2] the modelling of transverse conduction through cylindrical members,[3] the analysis of integrated circuits mounted to spreader plates, and the modelling of two-dimensional contact resistance problems.[4] Frequently, these problems are analyzed by modelling the heat inflow and outflow portions of the surface by an isothermal specification with the remaining surface being adiabatic.

THERMAL RESISTANCE OF TWO-DIMENSIONAL GEOMETRIES

For problems whose geometry conforms to a Cartesian coordinate system and for which one or both of the diabatic portions of the surface spans the entire extent of one of the coordinate directions, solutions can readily be obtained by application of the method of separation of variables.[5] However, if the diabatic portions of the boundary span only a portion of the range of the corresponding coordinate, if the contact spans a corner of the geometry, or if the inflow and outflow surfaces lie on adjacent boundaries or on the same boundary of the domain, the analysis becomes exceedingly complex, and, indeed, is not tractable for many of these problems. Moreover, if the geometry of the cross section is not that of a simple rectangle, further complications arise. Thus, the majority of mixed boundary value problems in circular cylinder coordinates, elliptic cylinder coordinates, and parabolic cylinder coordinates, for example, remain intractable.

The particular problem classification examined in this work is that for which the temperature distribution satisfies Laplace's equation in two dimensions. Moreover, the geometric boundaries of the domain are restricted to those which lie along lines of constant coordinate value in a general orthogonal curvilinear coordinate system. The boundary conditions are that the diabatic boundaries are isothermal at different temperatures and that the remaining surface is adiabatic. No restriction, however, is placed on the location of the two diabatic boundary portions. This description therefore defines a very large class of conduction heat transfer problems of practical interest to the heat transfer community.

A derivation will be presented which shows that for any coordinate system which is generated through the use of a conformal transformation, the thermal resistance can be determined through the solution of an appropriate corresponding problem whose domain is of rectangular shape in a Cartesian-like coordinate frame. In this regard, the exception of the circular cylinder coordinate system will be noted and the necessary modification to this system presented. The above mentioned derivation will therefore establish the uniqueness of the thermal resistance under conformal transformation of the domain.

Having established the uniqueness of the thermal resistance under conformal transformation, a detailed procedure will be provided to enable determination of the thermal resistance of the mixed boundary value problem posed in the above referenced Cartesian-like coordinate frame. This will be done through successive application of the Schwartz-Christoffel conformal transformation.[6] The application of the Schwartz-Christoffel transformation ultimately leads to a specific configuration whose solution is provided in greater detail and forms a basis solution necessary to the solution of each of the many inflow/outflow boundary configurations.

Finally, the procedure is demonstrated by evaluating the thermal resistance of two conduction heat transfer problems whose solution is not readily obtainable through the use of conventional solution techniques.

Analysis

Description of the Problem

The problem classification under consideration in this work can be defined by invoking the following assumptions:

1) Steady-state conduction.

2) Two-dimensional planar conduction.

3) No internal heat sources or sinks.

4) Geometric boundaries form part of an orthogonal curvilinear coordinate system.

5) Inflow and outflow boundaries are each isothermal.

6) Remaining surface is adiabatic.

Although the above list of assumptions may appear to be somewhat restrictive, a large number of problems are encompassed by the class of problems to which this work is addressed. For example, Fig. 1 illustrates nine specific configurations within a Cartesian coordinate system alone which are defined by the proposed problem classification. In the figure, the dark thicker portions of the boundary indicate the inflow and outflow isothermal portions of the boundary. In Fig. 2, only a few selected examples in a non-Cartesian coordinate system to which the method applies are presented. Through the use of symmetry arguments, where appropriate, certain multiple inflow/outflow boundary region problems also fall under the classification of problem that is addressed in this work.

Clearly, upon examination of the thermal configurations presented in Figs. 1 and 2, analytic solutions using conventional methods will be extremely complex, if indeed possible, to achieve. Before providing the procedure for obtaining the thermal resistance to these problems, however, it is constructive to examine the question pertaining to the uniqueness of the thermal resistance under conformal transformation of the domain. This question is addressed in the following section.

THERMAL RESISTANCE OF TWO-DIMENSIONAL GEOMETRIES 183

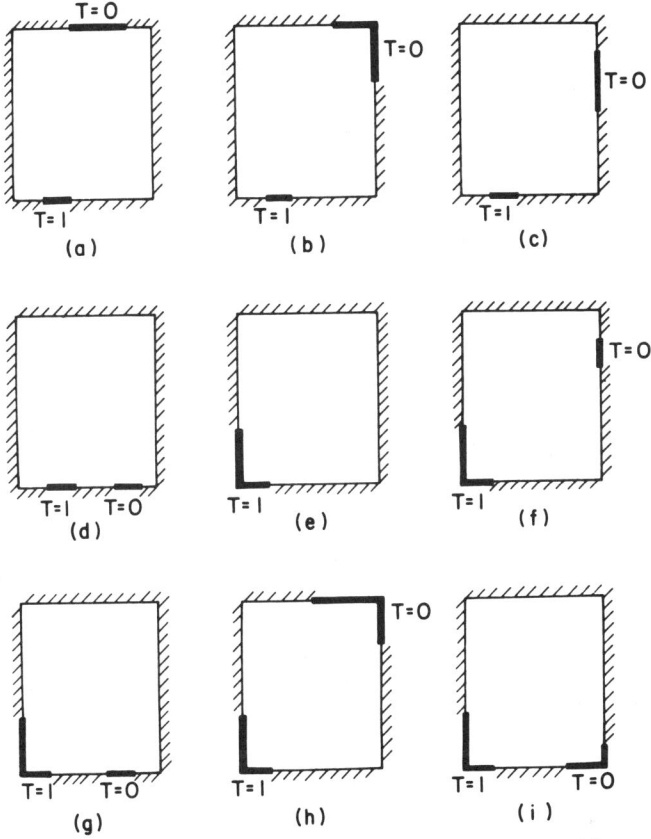

Fig. 1 Specific configuration of the thermal problem in a Cartesian frame.

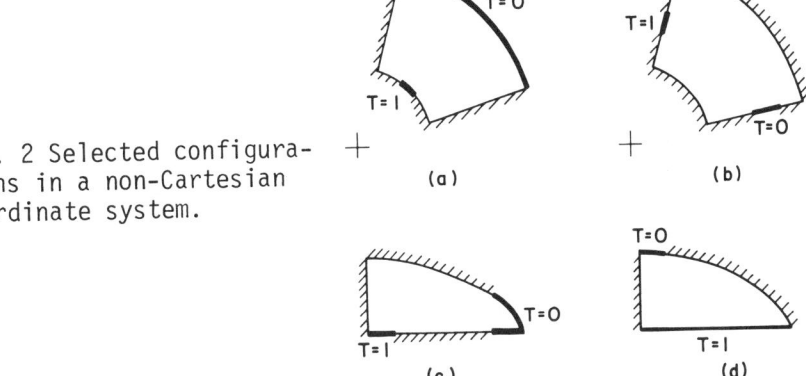

Fig. 2 Selected configurations in a non-Cartesian coordinate system.

Uniqueness of Resistance for Two-Dimensional Planar Problems

In addressing the uniqueness of the thermal resistance of geometries generated through the use of a conformal transformation from a Cartesian coordinate system, we examine coordinate systems generated through the transformation relation

$$\vec{u} = f(\vec{\rho}) \qquad (1)$$

which can be written as

$$\vec{\rho} = g(\vec{u}) \qquad (2)$$

where g is the inverse operator to f. In the above the Cartesian coordinates are given by

$$\vec{\rho} = x + iy \qquad (3)$$

and the transformed coordinates are given by

$$\vec{u} = u_1 + iu_2 \qquad (4)$$

It can readily be shown[7] that, through the transformation defined in this way, the Cauchy-Riemann conditions are valid. These conditions can be written as

$$\frac{\partial x}{\partial u_1} = \frac{\partial y}{\partial u_2} \qquad (5a)$$

$$\frac{\partial x}{\partial u_2} = -\frac{\partial y}{\partial u_1} \qquad (5b)$$

For planar problems, the third dimension is transformed according to the identity that

$$z = z \qquad (6)$$

where z is the third coordinate direction in both the original and the transformed plane. The metric or Lamé coefficients, which relate distances in the transformed plane to those in the original plane are defined by[4]

$$g_i = \left(\frac{\partial x}{\partial u_i}\right)^2 + \left(\frac{\partial y}{\partial u_i}\right)^2 + \left(\frac{\partial z}{\partial u_i}\right)^2, \quad i=1,2,3 \qquad (7)$$

where u_i is a general coordinate in the transformed plane. In this work we will reserve u_3 to denote the z direction so that

u_1 and u_2 can be identified as the coordinates lying in the plane of the cross section. Adopting the above coordinate association the metric or Lamé coefficients are

$$g_1 = (\frac{\partial x}{\partial u_1})^2 + (\frac{\partial y}{\partial u_1})^2 \qquad (8a)$$

$$g_2 = (\frac{\partial x}{\partial u_2})^2 + (\frac{\partial y}{\partial u_2})^2 \qquad (8b)$$

$$g_3 = 1 \qquad (8c)$$

Utilizing the Cauchy-Riemann conditions in Eq. (8b), it now becomes evident that

$$g_1 = g_2, \; g_3 = 1 \qquad (9)$$

for all coordinate systems defined in the manner indicated above.

Recalling that Laplace's equation in a general, orthogonal curvilinear coordinate system is written in two dimensions (u_1 and u_2) for constant properties as

$$\frac{\partial}{\partial u_1}\left[\frac{\sqrt{g}}{g_1}\frac{\partial T}{\partial u_1}\right] + \frac{\partial}{\partial u_2}\left[\frac{\sqrt{g}}{g_2}\frac{\partial T}{\partial u_2}\right] = 0 \qquad (10)$$

where

$$\sqrt{g} \equiv \sqrt{g_1 \, g_2 \, g_3} \qquad (11)$$

we find, using the result expressed in Eq. (9), that Laplace's equation becomes

$$\frac{\partial^2 T}{\partial u_1^2} + \frac{\partial^2 T}{\partial u_2^2} = 0 \qquad (12)$$

which is exactly of the same form as Laplace's equation when expressed in the original Cartesian coordinate system.

Two additional issues must now be addressed in order to establish the uniqueness of the thermal resistance of the proposed problem classification under conformal transformation. These are:

1) The statement of the boundary conditions.

2) The evaluation of the total heat flow rate, Q, for use in the definition of the thermal resistance, R, given by

$$R = \Delta T/Q \qquad (13)$$

To examine these issues, we consider the geometry of Fig. 3. In the original plane the boundary specification is

1) Over S_1, $T = T_1$ \hfill (14a)

2) Over S_2, $T = T_2$ \hfill (14b)

3) Over S_3, $\frac{\partial T}{\partial n} = 0$ \hfill (14c)

while for the transformed geometry, the boundary specification is

1) Over S_1', $T = T_1$ \hfill (15a)

2) Over S_2', $T = T_2$ \hfill (15b)

3) Over S_3', $\frac{\partial T}{\partial n} = 0$ \hfill (15c)

Thus, the form of the boundary condition specification is the same in both the original and transformed planes.

Finally, we examine the evaluation of the total heat flow rate. Since the domain is assumed in this work to be bounded by coordinate surfaces in the transformed domain, it is necessary to establish only that the integral

$$Q = \iint_{S_1} q \, dA \qquad (16)$$

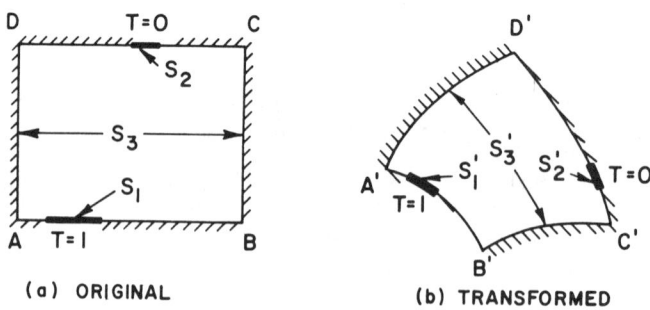

Fig. 3 Original and transformed problem geometry.

for example, is independent of the transformation metrics, and for this purpose it can be assumed that the surface lies exclusively along a single coordinate surface. The element of area, dA_1, formed by the u_2 and u_3 coordinates for a constant value of u_1 is given by[4]

$$dA_1 = \sqrt{g_2 g_3}\, du_2\, du_3 \qquad (17)$$

and the heat flux in the u_1 direction over the surface of constant u_1 is given by

$$q_1 = \left. \frac{-k}{\sqrt{g_1}} \frac{\partial T}{\partial u_1} \right|_{u_1 = \text{const}} \qquad (18)$$

The above integral then becomes

$$Q = \int\int_{u_2\, u_3} \frac{-k\sqrt{g_2 g_3}}{\sqrt{g_1}} \frac{\partial T}{\partial u_1} \left. du_2\, du_3 \right|_{u_1 = \text{const}} \qquad (19)$$

where the integration is performed over the appropriate limits defining S_1. In view of the results expressed by Eq. (9), Eq. (19) can be written as

$$Q = \int\int_{u_2\, u_3} -k \frac{\partial T}{\partial u_1} \left. du_2\, du_3 \right|_{u_1 = \text{const}} \qquad (20)$$

which, again, takes on exactly the same form as that utilized in a Cartesian frame, and is independent of the metric coefficients since it has already been shown that the temperature distribution, and hence its derivative, is independent of the metrics of scale defining the transformation.

It is therefore seen that the solution of the conduction problem in the transformed plane, subject to the assumptions described earlier, is identical to the solution of a conduction problem in a simple Cartesian coordinate system. The only criterion which must be imposed is that the geometry of the "equivalent Cartesian problem" must be specified by the values of the coordinates u_1 and u_2 which define the boundary surfaces in the transformed coordinate geometry. Indeed, the irregular, transformed geometry of Fig. 3b can be interpreted as simply the representation of a regular, Cartesian-like coordinate system on an actual set of Cartesian coordinate axes.

It can therefore be concluded that all mixed boundary value problems of the classification described earlier in this paper, for which the geometric boundaries of the domain are bounded by lines of constant coordinate value in an orthogonal curvilinear coordinate system, may be solved by solving an equivalent mixed boundary value problem in a Cartesian coordinate frame.

Basis Problem Solution

It was demonstrated in the preceding section that all mixed boundary value problems, of the classification considered in this work and whose cross-sectional geometry is formed by coordinate surfaces of a curvilinear orthogonal coordinate system, may be posed as equivalent problems in a pure Cartesian framework. Therefore, all such problems can be described by configurations of the type illustrated in Fig. 1. Further, it will be shown in the following section that each case of Fig. 1 can be reduced through conformal transformation to a single specific geometry. This geometric configuration is that illustrated in Fig. 4. Since this geometric configuration forms the basic solution for all possible configurations, the solution to this problem is presented in detail.

The solution to this basis problem is in itself obtained through a series of conformal transformations. The intermediate and final planes resulting from the transformations are shown in Fig. 5 for this basis problem. The transformations necessary for effecting the successive conformal mappings are presented below.

The first transformation maps the interior of the original rectangle from the z plane to the upper half-space of the s plane. This is effected through the Schwartz-Christoffel

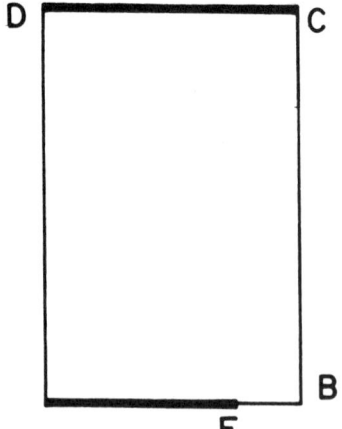

Fig. 4 Basis problem geometry.

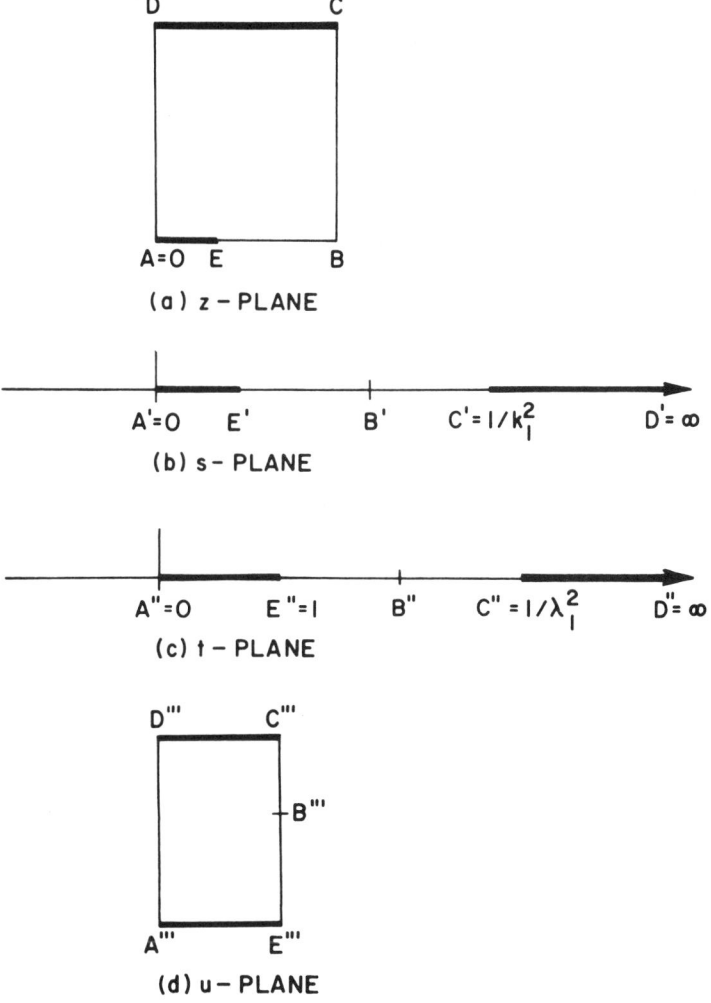

Fig. 5 Transformation of the basis problem geometry.

transformation[7] given in differential form by

$$\frac{dz}{ds} = C_1 (s - A')^{-\frac{1}{2}} (s - B')^{-\frac{1}{2}} (s - C')^{-\frac{1}{2}} \qquad (21)$$

where the rectangle is opened or "cut" through point D in the z plane. Integrating Eq. (21) subject to the conditions that

1) $z = 0$, $s = 0$ (22a)

2) $z = |AB|$, $s = 1$ (22b)

3) $z = |AB| + |BC|i$, $s = 1/k_1^2$ (22c)

leads to the transformation equation

$$s = sn^2 [z K(k_1)/|AB|, k_1] \quad (23)$$

where sn is the Jacobian elliptic sine amplitude function, $K(k_1)$ is the complete elliptic integral of the first kind of modulus k_1, and the modulus is determined by solving the transcendental equation

$$\frac{K'(k_1)}{K(k_1)} = \frac{|BC|}{|AB|} \quad (24)$$

in which $K'(k_1)$ is the complete elliptic integral of the first kind of complementary modulus

$$k_1' = \sqrt{1 - k_1^2} \quad \text{such that}$$

$$K'(k_1) = K(k_1') = \int_0^1 \frac{d\eta}{\sqrt{1-\eta^2}\sqrt{1-k_1'^2\eta^2}} \quad (25)$$

Rather than solving the transcendental equation, Eq. (24), a more direct evaluation of k_1 is possible utilizing the approximation of Schneider.[3] This approximation states that

$$K'(k_1)/K(k_1) \approx (2/\pi)\ln[2(1 + k_1')/k_1] \quad k_1 \leq 0.707 \quad (26a)$$

and

$$K'(k_1)/K(k_1) \approx (\pi/2)[\ln[2(1 + k_1)/k_1']]^{-1} \quad k_1 \geq 0.707 \quad (26b)$$

and is accurate to within 0.2% over the entire range of $0 \leq k_1 \leq 1$. In this manner, k_1 is evaluated from

$$k_1 = 4 \exp[\frac{\pi}{2} \frac{K'(k_1)}{K(k_1)}] \Big/ [\exp[\frac{\pi K'(k_1)}{K'(k_1)}] + 4] \quad k_1 \leq 0.707 \quad (27a)$$

or from

$$k_1 = [\exp[\frac{\pi K(k_1)}{K'(k_1)}] - 4] \Big/ [\exp[\frac{\pi K(k_1)}{K'(k_1)}] + 4] k_1 \geq 0.707 \quad (27b)$$

This, then, completes the transformation from the z plane to the s plane.

The transformation from the s plane to the t plane, in preparation for mapping the upper half-space of the t plane to the interior of the rectangle shown in the u plane, is simply given by

$$t = s/E' \quad (28)$$

where from the transformation, Eq. (23), to the s plane, E' is given by

$$E' = sn^2 [|AE| K(k_1)/|AB|, k_1] \quad (29)$$

This transformation, Eq. (28), has the influence of setting

1) $A'' = 0$ \quad (30a)

2) $E'' = 1$ \quad (30b)

3) $C'' = 1/\lambda_1^2$ \quad (30c)

where

$$\lambda_1 = k_1 \sqrt{E'} \quad (31)$$

The transformation which maps the t plane to the u plane is essentially the inverse of that used to map the z plane to the s plane. This transformation is given by

$$u = \frac{1}{K(\lambda_1)} sn^{-1} (\sqrt{t}, \lambda_1) \quad (32)$$

where for convenience $E''' = 1$ has been assigned.

The thermal resistance for the geometry of Fig. 5d is given simply by

$$R = |A''' \; D'''|/k \; L \quad (33)$$

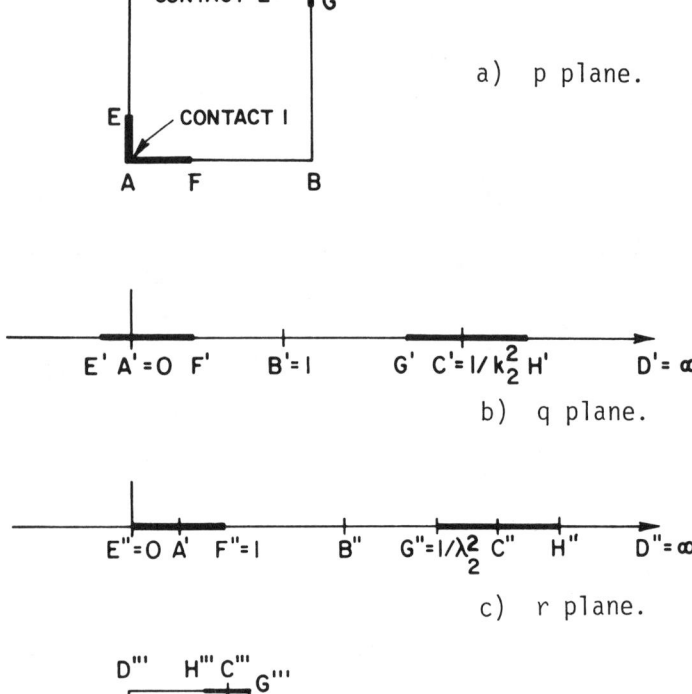

Fig. 6 Transformation of the general problem to the basis problem configuration.

where L is the thickness of the member in the z direction (i.e. into the page), and the magnitude $|A''' \ D'''|$ is given by

$$|A''' \ D'''| = [1/K(\lambda_1)]sn^{-1}[\infty, \lambda_1] = K'(\lambda_1)/K(\lambda_1) \quad (34)$$

where the ratio $K'(\lambda_1)/K(\lambda_1)$ can be determined using tables of the ratio of complete elliptic integrals of the first kind[8] or using the approximation of Schneider.[3]

Thus, it is now possible by following the above described procedure, to determine the thermal resistance of the "basis problem." It is shown below that the solution of all of the mixed boundary value problems described in Fig. 1 can be cast into the form of the basis problem described in this section.

Generalized Procedure for Two-Dimensional Planar Problems

The second requirement of the general analysis is to demonstrate that all mixed boundary value problems of the type addressed in this paper can be transformed to the basis problem geometry. To do this, the configuration of Fig. 6a is considered and the contact locations positioned such that contact 2 is always located counterclockwise from contact 1 relative to the point D. This convention, of course, poses no limitations whatsoever on the analysis, but simply adopts a relative orientation of contact 2 relative to contact 1. Furthermore, the contacts can be located on corners as shown or wholly contained within a side of the conducting member. The transformation steps corresponding to parts a to d of Fig. 6 are essentially those already discussed for application to the basis problem described in the previous section. The necessary transformation steps are described below.

The first transformation transforms the interior of the rectangle of Fig. 6a into the upper half space of Fig. 6b. As seen with the basis problem, this transformation is

$$q = sn^2 [p\ K(k_2)/|AB|,\ k_2] \tag{35}$$

where the modulus k_2 is determined from

$$K'(k_2)/K(k_2) = |AD|/|AB| \tag{36}$$

and where, again, the approximation of Schneider[3] may be used or tabulated values of the ratio $K'(k)/K(k)$.[7] In the q plane of Fig. 6 the identification has been made that $A' = 0$, $B' = 1$, and $C' = 1/k_2^2$ with k_2 as determined from Eq. (36).

The second transformation, from the q plane to the r plane is effected in order to make the r plane suitable for transformation to the basis problem geometry of Fig. 6d. This transformation is given simply by

$$r = (q - E')/|E'\ F'| \tag{37}$$

and has the influence of relocating the origin to the point E" and establishing F" = 1. For the final transformation, the identification is made that

$$G'' = 1/\lambda_2^2 \tag{38}$$

where from (37), the value of λ_2 is given by

$$2 = \sqrt{|E' \ F'|/(G' - E')} \tag{39}$$

where E', F', and G' are obtained from the transformation equation, Eq. (35), for which the argument may be complex. The actual values of E', F', and G', however, will be members of the set of real numbers.

The final transformation, transforming the r plane to the z plane, is essentially the inverse of the transformation which was used to map the p plane to the q plane. This transformation is

$$z = [1/K(\lambda_2)]\text{sn}^{-1}(\sqrt{r}, \lambda_2) \tag{40}$$

with λ_2 as determined earlier, and with F''' set equal to unity for convenience. The final geometric configuration, that of Fig. 6d, is seen to be identical to the basis problem geometry (although vertically and horizontally inverted) and hence the actual thermal resistance can be determined following the procedure described in the previous section.

Application

As demonstration of the procedure for effecting the above described transformations, two example problems will be examined. The first of these is the basis problem geometry and the second is a problem in circular cylinder coordinates which requires transformation to the basis geometry and then its solution is completed following the steps which will be executed for the first example problem. Prior to developing the examples, however, certain properties of the Jacobian elliptic sine amplitude function will be provided as these are useful to effecting the transformation. These properties are given below.

$$\text{sn}(K,k) = 1 \tag{41a}$$

$$\text{sn}(K + iK',k) = 1/k \tag{41b}$$

THERMAL RESISTANCE OF TWO-DIMENSIONAL GEOMETRIES   195

$$sn(x + iK',k) = 1/(k\ sn(x,k)) \quad (41c)$$

$$sn(K + iy,k) = 1/\sqrt{1 - k'^2\ sn^2(y,k')} \quad (41d)$$

$$sn(iy,k) = i\ sn(y,k')/cn(y,k') \quad (41e)$$

$$sn^{-1}(x,k) = F(x,k)\ ;\ 1 > x > 0 \quad (41f)$$

$$sn^{-1}(x,k) = K(k) + iF(\sqrt{(x^2 - 1)/x^2\ k'^2},k')\ ;$$
$$1/k > x > 1 \quad (41g)$$

$$sn^{-1}(x,k) = [K(k)-F(\sqrt{(x^2 - 1/k^2)/(x^2 - 1)},k)] + iK'(k);$$
$$x > \frac{1}{k} \quad (41h)$$

$$sn^2(x,k) + cn^2(x,k) = 1 \quad (41i)$$

Basis Problem Example

The geometry chosen for the basis problem example is shown in Fig. 7. The dimensions are those shown and are chosen to be the values shown since the authors have previously performed a finite element solution for this configuration. The appropriate transformed locations on the t plane of Fig. 5 are

$$A' = 0$$
$$B' = sn^2(K(k_1),k_1) = 1$$

Fig. 7 Example of basis problem solution.

LOCATION	COORDINATES (x,y)
A	(0,0)
B	(1,0)
C	(1, 1.4737)
D	(0, 1.4737)
E	(0.75, 0)

$$C' = sn^2(K + iK', k_1) = \frac{1}{k_1^2} = 6.9143$$

$$D' = sn^2(iK', k_1) = \infty$$

$$E' = sn^2(0.75K, k_1) = 0.8627$$

where $k_1 = 0.3803$ was determined using Eq. (27a), and $K(k_1)$ subsequently evaluated to be 1.6327. The appropriate locations in the t plane of Fig. 5 are then given by

$$A'' = 0$$

$$B'' = 1.1592$$

$$C'' = 8.0147$$

$$D'' = \infty$$

$$E'' = 1$$

From the value of C'', we determine λ_1 to be $\lambda_1 = 0.3532$. Finally, the value of $RkL = |A''' \ D'''|$ in the u plane is given by

$$|A''' \ D'''| = K'(\lambda_1)/K(\lambda_1) = 1.5242$$

which compares favorably, to within the accuracy of the finite element result, to the numerically obtained value of 1.5268.

Circular Cylinder Example

The circular cylinder system is one of the few for which the laws governing conformal transformation are conventionally not applied in a consistent fashion. The logarithmic transformation is used to generate this coordinate system and the coordinates, conformally consistent, should be (ln r, θ) rather than the conventional usage of (r,θ). The particular problem considered is shown in the x-y plane in Fig. 8a and in the ln r-θ space in Fig. 8b. Now according to our previous proof of the uniqueness of the thermal resistance, a shift in origin will not affect the evaluation of the resistance provided the shift is made in ln r-θ space. Shifting the origin to coincide with point A yields the configuration shown in Fig. 8c with coordinates as shown. The problem shown in Fig. 8c is therefore the problem which will be solved.

THERMAL RESISTANCE OF TWO-DIMENSIONAL GEOMETRIES 197

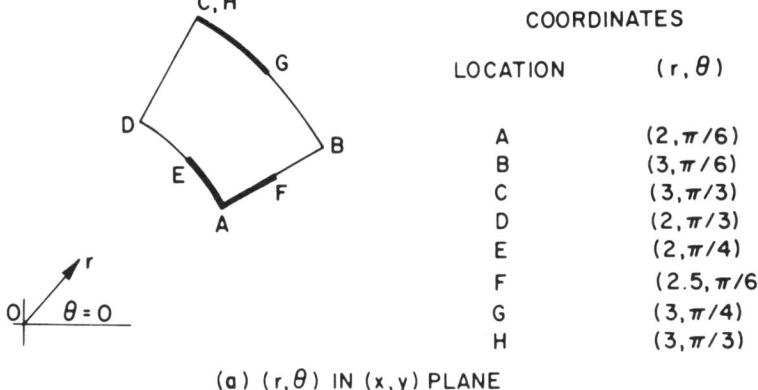

(a) (r,θ) IN (x,y) PLANE

LOCATION	(r,θ)
A	$(2, \pi/6)$
B	$(3, \pi/6)$
C	$(3, \pi/3)$
D	$(2, \pi/3)$
E	$(2, \pi/4)$
F	$(2.5, \pi/6)$
G	$(3, \pi/4)$
H	$(3, \pi/3)$

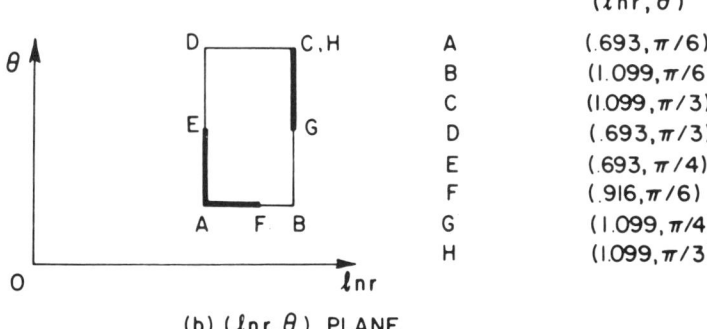

(b) $(\ln r, \theta)$ PLANE

	$(\ln r, \theta)$
A	$(.693, \pi/6)$
B	$(1.099, \pi/6)$
C	$(1.099, \pi/3)$
D	$(.693, \pi/3)$
E	$(.693, \pi/4)$
F	$(.916, \pi/6)$
G	$(1.099, \pi/4)$
H	$(1.099, \pi/3)$

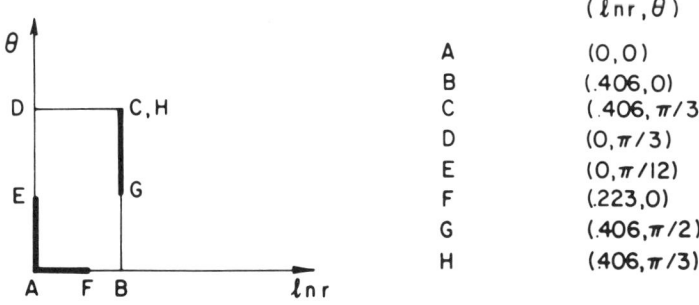

(c) TRANSLATED $(\ln r, \theta)$ PLANE

	$(\ln r, \theta)$
A	$(0, 0)$
B	$(.406, 0)$
C	$(.406, \pi/3)$
D	$(0, \pi/3)$
E	$(0, \pi/12)$
F	$(.223, 0)$
G	$(.406, \pi/2)$
H	$(.406, \pi/3)$

Fig. 8 Example of general problem for the case of circular cylinder coordinates.

The appropriate transformed locations on the q plane of Fig. 6 are

$$A' = sn^2(0, k_2) = 0$$

$$B' = sn^2(K, k_2) = 1$$

$$C' = H' = sn^2(K + iK', k_2) = 1/k_2^2 = 4.1311$$

$$D' = sn^2(iK', k_2) = \infty$$

$$E' = sn^2(0.643iK, k_2) = sn^2(1.082i, k_2) = -2.033$$

$$F' = sn^2(0.549K, k_2) = sn^2(0.9238, k_2) = 0.6115$$

$$G' = sn^2(K+0.643iK, k_2) = sn^2(K+1.082i, k_2) = 2.033$$

where $k_2 = 0.492$ was determined using Eq. (27a) and $K(k_2)$ subsequently evaluated to be 1.6815. The properties given in Eqs. (41) have been used extensively in evaluating the above transformed locations.

The transformed locations in the r plane of Fig. 6, using $|E' \; R'| = 2.6445$, are given by

$$A'' = 0.7688$$

$$B'' = 1.1469$$

$$C'' = H'' = 2.3309$$

$$D'' = \infty$$

$$E'' = 0$$

$$F'' = 1.0$$

$$G'' = 1.5375$$

From the value of G'', we determine λ_2 to be $\lambda_2 = 0.8065$ and this leads to the value of $|E''' \; D'''|$ in the z plane of

$$|E''' \; D'''| = K'(\lambda_2)/K(\lambda_2) = 0.8685$$

and the value $K(\lambda_2) = 2.0081$.

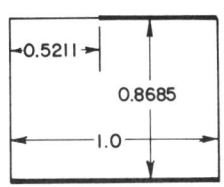
(a) BASIS GEOMETRY FOR EXAMPLE 2

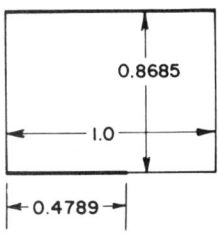
(b) BASIS GEOMETRY AFTER ROTATION

Fig. 9 Equivalence of the transformed general problem and the basis problem.

The transformed locations in the z plane are given by

$$A''' = 0.4980 \, sn^{-1}(\sqrt{.7688}, \lambda_2) = 0.6039$$

$$B''' = 0.4980 \, sn^{-1}(\sqrt{1.1469}, \lambda_2) = 1 + 0.4912 \, i$$

$$C''' = H''' = 0.4980 \, sn^{-1}(\sqrt{2.3309}, \lambda_2) = 0.52106 + 0.8685i$$

$$D''' = 0.4980 \, sn^{-1}(\infty, \lambda_2) = 0.8685 \, i$$

$$E''' = 0.4980 \, sn^{-1}(0, \lambda_2) = 0$$

$$F''' = 0.4980 \, sn^{-1}(1, \lambda_2) = 1$$

$$G''' = 0.4980 \, sn^{-1}(\sqrt{1.5375}, \lambda_2) = 1 + 0.8685 \, i$$

The above locations are shown schematically in Fig. 9a. The basic configuration shown in Fig. 9a corresponds with the basis problem geometry considered earlier. Since a pure rotation of the geometry will not alter its thermal resistance, provided that the boundary conditions are rotated together with the geometry, the identification of the transformed geometry with that of the basis problem is more evident upon examination of Fig. 9b which reflects this rotation. Performing the transformations required to solve this particular configuration of the basis problem geometry leads to a nondimensional thermal resistance of RkL = 1.1096 which agrees well with the result of 1.1214 obtained using the method of finite differences.

Thus, it is observed from the above applications that, although some familiarity with the Jacobian elliptic functions and elliptic integrals of the first kind is required, the procedure outlined in this paper can be readily applied to solve problems which are highly complex, if not intractable, when using conventional solution techniques.

Discussion and Conclusions

The classification of thermal problem addressed in this work is that for which heat is conducted from one isothermal contact through a homogeneous medium having constant thermal conductivity to a second isothermal contact. The heat flow is steady-state with no internal heat generation and the geometric configuration is such that the domain is bounded by coordinate surfaces of a two-dimensional, planar, orthogonal curvilinear coordinate system. The boundary of the domain, with the exception of the two isothermal contacts, is adiabatic.

It was necessary, before providing the solution procedure for solving the above problem classification, to demonstrate the uniqueness of the thermal resistance for such geometric and thermal configurations under the influence of a conformal transformation. This was established by considering the governing equation, the boundary condition specification, and the evaluation of the total heat flow rate in both the original and transformed coordinate system.

In anticipation of the sequence of transformations which form a general solution procedure, a basis problem configuration was examined. A detailed description of the procedure used to solve this basic problem classification has been presented. The procedure consists of four successive applications of the Schwartz-Christoffel conformal transformation.

The general procedure for solving problems of the general classification addressed in this work was then described. It was illustrated that all problems of the general classification considered in this paper can be reduced to the basis problem configuration through four additional applications of the Schwartz-Christoffel transformation. The details of this procedure have been presented.

Finally, two example problems were solved utilizing the procedures outlined in this paper. The purpose the examples serve is to provide guidance to the reader in the application of the transformation procedures, and to provide the reader with the necessary familiarity with the functions involved in effecting the transformations.

The procedure presented in this paper is a very powerful one for solving problems of the classification to which this work has been addressed. This procedure will therefore vastly expand the repertoire of solutions available to the heat transfer analyst.

Acknowledgments

The authors thank the Natural Sciences and Engineering Research Council of Canada for their support of this project in the form of an operating grant to G.E. Schneider. The authors also thank Mr. M. Zedan for obtaining the finite difference solution for the second example cited in this work.

References

[1] Schneider, G.E., Yovanovich, M.M., and Cane, R.L.D., "Thermal Constriction Resistance of a Convectively Cooled Flat Plate with Non-Uniform Flux Over Its Opposite Face," AIAA Paper No. 78-86, Jan. 1978, Huntsville.

[2] Oliveira, H.Q., and Forslund, R.P., "The Effect of Thermal Constriction Resistance in the Design of Channel Plate Heat Exchangers," ASME Journal of Heat Transfer, Aug. 1974, p. 292.

[3] Schneider, G.E., "Thermal Resistance of a Cylinder with Two Diametrically Opposite Symmetric Isothermal Caps," ASME Journal of Heat Transfer, Series C, Vol. 97, No. 3, Aug. 1975, pp. 465-466.

[4] Yovanovich, M.M., Advanced Heat Conduction, Hemisphere Publishing Company, to be published.

[5] Hildebrand, F.B., Advanced Calculus for Applications, Prentice-Hall, Englewood Cliffs, 1976.

[6] Binns, K.G., and Lawrenson, P.J., Analysis and Computation of Electric and Magnetic Field Problems, Pergamon Press, New York, 1973.

[7] Gibbs, W.J., Conformal Transformations in Electrical Engineering, Chapman and Hall Ltd., 1958.

[8] Abramowitz, M., and Stegun, I.A., Handbook of Mathematical Functions, Dover Publications, New York, 1965.

SOME BASIC THREE-DIMENSIONAL INFLUENCE COEFFICIENTS FOR THE SURFACE ELEMENT METHOD

M. M. Yovanovich[*] and K. A. Martin[+]

University of Waterloo,
Waterloo, Ontario, Canada

Abstract

Expressions are developed for the temperature rise and the heat flux vector at an arbitrary field point due to distributed volumetric, surface, and line heat sources utilizing the integral form of the solution to Laplace's equation. The application of the integral solution to the problem of thermal constriction resistance of contact areas of arbitrary shape subjected to the boundary condition of the first, second, or third kind is considered. The importance of influence coefficients in the efficient and accurate numerical solution of thermal constriction problems using the surface element method is discussed. A list of basic three-dimensional solutions for several important geometric source areas is presented for future reference.

Nomenclature

A	= position; area
$\vec{a}_1, \vec{a}_2, \vec{a}_3$	= unit vectors
a	= rectangular side
B	= position
$B(\ )$	= complete elliptic integral
b	= rectangular side
C	= position

Presented as Paper 80-1491 at the AIAA 15th Thermophysics Conference, Snowmass, Colo., July 14-16, 1980. Copyright © by M. M. Yovanovich. Published by the American Institute of Aeronautics and Astronautics with permission.

[*]Professor, Thermal Engineering Group, Department of Mechanical Engineering.

[+]Graduate Research Assistant, Thermal Engineering Group, Department of Mechanical Engineering.

INFLUENCE COEFFICIENTS FOR SURFACE ELEMENT METHOD

C_{ij}	= influence coefficient
D	= position
$D(\ )$	= complete elliptic integral
d_1, d_2, d_3	= distances
$E(\ )$	= complete elliptic integral of the second kind
$F(\ )$	= incomplete elliptic integral of the first kind
G_{ij}	= geometric coefficient
h	= contact conductance or film coefficient
$I_0(\ )$	= modified Bessel function of the first kind of order zero
$J_0(\ ), J_1(\ )$	= Bessel functions of the first kind of order zero and one
i, j	= field and source point indices
$K_0(\ )$	= modified Bessel function of the second kind of order zero
$K(\ )$	= complete elliptic integral of the first kind
k	= thermal conductivity
ℓ	= length
m	= source strength per unit length
$m(\vec{s})$	= source strength per unit length
P	= field point location; perimeter
$P_n(\cos\theta)$	= Legendre polynomial of order n
Q	= heat flow rate; total source strength
q	= source strength per unit area
$q(\vec{s})$	= source strength per unit area
R_c	= constriction resistance
r	= radial coordinate
$\vec{r}$	= field point position vector
$\vec{s}$	= source point position vector
T	= temperature rise
$T(\vec{r})$	= temperature rise at field point
T_f	= external source temperature
T_s	= source temperature
t	= dummy variable
V	= volume
w	= dummy variable
x, y, z	= Cartesian coordinates
β	= dummy variable
δ	= perpendicular in right triangle
ζ	= source point coordinate
η	= source point coordinate
θ	= polar coordinate; dummy variable
κ	= modulus of complete elliptic integrals
μ	= argument of Legendre polynomials
ξ	= source point coordinate
π	= pi
ρ	= volumetric source strength; polar coordinate
$\rho(\vec{s})$	= volumetric source strength

ρ = effective ring radius
Ω = right triangle influence parameter
ω_o = vertex angle of right triangle
∇ = Laplacian operator

Introduction

During the past several years Yovanovich and his co-workers[1-7] have demonstrated that the surface element method (SEM) or the boundary integral equations method (BIEM) are practical and efficient methods for obtaining numerical solutions to Laplace's equation. By means of these methods they determined the thermal constriction resistances of singly and doubly connected, planar contact areas of arbitrary shape on insulated, isotropic half-spaces. Isothermal contacts[7] as well as contacts subjected to a uniform flux distribution[1-5] have been considered. Recently Martin[6] has shown how the SEM can be applied to singly and doubly connected, planar contact areas subjected to the boundary condition of the third kind. He has demonstrated that numerical solutions for the boundary condition of the third kind will yield under certain conditions the two limiting cases: boundary conditions of the first and second kinds. The numerical solutions have been found to be efficient and very accurate (errors less than 1% when compared with known exact solutions).

These techniques as applied to thermal constriction problems are based upon the integral form of the solution to Laplace's equation:

$$\nabla^2 T = \frac{\partial^2 T}{\partial x^2} + \frac{\partial^2 T}{\partial y^2} + \frac{\partial^2 T}{\partial z^2} = 0 \qquad (1)$$

The solution can be written as[8-12]

$$T(\vec{r}) = \frac{1}{2\pi k} \iint_A \frac{q(\vec{s}) dA}{|\vec{r} - \vec{s}|} \qquad (2)$$

where $T(\vec{r})$ is the temperature excess related to some arbitrary ambient reference temperature, k is the thermal conductivity of the isotropic conductor, $q(\vec{s})$ is the heat flux distribution over the contact area of interest. The position vector to the arbitrary field point (x,y,z) is $\vec{r}$ while the position vector to an arbitrary source point (ξ,η,ζ) is denoted by $\vec{s}$. These position vectors are defined as

$$\vec{r} = x\vec{a}_1 + y\vec{a}_2 + z\vec{a}_3 \qquad (3)$$

$$\vec{s} = \xi\vec{a}_1 + \eta\vec{a}_2 + \zeta\vec{a}_3 \tag{4}$$

where $\vec{a}_1$, $\vec{a}_2$, and $\vec{a}_3$ are unit vectors.

The distance between an arbitrary source point (ξ,η,ζ) and a field point (x,y,z) as depicted in Fig. 1 is

$$|\vec{r} - \vec{s}| = \sqrt{(x-\xi)^2 + (y-\eta)^2 + (z-\zeta)^2} \tag{5}$$

In its simplest form the SEM when applied to thermal constriction problems consists of dividing the contact area into a finite number, N, of surface elements, A_j, over each of which the heat flux, q_j, is assumed to be uniform. The centroid of a typical surface element is designated by coordinates (x_i, y_i, z_i) and by the position vector $\vec{r}_i$ from the origin. The temperature excess T_i (or simply temperature rise if the reference temperature is taken to be zero) at the centroid of the typical surface element A_i is

$$\begin{aligned} T_i &= \frac{1}{2\pi k} \iint_A \frac{q(\vec{s})dA}{|\vec{r}_i - \vec{s}|} \\ &= \sum_{j=1}^{N} \frac{1}{2\pi k} \iint_{A_j} \frac{q_j \, dA_j}{|\vec{r}_i - \vec{s}|} \\ &= \sum_{j=1}^{N} q_j \left[\frac{1}{2\pi k} \iint_{A_j} \frac{dA_j}{|\vec{r}_i - \vec{s}|} \right] \end{aligned} \tag{6}$$

Equation (6) can be conveniently expressed in the following manner:

$$T_i = \sum_{j=1}^{N} C_{ij} q_j \tag{7}$$

where $C_{ij} q_j$ represents the temperature rise at the point (x_i, y_i, z_i) due to the surface element A_j. C_{ij} represents the temperature rise at the centroid of A_i due to thermal sources of unit strength distributed over the surface element A_j. The influence coefficients C_{ij} are known; they consist of the integrals in Eq. (6).

Equation (7) can be written in more compact form using matrix notation:

$$\{T\} = [C]\{q\} \tag{8}$$

For boundary conditions of the second kind, Eq. (8) can be solved directly for T_i because q_i is known. For boundary conditions of the first kind T_i is known and Eq. (8) must be solved for q_j. In matrix form we have

$$\{q\} = [C]^{-1} \{T\} \quad (9)$$

where $[C]^{-1}$ is the inverse matrix of $[C]$.

For the contact areas subjected to the boundary condition of the third kind the solution to Eq. (1) can be expressed as

$$T(\vec{r}) = \frac{1}{2\pi k} \iint_A h(\vec{s}) \left[T_f - T(\vec{s}) \right] \frac{dA}{|\vec{r}-\vec{s}|} \quad (10)$$

where $h(\vec{s})$ is the thermal contact conductance (or heat transfer coefficient in convective problems), and T_f is the external source temperature.

For uniform T_f and uniform h over each surface element Martin[6] has shown that the temperature rise at any point (x_i, y_i, z_i) is given by

$$T_i = \frac{1}{2\pi k} \sum_{j=1}^{N} h_j (T_f - T_{sj}) \iint_{A_j} \frac{dA_j}{|\vec{r}_j - \vec{s}|} \quad (11)$$

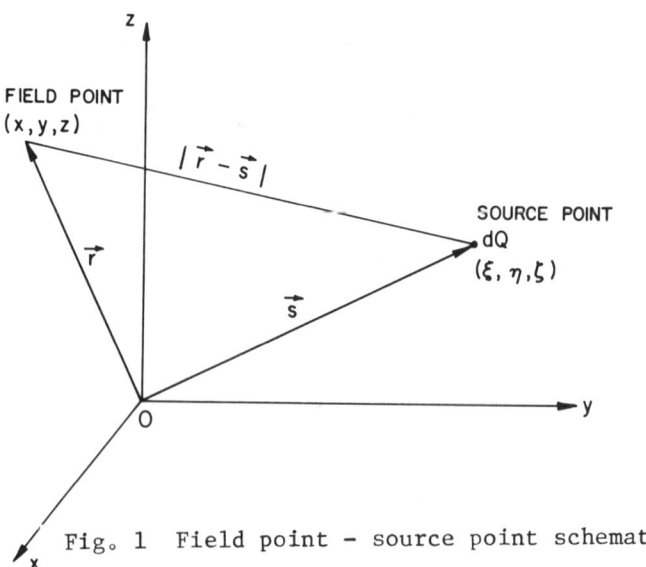

Fig. 1 Field point – source point schematic.

where T_{sj} is the temperature of the surface element A_j over which the external source is applied.

Using matrix notation Eq. (11) becomes

$$\frac{1}{k}[G]\{h(T_f-T_s)\} = \{T_s\} \qquad (12)$$

where

$$G_{ij} \equiv \frac{1}{2\pi}\iint_{A_j} \frac{dA_i}{|\vec{r}_i-\vec{s}|} = k\,C_{ij} \qquad (13)$$

The geometric coefficients G_{ij} are related to the previously discussed influence coefficients C_{ij} by means of Eq. (13).

Solving Eq. (12) for the unknown contact area temperature rise T_s we obtain

$$\left[[G] + k[H]\right]\{T_s\} = [G]\{T_f\} \qquad (14)$$

or simply

$$[G']\{T_s\} = \{T_f\} \qquad (15)$$

In Eq. (14) $[H]$ is the identity matrix. Martin[6] has demonstrated that the general solution for the boundary condition of the third kind, Eq. (11) and (14) reduced to the solution for the boundary condition of the first kind when $k/h \to 0$ and reduces to the solution for the boundary condition of the second kind when $k/h \to \infty$.

The thermal constriction resistance[1-7]

$$R_c \equiv \frac{\text{average contact temperature rise}}{\text{total heat flow rate}}$$

$$= \frac{1}{A}\sum_{i=1}^{N} T_i\,dA_i \bigg/ \sum_{j=1}^{N} q_j\,dA_j \qquad (16)$$

For boundary conditions of the first kind, T_i in Eq. (16) is known and the unknown q_i must be determined by means of Eq. (9). On the other hand for boundary conditions of the second kind, q_i in Eq. (16) is known and the unknown T_i can be determined by Eq. (8).

When the boundary condition of the third kind is specified, both T_i and q_i in Eq. (16) are unknown. The surface temperature T_i is determined by means of Eq. (14) and q_i is determined from $q_i = h(T_f-T_i)$. In all cases the efficient and accurate solutions will depend upon analytical or numerical evaluation of the geometric and influence coefficients G_{ij} and C_{ij}, respectively.

The influence coefficients appear in integral solutions of Laplace's equation in several different physical areas: Newtonian potential[8,11,16], electrostatics[9,10,12,19-21] and elastostatics[11,18] for example; some special cases have been considered as examples in mathematical treatises.[13-16]

The purpose of this paper is to compile a list of influence coefficients, noting their characteristics, and their applicability to a variety of thermal problems. This list should be useful to the thermal analyst who is interested in solving thermal constriction problems as they appear in several technological areas.

Influence Coefficients for Point Sources

Point Source

The temperature rise at an arbitrary point (x,y,z) due to a thermal source of strength **dQ located at** (ξ,υ,ζ) as shown

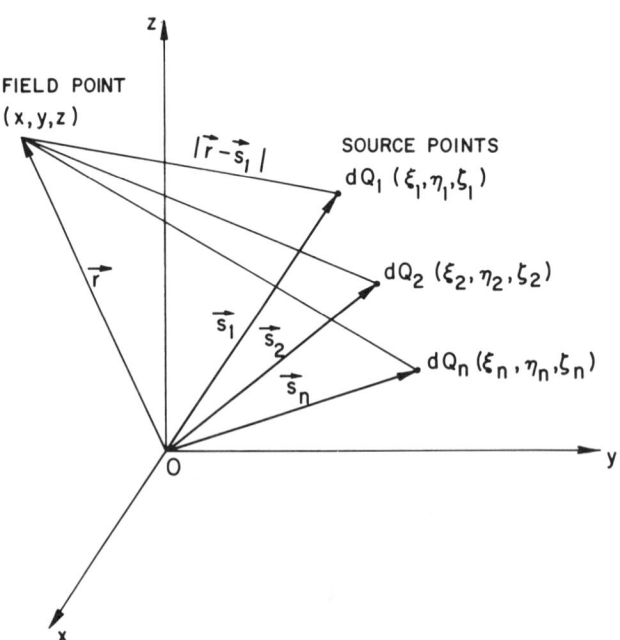

Fig. 2 Field point - multiple source point schematic.

in Fig. 1 is given by

$$4\pi k T(\vec{r}) = \frac{dQ}{|\vec{r} - \vec{s}|} \qquad (17)$$

where the position vectors $\vec{r}$ and $\vec{s}$ are defined by Eq. (3) and (4). It can be shown that Eq. (17) is the solution to Eq. (1).

Multiple Point Sources

Next consider the set of point sources dQ_i located at (ξ_i, η_i, ζ_i) with position vectors s_i as shown in Fig. 2. The total temperature rise at the field point (x,y,z) due to n point sources is obtained by the superposition of the effects of the individual point sources acting alone. Therefore,

$$4\pi k T(\vec{r}) = \frac{dQ_1}{|\vec{r}-\vec{s}_1|} + \frac{dQ_2}{|\vec{r}-\vec{s}_2|} + \frac{dQ_3}{|\vec{r}-\vec{s}_3|} + \cdots + \frac{dQ_n}{|\vec{r}-\vec{s}_n|}$$

$$= \sum_{j=1}^{n} \frac{dQ_j}{|\vec{r}-\vec{s}_j|} \qquad (18)$$

When the point sources can be modelled as a uniform flux q_j emitted by a differential area dA_j, Eq. (18) becomes

$$T(x,y,z) = \frac{1}{4\pi k} \sum_{j=1}^{n} \frac{2 q_j\, dA_j}{|\vec{r}-\vec{s}_j|} \qquad (19)$$

The factor of 2 in Eq. (19) appears because heat is emitted from both sides of the differential area. This was taken into account when the integral solution, Eq. (2), was developed for contact areas situated on half-spaces.

Distributed Sources

We next consider the temperature rise due to an arbitrary volumetric distribution of thermal sources ρ as depicted in Fig. 3. The volume V can be approximated as the sum of n cubes; the typical cube has volume ΔV_j and source density ρ_j. The temperature rise at the arbitrary point (x,y,z) can be expressed as

$$4\pi k T(\vec{r}) = \lim_{\substack{n\to\infty \\ V_j \to 0}} \sum_{j=1}^{n} \frac{\rho_j\, \Delta V_j}{|\vec{r}-\vec{s}_j|} \qquad (20)$$

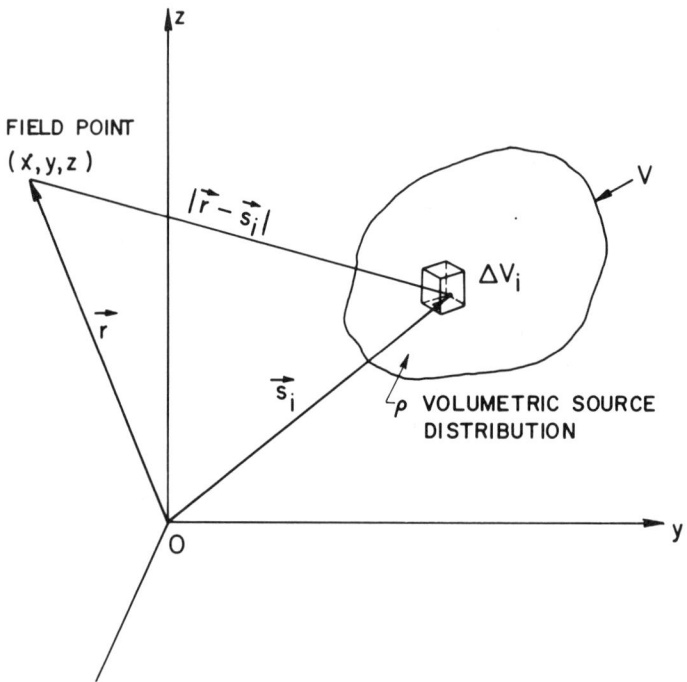

Fig. 3 Field point - volumetric source point configuration.

where $\vec{s}_j$ is the position vector from the common origin to the typical volume element. In the limit the summation becomes a volume integral; therefore,

$$4\pi k T(\vec{r}) = \iiint_V \frac{\rho(\vec{s})\, dV}{|\vec{r} - \vec{s}|} \qquad (21)$$

In Cartesian coordinates Eq. (21) becomes

$$4\pi k T(x,y,z) = \iiint_V \frac{\rho(\xi,\eta,\zeta)\, d\xi\, d\eta\, d\zeta}{\sqrt{(x-\xi)^2 + (y-\eta)^2 + (z-\zeta)^2}}$$

The integrand is a function of six variables (x,y,z) and (ξ,η,ζ). Integration eliminates the dummy variables (ξ,η,ζ), therefore the temperature rise is a function of (x,y,z) only.

Flux Distribution

The spatial flux distribution $\vec{q}$ is often of interest to the thermal analyst. It may be determined at any point

(x,y,z) by means of Fourier's rate equation

$$\vec{q} = -k \text{ grad } T = -k \nabla T \tag{23}$$

Substitution of Eq. (21) into Eq. (23) yields

$$\vec{q}(\vec{r}) = -k\nabla \left[\frac{1}{4\pi k} \iiint_V \frac{\rho(\vec{s}) \, dV}{|\vec{r} - \vec{s}|} \right] \tag{24}$$

Using Leibnitz's rule for the differentiation of an integral we obtain

$$\vec{q}(\vec{r}) = -\frac{1}{4\pi} \iiint_V \nabla \left[\frac{\rho(\vec{s})}{|\vec{r} - \vec{s}|} \right] dV \tag{25}$$

But,

$$\nabla \left[\frac{\rho(\vec{s})}{|\vec{r} - \vec{s}|} \right] = \rho(\vec{s}) \, \nabla \left[\frac{1}{|\vec{r} - \vec{s}|} \right] \tag{26}$$

It can be shown that

$$\nabla \left[\frac{1}{|\vec{r} - \vec{s}|} \right] = \frac{(\vec{r} - \vec{s})}{|\vec{r} - \vec{s}|^3} \tag{27}$$

Combining Eqs. (26) and (27) with Eq. (25) gives for the flux vector

$$\vec{q}(\vec{r}) = \frac{1}{4\pi} \iiint_V \frac{\rho(\vec{s})(\vec{r} - \vec{s}) dV}{|\vec{r} - \vec{s}|^3} \tag{28}$$

Using Cartesian coordinates the three flux components may be written as

$$q_x(x,y,z) = \frac{1}{4\pi} \iiint_V \frac{\rho(\xi,\eta,\zeta)(x-\xi) d\xi \, d\eta \, d\zeta}{\left[(x-\xi)^2 + (y-\eta)^2 + (z-\zeta)^2\right]^{3/2}} \tag{29}$$

$$q_y(x,y,z) = \frac{1}{4\pi} \iiint_V \frac{\rho(\xi,\eta,\zeta)(y-\eta) d\xi \, d\eta \, d\zeta}{\left[(x-\xi)^2 + (y-\eta)^2 + (z-\zeta)^2\right]^{3/2}} \tag{30}$$

$$q_z(x,y,z) = \frac{1}{4\pi} \iiint_V \frac{\rho(\xi,\eta,\zeta)(z-\zeta) d\xi \, d\eta \, d\zeta}{\left[(x-\xi)^2 + (y-\eta)^2 + (z-\zeta)^2\right]^{3/2}} \tag{31}$$

Two methods are available for obtaining the flux distribution:
1) calculate the temperature rise by means of Eq. (22), then determine the flux by means of Fourier's rate equation; and
2) calculate the flux directly by means of Eq.(28).

Influence Coefficients for Distributed Sources

Surface and Line Sources

For analytical and computational purposes we may regard surface source distributions and line source distributions as special cases of volume source distributions.

For example the volume source distribution, Eq. (21), reduces to the surface source distribution

$$4\pi k T(\vec{r}) = \iint_A \frac{q(\vec{s}) \, dA}{|\vec{r} - \vec{s}|} \qquad (32)$$

where $q(\vec{s})$ is the surface strength per unit area; and to the line source distribution

$$4\pi k T(\vec{r}) = \int_\ell \frac{m(\vec{s}) \, d\ell}{|\vec{r} - \vec{s}|} \qquad (33)$$

where $m(\vec{s})$ is the line source strength per unit length.

These concepts will be used to develop additional solutions in the subsequent sections.

Uniform Finite Line Source

Consider a finite line source of length 2a, strength m watts/unit length and the total strength is $Q = 2ma$ watts. By means of Eq. (33) the temperature rise at any field point (ρ, z), Fig. 4 is

$$T(\rho, z) = \frac{1}{4\pi k} \cdot \frac{Q}{2a} \int_{\zeta=-a}^{\zeta=+a} \frac{d\zeta}{\sqrt{\rho^2 + (z-\zeta)^2}}$$

$$= \frac{Q}{8\pi ka} \ln \frac{(z+a) + r_1}{(z-a) + r_2} \qquad (34)$$

where r_1 and r_2 are the distances from the ends of the finite line source to the point (ρ, z). The isothermal surfaces are confocal rotational (prolate) ellipsoids.

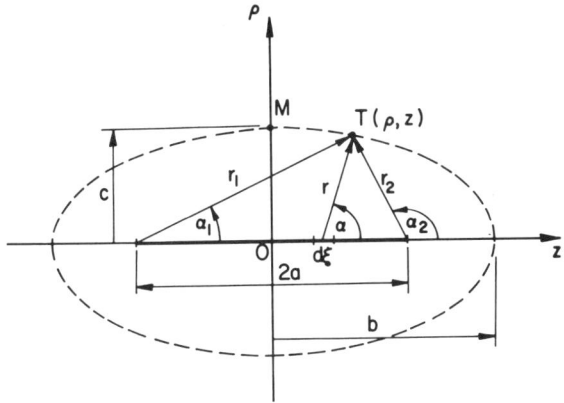

Fig. 4 Finite line source temperature rise field configuration.

The heat flux components can be determined by means of Fourier's rate equation. Thus

$$q_z = -k \frac{\partial T}{\partial z} = \frac{Q}{8\pi a} \frac{1}{\rho} (\sin\alpha_2 - \sin\alpha_1)$$

$$= \frac{Q}{8\rho a} (\frac{1}{r_2} - \frac{1}{r_1}) \quad (35)$$

and

$$q_\rho = -k \frac{\partial T}{\partial \rho} = \frac{Q}{8\pi a} \frac{1}{\rho} (\cos\alpha_1 - \cos\alpha_2) \quad (36)$$

The isothermal surfaces have semi-major and minor axes b and c respectively, Fig. 4. Also $a = \sqrt{b^2 - c^2}$. If we select the point M on the isothermal ellipsoidal surface where $z = 0$, $r_1 = r_2 = b$, then

$$T = \frac{Q}{8\pi ka} \ln \frac{b + a}{b - a}$$

$$= \frac{Q}{8\pi k \sqrt{b^2 - c^2}} \ln \frac{b + \sqrt{b^2 - c^2}}{b - \sqrt{b^2 - c^2}} \quad (37)$$

The influence coefficient for a finite line source can be determined from Eq. (37).

Uniform Circular Ring Source

Consider the case of a total source strength Q uniformly distributed over a circular ring of radius a. The line source strength is $m = Q/2\pi a$ watts/unit length. Using Eq. (33) the temperature rise at an arbitrary field point (ρ,z), Fig. 5, is therefore

$$T(\rho,z) = \frac{m}{4\pi k} \int_{\psi=0}^{\psi=2\pi} \frac{a\,d\psi}{r}$$

$$= \frac{m}{4\pi k} \int_{\psi=0}^{\psi=2\pi} \frac{a\,d\psi}{\sqrt{(\rho-a\cos\psi)^2 + (a\sin\psi)^2 + z^2}} \quad (38)$$

The following transformation:

$$\cos\psi = 2\sin^2 t - 1, \qquad d\psi = -2\,dt$$

reduces Eq. (38) to the following integral:

$$T(\rho z) = \frac{Q}{4\pi k} \times \frac{2}{\pi} \frac{1}{\sqrt{(\rho+a)^2 + z^2}} \int_0^{\pi/2} \frac{dt}{\sqrt{1 - \kappa^2 \sin^2 t}}$$

$$= \frac{Q}{4\pi k} \times \frac{2}{\pi} \frac{1}{\sqrt{(\rho+a)^2 + z^2}} K(\kappa) \quad (39)$$

where $K(\kappa)$ is the complete elliptic integral of the first kind of modulus κ where

$$\kappa^2 = \frac{4\rho a}{(\rho+a)^2 + z^2} \quad (40)$$

Along the axis $\rho = 0$, $\kappa^2 = 0$ and $K(0) = \pi/2$; therefore,

$$T(o,z) = \frac{Q}{4\pi k} \times \frac{1}{\sqrt{a^2 + z^2}} \quad (41)$$

This result can be obtained directly by a simple integration of Eq. (38).

In the plane of the ring source $z = 0$, the temperature rise is

$$T(\rho,o) = \frac{Q}{4\pi k} \times \frac{2}{\pi} \times \frac{1}{(\rho+a)} K(\kappa) \quad (42)$$

where $\kappa^2 = 4\rho a/(\rho+a)^2$.

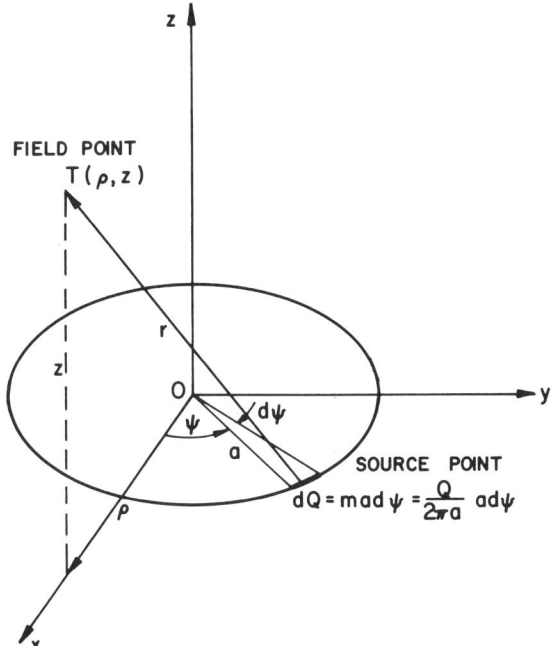

Fig. 5 Circular ring source configuration.

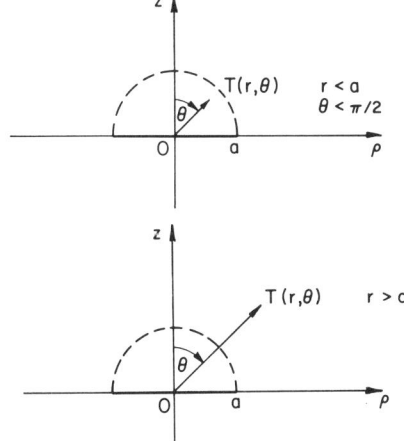

Fig. 6 Inner and outer field points for ring and circular sources.

Alternate expressions are available[9,12,16] for the temperature rise at any field point (r,θ) where $z = r\cos\theta$, $\rho = r\sin\theta$, and $\mu = \cos\theta$, Fig. 6,

$$T(r,\theta) = \frac{Q}{4\pi ka}\left(1 - \frac{1}{2}(\frac{r}{a})^2 P_2(\mu) + \frac{1\cdot 3}{2\cdot 4}(\frac{r}{a})^4 P_4(\mu) - \ldots\right) \qquad (43)$$

for $r < a$, or

$$T(r,\theta) = \frac{Q}{4\pi ka}\left(\frac{a}{r} - \frac{1}{2}(\frac{a}{r})^2 P_2(\mu) + \frac{1.3}{2.4}(\frac{a}{r})^5 P_4(\mu) - \ldots\right) \quad (44)$$

for $r > a$. $P_2(\mu)$, $P_4(\mu)$, etc. are the even order Legendre polynomials. It can be seen from Eq. (44) that when $a/r \to 0$, $T(r,\theta) \to Q/4\pi kr$ independent of θ; the ring appears to be a point source of strength Q located at the origin.

Other alternate expressions are presented in the text by Budak et al.[14]

$$T(\rho,z) = \frac{2Q}{\pi k}\int_0^\infty K_o(\omega a)I_o(\omega\rho)\cos\omega z\, d\omega \quad \text{for } \rho < a \quad (45)$$

$$T(\rho,z) = \frac{2Q}{\pi k}\int_0^\infty I_o(\omega z)K_o(\omega\rho)\cos\omega z\, d\omega \quad \text{for } \rho > a \quad (46)$$

where $I_o(\ )$ and $K_o(\ )$ are modified Bessel functions of the first and second kind, respectively, of order zero.

Uniform Source Distribution over a Circular Area

Suppose that heat is supplied uniformly over the circular area $0 < \rho < a$ in the plane $z = 0$. The total source strength $Q = q 2\pi a^2$ results in a temperature rise at the field point $P(\rho,z)$ which is[17]

$$T(\rho,z) = \frac{qa}{k}\int_0^\infty e^{-w|z|} J_o(w\rho)\, J_1(wa)\, \frac{dw}{w} \quad (47)$$

where $J_o(\ )$ and $J_1(\ )$ are Bessel functions of the first kind of order zero and unity, and w is a dummy variable.

Along the axis $\rho = 0$, $|z| > 0$, the temperature rise can be determined directly by means of the following double integration:

$$4\pi k T(o,z) = 2q \int_0^{2\pi}\int_0^a \frac{\rho\, d\rho\, d\phi}{\sqrt{a^2+z^2}} \quad (48)$$

$$= 4\pi q\left[\sqrt{a^2 + z^2} - z\right]$$

In contact plane $z = 0$, the temperature rise for internal points $0 < \rho < a$ is given[1]:

$$T(\rho) = \frac{2}{\pi} \frac{qa}{k} E(\kappa) \qquad (49)$$

where $E(\ )$ is the complete elliptic integral of the second kind,

$$E(\kappa) = \int_0^{\pi/2} \sqrt{1 - \kappa^2 \sin^2 t}\ dt \qquad (50)$$

of modulus $\kappa = \rho/a$ and $0 < \kappa < 1$.

External to the contact area, the temperature rise is[1]

$$T(\rho) = \frac{2}{\pi} \frac{qa}{k} \kappa B(\kappa) \qquad (51)$$

where $\kappa = a/\rho \lesssim 1$, and

$$B(\kappa) = K(\kappa) - D(\kappa) \text{ and } D(\kappa) = \left[K(\kappa) - E(\kappa)\right]/\kappa^2 \qquad (52)$$

In Eqs. (52), $K(\ )$ and $E(\ )$ are complete elliptic integrals of the first and second kind respectively. $B(\ )$ is also a complete elliptic integral, $B(1) = 1$ and $B(0) = \pi/4$.

Martin[6] has developed the following polynomial approximation for $B(\kappa)$:

$$B(\kappa) = 0.7854 + 0.1072\kappa^2 + 0.081749\kappa^{6.8154} + 0.024619\kappa^{48.47} \qquad (53)$$

with a maximum error of 0.10% when $\kappa = 1$.

For arbitrary field points (ρ,z) or (r,θ), Fig. 6, where $\rho = r\sin\theta$, $z = r\cos\theta$, we have[12,16]

$$T(r,\theta) = \frac{qa}{k} \sum_{n=1}^{\infty} A_{2n} \left(\frac{a}{r}\right)^{2n-1} P_{2n-2}(\cos\theta) \qquad (54)$$

for $r > a$, and

$$T(r,\theta) = \frac{qa}{k}\left[1 - \left(\frac{r}{a}\right)\cos\theta + \sum_{n=1}^{\infty} A_{2n}\left(\frac{r}{a}\right)^{2n} P_{2n}(\cos\theta)\right] \qquad (55)$$

for r < a. In both equations we have

$$A_2 = \frac{1}{2}$$

$$A_{2n} = (-1)^{n+1} \frac{1 \cdot 3 \cdot 5 \cdot \ldots \cdot (2n-3)}{2 \cdot 4 \cdot 6 \cdot \ldots \cdot (2n)} \qquad (56)$$

$$A_{2n+1} = 0$$

It is seen from an examination of Eq. (48) and Eq. (51) with Eq. (53) that the temperature rise produced by a circular area with uniformly distributed sources approaches 99% of the point source temperature rise $T = Q/4\pi kr$ within 1.5 diameters.

Uniform Source Distribution Over a Circular Annulus

Consider the circular annulus having radii a, b with a < b lying in the z = 0 plane, Fig. 7. The total source strength is $Q = q\pi 2 \, (b^2 - a^2)$. By means of Eq. (32) using polar coordinates (ρ, θ, z), the temperature rise at any point $P(\rho,z)$ is[6]

$$T(\rho,z) = \frac{2}{\pi} \frac{q}{k} \int_a^b \frac{\beta K(\kappa) \, d\beta}{\sqrt{(\rho+\beta)^2 + z^2}} \qquad (57)$$

where K( ) is the complete elliptic integral of the first kind with modulus

$$\kappa^2 = \frac{4\rho\beta}{(\rho+\beta)^2 + z^2} \qquad (58)$$

and β is a dummy variable.

A closed form expression for the integral in Eq. (57) is presently not available. One can however obtain expressions for the temperature rise along the axis and in the plane of the annulus by means of the superposition principle. Yovanovich[1] superposed a uniformly distributed source (+q over 0 < ρ < b) and a uniformly distributed sink (-q over 0 < ρ < a < b) to give the circular annulus solution. He obtained for the temperature rise along the axis $\rho = 0$, $z \geqslant 0$.

$$T(o,z) = \frac{q}{k} \left[\sqrt{b^2 + z^2} - \sqrt{a^2 + z^2} \right] \qquad (59)$$

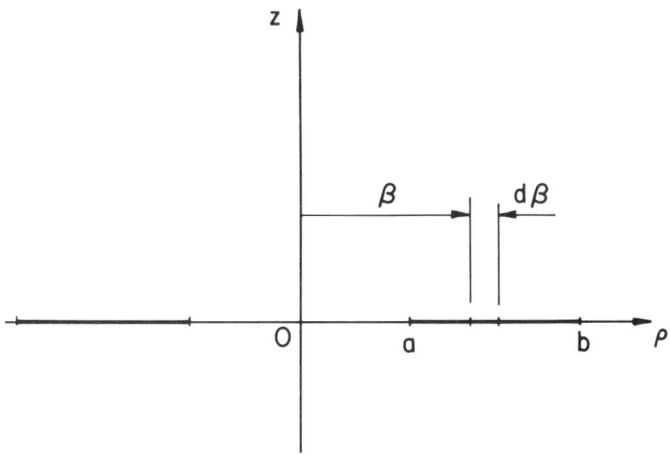

Fig. 7 Circular annulus source configuration.

and for points in the contact plane z = 0, he gave three expressions:

1) $0 \leq \rho < a < b$

$$T(\rho) = \frac{2}{\pi} \frac{qb}{k} E(\frac{\rho}{b}) - \frac{2}{\pi} \frac{qa}{k} E(\frac{\rho}{a}) \qquad (60)$$

2) $a \leq \rho \leq b$

$$T(\rho) = \frac{2}{\pi} \frac{qb}{k} E(\frac{\rho}{b}) - \frac{2}{\pi} \frac{qa}{k} \kappa B(\kappa) \qquad (61)$$

3) $\rho \geq b$

$$T(\rho) = \frac{2}{\pi} \frac{qb}{k} \frac{b}{\rho} B(\frac{b}{\rho}) - \frac{2}{\pi} \frac{qa}{k} \frac{a}{\rho} B(\frac{a}{\rho}) \qquad (62)$$

where E( ) and B( ) are complete elliptic integrals defined by Eqs. (50) and (52) respectively.

Alternate expressions for the temperature rise at an arbitrary point $P(\rho,z)$ or $P(r,\theta)$, Fig. 6, can be obtained by the superposition of the circular source solutions, Eqs. (54-56). The region above or below a circular annulus can be separated into three zones corresponding to $r < a$, $a < r < b$, and $r > b$. The temperature rise within each zone can be developed from Eqs. (54-56). They are

1) $r < a$

$$T(r,\theta) = \frac{qb}{k}\left[1 - \frac{r}{b}\cos\theta + \sum_{n=1}^{\infty} A_{2n}(\frac{r}{b})^{2n} P_{2n}(\cos\theta)\right]$$

$$-\frac{qa}{k}\left[1 - \frac{r}{a}\cos\theta + \sum_{n=1}^{\infty} A_{2n}\left(\frac{r}{a}\right)^{2n} P_{2n}(\cos\theta)\right] \quad (63)$$

2) $a < r < b$

$$T(r,\theta) = \frac{qb}{k}\left[1 - \frac{r}{b}\cos\theta + \sum_{n=1}^{\infty} A_{2n}\left(\frac{r}{b}\right)^{2n} P_{2n}(\cos\theta)\right]$$

$$-\frac{qa}{k}\sum_{n=1}^{\infty} A_{2n}\left(\frac{a}{r}\right)^{2n-1} P_{2n-2}(\cos\theta) \quad (64)$$

3) $r > b$

$$T(r,\theta) = \frac{qb}{k}\sum_{n=1}^{\infty} A_{2n}\left(\frac{b}{r}\right)^{2n-1} P_{2n-2}(\cos\theta)$$

$$-\frac{qa}{k}\sum_{n=1}^{\infty} A_{2n}\left(\frac{a}{r}\right)^{2n-1} P_{2n-2}(\cos\theta) \quad (65)$$

Martin[6] has shown that the temperature rise at any point $P(\rho,z)$ can be computed with an error less than 1% when the circular annulus is replaced by an equivalent ring source. The strength per unit length of the ring is obtained from

$$Q = 2\pi\bar{\rho}m = q\,2\pi(b^2 - a^2) \quad (66)$$

If the effective ring radius is chosen to be

$$\bar{\rho} = (a+b)/2 \quad (67)$$

then the strength per unit length becomes

$$m = 2(b-a)q \quad (68)$$

When these equations are substituted into Eq. (39), we obtain the following equivalent ring expression:

$$T(\rho,z) = \frac{q(b-a)}{\pi k}\sqrt{\frac{\bar{\rho}}{\rho}}\,\kappa\,K(\kappa) \quad (69)$$

where the modulus of $K(\ )$ is defined as

$$\kappa^2 = \frac{4\bar{\rho}\rho}{(\bar{\rho}+\rho)^2 + z^2} \quad (70)$$

INFLUENCE COEFFICIENTS FOR SURFACE ELEMENT METHOD   221

The difference between the exact solution given by Eqs. (57) and (58) and the approximate solution given by Eqs. (69) and (70) will be less than 1% provided the arbitrary point does not lie within the volume $(2a-b) \leq \rho \leq (2b-a)$, $0 \leq z \leq 3(b-a)$ adjacent to the circular annulus.

Uniform Source Distribution over a Right Triangular Area

Suppose a uniform heat source is distributed over the right triangle with vertices $A(0,0)$, $B(\delta,0)$ and $C(\delta, \delta\tan\omega_o)$ with the vertex angle at A denoted by ω_o and the perpendicular from A to B designated δ. The triangular area lies in the $z = 0$ plane and has the total strength $Q = q\delta^2 \tan\omega_o$.

The temperature rise at the point $P(0,0,z)$ located directly above the vertex A is obtained from

$$T(0,0,z) = \frac{q}{4\pi k} \int_0^{\omega_o} \int_0^{\rho_o} \frac{2\rho \, d\rho \, d\theta}{\sqrt{\rho^2 + z^2}} \qquad (71)$$

where $\rho_o = \delta/\cos\theta$. The first integration with respect to ρ gives

$$T(0,0,z) = \frac{q}{2\pi k} \left\{ \int_0^{\omega_o} \sqrt{\delta^2 + z^2 - z^2 \sin^2\theta} \, \frac{d\theta}{\cos\theta} - z\omega_o \right\} \qquad (72)$$

Yovanovich[22] has obtained the following closed form expression for Eq.(72):

$$\frac{2\pi k T}{q\delta} = \frac{1}{2} \ln \frac{\sqrt{1 + \zeta^2 \cos^2\omega_o} + \sin\omega_o}{\sqrt{1 + \zeta^2 \cos^2\omega_o} - \sin\omega_o}$$

$$+ \zeta \sin^{-1}\left[\frac{\zeta\sin\omega_o}{\sqrt{1+\zeta^2}}\right] - \zeta\,\omega_o \qquad (73)$$

with $\zeta = z/\delta$. The function on the right hand side of Eq. (73) is defined to be $\Omega(\zeta,\omega_o)$, therefore Eq. (73) can be written as

$$T = \frac{q\delta}{2\pi K} \Omega(\zeta,\omega_o) \qquad (74)$$

when z = 0, the omega function reduces to the following equivalent expressions:

$$\Omega(0,\omega_o) = \ln\tan(\frac{\pi}{4} + \frac{\omega_o}{2})$$

$$= \frac{1}{2}\ln\frac{1 + \sin\omega_o}{1 - \sin\omega_o}$$

$$= F(\omega_o, 1) \quad (75)$$

where $F(\omega_o,1)$ is the incomplete elliptic integral of the first kind of unit modulus,

$$F(\omega_o,1) = \int_0^{\omega_o} \frac{d\theta}{\sqrt{1-\sin^2\theta}} \quad (76)$$

If $\omega_o < 85°$ and $\zeta > P/\delta$ where the perimeter P of the triangular area is

$$P = \delta\left[1 + \tan\omega_o + \sqrt{1 + \tan^2\omega_o}\right] \quad (77)$$

the exact expression, Eq. (73), can be approximated by

$$\frac{2\pi kT}{q\delta} = \frac{A}{r_o} = \frac{1.5\tan\omega_o}{\sqrt{4 + 9\zeta^2 + \tan^2\omega_o}} \quad (78)$$

where A is the area of the triangle and r is the distance from the centroid of the triangle to the field point $(0,0,z)$. The error is less than 1.0% provided $\omega_o < 85°$ and $\zeta > P/\delta$.

By the superposition principle, Eq. (74) can be used to obtain the temperature rise at points which are located directly above any vertex of any arbitrary triangular area.

Uniform Source Distribution over a Rectangular Area

A rectangular area lying in the first quadrant $(x > 0, y > 0)$ has corners located at $A(x_1,y_1)$, $B(x_2,y_1)$, $C(x_2,y_2)$, $D(x_1,y_2)$ with a uniform surface source density q. The temperature rise at the field point $P(0,0,z)$ is obtained by means of Eq. (32) where the integration is taken over both faces of the rectangle. Thus,

$$\frac{2\pi kT}{q}(0,0,z) = x_1 \ln \frac{y_1 + \sqrt{x_1^2 + y_1^2 + z^2}}{y_2 + \sqrt{x_1^2 + y_2^2 + z^2}}$$

$$+ x_2 \ln \frac{y_2 + \sqrt{x_2^2 + y_2^2 + z^2}}{y_1 + \sqrt{x_2^2 + y_1^2 + z^2}}$$

$$+ y_1 \ln \frac{x_1 + \sqrt{x_1^2 + y_1^2 + z^2}}{x_2 + \sqrt{x_2^2 + y_1^2 + z^2}}$$

$$+ y_2 \ln \frac{x_2 + \sqrt{x_2^2 + y_2^2 + z^2}}{x_1 + \sqrt{x_1^2 + y_2^2 + z^2}}$$

$$- z \arctan \frac{x_1 y_1}{z\sqrt{x_1^2 + y_1^2 + z^2}}$$

$$- z \arctan \frac{x_2 y_2}{z\sqrt{x_2^2 + y_2^2 + z^2}}$$

$$+ z \arctan \frac{x_1 y_2}{z\sqrt{x_1^2 + y_2^2 + z^2}}$$

$$+ z \arctan \frac{x_2 y_1}{z\sqrt{x_2^2 + y_1^2 + z^2}} \quad (79)$$

When the field point lies in the plane of the rectangle, Eq. (79) becomes

$$\frac{2\pi kT}{q}(0,0,0) = x_1 \ln \frac{y_1 + \sqrt{x_1^2 + y_1^2}}{y_2 + \sqrt{x_1^2 + y_2^2}}$$

$$= x_2 \ln \frac{y_2 + \sqrt{x_2^2 + y_2^2}}{y_1 + \sqrt{x_2^2 + y_1^2}}$$

$$+ y_1 \ln \frac{x_1 + \sqrt{x_1^2 + y_1^2}}{x_2 + \sqrt{x_2^2 + y_2^2}}$$

$$+ y_2 \ln \frac{x_2 + \sqrt{x_2^2 + y_2^2}}{x_1 + \sqrt{x_1^2 + y_2^2}} \qquad (80)$$

For the field point at the center of the rectangular area with sides $a = (x_2 - x_1)$, $b = (y_2 - y_1)$, Eq. (80) reduces to

$$\frac{2\pi kT}{q} = a \ln \left(\frac{b + \sqrt{a^2 + b^2}}{-b + \sqrt{a^2 + b^2}}\right) + b \ln \left(\frac{a + \sqrt{a^2 + b^2}}{-a + \sqrt{a^2 + b^2}}\right) \qquad (81)$$

An alternate expression can be developed for Eq. (81) using the right triangular area solution, Eq. (74) with $\zeta = 0$. By symmetry there are two sets of four identical triangles with

$$\delta_1 = \frac{a}{b}, \; \omega_1 = \tan^{-1} \frac{b}{a}, \; \delta_2 = \frac{b}{a}, \; \omega_2 = \tan^{-1} \frac{a}{b} \qquad (82)$$

Superimposing solutions we obtain for the temperature rise at the centroid of the rectangular area (a×b),

$$\frac{\pi kT}{q} = a \, \Omega(0, \tan^{-1} \frac{b}{a}) + b \, \Omega(0, \tan^{-1} \frac{a}{b}) \qquad (83)$$

which is equivalent to Eq. (81).

INFLUENCE COEFFICIENTS FOR SURFACE ELEMENT METHOD   225

For the field point which lies directly above the vertex A, Kellogg[8] gives the following simple expression for the temperature rise:

$$\frac{2\pi kT}{q} = b \ln \frac{a + d_3}{d_2} + a \ln \frac{b + d_3}{d_1} - z \arctan \frac{ab}{zd_3} \quad (84)$$

with $z = PA$, $d_1 = PB$, $d_2 = PC$, and $d_3 = PD$.

The temperature fields produced by a square source and a circular source having the same total area are similar for field points which are not too close to the source. For example a 2 x 2 square source will raise the temperature of a field point which lies in the plane of the source at a distance of two units from the centroid,

$$\frac{kT}{q\sqrt{A}} = 0.1652 \quad (85)$$

where A is the area of the square.

A circular source of radius $a = 2/\sqrt{\pi}$ will raise the temperature of the same field point, according to Eq. (51) with $\kappa = 1/\sqrt{\pi}$,

$$\frac{kT}{q\sqrt{A}} = 0.1664 \quad (86)$$

The difference between the square source and the equivalent circular source is less than 1% for the relatively near field point considered above. For field points whose distance from the centroid of the square source r_o is equal to or greater than the square root of the area (one side only), i.e., $r_o/\sqrt{A} > 1$, its temperature rise can be computed with negligible error using Eqs. (51) and (53) for an equivalent circular source.

Uniform Source Distribution over an Infinite Strip

Consider an infinite strip of width $-a < x < a$, lying in the $y = 0$ plane with uniform surface density q. The temperature rise at any point $P(x,y)$ can be obtained by means of the following integral:

$$4\pi kT(x,y) = 2q \int_{-a}^{a} \int_{-\infty}^{\infty} \frac{d\xi \, d\zeta}{\sqrt{(x-\xi)^2 + y^2 + \zeta^2}}$$

$$= -2q \int_{-a}^{a} \ln \sqrt{(x-\xi)^2 + y^2} \, d\xi \qquad (87)$$

A second integration yields[14]

$$\frac{2\pi \, kT(x,y)}{q} = 2a - y \arctan\left[\frac{2ay}{x^2+y^2-a^2}\right]$$

$$- \frac{(a-x)}{2} \ln\left[y^2 + (a-x)^2\right]$$

$$- \frac{(a+x)}{2} \ln\left[y^2 + (a+x)^2\right] \qquad (88)$$

Summary and Conclusions

Several basic integral solutions of Laplace's equation for point sources and distributed sources have been considered. Expressions for the temperature rise at arbitrary field points due to a finite line source, a circular ring source, a circular source, a circular annular source, a rectangular source, and a right triangular source have been developed. Alternate expressions are also presented for certain geometries. Temperature rise expressions along the axis and in the plane of the source are also developed and discussed.

The fundamental solutions presented in this paper form the basis of the surface element method which is an efficient and accurate numerical technique for the solution of Laplace's equation with complex boundary conditions. Martin[6] has demonstrated the power of this numerical method when applied to certain important thermal constriction problems such as singly and doubly connected, planar contact areas subjected to the boundary condition of the third kind with uniform and nonuniform contact conductance. In his solutions he employed several of the basic solutions given here.

Acknowledgments

The authors gratefully acknowledge the financial support of the Natural Sciences and Engineering Research Council of Canada under Grant No. A7445 to Dr. M. M. Yovanovich.

References

[1] Yovanovich, M. M., "Thermal Constriction Resistance of Contacts on a Half-Space Integral Formulation," Radiative Transfer and Thermal Control, Progress in Astronautics and Aeronautics, Vol. 49, edited by A. M. Smith, AIAA, New York, 1976, pp. 397-418.

[2] Yovanovich, M. M., Burde, S. S., and Thompson, J. C., Thermal Constriction Resistance of Arbitrary Contacts with Constant Flux," Thermophysics of Spacecraft and Outer Planet Entry Probes, Progress in Astronautics and Aeronautics, Vol. 56, edited by A. M. Smith, AIAA, New York, 1977, pp. 127-139.

[3] Yovanovich, M. M. and Schneider, G. E., "Thermal Constriction Resistance due to a Circular Annular Contact," Thermophysics of Spacecraft and Outer Planet Entry Probes, Progress in Astronautics and Aeronautics, Vol. 56, edited by A. M. Smith, AIAA, New York, 1977, pp. 141-154.

[4] Yovanovich, M. M. and Burde, S. S., "Centroidal and Area Average Resistances of Nonsymmetric, Singly-Connected Contacts," AIAA Journal, Vol. 15, No. 10, Oct. 1977, pp. 1523-1525.

[5] Yovanovich, M. M., Martin, K. A., and Schneider, G. E., "Constriction Resistance of Doubly-Connected Contact Areas Under Uniform Flux," AIAA Paper No. 79-1070, June 1979, Orlando.

[6] Martin, K. A., M.A. Sc. thesis, Department of Mechanical Engineering, University of Waterloo, Aug. 1980.

[7] Schneider, G. E., "Thermal Resistance Due to Arbitrary Dirichlet Contacts on a Half-Space," Thermophysics and Thermal Control, Progress in Astronautics and Aeronautics, Vol. 65, edited by R. Viskanta, AIAA, New York, 1979, pp. 103-119.

[8] Kellogg, O.D., Foundations of Potential Theory, Dover Publications, Inc., New York, 1964.

[9] Jeans, J. H., The Mathematical Theory of Electricity and Magnetism, 4th ed. Cambridge University Press, London, 1963.

[10] Stratton, J. A., *Electromagnetic Theory*, McGraw-Hill, New York, 1941.

[11] Jaswon, M. A. and Symm, G. T., *Integral Equation Methods in Potential Theory and Elastostatics*, Academic Press, New York, 1977.

[12] Durand, E., *Electrostatique et Magnetostatique*, Masson et Cie, Paris, France, 1953.

[13] Tikhonov, A. N. and Samarskii, A. A., *Equations of Mathematical Physics*, The MacMillan Company, New York, 1963.

[14] Budak, B. M., Samarskii, A. A., and Tikhonov, A. N., *A Collection of Problems on Mathematical Physics*, The MacMillan Company, New York, 1964.

[15] Lebedev, N. N., Skalskaya, I. P., and Uflyand, Y. S., *Worked Problems in Applied Mathematics*, Dover Publications, Inc., New York, 1979.

[16] Byerly, W. E., *Fourier Series and Spherical, Cylindrical, and Ellipsoidal Harmonics*, Dover Publications, Inc., New York, 1959.

[17] Carslaw, H. S. and Jaeger, J. C., *Conduction of Heat in Solids*, Oxford University Press, London, 1959.

[18] Conway, H. D. and Farnham, K. A., "The Relationship Between Load and Penetration for a Rigid, Flat-Ended Punch of Arbitrary Cross Section," *International Journal of Engineering Science*, Vol. 6, Dec. 1968, pp. 489-496.

[19] Reitan, D. K. and Higgins, T. J., "Accurate Determination of Capacitance of a Thin Rectangular Plate," *AIEE Transactions*, Vol. 75, 1957, pp. 761-766.

[20] Higgins, T. J. and Reitan, D. K., "Calculation of the Capacitance of a Circular Annulus by the Method of Subareas," *AIEE Transactions*, Vol. 70, 1951, pp. 926-933.

[21] Reitan, D. K. and Higgins, T. J., "Calculation of the Electrical Capacitance of a Cube," *Journal of Applied Physics*, Vol. 22, No. 2, 1950, pp. 223-226.

[22] Yovanovich, M. M., unpublished work on influence coefficients.

APPROXIMATE SOLUTIONS OF TRANSIENT HEAT CONDUCTION IN A FINITE SLAB

T. F. Zien[*]

Naval Surface Weapons Center, Silver Spring, Md.

Abstract

An integral procedure suitable for approximate solutions of two-point boundary value problems is presented here with the specific application to the transient heat conduction in a finite slab. The method is based on a further generalization of the ideas used in the author's earlier studies of viscous boundary-layer flows and transient heat conduction in semi-infinite slabs. The method is particularly suitable for accurate calculations of boundary heat flux and temperature. Only linear problems are treated in this paper, although the basic ideas are believed to be applicable to nonlinear problems as well. Results are compared with exact solutions and other approximate solutions wherever appropriate. Simplicity, accuracy, and weak profile dependence of the solution are the principal merits of the present method.

Nomenclature

a, a_0, a_1, a_2	=	coefficients of polynomial profile for temperature
b, b_0, b_1, b_2	-	coefficients of polynomial profile for temperature
f	=	temperature profile
HBI	=	heat balance integral
k	=	thermal conductivity
L	=	latent heat of ablation per unit mass
ℓ	=	width of the slab
Q	=	dimensionless heat flux
q	=	heat flux

Presented as Paper 80-1472 at the AIAA 15th Thermophysics Conference, Snowmass, Colo., July 14-16, 1980. This paper is declared a work of the U. S. Government and therefore is in the public domain.

[*]Head, Applied Mathematics Branch.

T	=	temperature
T^*	=	ablation temperature (in excess of the initial temperature)
t	=	time
x	=	space coordinate
α	=	thermal diffusivity
δ	=	thermal penetration depth
η	=	x/δ
θ	=	dimensionless temperature = $kT/q_0 \ell$
Θ	=	dimensionless temperature = $kT/q_0 \sqrt{\alpha t}$
ν	=	ablation parameter = L/CT^*
ξ	=	dimensionless coordinate, x/ℓ
τ	=	dimensionless time

Subscripts

o	=	$x = 0$
1	=	$\xi = 1$
p	=	at time of penetration, i.e., when $\delta = \ell$
s	=	steady-state values
T	=	total ablation

Superscript

*	=	ablation

Introduction

Transient heat conduction in a finite slab represents a problem of great importance in many technological areas. Its applications include heat-transfer gages, thermal insulations, metal casting, ice formation, thermal control of space vehicles, etc., to name only a few. The problem differs from that of a semi-infinite solid mainly in the existence of a physical length scale which precludes the possibility of similarity in the solution even for the simple case of linear heat conduction in a single phase. Exact solution is generally difficult, and would necessarily require considerable computational efforts. Therefore, approximate methods of calculation of such problems are desirable for practical purposes.

This paper presents an integral method suitable for approximate solution of two-point boundary-value problems with specific application to the transient heat conduction problem in a finite slab. The method represents a further generalization of the ideas used in the author's earlier works (e.g., Refs. 1-4), in viscous boundary-layer and transient heat conduction studies. The method is particularly suitable for accurate calculations of the boundary properties, e.g., boundary heat flux and temperature. The real potential of the method lies in its application to nonlinear problems such as melting, temperature dependent thermal properties, etc., where exact solutions are difficult. However, only linear problems are treated in this paper in an effort to illustrate the solution procedure with a minimum complexity. The nature of the method allows nonlinear problems to be treated without additional great difficulties, as amply demonstrated in the earlier papers, (e.g., Refs. 3,4). Therefore, the present method is expected to work for the problem of melting of finite slabs by inference from the results of Ref. 3 where both linear and nonlinear problems were solved for the semi-infinite slab configuration.

The Method and Its Application

Following Goodman,[5] we base the consideration of the problem on the concept of thermal penetration length, and divide the problem of heat conduction in a finite slab into two parts, a prepenetration period and a postpenetration period. In the prepenetration period, the thermal penetration length, $\delta(t)$, is less than the width of the slab, ℓ, so that the boundary condition specified at $x = \ell$ is not yet effective and the solid behaves essentially like a semi-infinite solid. As soon as the penetration length exceeds the slab width, the boundary condition at $x = \ell$ will influence the solution, and must be properly accounted for in the solution process. This period is referred to as the postpenetration period. The two parts of the solution are joined by requiring the continuity of temperature distributions in the solid at the time of thermal penetration, t_p.

Another characteristic feature of the finite slab problem is related to the presence of two physical boundary surfaces on which the boundary values are of interest. This is in contrast to the problem with semi-infinite solids, viscous boundary-layer flows, etc., where the boundary data at infinity are generally of little practical interest. In order to calculate the boundary values or their derivatives at both boundaries, an additional auxiliary equation is to be generated, and this forms the basis of the present combined moment scheme, or the θ-ξ scheme referred to in this paper.

Some examples are worked out in the following to illustrate the method. Results are compared to those of the previous ξ moment scheme and the classical heat balance integral (HBI) wherever appropriate. It is assumed in all the examples that the slab is maintained at a uniform temperature, $T_i = 0$ initially.

Given q_0 = Const and $T_1 = T_i = 0$

Prepenetration Solution. The equation and boundary conditions are listed below

$$\frac{\partial T}{\partial T} = \alpha \frac{\partial^2 T}{\partial x^2} \tag{1}$$

$$-k \left.\frac{\partial T}{\partial x}\right|_0 = q_0 \tag{2}$$

$$T(\delta, t) = 0 \tag{3}$$

Since the solution in this period is expected to behave like that for a semi-infinite solid, the length, ℓ, should not appear in the solution. Therefore, an approximate temperature profile, f, having similarity structure in the form

$$T = f = (q_0 \delta/k) \, a \, (1-x/\delta)^3 \tag{4}$$

is assumed where a is to be determined along with the thermal penetration depth $\delta(t)$. Note here that a is taken to be an independent parameter and $-k \left.\frac{\partial f}{\partial x}\right|_0 = q_0$ is <u>not</u> required.

We will obtain the solution by using the x-moment scheme.[3] Thus, the heat balance integral yields

$$\frac{d}{dt} \int_0^1 a\delta^2 (1-\eta)^3 d\eta = \alpha \tag{5}$$

where $\eta \equiv x/\delta$ and the exact boundary condition of $(\partial T/\partial x)_0 = -q_0/k$ is used.

The x-moment equation leads to

$$\frac{d}{dt} \int_0^1 a\delta^3 \eta (1-\eta)^3 d\eta = \alpha a \delta \tag{6}$$

Equations (5) and (6) form the required system for the unknowns $\delta(t)$ and a. The solutions are easily obtained as

$$\delta/\sqrt{\alpha t} = \sqrt{40/3} \tag{7a}$$

and
$$a = 3/10 \text{ (const)} \tag{7b}$$

$$\Theta_0 \equiv kt_0/q_0\sqrt{\alpha t} = 1.0954 \tag{7c}$$

compared to the exact value of $\Theta_0 = 1.1284$ for a semi-infinite solid.[6]

The penetration time, t_p, is defined as

$$\delta(t_p) = \ell \tag{8}$$

Therefore, t_p is determined to be

$$\alpha t_p/\ell^2 = \tau_p = 3/40 \tag{9}$$

which will be used in the postpenetration solution to be discussed below.

Postpenetration Solution. The equation and boundary conditions are

$$\frac{\partial \theta}{\partial \tau} = \frac{\partial^2 \theta}{\partial \xi^2} \tag{10}$$

$$\theta(1,\tau) = 0 \tag{11a}$$

$$\theta'(0,\tau) = -1 \tag{11b}$$

where the problem is formulated in dimensionless variables with

$$\theta \equiv \frac{kT}{q_0\ell}, \quad \xi \equiv \frac{x}{\ell}, \quad \tau \equiv \frac{\alpha t}{\ell^2} \tag{12}$$

The approximate profile, f, is also chosen in the form of a third degree polynomial so that matching with the prepenetration profile is possible. Thus,

$$f = a_0(\tau) + a_1(\tau)\xi + a_2(\tau)\xi^2 + a_3(\tau)\xi^3 \tag{13}$$

The basic boundary condition, Eq. (11a), implies

$$\frac{\partial \theta}{\partial \tau}\bigg|_{\xi=1} = 0 \tag{14a}$$

Equation (14a) then leads to the additional boundary condition

$$\frac{\partial^2 \theta}{\partial \xi^2}(1,\tau) = 0 \tag{14b}$$

by virtue of the heat equation, (10).

Equations (11a) and (14b) when applied to the profile, Eq. (13), give

$$a_3 = (a_0 + a_1)/2 \tag{15a}$$

and

$$a_2 = -(3/2)(a_0 + a_1) \tag{15b}$$

Note again that $\frac{\partial f}{\partial \xi}(0,\tau) = -1$ is not required.

Therefore we have

$$f = a_0 + a_1\xi - \frac{3}{2}(a_0 + a_1)\xi^2 + \frac{1}{2}(a_0 + a_1)\xi^3 \tag{16}$$

At $t = t_p$, the prepenetration profile, Eq. (4), gives

$$kT_p/q_0\ell = (3/10)(1 - 3\xi + 3\xi^2 - \xi^3) \tag{17}$$

whereas the postpenetration profile, Eq. (16), gives

$$kT_p/q_0\ell = a_{op} + a_{1p}\xi - (3/2)(a_{op} + a_{1p})\xi^2 + (1/2)(a_{op} + a_{1p})\xi^3 \tag{18}$$

Equation (17) and (18) can be made identical if

$$a_{op} = \frac{3}{10}, \quad a_{1p} = -\frac{9}{10} \tag{19}$$

Thus, the continuity of temperature distributions at t_p can be achieved if Eq. (19) is satisfied. Equation (19) provides the initial conditions for the postpenetration problem.

The quantities of particular interest are the boundary temperature, $a_0(\tau)$, and the heat flux at the other boundary, $\theta'(1,\tau)$ which is <u>not</u> to be taken as $f'(1,\tau)$. Therefore, the three quantities, $a_0(\tau)$, $a_1(\tau)$, and $\theta'(1,\tau)$ constitute the basic unknowns of the problem and three equations are to be generated for their solution. These equations are to be provided by the heat balance integral (HBI), the ξ-moment equation, and the θ-moment equation. This is the basis of the present method, i.e., ξ-θ scheme.

The HBI equation gives

$$\frac{d}{d\tau} \int_0^1 f d\xi = \frac{\partial \theta}{\partial \xi}\bigg|_1 + 1 \qquad (20)$$

the ξ-moment equation gives

$$\frac{d}{d\tau} \int_0^1 \xi f d\xi = a_0 + \frac{\partial \theta}{\partial \xi}\bigg|_1 \qquad (21)$$

and the θ_d moment equation gives

$$\frac{d}{d\tau} \int_0^1 f^2 d\xi = 2a_0 - 2 \int_0^1 \left(\frac{\partial f}{\partial \xi}\right)^2 d\xi \qquad (22)$$

Note here that θ_d is used to denote the use of the direct derivative for the term $\partial \theta/\partial \xi$ inside the right-hand side integral of Eq. (22), i.e., $\partial \theta/\partial \xi = \partial f/\partial \xi$ instead of using an integral expression for $\partial \theta/\partial \xi$ which can be obtained from a partial integration of the original heat equation.

Substituting f by the expression, Eq. (16), we have, from Eqs. (20, 21, and 22) respectively

$$5 \frac{da_0}{d\tau} + \frac{da_1}{d\tau} = 8(1 + \theta'_1) \qquad (23)$$

$$27 \frac{da_0}{d\tau} + \frac{da_1}{d\tau} = 120(a_0 + \theta'_1) \qquad (24)$$

$$(51a_0 + 9a_1) \frac{da_0}{d\tau} + (9a_0 + 2a_1)\frac{da_1}{d\tau} = 105a_0 \\ - 126a_0^2 - 42a_0 a_1 - 21a_1^2 \qquad (25)$$

which are to be solved for the three unknowns $a_0(\tau)$, $a_1(\tau)$, and θ'_1.

Eliminating θ'_1 from Eqs. (23) and (24), we obtain the following equation:

$$6 \frac{da_0}{d\tau} + \frac{da_1}{d\tau} = 15(1 - a_0) \qquad (26)$$

Equations (25) and (26) are to be solved with the initial conditions $a_0(\tau_p) = 3/10$ and $a_1(\tau_p) = -9/10$ where $\tau_p = 3/40$ [Eq. (9)].

The apparently nonlinear system of equations is reducible to the following linear system, $a_0 = -a_1$ being ruled out as a possible solution in view of their different initial values

$$\frac{da_0}{d\tau} = -3a_0 + 7a_1 + 10 \tag{27a}$$

$$\frac{da_1}{d\tau} = 3a_0 - 42a_1 - 45 \tag{27b}$$

The system, Eqs. (27a) and (27b) is readily solvable with standard methods. The solutions are given below:

$$\theta_0 = a_1 = 1 - 0.81019e^{-2.4688\tau} - 0.64975e^{-42.5312\tau} \tag{28a}$$

$$a_1 = -1 - 0.06149e^{-2.4688\tau} + 3.6693e^{-42.5312\tau} \tag{28b}$$

and θ_1' follows from Eq. (23) as

$$-\theta_1' = 1 - 1.2691e^{-2.4688\tau} + 2.2359e^{-42.5312\tau} \tag{28c}$$

where $-\theta_1' = q_1/q_0$ = dimensionless heat flux at $x = \ell$.

The exact solution of the problem can be expressed in terms of the following series[6]:

$$(\theta_0)_{exact} = 1 - 0.8113e^{-2.4674\tau} - 0.0901e^{-22.2167\tau} - \cdots \tag{29a}$$

$$-(\theta_1')_{exact} = 1 - 1.2732e^{-2.4674\tau} + 0.4244e^{-22.2167\tau} - \cdots \tag{29b}$$

In the postpenetration period where $\tau > \tau_p$, only the first two terms in Eqs. (29a) and (29b) are significant, and a comparison with Eqs. (28a) and (28c) clearly demonstrates the remarkable accuracy of the present approximate solution.

Solution by the ξ- moment Scheme. In the previous method of calculation, only the ξ-moment (or θ-moment) equation is

TRANSIENT HEAT CONDUCTION IN A FINITE SLAB

generated and used in conjunction with the HBI equation. This is referred to as the ξ-moment (or θ-moment) scheme. The ξ-moment scheme will now be used to solve the same problem with the same form of temperature profile so that the merit of the presently proposed θ-ξ scheme can be assessed.

Of course, the prepenetration solution as obtained earlier, Eqs. (7), is still the same. The postpenetration profile is again assumed to be in the form of Eq. (16). However, since only one auxiliary equation is available, only one of the two coefficients, $a_o(\tau)$, can be treated as unknown, and the other must be specified as a constant. $a_o(\tau)$ is the boundary temperature at the heating side, i.e., $a_o(\tau) = \theta_o(\tau)$, and is obviously not a constant. Therefore, we choose

$$a_1 = \text{const} = a_{1p} = -9/10 \tag{30a}$$

and

$$f = a_o - (9/10)\xi - (3/2)(a_o - 9/10)\xi_+^2 (a_o - 9/10)\xi^3/2 \tag{30b}$$

Note that the continuity of temperature profiles at τ_p is still insured.

The system of equation will consist of only Eqs. (20) and (21), and they are reduced to the following:

$$\frac{5}{8} \frac{da_o}{d\tau} = \frac{\partial \theta}{\partial \xi}\bigg|_1 + 1 \tag{31a}$$

$$\frac{9}{40} \frac{da_o}{d\tau} = a_o + \frac{\partial \theta}{\partial \xi}\bigg|_1 \tag{31b}$$

with $a_o(3/40) = 3/10$.

Eqs. (31) are easily solved to give

$$\theta_o = a_o = 1 = 0.8444e^{-2.5\tau} \tag{32a}$$

$$-\frac{\partial \theta}{\partial \xi}\bigg|_1 = 1 - 1.3194e^{-2.5\tau} \tag{32b}$$

A comparison of Eqs. (29) and (32) again shows the good accuracy of the solution, although it is somewhat inferior to the one obtained by the presently proposed ξ-θ_d scheme [Eqs. (28)].

HBI Solution. The same problem can also be solved approximately using the classical HBI method.[5] Here, the prepenetration profile is taken to be

$$T = f = (1/3)(q_0/k)\delta(1-\eta)^3 \qquad (33a)$$

and the postpenetration profile is assumed to be

$$\theta \equiv kT/q_0\ell = f = \theta_0^-\xi + (3/2)(1-\theta_0)\xi^2 - (1-\theta_0)\xi^{3/2} \qquad (33b)$$

where, in both profiles, the condition $-k\left.\frac{\partial T}{\partial x}\right|_0 = q_0$ is imposed.

The method uses only the HBI equation (20), with the derivatives taken directly from the assumed profile, i.e., $(\partial\theta/\partial\xi)_1 = (\partial f/\partial\xi)_1$ and $(\partial\theta/\partial\xi_0) = (\partial f/\partial\xi)_0$. The profile thus contains only one parameter, δ (prepenetration) or θ_0 (postpenetration), which is to be determined from the solution of the HBI equation. The solution is summarized below:

prepenetration:

$$kT_0/q_0\sqrt{\alpha t} = 1.1547; \quad \tau_p = 1/12$$

postpenetration

$$\theta_0 = 1 = 0.8143e^{-2.4\tau} \qquad (34a)$$

$$-\theta_1' = 1 - 1.2215e^{-2.4\tau} \qquad (34b)$$

The accuracy is seen to be not as good as that of the solutions by the present method.

Profile Dependence of the Approximate Solutions. To test the present method further, we use a simpler second-degree polynomial profile for the solution. Omitting the details, we will only present the solutions in the following:

Prepenetration solution (ξ-moment scheme):

$$T = f = bo(q_0\delta/k)(1-\eta)^2 \qquad (35a)$$

$$bo = 3/8, \quad \delta = \sqrt{8\alpha t}, \quad \tau_p = 1/8, \quad kT_0/q_0\sqrt{\alpha t} = 1.0607 \qquad (35b)$$

Postpenetration solution (ξ-θ_d scheme):

$$\theta \equiv kT/q_0 = f = \theta_0 + a_1\xi - (a_0 + a_1)\xi^2 \qquad (36a)$$

$$\theta_0 = 1 - 0.8093e^{-2.4860\tau} - 1.7796e^{-32.1807\tau} \qquad (36b)$$

$$-\theta_1' = 1 - 1.2833e^{-2.4860\tau} + 5.9854e^{-32.1807\tau} \qquad (36c)$$

The HBI solution of the same problem with a second-degree polynomial profile is also worked out, and is indicated for comparison:

Prepenetration solution (HBI):

$$T = f = (1/2)(q_0\delta/k)(1-\eta)^2 \qquad (37a)$$

$$\delta = \sqrt{6\alpha t}, \quad \tau = 1/6, \quad kT_0/q_0\sqrt{\alpha t} = 1.2247 \qquad (37b)$$

Postpenetration solution (HBI):

$$\theta \equiv kT/q_0\ell = f = a_0 - \xi + (1-a_0)\xi^2 \qquad (38a)$$

$$\theta_0 = 1 - 0.8244e^{-3\tau} \qquad (38b)$$

$$-\theta_1' = 1 - 1.6488e^{-3\tau} \qquad (38c)$$

A comparison of these approximate solutions with the exact solution reveals the relatively weak dependence of the present solution to the assumed profile, a highly desirable property of the solution discovered previously in the application of the early version of the method.[1-4] Furthermore, the need for improvements on the classical HBI solution becomes more obvious in view of its strong dependence on the profile.

These results are plotted in Figs. 1 and 2 for easy comparison.

Given T_0 = Const and $T_1 = 0$

For this problem, the boundary heat fluxes at both surfaces, q_0 and q_1, are the main quantities of interest in the solution.

In the prepenetration period, we use the profile

$$\theta \equiv T/T_0 = f = (1-\eta)^3 \qquad (39a)$$

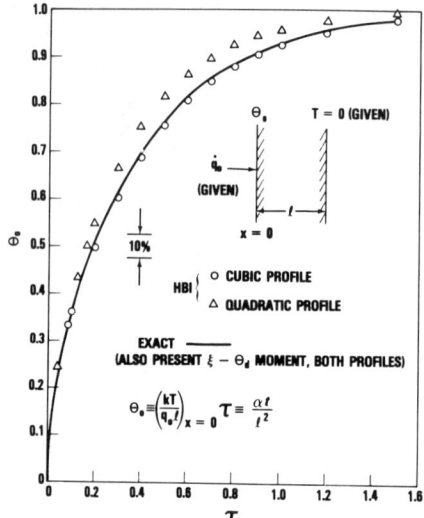

Fig. 1 Boundary temperature at x = 0.

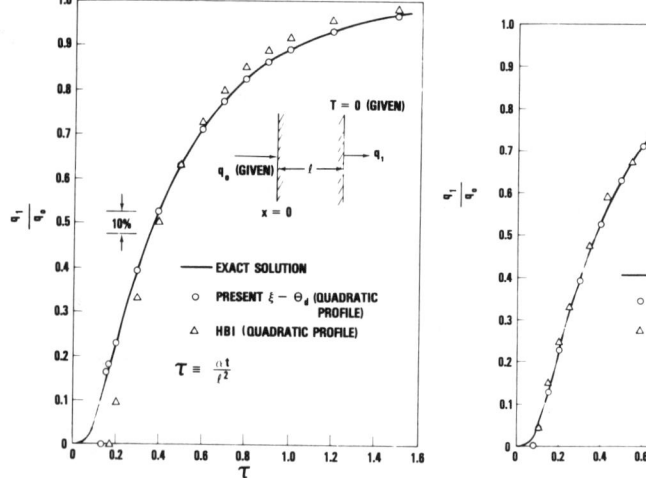

Fig. 2a Heat flux at $x = \ell$ – quadratic profile.

Fig. 2b Heat flux at $x = \ell$ – cubic profile.

where, as before, $\eta \equiv x/\delta$.

The x-moment scheme is used to give the following approximate solution

$$\delta = \sqrt{20\alpha t}, \quad \tau_p = 1/20, \quad Q_0 \equiv q_0\sqrt{\alpha t}/kT_0 = 0.5590 \qquad (39b)$$

compared to the exact solution (semi-infinite slab) of $Q_0 = 0.5642$.

In the postpenetration period, a profile

$$\theta \equiv T/T_0 = f = 1+b_1\xi-(3/2)(1+b_1)\xi^2+(1+b_1)\xi^3/2 \quad (40)$$

is assumed which ensures continuity at τ_p.

The $\xi-\theta_d$ scheme leads to the following system of ordinary differential equations for the unknowns, θ_0', θ_1' and $b_1(\tau)$,

$$\frac{1}{8}\frac{db_1}{d\tau} = \theta_1' - \theta_0' \quad (41a)$$

$$\frac{7}{120}\frac{db_1}{d\tau} = 1 + \theta_1' \quad (41b)$$

$$3\frac{db_1}{d\tau} + \frac{2}{3}b_1\frac{db_1}{d\tau} = -35\theta_0' - 7(6+2b_1+b_1^2) \quad (41c)$$

with $b_1(\tau_p) = -3$. The prime denotes, as before, differentiation with respect to ξ.

Solution to Eqs. (41) is easily obtained as

$$-\theta_0' = 1 + 2.3666e^{-10.5\tau} \quad (42a)$$

$$-\theta_1' = 1 - 2.0708e^{-10.5\tau} \quad (42b)$$

$$b_1 = -1 - 3.3809e^{-10.5\tau} \quad (42c)$$

The exact solution has the power series expression[6]:

$$-(\theta_0')_{exact} = 1 + 2e^{-9.8696\tau} + 2e^{-39.4784\tau} + \cdots \quad (43a)$$

$$-(\theta_1)_{exact} = 1 - 2e^{-9.8696} + 2e^{-39.4784\tau} - \cdots \quad (43b)$$

Introducing the steady-state heat flux

$$q_s = kT_0/\ell, \quad (44)$$

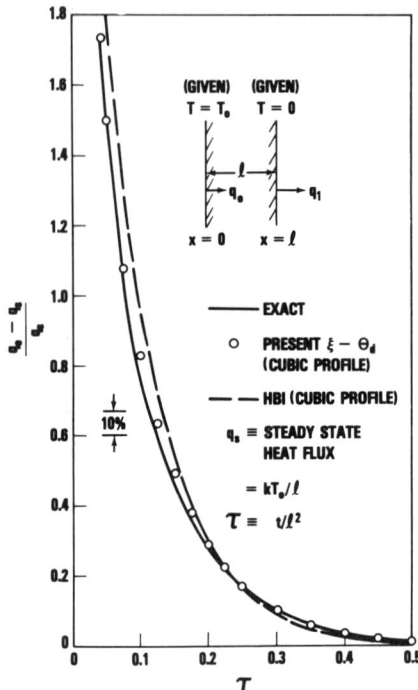

Fig. 3 Boundary heat flux at $x = 0$.

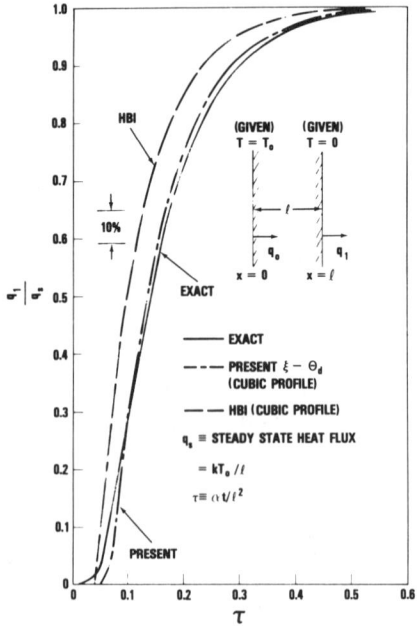

Fig. 4 Boundary heat flux at $x = \ell$.

we can write

$$-\theta'_0 = q_0/q_s \qquad (45a)$$

and

$$-\theta'_1 = q_1/q_s \qquad (45b)$$

The classical HBI solution with the same third-degree polynomial profile is also presented below for comparison:

Prepenetration period:

$$q_0\sqrt{\alpha t}/kT_0 = 0.6124, \quad \tau_p = 1/24 \qquad (46)$$

Post-penetration period:

$$-\theta'_0 = 1 + 3.2974 e^{-12\tau} \qquad (47a)$$

$$-\theta'_1 = 1 - 1.6487 e^{-12\tau} \qquad (47b)$$

The superiority of the present solution is thus clearly demonstrated. These results are shown in Figs. 3 and 4.

Given q_0 = Const and $q_1 = 0$
_____

For this problem, a second-degree polynomial profile is more suitable than a third-degree polynomial. [Consider the postpenetration period. In view of the boundary condition of $q_1 = 0$, we have $(\partial\theta/\partial\xi)_1 = 0$ for all τ. Thus $(\partial^2\theta/\partial\xi\partial\tau)_1 = 0$ which leads to $(\partial^3\theta/\partial\xi^3)_1 = 0$. This requires that the coefficient of the cubic term be zero in a third-degree polynomial temperature profile. Therefore, the post-penetration profile reduces to a second-degree polynomial.] Thus, for the prepenetration period, we use

$$T = f = b_0 (q_0 \delta/k)(1-\eta)^2 \qquad (48a)$$

The x-moment solution is readily obtained as

$$b_0 = 3/8, \quad \delta = \sqrt{8\alpha t}, \quad kT_0/q_0\ell = 1.0607\sqrt{\tau} \qquad (48b)$$

where $\tau \equiv \alpha t/\ell^2$. The exact solution (semi-infinite slab problem) gives $kT_0/q_0\ell = 1.0954\sqrt{\tau}$ [see Eq. (7c)].

In the postpenetration period, a profile

$$\theta \equiv kT/q_0\ell = \theta_0(\tau) + a_1\xi + a_2\xi^2 \qquad (49)$$

is assumed which can ensure continuity with the prepenetration profile at $\tau_p = 1/8$.

The ξ-θ_d scheme then leads to the following system of equations for $\theta_0(\tau)$, $a_1(\tau)$ and $a_2(\tau)$:

$$\frac{d\theta_0}{d\tau} + \frac{1}{2}\frac{da_1}{d\tau} + \frac{1}{3}\frac{da_2}{d\tau} = 1 \qquad (50a)$$

$$\frac{1}{2}\frac{d\theta_0}{d\tau} + \frac{1}{3}\frac{da_1}{d\tau} + \frac{1}{4}\frac{da_2}{d\tau} = -a_1 - a_2 \qquad (50b)$$

$$(2\theta_0 + a_1 + \frac{2}{3}a_2)\frac{d\theta_0}{d\tau} + (\theta_0 + \frac{2}{3}a_1 + \frac{1}{2}a_2)\frac{da_1}{d\tau}$$
$$+ (\frac{2}{3}\theta_0 + \frac{1}{2}a_1 + \frac{2}{5}a_2)\frac{da_2}{d\tau} = 2\theta_0 - 2a_1^2 - 4a_1 a_2 - \frac{8}{3}a_2^2 \qquad (50c)$$

with

$$\theta_0(\tau_p) = a_2(\tau_p) = 3/8 \qquad (51a)$$

and

$$a_1(\tau_p) = 3/4 \qquad (51b)$$

The system of nonlinear differential equations is again reducible to a system of linear equations as follows:

$$\frac{d\theta_0}{d\tau} = 9 + 6a_1 - 4a_2 \qquad (52a)$$

$$\frac{da_1}{d\tau} = 12(-3 + 4a_2 - a_1) \qquad (52b)$$

$$\frac{da_2}{d\tau} = 30(1 - 2a_2) \qquad (52c)$$

In the reduction, the other solution corresponding to $a_2(\tau) \equiv 0$ is discarded in view of the boundary condition, Eq. (51a).

Closed form solution is easily obtained as

$$\theta_0(\tau) = \tau + \frac{1}{3} - \frac{1}{16}e^{-12(\tau-\tau_p)} - \frac{1}{48}e^{-60(\tau-\tau_p)} \qquad (53a)$$

$$a_1(\tau) = -1 + \frac{1}{8}e^{-12(\tau-\tau_p)} + \frac{1}{8}e^{-60(\tau-\tau_p)} \qquad (53b)$$

$$a_2(\tau) = \frac{1}{2} - \frac{1}{8} e^{-60(\tau-\tau_p)} \quad (53c)$$

from which we find

$$\theta_0 = \tau + \frac{1}{3} - 0.2801 e^{-12\tau} - 37.67 e^{-60\tau} \quad (54a)$$

$$\theta_1 = \tau - \frac{1}{6} + 0.2801 e^{-12\tau} - 37.67 e^{-60\tau} \quad (54b)$$

The exact solution of the problem gives[6]

$$(\theta_0)_{exact} = \tau + \frac{1}{3} - 0.2026 e^{-\pi^2 \tau} - 0.0507 e^{-4\pi^2 \tau} - \cdots \quad (55a)$$

and

$$(\theta_1)_{exact} = \tau - 1/6 + 0.2026 e^{-\pi^2 \tau} - 0.0507 e^{-4\pi^2 \tau} + \cdots \quad (55b)$$

The following second-degree polynomial profiles are used in conjunction with the classical HBI method:

Prepenetration:

$$T = f = (1/2)(q_0 \delta/k)(1-\eta)^2 \quad (56a)$$

Postpenetration

$$\theta \equiv kt/q_0 \ell = f = \theta_0(\tau) - \xi + \xi^2/2 \quad (56b)$$

The results of HBI method are indicated below:

$$kT_0/q_0\sqrt{\alpha t} = 1.2247 \quad (\tau < \tau_p = 1/6) \quad (57a)$$

and

$$\theta_0 = \tau + 1/3, \quad \theta_1 = \tau - 1/6 \quad (\tau > \tau_p = 1/6)$$

Note that the exponential decay terms in the postpenetration solution are totally missing from the HBI solution. The accuracy of the solution is thus expected to be rather poor for moderate times.

Ablation Problem - A Brief Discussion

The ablation of a finite slab can be treated in a manner similar to that for the semi-infinite solid.[3,4] As was amply demonstrated in Refs. 3 and 4, the presence of phase transition introduces only additional complexity of the problem but not essential difficulties in the application of the present integral method. However, the finite thickness of the slab introduces a new time scale into the problem, i.e., the thermal penetration time, t_p, discussed in the previous sections. Therefore, a discussion of the ablation problem of a finite slab in the framework of the present integral method will inevitably involve the use of two time scales, t_p and t^*. t^* is the preablation time [3,4] defined as the time required for the boundary to reach the ablation temperature T^*, i.e., $T_o(t^*) = T^*$ = ablation temperature (in excess of the initial temperature). Of course, in the problem where total ablation of the slab is possible, such as the one with an insulated boundary (previous section), another characteristic time t_T would appear which is defined as the time required for the total ablation of the slab.

To fix the idea, we consider the problem of ablation with boundary conditions as specified in the previous section. Suppose we use the second-degree polynomial profile for the temperature, so that the penetration time is given by $\tau_p = 1/8$. Depending on the intensity of external heating, q_o^*, the preablation time could be either less or greater than the penetration time

$$i \quad \tau^* < \tau_p < \tau_T$$

This case corresponds to a sufficiently large q_o so that ablation begins before total thermal penetration. Here, the preablation period is the same as the corresponding case of a semi-infinite solid, and τ^* can be determined from the prepenetration solution given by Eq. (48b), i.e.,

$$kt^*/q_o \ell = (9\tau^*/8)^{\frac{1}{2}}$$

or

$$\tau^* = (8/9)(1/Q_o^*)^2 \tag{59}$$

where $Q_o^* \equiv q_o \ell / kT^*$ = dimensionless measure of the applied heat flux.

The ablation solution would then be divided into prepenetration and postpenetration periods, with the prepenetration period being identical to that associated with the semi-infinite solid

$$ii \quad \tau_p < \tau^* < \tau_T$$

In this case, total thermal penetration into the slab preceeds the onset of ablation, and the preablation time,

τ^*, should be determined by the postpenetration solution, Eqs. (54a). We have

$$\tau^* \approx 1/Q_0^* - 1/3 \tag{60}$$

The ablation solution will then be based entirely on the finite slab model.

In either case, the exact time for total ablation, t_T, can be obtained from a consideration of energy balance. For the problem under consideration, Citron[7] gives

$$\tau_T = \rho \ell (L + CT^*)/q_0 \tag{61}$$

where ρ is the solid density, L is the latent heat of ablation per unit mass of the solid, and C is the specific heat. In dimensionless form, Eq. (61) becomes

$$\tau_T \equiv \alpha t_T/\ell^2 = (1+\nu)/Q_0^* \tag{62}$$

where $\nu \equiv L/CT^*$ is the ablation parameter.[3]

Concluding Remarks

The so-called ξ-θ_d scheme of the integral method as illustrated in this paper appears capable of producing accurate approximate solutions for two-point boundary-value problems, as evidenced by the sample solutions of some transient heat conduction problems for a finite slab. The method can predict accurate boundary derivatives (heat flux) or boundary values (temperature), depending on the nature of the problem posed. Furthermore, the approximate solutions seem to be only weakly dependent on the temperature profile assumed in the solution process, and this represents an additional significant improvement over the classical HBI method.

In view of the fact that the method can treat nonlinear and linear problems alike, the problem of melting or ablation of a finite slab should also be amenable to the present method.

Acknowledgment

The research was supported jointly by the NSWC Independent Research Program and the Naval Air Systems Command through the Strike Warfare Technology Block (Naval Weapons Center). The author gratefully acknowledges some useful discussion with his colleagues John Bell, Alan Berger, and Jeffrey Youngs on certain mathematical solutions.

References

[1] Zien, T. F., "Skin Friction on Porous Surfaces Calculated by a Simple Integral Method," AIAA Journal, Vol. 10, No. 10, Oct. 1972, pp. 1267-1268.

[2] Zien, T. F., "Approximate Analysis of Heat Transfer in Transpired Boundary Layer with Effects of Prandtl Number," International Journal of Heat and Mass Transfer, Vol. 19, No. 5, May 1976, pp. 513-521.

[3] Zien, T. F., "Study of Heat Conduction with Phase Transition Using an Integral Method," Thermophysics of Spacecraft and Outer Planet Entry Probes, Progress in Astronautics and Astronautics, Vol. 56, edited by A. M. Smith, AIAA, New York, 1977, pp. 87-111.

[4] Zien, T. F., "Integral Solutions of Ablation Problems with Time-Dependent Heat Flux," AIAA Journal, Vol. 16, No. 12, Dec. 1978, pp. 1287-1295.

[5] Goodman, T. R., "Application of Integral Methods to Transient Nonlinear Heat Transfer," Advances in Heat Transfer, Vol. 1, 1964, pp. 51-122.

[6] Carslaw, H. S. and Jaeger, J. C., Conduction of Heat in Solids, 2nd ed. Oxford University Press, London, 1959, Chap.3.

[7] Citron, S. J., "Heat Conduction in a Melting Solid," Journal of the Aerospace Sciences, Vol. 27, No. 3, March 1960, pp. 219-228.

FINITE-ELEMENT ANALYSIS FOR CONDUCTION AND ABLATION MOVING BOUNDARY

Jin H. Chin*

Lockheed Missiles & Space Company, Inc., Sunnyvale, Calif.

Abstract

A computerized finite-element procedure is developed to determine the two-dimensional and axisymmetric thermal response of ablating bodies. The coupling of nonlinear conduction and ablation moving boundary in an anisotropic multimaterial region is analyzed with a fixed finite-element grid. An extended integral method is formulated for evaluation of the properties of partially ablated elements. The net conductive flux into the solid, given by a nonlinear energy balance at the ablating surface, is simulated by a moving surface source. The finite-element results for the problems of one-dimensional and axisymmetric ablation of graphite compare favorably with that obtained by finite-difference methods. Application of the procedure to solid rocket nozzle analysis is demonstrated.

Nomenclature

B' = mass-transfer parameter $\dot{m}/\bar{C}_H$
C = volumetric heat capacitance matrix
C_H = zero blowing heat-transfer coefficient
$\bar{C}_H$ = blowing heat-transfer coefficient
c = specific heat
F = radiation form factor
H_r = recovery enthalpy

Presented as Paper 80-1488 at the AIAA 15th Thermophysics Conference, Snowmass, Colo., July 14-16, 1980. Copyright © American Institute of Aeronautics and Astronautics, Inc., 1980. All rights reserved.
*Staff Engineer Senior, Aero/Thermodynamics & Vulnerability, Engineering Technology.

h	= heat-transfer coefficient
h_c	= enthalpy of solid at wall temperature
h_w	= enthalpy of gas species at wall temperature
K	= thermal-conductance matrix
k	= thermal conductivity
L	= distance or length
L_N	= normalizing length = 0.1 in (2.54 mm)
$\dot{m}$	= ablation rate per unit area
N	= element shape function
n	= unit outward normal
p_e	= boundary-layer edge pressure
Q	= heat flux per unit volume
q	= heat rate per unit area
q_N	= normalizing $q = C_H H_r$ at stagnation point
q_{rad}	= incident radiative q
r	= radius
r_o	= outer radius of sphere shell segment
S	= instantaneous total ablation
T	= absolute temperature
T_N	= normalizing temperature = 10000 R (5555.55 K)
t	= time
V	= temperature integral of v
v	= material property
x	= Cartesian coordinates
α	= nodal point index
α_w	= wall absorptivity for q_{rad}
β	= nodal point index
Γ	= domain boundary surface
ε	= emissivity
Θ	= angle to stagnation axis
σ	= Stefan-Boltzmann constant
ρ	= density
ω	= time integration parameter
Ω	= domain volume

Subscripts

av	= average
f	= phase change, ablation
f±Δ	= at $T_f \pm \Delta T$ location
i	= coordinate direction
j	= coordinate direction
n	= at time step t_n

out = leaving the domain surface to surroundings
w = at wall conditions
α = nodal point
β = nodal point
* = at weighted average time and temperature conditions
∞ = ambient environment conditions

Superscripts

e = element
i = ith iteration

Convention

$\dot{T} = \partial T/\partial t$

$T_{,i} = \partial T/\partial x_i$

$N_\alpha T_\alpha = \sum_{\alpha=1}^{N_p} N_\alpha T_\alpha$, N_p = number of element nodes

Introduction

Interest in the moving boundary problem is evident through many recent publications.[1-3] The author's earlier interest in moving boundary problems concerns the ablation and thermal response of externally heated nosetips of re-entry vehicles.[4,5] With fixed rectangular meshes, the finite-difference space discretization of the Fourier conduction equation yields relatively simple expressions for the regular interior node; however, the expressions become complicated, and the solution accuracy decreases when the nodes adjacent to the ablation moving boundary are considered.[4] For single material regions, a transformed, deforming coordinate system coupled with a finite-difference space discretization simplifies the boundary motion calculation considerably and may reduce the total number of nodal points required for a given problem.[5] However, the advantage of the deforming coordinate formulation decreases when multimaterial regions are considered and when the ablation moving surface crosses the material boundary. This observation is also valid for the finite-element method formulated in terms of a deforming coordinate system, although this approach has been shown to be effective in solving certain simple heat conduction problems with and without phase change.[6] To maintain the advantage of the finite-element method, particularly for application to multimaterial regions with complex geometries, fixed or time invariant grids are generally preferred.[2,7] With fixed grids, the regular finite-element numerical procedures may be used for the interior elements which contain no part of the moving boundary. Special methods have been devised to treat elements which undergo a phase change at a given temperature, with or without removal of the new phase.[2,8,7] However, the ablation temperature is not a constant for materials such as graphites subjected to a rocket nozzle or re-entry heating environment. In fact, the ablation temperature, governed by a nonlinear energy balance at the ablating surface, must be solved together with the internal conduction equation. This paper presents an algorithm for finite-element analysis for conduction and ablation. Examples of computed results are given.

Analysis

Finite-Element Formulation

In terms of an orthogonal Cartesian coordinate x_i and using the tensor index notation and summation convention, the law of conservation of energy is given by

$$\rho c \dot{T} + q_{i,i} - Q = 0 \quad \text{in } \Omega \tag{1}$$

subjected to boundary conditions

$$T - T_w(x, t) = 0 \quad \text{on } \Gamma_1 \tag{2}$$

and

$$n_i q_i - f(x, t) = 0 \quad \text{on } \Gamma_2 \tag{3}$$

The heat flux vector is given by the Fourier law

$$q_i = -k_{ij} T_{,j} \quad \text{in } \Omega \tag{4}$$

Spacewise discretization of Eq. (1), subjected to boundary conditions (2) and (3), may be performed using the Galerkin method.[9] This method leads to the system of differential equations for the element Ω^e

$$C^e_{\alpha\beta} \dot{T}_\beta + K^e_{\alpha\beta} T_\beta - Q^e_\alpha + F^e_\alpha = 0 \quad \text{in } \Omega^e \tag{5}$$

where

$$C^e_{\alpha\beta} = \int_{\Omega^e} \rho c\, N_\alpha N_\beta\, d\Omega, \quad K^e_{\alpha\beta} = \int_{\Omega^e} k_{ij} N_{\alpha,i} N_{\beta,j}\, d\Omega$$

$$Q^e_\alpha = \int_{\Omega^e} Q N_\alpha\, d\Omega, \quad F^e_\alpha = \int_{\Gamma^e_2} f N_\alpha\, d\Gamma \tag{6}$$

Element shape functions $N_\beta(x)$[9,10] have been used to approximate the temperature field within Ω^e as linearly dependent upon the element node temperatures T_β

$$T(x) = N_\beta(x) \, T_\beta \tag{7}$$

Matrix and vector coefficients of Eq. (6) are evaluated for each element and then assembled to obtain the matrix and vector coefficients for the whole domain Ω. Generally, F_α^e needs to be evaluated only for elements having one or more boundaries as part of Γ_2. The evaluation of the integrals in Eq. (6) is generally by numerical quadratures. For instance, for a linear quadrilateral element with four corner nodes, the optimum method is the Gauss-Legendre quadrature with four integration points.[10]

Assembled equations for the whole domain are identical to Eqs. (5) and (6) except the superscript e is removed from all expressions. In matrix form, the assemblage of Eq. (5) may be written as

$$C\dot{T} + KT - Q + F = 0 \quad \text{in } \Omega \tag{8}$$

which represents a system of n equations for the unknown temperatures.

Linear and Nonlinear Boundary Conditions

For linear Neumann boundary conditions, the function f in Eq. (3) may be represented by

$$f(x,t) = a(x,t) \, T(x,t) + b(x,t) \tag{9}$$

Equation (9) covers two special cases:

Convective cooling

$$f = h(T - T_\infty), \quad a = h, \quad b = -h \, T_\infty \tag{10}$$

Applied heat flux removal

$$f = q_{out}, \quad a = 0, \quad b = q_{out} \tag{11}$$

Substituting Eq. (10) into the fourth member of Eq. (6) and using Eq. (7), the following expression is obtained

$$F_\alpha^e = H_{\alpha\beta}^e \, T_\beta - \int_{\Gamma_2^e} h \, T_\infty \, N_\alpha \, d\Gamma \tag{12}$$

where

$$H^e_{\alpha\beta} = \int_{\Gamma^e_2} h\, N_\alpha\, N_\beta\, d\Gamma \qquad (13)$$

Convection matrix $H^e_{\alpha\beta}$ may be combined with the conductance matrix $K^e_{\alpha\beta}$ to form a new effective conductance matrix for use in Eq. (5).

For thermal ablation of materials such as graphite in a rocket nozzle or re-entry heating environment, the function f in Eq. (3) may be written as

$$-f(x,t,T) = \overline{C}_H(H_r - h_w) - \dot{m}(h_w - h_c) + \alpha_w q_{rad} - \sigma F \epsilon_w T_w^4 \qquad (14)$$

where for simplifying the discussions, equal diffusion coefficients and equal heat and mass-transfer coefficients have been assumed. Equation (14) is highly nonlinear in T. Generally, the material properties ρc and k also are functions of T.

Time Integration

A mixed explicit-implicit iterative scheme for time integration is used. Let

$$\Delta t = t_{n+1} - t_n$$
$$t_* = (1-\omega) t_n + \omega\, t_{n+1} \qquad (15)$$
$$T^i_* = (1-\omega) T_n + \omega\, T^i_{n+1}$$

where ω $(0 \le \omega \le 1)$ is a weighting factor: $\omega = 0$ for the explicit scheme, $\omega = 0.5$ corresponding to the Crank-Nicloson method, and $\omega = 1$ for the fully implicit scheme.

Replacing $\dot{T}$ by $(T_{n+1} - T_n)/\Delta t$, using the expressions given by Eq. (15), and by adding and subtracting T^i_{n+1} from T^{i+1}_{n+1}, Eq. (8) may be manipulated to yield the following matrix equation for the solution of the corrections

$$\left(\frac{C^i_*}{\Delta t} + \omega K^i_*\right)\left(T^{i+1}_{n+1} - T^i_{n+1}\right) = -K^i_* T^i_* - \frac{C^i_*}{\Delta t}\left(T^i_{n+1} - T_n\right) + Q^i_* - F^i_* \qquad (16)$$

Successive iteration with Eq. (16) may be continued until $(T^{i+1}_{n+1} - T^i_{n+1})$ is reduced to a specified small value.

An alternate time integration scheme may be the three level Lees[11,2,8] algorithm.

Ablation Moving Boundary, Extended Integral Average

Figure 1 shows a partially ablated element with local node numbers 1, 2, 3, and 4, and locations of the integration points and surface points along the ablation moving boundary. Material properties change abruptly across the moving boundary ($\rho c = k = 0$ at integration points 2 and 4). The numerical integration of the element matrix and vector coefficients given by Eq. (6) becomes inaccurate. Increasing the number of integration points may help but with corresponding increase in computation efforts and costs.

The fact that the element shape functions are introduced to approximate the element temperature field suggests that the solution sought is of an integral average nature rather than concerning the fine details of variation from point to point. The integrations represented by Eq. (6) may be viewed as a process "to obtain an integral of the contributions within an element and to distribute these contributions to the element nodal points proportionally." Evaluation of the integrals of Eq. (6) by numerical quadrature with a relatively small number of integration points requires that the quantities ρc, k_{ij}, Q, and f be relatively well behaved. For situations where these quantities change abruptly within an element, an integral average for these quantities is thus suggested. The enthalpy average of ρc for a phase-change problem is such an integral average.[2,8]

The relationship between temperature integration and geometry integration is recently illustrated.[7] Consider a one-dimensional linear element with nodes at x_1 and x_2. A phase change interface occurs at x_f within the element, separating the two phases with thermal conductivities k_1, k_2 and volumetric heat capacities $(\rho c)_1$, $(\rho c)_2$. Per unit cross-sectional

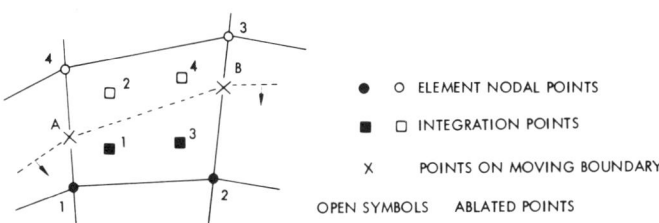

Fig. 1 Partially ablated element.

area, the overall conduction resistance for this element is given by

$$R = \frac{x_f - x_1}{k_1} + \frac{x_2 - x_f}{k_2} = \frac{x_2 - x_1}{k_{av}} \qquad (17)$$

where k_{av} is an average thermal conductivity within the element. Similarly, the overall volumetric thermal capacitance for this element is given by, assuming a small phase change temperature interval $2\Delta T$

$$C = (x_{f-\Delta} - x_1)(\rho c)_1 + (x_{f+\Delta} - x_{f-\Delta})(\rho c)_f$$

$$+ (x_2 - x_{f+\Delta})(\rho c)_2 = (x_2 - x_1)(\rho c)_{av} \qquad (18)$$

where $(\rho c)_{av}$ is an average element volumetric heat capacity. For a linear temperature field within the element, Eqs. (17) and (18) may be written, respectively, as follows

$$\frac{1}{k_{av}} = \frac{x_f - x_1}{x_2 - x_1} \frac{1}{k_1} + \frac{x_2 - x_f}{x_2 - x_1} \frac{1}{k_2}$$

$$= \frac{T_f - T_1}{T_2 - T_1} \frac{1}{k_1} + \frac{T_2 - T_f}{T_2 - T_1} \frac{1}{k_2} \qquad (19)$$

$$(\rho c)_{av} = \frac{x_{f-\Delta} - x_1}{x_2 - x_1}(\rho c)_1 + \frac{x_{f+\Delta} - x_{f-\Delta}}{x_2 - x_1}(\rho c)_f + \frac{x_2 - x_{f+\Delta}}{x_2 - x_1}(\rho c)_2$$

$$= \frac{T_{f-\Delta} - T_1}{T_2 - T_1}(\rho c)_1 + \frac{\lambda}{T_2 - T_1} + \frac{T_2 - T_{f+\Delta}}{T_2 - T_1}(\rho c)_2 \qquad (20)$$

where $2\Delta T(\rho c)_f$ has been replaced by the volumetric latent heat λ. Equations (20) and (19) show that the temperature integral average is appropriate for the volumetric heat capacity and for the inverse of thermal conductivity.

The above approach leads to the algorithm of "extended integral averages." Let v represent material properties which may depend upon temperature and possibly other variables such as the pressure. The values of v at the element nodal points β are first evaluated. It is assumed that the variation of v with T is linear within the element in a least square sense

$$v_\beta = a + b(T_\beta - T_{\beta min}) \qquad (21)$$

where $T_{\beta min}$ is the minimum of all T_β for the element and the constants a and b are determined by the least square method. Then the integral of v with respect to T is calculated as follows

$$V_\beta = \int_{T_{\beta min}}^{T_\beta} [a + b(T - T_{\beta min})] \, dT \qquad (22)$$

where $T_\beta \leq T_{w\beta}$, the temperature of the ablating wall <u>associated with node β</u>. Now, a small increment of temperature, ΔT, is used for the ablation temperature range: $T_{w\beta} \leq T_\beta \leq T_{w\beta} + \Delta T$. Define $V_{w\beta}$ as the value of V_β when $T_\beta = T_{w\beta}$

$$V_{w\beta} = (T_{w\beta} - T_{\beta min})[a + b(T_{w\beta} - T_{\beta min})/2] \qquad (23)$$

It is assumed that v_β drops to zero linearly within the ablation temperature range. Then, for $T_\beta \geq T_{w\beta}$

$$V_\beta = \begin{cases} V_{w\beta} + v_{w\beta}(T_\beta - T_{w\beta})(1-(T_\beta - T_{w\beta})/2\Delta T), & T_{w\beta} \leq T_\beta \leq T_{w\beta} + \Delta T \\ V_{w\beta} + v_{w\beta}\Delta T/2, & T_\beta \geq T_{w\beta} + \Delta T \end{cases} \qquad (24)$$

Using the derivatives of the element shape functions, the extended integral average, v_{av}, at the integration points of the element is then calculated as follows

$$v_{av} = (V_{,i} V_{,i})^{\frac{1}{2}} / (T_{,i} T_{,i})^{\frac{1}{2}} \qquad (25)$$

The extended integral average is not rigorous. As long as the variation of v with location within an element may be approximated by its variation with temperature, then the temperature may be reasonably used as the variable of integration.

The ablating surface energy balance Eq. (14) indicates that the ablation temperature is not constant. Thus, for the partially ablated element shown in Fig. 1, generally $T_{wA} \neq T_{wB}$. What is the appropriate $T_{w\beta}$ for use in Eqs. (23) and (24) when β represents the ablated node 4 or 3? Lacking a rigorous basis, the effective ablation temperature $T_{w\beta}$ for ablated node β is determined from a geometrical consideration (Fig. 1)

$$T_{w\beta} = \left(\frac{T_{wA}}{L^2_{\beta A}} + \frac{T_{wB}}{L^2_{\beta B}}\right) \Big/ \left(\frac{1}{L^2_{\beta A}} + \frac{1}{L^2_{\beta B}}\right) \qquad (26)$$

where $L_{\beta A}$ and $L_{\beta B}$ are the distances between ablated node β (4 or 3) and surface points A and B, respectively.

The algorithm of the extended integral averages, Eqs. (21-26), is thus an extension of the method of one-dimensional averages, Eqs. (17-20). For partially ablated elements, the thermal properties of ablated region become zero: $k_2 = (\rho c)_2 = 0$ in Eqs. (19) and (20), respectively. Because $1/k_2 \to \infty$ as $k_2 \to 0$, Eq. (19) cannot be used, and, correspondingly, Eq. (21) is a poor approximation for $v = 1/k$. The extended integral average on $1/k$ is expected to yield poor numerical behavior. It will be shown later that the extended integral averages applied to both ρc and k give good results.

Moving Surface Source

The net conductive heat flux into the solid is given by the right-hand side of Eq. (14). This heat flux at the ablating surface may be simulated by a moving surface source within the surface or boundary element. Consider a point source $g(x_a) = g\,\delta(x-x_a)$ at location x_a within an element, where δ is the Dirac unit impulse function. The effect of this point source is distributed to the element nodal points by means of the element shape functions [e.g., third of Eq. (6)]

$$G^e_{x_a \alpha} = \int_{\Omega^e} g\,\delta(x - x_a)\, N_\alpha\, d\Omega$$

$$= g\, N_\alpha(x_a) \qquad (27)$$

The effect of a surface source distributed to the element nodal points is then obtained by integration of Eq. (27) over the ablating surface Γ^e_a

$$G^e_\alpha = \int_{\Gamma^e_a} g(x_a)\, N_\alpha(x_a)\, d\Gamma \qquad (28)$$

To simplify the numerical bookkeeping procedure, the movement of the surface points is constrained to be along fixed straightlines coincident with the element boundaries. Linear quadrilateral isoparametric elements with four corner nodes are used. Second-order Gauss-Legendre quadrature is used for evaluation of the element integrals in Eqs. (6) and (28). Completely ablated elements are dropped numerically. The temperature of the ablated nodes of a partially ablated element continues to contribute to the numerical solution.

Results and Conclusion

To validate the analysis methods, a one-dimensional problem is first analyzed. A graphite column is subjected to a

CONDUCTION AND ABLATION BY FINITE ELEMENT

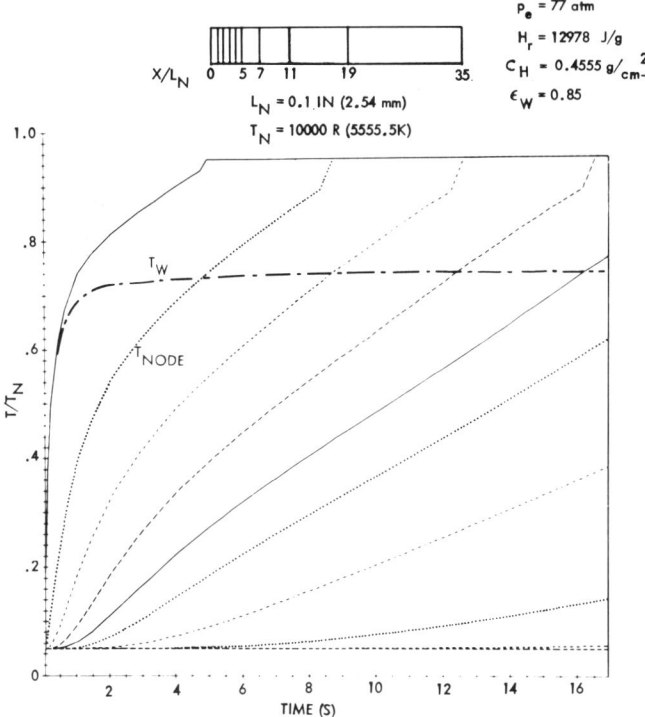

Fig. 2 One-dimensional ablation of graphite, temperatures.

re-entry heating at the left face with the remaining boundaries insulated. A simplified graphite-air chemistry is used for the surface energy balance. The equations used are as follows

$$\xi = \exp(29.994137 - 101055.55/T_w)$$

$$B' = (0.0371\,\xi + 0.174\,p_e) / (p_e - 0.02778\,\xi)$$

$$h_w = (-5454 + (0.8478 + 9.656 \times 10^{-5}\,T_w)\,T_w + (22115 + 1.548\,T_w)\,B') / (1 + B')$$

with T_w in K, p_e in atm, and h_w in J/g (kW-s/kg). To simulate the kinetic effects at low surface temperatures, a linear variation of B' versus T_w is assumed between 1000 K and the temperature corresponding to B' = 0.175, given by

$$T_w(B'=0.175) = 101055.55 / (33.607573 - \ln p_e)$$

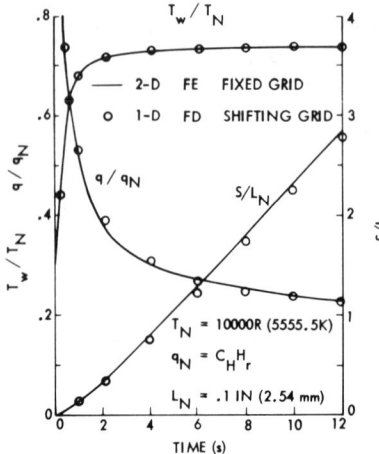

Fig. 3 One-dimensional ablation of graphite, comparison of 2-D FE and 1-D FD results.

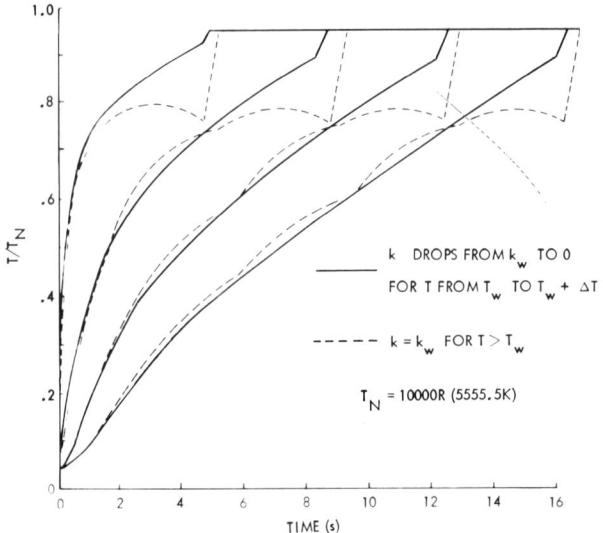

Fig. 4 One-dimensional ablation of graphite, effects of k averaging scheme.

An ablation temperature range $\Delta T = 1$ R (0.556 K) and a time integration parameter $\omega = 0.6$ are used. When an element is dropped because of its being completely ablated, the temperature of the dropped nodes is set to 9500 R (5278 K) for computer plotting purposes. The extended integral averages defined by Eqs. (21-25) are used for both ρc and k (not inverse k).

Figure 2 shows the temperature history at the ablating wall and at the nodal points, with the finite-element grid and pertinent environment variables at the top of the figure. To validate the two-dimensional (2-D) finite-element (FE) solution, the same problem is analyzed by a one-dimensional (1-D) finite-difference (FD) conduction code using a shifting coordinate following the ablating surface.[12] Figure 3 indicates a favorable agreement between the 2-D FE and 1-D FD results.

Changing the ablation temperature range ΔT from 1 R (0.556 K) fo 50 R (27.8 K) produces only a very small effect.

Instead of letting k drop from k_w at T_w to 0 at $T_w + \Delta T$, a separate calculation is made assuming $k = k_w$ for $T > T_w$. The result shows unacceptable wavy temperature histories compared with that obtained by using the extended integral average on k, as indicated in Fig. 4.

Another calculation is made using the extended integral average on 1/k. The results are similar to that given in Fig. 4, with unacceptable wavy temperature histories. Thus, the comparison of the results of Figs. 2 and 4 shows that the extended integral average on k, in contrast to 1/k, should be used for further analysis.

The validity of the analysis method is further assessed with an axisymmetric problem. A 60 deg sphere shell segment

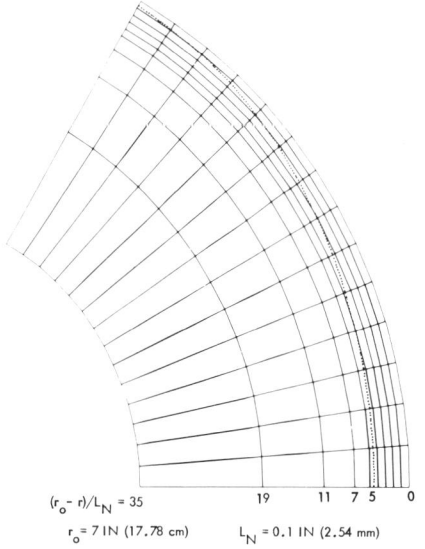

Fig. 5 Sphere shell segment FE grid and ablating surface at 18 s.

of graphite is subjected to the same heating environment, at the stagnation point, as in the first problem. A uniform H_r and a $\cos^2\theta$ distribution of p_e and C_H are imposed on the heating surface. The FE grid and the ablating surface at 18 s are shown in Fig. 5. The same problem is analyzed by a FD code using a deforming grid.[5] Figures 6, 7, and 8 show that the current FE results agree well with the FD solution.

The advantage of the fixed grid FE formulation is demonstrated by considering the ablation and thermal response of an experimental solid rocket nozzle. The FE mesh of the forward movable section of the submerged solid rocket nozzle is given in Fig. 9. Figure 10 shows a computer isotherm plot at 20 s, including a special 600 R (333 K) isotherm. A row of contact resistance elements separating the throat region and the forward transition ring causes a temperature discontinuity there. The internal material boundaries and the location of the in-

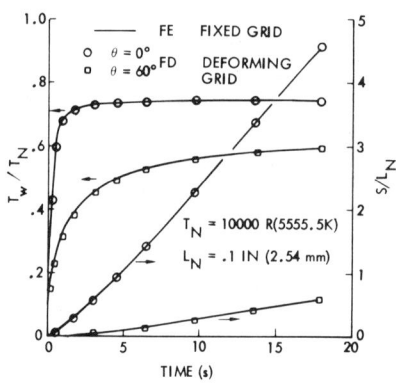

Fig. 6 Sphere shell segment surface temperature and ablation history.

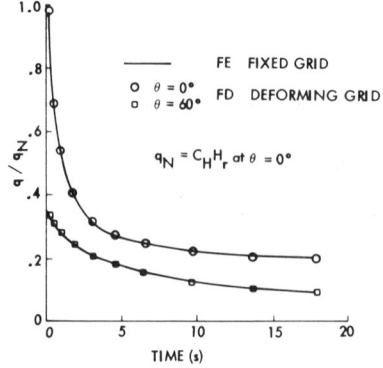

Fig. 7 Sphere shell segment net wall conductive heat fluxes.

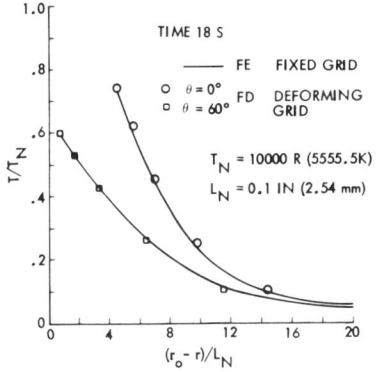

Fig. 8 Sphere shell segment temperature distributions at 18 s.

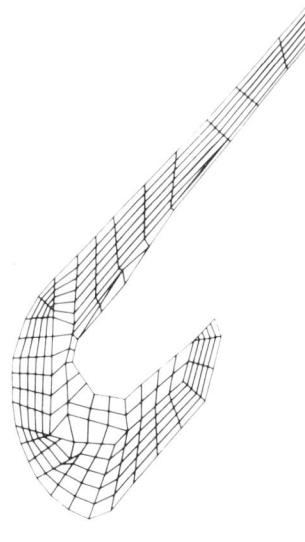

Fig. 9 Nozzle forward section, FE mesh.

Fig. 10 Nozzle forward section, isotherms, and ablating surface at 20 s.

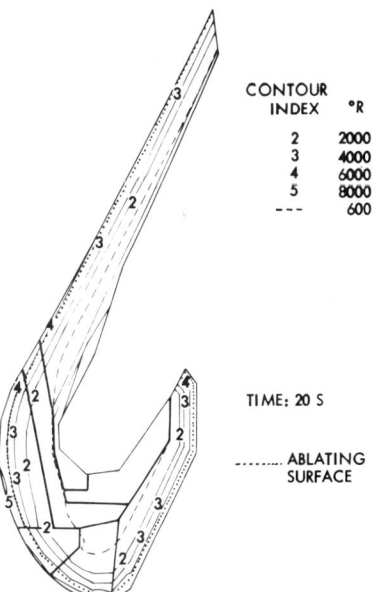

stantaneous surface are noted. Isotherms between the initial and the instantaneous surfaces are fictitious.

It is concluded from this study that the algorithm of extended integral average and moving surface source, in conjunction with a fixed grid finite-element formulation, enables the analysis of ablation and thermal response of problems with multimaterial regions and complex geometries.

Acknowledgments

This work was performed under LMSC Independent Development Program.

References

[1] Ockendon, J. R. and Hodgkins, W. R., Eds., Moving Boundary Problems in Heat Flow and Diffusion, Proceedings of Conference held at University of Oxford, 25-27 March 1974, Clarendon Press, Oxford 1975.

[2] Comini, G., Del Guidice, S., Lewis, R. W., and Zienkiewicz, O. C., "Finite Element Solution of Non-Linear Conduction Problems with Special Reference to Phase Change," International Journal for Numerical Methods in Engineering, Vol. 8, No. 3, 1974, pp. 613-624.

[3] Lewis, R. W. and Morgan, K., Eds., Numerical Methods in Thermal Problems, Proceedings of the First International Conference, University College, Swansea, July 2-6, 1979. Pineridge Press, Swansea, United Kingdom, 1979; Section 2 contains eleven papers on Phase Change.

[4] Chin, Jin H., "Coupling of Shape Change, Heating Distribution and Internal Conduction for Ablating Bodies," AIAA Progress in Astronautics and Aeronautics: Fundamentals of Spacecraft Thermal Design, Vol. 29, edited by John W. Luccas, MIT Press, Cambridge, Mass., 1972, pp. 333-347.

[5] Chin, Jin H., "Shape Change and Conduction for Nosetip at Angle of Attack," AIAA Journal, Vol. 13, No. 5, 1975, pp. 599-604.

[6] O'Neill, K. and Lynch, D. R., "A Finite Element Solution for Porous Medium Freezing, Using Hermite Basis Functions and a Continuously Deforming Coordinate System," Numerical Methods in Thermal Problems, edited by R. W. Lewis and K. Morgan, Pineridge Press, Swansea, United Kingdom, 1979, pp. 548-559.

[7] Lemmon, E. C., "Multidimensional Integral Phase Change Approximations for Finite Element Conduction Codes," to appear in *Recent Numerical Advances in Thermal Problems*, edited by O. C. Zienkiewicz, K. Morgan, and R. W. Lewis.

[8] Morgan, K., Lewis, R. W., and Zienkiewicz, O. C., "An Improved Algorithm for Heat Conduction Problems with Phase Change," *International Journal for Numerical Methods in Engineering*, Vol. 12, No. 7, 1978, pp. 1191-1195.

[9] Zienkiewicz, O. C., *The Finite Element Method*, Third Edition, McGraw-Hill Book Company (UK) Limited, London, 1977.

[10] Bathe, K-J. and Wilson, E. L., *Numerical Methods in Finite Element Analysis*, Prentice-Hall, Inc., Englewood Cliffs, 1976.

[11] Lees M., "A Linear Three-Level Difference Scheme for Quasilinear Parabolic Equations," *Mathematics of Computation*, Vol. 20, No. 96, 1966, pp. 516-622.

[12] "User's Manual, Aerotherm Charring Material Thermal Response and Ablation Program, Version 3," Vol. 1, AFRPL-TR-70-92, April 1970.

THERMAL RESISTANCE OF CONTACTS: INFLUENCE OF OXIDE FILMS

F.R. Al-Astrabadi*
British Aerospace Dynamics Group
Stevenage, Hertfordshire, U.K.

and

P.W. O'Callaghan[†] and S.D. Probert[‡]
Cranfield Institute of Technology,
Cranfield, Bedfordshire, U.K.

Abstract

A theoretical prediction for the thermal resistance of a contact between oxidized nominally flat randomly rough metallic surfaces is developed. Stochastic representations of microtopographies are used and uniform film thicknesses are assumed. The microcontacts produced are then considered as two types: a) metal-to-metal bridges surrounded by contacting annular oxide areas, and b) oxide-to-oxide bridges. The thermal resistance vs applied loading variations are compared with those resulting from an alternative analysis. The estimates are evaluated using newly obtained experimental data for oxidized contacts between EN3B mild steel specimens. For these, measurements of surface topographies and oxide film thicknesses were obtained. The theoretical predictions were found to be in reasonable agreement with experimental measurements.

Nomenclature

a = radius of microcontact spot, m
A = contact area, m^2
b = radius of heat flow channel, m
$g( )$ = constriction alleviation function described by Eq.(20)
k = thermal conductivity, $Wxm^{-1}K^{-1}$
M = micro indentation hardness, Nxm^{-2}

Presented as Paper 80-1467 at the AIAA 15th Thermophysics Conference, Snowmass, Colo., July 14-16, 1980. Copyright © American Institute of Aeronautics and Astronautics, Inc., 1980. All rights reserved.
*Senior Thermal Design Engineer.
†Senior Lecturer in Energy Management.
‡Professor of Applied Energy.

N_A = mean number of microcontacts per unit area, m^{-2}
P = contact pressure, P_a
Q = rate of heat flow crossing the interface, W
R = thermal contact resistance of unit area, $m^2 \times kW^{-1}$
t = normalized mean plane separation, $(=u/\sigma)$
u = mean plane separation, m
W = normal load, N
x = lateral spread of oxide surrounding each metal-to-metal contact spot, Eq. (8), m
y = vertical thickness of oxide, Eq. (0), m
Z = defined by Eq. (31)
ΔT = apparent temperature discontinuity across the interface, K
δ = thickness of surface coating, m
$\Phi(t) = (= \int_0^t \phi(t)\, dt)$
$\phi(t)$ = Gaussian probability function described by Eq.(1)
σ = rms roughness (standard deviation of surface heights), m
Ψ = $(= |\bar{\psi}|/\sigma)$, rad $\times m^{-1}$
ψ = asperity flank slope, rad

Subscripts

1,2 = surfaces 1 and 2 respectively
an = annular type micro contact spots
c = surface coating
A = per unit nominal area
M = metal
N = nominal
0 = oxide
00 = oxide-to-oxide
R = real
S = effective for a contact between dissimilar surfaces
tot = total

The Surface Oxide Problem

There are numerous predictions of the steady state thermal contact resistance (TCR) ensuing when two nominally flat, clean metallic surfaces (having Gaussian distributions of surface heights) are brought into contact under a constant normal applied mechanical load.[1-5]

In general, a reasonably accurate estimate of TCR may be made, providing that:
1) Appropriate descriptions of the surfaces can be obtained.
2) The predominant mode (i.e. either elastic, elasto-plastic

or entirely plastic) of asperity deformation can be established.
3) The thermophysical and mechanical properties of the materials are known.

The actual area of contact occurring under a specified mechanical load may then be estimated, and also the effects of constricting the heat to flow through the resulting array of microcontact bridges may be ascertained.

The presence of surface films, coatings, contaminant layers or interfacial inserts, however, complicates the predictive analysis. Soft, high conductivity inserts have been used to enhance the thermal and electrical conductance of single contacts.[6-13] Multiple layers of hard, thin, low conductivity materials, when inserted at structural joints, produce highly insulative supports.[14-17]

An oxide is, in general, harder and less ductile than its parent metal.[18] Thus the formation of oxides tends to reduce the true metal-to-metal contact for freshly assembled joints, and so increases TCR. Usually such interfaces suffer from oxide contamination, especially at higher temperatures. Thus, in order to attempt to be realistic, it is desirable that the presence of surface oxides should be considered in any prediction of thermal behavior.

Classifications

I) The TCR of metal-to-metal joints between clean surfaces, assembled in a high vacuum and then subjected to a constant normal applied loading, with an invariant thermal current passing across the contact, is considered. Due to the presence of an oxidizing atmosphere, the TCR should decrease as a result of oxide film buildup around the contacting asperities, thereby leading to enhanced annular areas of oxide-to-oxide contact, as well as additional oxide-to-oxide contacts. So the initial array of metal-to-metal microcontacts is supplemented by additional oxide-to-oxide bridges. There exists, however, no previous experimental or other practical evidence to support this contention. The previously published papers (e.g., Ref. 18) imply that TCR <u>always increases</u> with oxide film growth. This appeared reasonable because of the following factors:

i) The contact is seldom subjected to a constant load and heat flux.
ii) Such mechanical and thermal fluctuations result in a 'make-

and-break' contact behavior, allowing the growth of surface oxides to disrupt the metallic contact bridges.
iii) The buildup of oxide in the noncontact regions could prise the surfaces apart and hence break metallic bridges.
iv) The oxide and contaminant formation induces passive transient behavior, encouraging factors ii) and iii) above.

II) When coated surfaces are pressed together, the resulting microcontact spot population differs from that which would occur when clean surfaces are brought into contact under identical conditions. The following ratios influence the thermal contact behavior:

$$\frac{\text{coating hardness}}{\text{metal hardness}} = M_c/M_M$$

If the coating, or interfacial shim, is softer than the base metal, a greater real area of contact will be produced under a specified, loading pressure.

$$\frac{\text{coating thermal conductivity}}{\text{metal thermal conductivity}} = k_c/K_M$$

$$\frac{\text{thickness of coating (or shim)}}{\text{rms roughness of surface}} = \delta/\sigma$$

a) When $\delta \gg \sigma$, assuming that the coating material is incompressible, then the sign of the change in TCR is independent of the value of k_C/k_M. At all loads the TCR exceeds that which results for the bare surfaces (providing that the interfacial film is not extruded out of the contact plane).

b) Provided δ is of the order of or less than σ, then the following apply:

i) If $M_c \ll M_M$, a decrease in TCR will occur for all loads such that the mean plane separation, u, is greater than δ, regardless of the value of k_c/k_M. The conductance at each metal-to-metal contact is enhanced by the additional annular coating-to-coating bridges, as classification I suggests.

Under higher loads, if $u < \delta$, providing that the insert is fully ductile but incompressible, then the TCR should remain invariant as full contact is made when $u = \delta$. In this case, the contact configuration is largely dictated by the hardness of the metal surfaces (except under very light loads when the coating-to-coating contact occurs in isolation), i.e., the insert acts as a filler or gasket.

ii) If $M_C > M_M$, the mean plane separation, and consequently the contact configuration under a particular applied loading pressure, should be determined according to the surface topography and the hardness of the coating. In practice, however, because the coating is often thin and brittle, it can be ruptured during loading, so that metal-to-metal contacts also form. This mechanism is abetted by the work hardening of the metal asperities under stress. Any debris, resulting from relative shearing movements of the contacting surfaces, may increase or reduce the TCR depending upon the relative opposing effects of the ratios M_C/M_M and k_M/k_C, i.e., the ratio $(M_C/M_M)/(k_C/k_M)$. For oxide films, both these factors lead to an increase in the TCR, i.e., $(M_C/M_M)/(k_C/k_M) > 1$. Brittle fracture may result in a reduction of the TCR and this can be somewhat accounted for in level crossing[18] analyses by assuming plastic penetration of the metal asperities through the oxide film. Mean plane separations are however greater than those which would ensue for contacts between clean surfaces, because the separations depend upon both metal and oxide hardnesses and the relative amounts of metal-to-metal and oxide-to-oxide contact areas produced.

The Present Predictions

An existing stochastic prediction[1,15] for TCR is extended here to cope with the presence of interfacial insert materials, particularly in this instance oxide films. The resulting estimates are then compared with those produced by the prediction due to Yip[18] as well as with experimentally measured data.

Assumptions of The Present Analysis

In order to develop even a mathematical model of the contact assembly behavior the following assumptions, although often only crude approximations, are introduced:

i) All microcontact regions are taken to be circular.

ii) The thickness of the oxide film is uniform. Thus the presence of the oxide is assumed not to alter the surface topography from that observed prior to oxidation.

iii) The height distributions of the contacting surfaces are such that they may be described by Gaussian probability functions.

iv) The surface asperities are assumed to be right circular cones which deform in an ideal plastic manner. (The analysis assumes that all the material within intersections of the oxide surface is annihilated as normal loading proceeds.)

v) The oxide films are relatively thin and brittle so that metal-to-metal microcontacts form at mating asperities.

vi) The effective thermal conductivity of an oxide-to-oxide microcontact is given by the harmonic mean of the conductivities of the oxide and metal.

vii) The effective thermal conductivity of a metal-to-metal contact surrounded by an annulus of oxide is given by the arithmetic mean of the conductivities of oxide and metal.

viii) The theory neglects any contact resistance at the oxide metal substrate interface.

Analysis

Figure 1 represents a section through an oxidized surface having a roughness σ and a mean asperity slope $|\bar{\psi}|$: ordinate heights have been normalized using the roughness σ. The mean plane through the oxide surface has an equal spread of ordinates above and below it (i.e., statistically $\pm \sim 3$). An imaginary perfectly flat surface displaced a distance t from

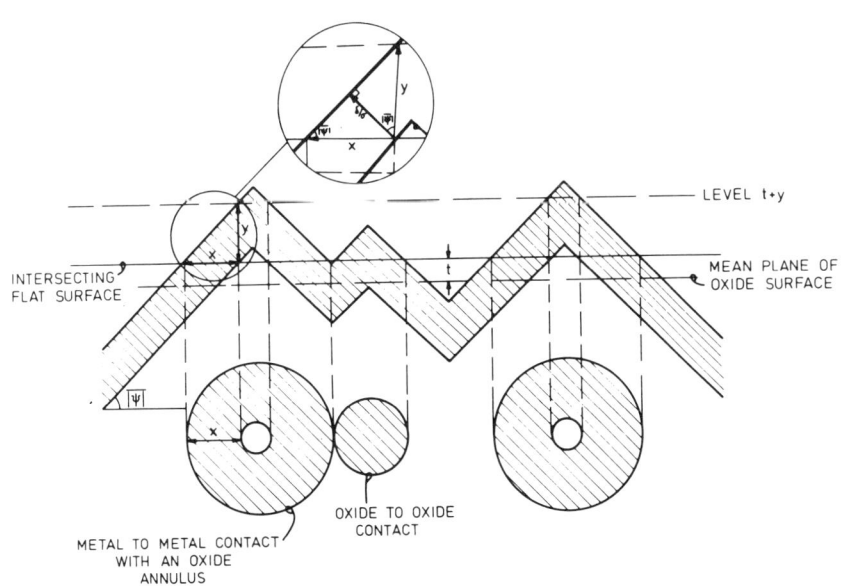

Fig. 1 Idealized contact mechanism.

this mean plane results in circular contact areas at the intersections of the two surfaces.

The total area of real contact is calculated from the probability that $\phi(t)$ ordinates of the rough surface lie at distances greater than t from the mean plane.

If
$$\phi(t) = 1/\sqrt{2\pi}\ e^{-t^2/2} \tag{1}$$

represents the random ordinate distribution, then the area of contact (i.e., the intersecting area) is given by:

$$A_R = \int_t^\infty \phi(t)dt \Big/ \int_{-\infty}^\infty \phi(t)dt \tag{2}$$

and because the ordinates are normalized

$$\int_{-\infty}^\infty \phi(t)dt = 1 \tag{3}$$

$$\therefore A_R = \int_t^\infty \phi(t)dt = \int_0^\infty \phi(t)dt - \int_0^t \phi(t)dt$$

i.e.,
$$A_R = (1/2 - \Phi(t)) \tag{4}$$

where
$$\Phi(t) = \int_0^t \phi(t)dt$$

For a purely plastic deformation, the flow stress or microindentation hardness of the material at the intersections is given by:

$$M = W/A_R\ A_N = P/A_R \tag{5}$$

$$A_R = P/M \tag{6}$$

Substituting Eq.(4) into Eq.(6)

$$\Phi(t) = 1/2 - P/M \tag{7}$$

Referring to Fig.1

$$x = \delta/\sigma\ \sin|\bar{\psi}| \tag{8}$$

and
$$y = \delta/\sigma \cos |\overline{\psi}| \quad (9)$$

The intersecting area at level t,
$$A_{M+O} = 1/2 - \Phi(t) \quad (10)$$

The intersecting area at level $t + y$,
$$A_M = 1/2 - \Phi(t + y) \quad (11)$$

Thus the total area of metal-to-metal microcontacts, A_M. The total area of oxide-to-oxide contacts can then be calculated from
$$A_O = A_{M+O} - A_M \quad (12)$$

The Normalized Mean Plane Separation t

The normal load is distributed over the metal and the oxide microcontact zones. Then
$$W = W_O + W_M = PA_N \quad (13)$$

but
$$A_O = W_O/M_O \text{ and } A_M = W_M/M_M \quad (14)$$

where W_O and W_M are the partial loads borne by the oxide and metal respectively. Substituting from Eq.(14) into Eq.(13) gives
$$A_O M_O + A_M M_M = PA_N \quad (15)$$

Substituting from Eq.(12) into Eq.(15) reveals that
$$(A_{M+O} - A_M) M_O + A_M M_M = PA_N \quad (16)$$

Substituting from Eqs. (10) and (11) into Eq. (16), simplifying and rearranging, results in
$$\left[1/2 - \Phi(t+y) \right] + \left[\Phi(t+y) - \Phi(t) \right] M_O/M_M = P/M_M \quad (17)$$

Thus the normalized and mean plane separation t may be obtained in terms of the nominal applied normal loading pressure P for a given oxide film thickness provided the hardnesses of the oxide and base metal are known, and the surface characteristics are specified.

The Microcontact Parameters

Having ascertained the predicted mean plane separation, the microcontact parameters may be calculated[3,5,15] as follows:

The total mean number of microcontacts per unit nominal area at level t, is

$$\overline{N}_{Atot} = \Psi^2 \pi t \, \phi(t)/8 \qquad (18)$$

The total number of metal-to-metal microcontacts per unit nominal area at level t, is

$$\overline{N}_{AM} = \Psi^2 \pi(t+y) \, \phi(t+y)/8 \qquad (19)$$

Each metal microcontact is enclosed by an oxide region of annular thickness, x. Thus $\overline{N}_{AM}$ is also the number of these annular contacts at level t. The total number of oxide-to-oxide microcontacts per unit nominal area at level t is given by

$$\overline{N}_{AOO} = \overline{N}_{Atot} - \overline{N}_{AM} \qquad (20)$$

The mean radius of all the microcontacts at level t is

$$\overline{a}_{tot} = 2/\pi\Psi t \qquad (21)$$

The mean radius of the metal-to-metal microcontacts at level t is

$$\overline{a}_M = 2/\pi\Psi \, (t+y) \qquad (22)$$

The mean radius of the metal-to-metal microcontacts surrounded by annular oxide load bearing areas at level t is

$$\overline{a}_{an} = \overline{a}_M + x\sigma \qquad (23)$$

Now, within a linear intersection of the surface

$$\overline{N}_{Atot} \, \overline{a}_{tot} = \overline{N}_{AM} \, \overline{a}_{an} + \overline{N}_{AOO} \, \overline{a}_{OO}$$

from which the mean radius of the oxide-to-oxide microcontacts at level t may be deduced,

i.e.,

$$\overline{a}_{OO} = \left(\overline{N}_{Atot} \, \overline{a}_{tot} - \overline{N}_{AM} \, \overline{a}_{an} \right) / \overline{N}_{AOO} \qquad (24)$$

The Thermal Contact Resistance

Having developed the component parts, the estimate of overall thermal contact resistance is relatively easy to synthesize from constrictional analysis, viz

$$R_{00} = (2 \bar{a}_{00} \bar{N}_{A00} k_{00})^{-1} \qquad (25)$$

for oxide-to-oxide microcontacts where, because metal and oxide appear in thermal series

$$2/k_{00} = 1/k_M + 1/k_O \qquad (26)$$

Also

$$R_{an} = (2 \bar{a}_{an} \bar{N}_{AM} k_{an})^{-1} \qquad (27)$$

for the metal-to-metal contacts surrounded by oxide annuli, where, because metal and oxide appear in parallel thermal circuits,

$$k_{an} = (k_M + k_O)/2 \qquad (28)$$

The total thermal contact resistance of unit nominal area is then the result of both the oxide-to-oxide and annular type composite microcontacts in parallel, i.e.

$$1/R_{tot} = 1/R_{00} + 1/R_{an} \qquad (29)$$

The constriction alleviation factor,[19] $g(Z)$, should be applied to R_{tot} to allow for interference to the heat flow among adjacent channels, i.e.,

$$g(Z) = 1 - 1.409Z + 0.296Z^3 - \ldots \qquad (30)$$

where

$$Z = \bar{a}_{tot}/\bar{b}_{tot} \qquad (31)$$

and

$$\bar{b}_{tot} = (A_N/\pi \bar{N}_{Atot} A_N)^{0.5} = (\pi N_{Atot})^{-0.5} \qquad (32)$$

Parameters Describing the Contact

In order to explain the behavior of two surfaces 1 and 2 of the same material, but with differing surface topographies and oxide film thicknesses, which are pressed together, certain compound parameters σ_s, $|\overline{\psi}|_s$, t_s, and δ_s are used in the prediction. This permits the actual contact configuration to be represented as the statistically equivalent assembly between a perfectly flat smooth surface and a rough oxidized surface.

The effective compound rms roughness

$$\sigma_s = (\sigma_1^2 + \sigma_2^2)^{0.5} \tag{33}$$

and the effective compound mean absolute surface slope

$$|\overline{\psi}|_s = (|\overline{\psi}|^2 + |\overline{\psi}|^2)^{0.5} \tag{34}$$

The normalized mean plane separation

$$t_s = u/\sigma_s \tag{35}$$

and the effective compound oxide film thickness

$$\delta_s = \delta_1 + \delta_2 \tag{36}$$

Experimental Measurements

Six contact assemblies were formed between the flat faces of twelve cylindrical, 25 mm diam, 30 mm long specimens of EN3B mild steel. These faces were ground, lapped, and polished in guard holders to optical flatness, cleaned with acetone on a Soni machine, and then with isopropyl alcohol. A controlled degree of random surface roughness was achieved by sand blasting the surfaces in the guard holders. Sets of three identical specimen pairs (two for the purpose of measuring TCR and a third for oxide film thickness measurements) were then oxidized in a) a high temperature furnace, or b) an environmental chamber. Details of the specimen assemblies are given in Table 1.

Surface Topography Measurements

Surface roughnesses, asperity peak curvatures, auto correlations and mean flank slopes of surface asperities were obtained before and after oxidation by digitizing the surface heights information at 2.82 μm intervals using a Talysurf 4 stylus profilometer (see, for example, Figs.2 and 3).

THERMAL RESISTANCE OF CONTACTS INFLUENCE OF OXIDE FILMS 277

Table 1 The specimen assemblies [a]

Assembly	Oxide thickness (μm)	Roughness before oxidation (rms μm)	Roughness after oxidation (rms μm)	Mean surface slope before oxidation (rad)	Mean surface slope after oxidation (rad)	Symbol used in Fig. 5
1. Upper specimen	0.118	1.45	1.57	0.127	0.132	×
Lower specimen	0.118	1.51	1.67	0.127	0.135	
2. Upper specimen	0.0546	1.50	1.57	0.123	0.130	+
Lower specimen	0.0546	1.53	1.67	0.120	0.134	
3. Upper specimen	0.118	1.85	1.98	0.17	0.18	
Lower specimen	0.118	1.90	2.03	0.17	0.19	
4. Upper specimen	0.1	0.12	0.12	0.032	0.038	*
Lower specimen	0.1	0.12	0.123	0.030	0.040	
5. Upper specimen	0 (clean)	0.46	-	0.04	-	●
Lower specimen	Thick, oxidized at 70° and 80% humidity for 72h	0.45	2.28	0.04	0.21	
6. Upper specimen	0 (clean)	0.45	-	0.041	-	○
Lower specimen	0 (clean)	0.44	-	0.042	-	

a. The specimens used in assemblies 1 to 4 were oxidized in a high temperature furnace to produce uniform thicknesses of oxide film. Material) mild steel EN3B; thermal conductivity of the mild steel) 47 W × m^{-1}K^{-1}; thermal conductivity of oxide) 0.875 W × m^{-1}K^{-1}; hardness of the mild steel) 1.67 × 10^9 N × M^{-2}; hardness of oxide) 3.51 × 10^9 N × m^{-2}; diameter of specimens) 25.0 ± 0.1 mm; length of specimen) 30.0 ± 0.1 mm.

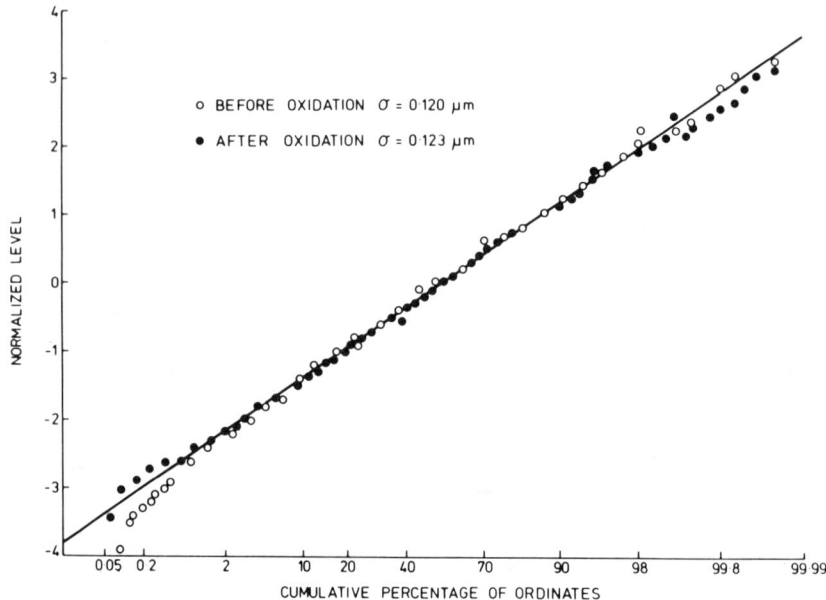

Fig.2 Ordinate height distribution for a typical contacting surface having a thin oxide coating.

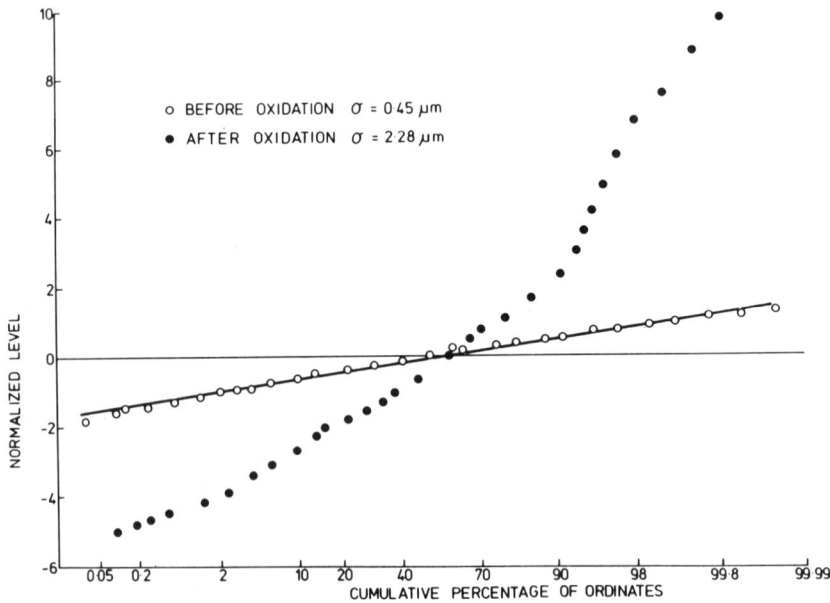

Fig. 3 Ordinate height distribution for a heavily oxidized surface.

THERMAL RESISTANCE OF CONTACTS INFLUENCE OF OXIDE FILMS

The data were transferred to punch tape using a Solartron data logger, and analyzed using a computer program described elsewhere.[15] Distributions and mean values for the surface parameters were thus obtained. Values are listed in Table 1.

Oxide Film Thicknesses

Film thicknesses were measured by cutting 2.5 mm x 15 mm samples from one specimen of each set using a spark erosion

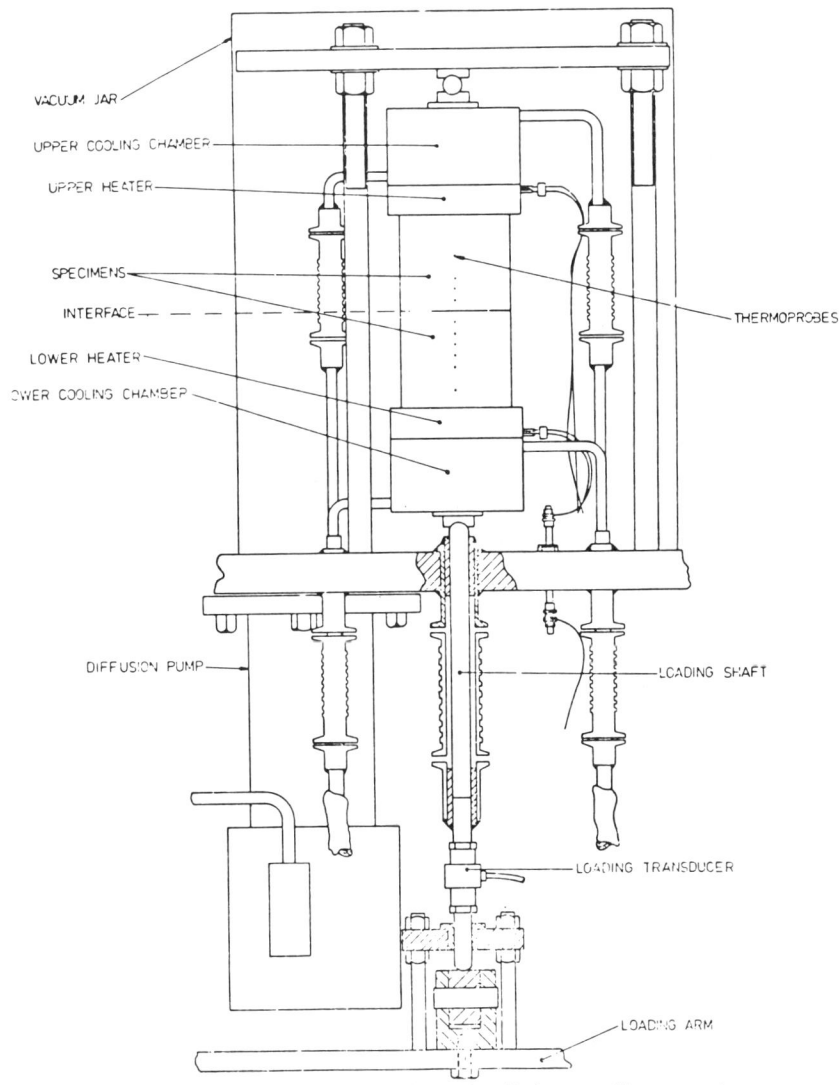

Fig.4 The evacuated longitudinal heat flow system.

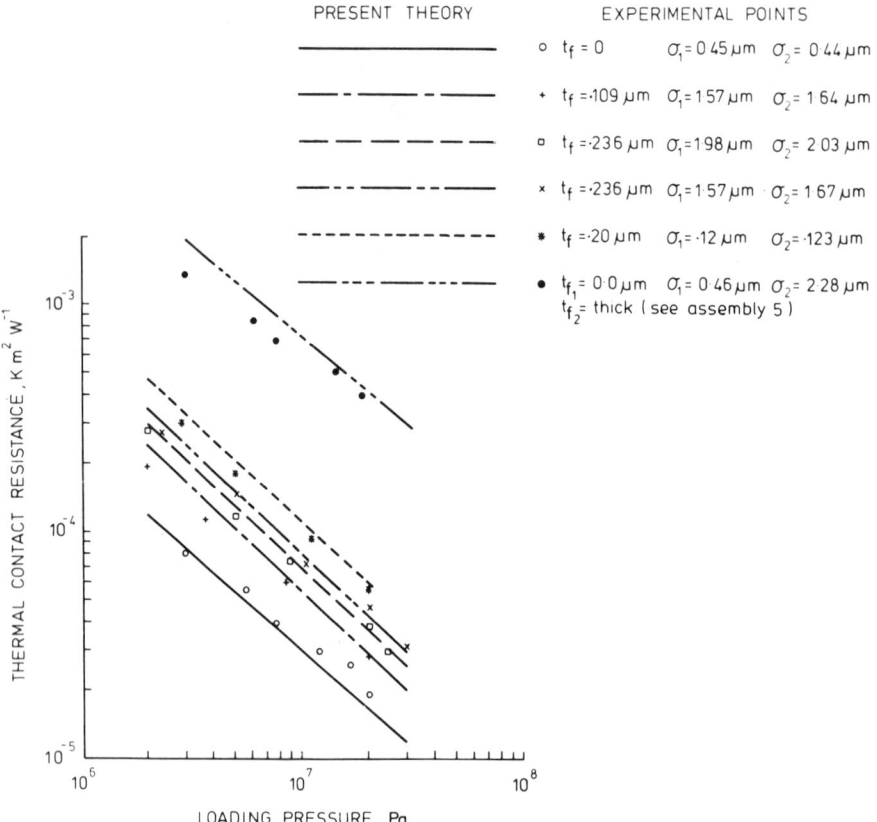

Fig.5 Comparison of the predictions with measured data.

machine. These samples were then viewed with 1) a Stereoscan electron microscope for the smoother surfaces or 2) a Reichart metallographic microscope for the rougher surfaces.

The oxide layers formed under high humidity conditions in the environmental chamber were grossly nonuniform and for this reason it was not possible to estimate their mean thicknesses using the metallographic microscope alone. The measured thicknesses are provided in Table 1.

Thermal Contact Resistance

The thermal interface resistances at the contact planes were measured using an evacuated, longitudinal, one-dimensional, heat flow system (Fig.4), which has been described fully elsewhere.[5,15] Normal loads were applied via a dead-

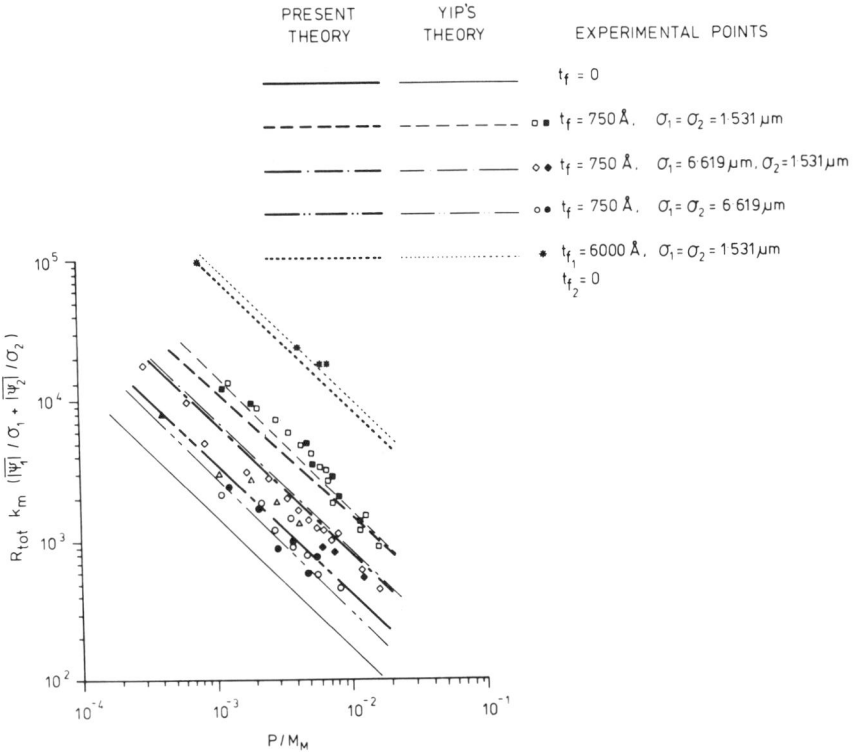

Fig.6 Comparison of the present prediction with that due to Yip.[18]

weight lever system. The total interface resistance for each contact under a particular applied load was calculated from an estimate of the steady state temperature difference across the interface by extrapolating axially placed thermojunction indications to the surface plane, and averaging the heat flows through each contacting specimen, i.e.,

$$R_{tot} = A_N \Delta T/Q \qquad (37)$$

The measured values are plotted in Fig.5 together with predictions from the theory. The accuracy of the thermal contact resistance measurements was within 5%.[5,20]

Discussion and Conclusions

The oxide films formed in the high temperature furnace had little effect on the surface topographies measured prior to oxidation. It may be seen from Fig.2 that the ordinate height distribution remained Gaussian after oxidation. This was so for all the surfaces prepared in the furnace. The thick and irregular oxide film grown in the environmental chamber, however, caused a redistribution of surface heights which led to severe skewness (see Fig.3), manifested in a fivefold increase in surface roughness.

The presence of oxide films always increased the thermal contact resistance in accordance with the prediction outlined (see Fig.5). The prediction analysis for assembly 5 differed slightly in that the rough, hard, thick oxide surface was assumed to penetrate the relatively softer surface without brittle fracture occurring.

Figure 6 superimposes predictions from the present theory upon the experimental results and analytical estimates due to Yip. It may be deduced that either theory is capable of predicting the phenomenon to an acceptable degree. The present theory does not, however, require an iterative computational procedure. It should be noted that the experimental data were obtained for the first loading of fresh contacts. Thus no elastic effects were isolated.

References

[1] Al-Astrabadi, F.R., O'Callaghan, P.W., and Probert, S.D., "Effects of Surface Finish on Thermal Contact Resistance between Different Materials," AIAA Paper No.79-1065, June 1979, Orlando.

[2] Jones, A.M., O'Callaghan, P.W., and Probert, S.D., "Prediction of Contact Parameters from the Topographies of Contacting Surfaces," Wear, Vol. 31, May 1975, pp 89-107.

[3] Tsukizoe, T. and Hisakado, T., "On the Mechanism of Contact Between Metal Surfaces," Transactions of the ASME, Vol. 87D, Sept. 1965, pp. 666-674.

[4] Yovanovitch, M.M., "Thermal Contact Conductance of Turned Surfaces," AIAA Paper 71-80, Jan. 1971,

[5] O'Callaghan, P.W. and Probert, S.D., "Thermal Resistance and Directional Index of Pressed Contacts between Smooth, Non-wavy Surfaces," Journal of Mechanical Engineering Science, Vol. 16, 1974, pp 41-55.

[6] Yovanovitch, M.M., "Effects of Foils upon Joint Resistance - Evidence of Optimal Thickness," AIAA Paper 72-282, 1972, San Antonio.

[7] Barzelay, M.E., Tong, K.N., and Holloway, G.F., "Effect of Pressure on the Thermal Conductance of Contact Joints," NASA Report No. TN-3295, May 1955.

[8] Cunnington, G.R., "Thermal Conductance of Filled Aluminium and Magnesium Joints in a Vacuum Environment," ASME Paper No. 64WA/HT - 40, 1964.

[9] Fried, E., "Thermal Joint Conductance in a Vacuum," ASME Paper No. 63-AHGT-18, 1963.

[10] Fried, E., "Study of Interface Thermal Contact Conductance," Summary Report, Contract No. NAS8-11247, GE Document No. 655D 4395, 1965.

[11] Fried, E., and Kelley, M.J., "Thermal Conductance of Metallic Contacts in a Vacuum," AIAA Paper No. 65-661, Monterey.

[12] Fry, E.M., "Measurements of Contact Coefficients of Thermal Conductance," AIAA Paper no. 65-662, 1965, Monterey.

[13] Cetinkale, T.N. and Fishenden, M., "Thermal Conductance of Metal Surfaces in Contact," *Proceedings of the International Conference on Heat Transfer,* 1951, pp 271-275.

[14] Al-Astrabadi, F.R., O'Callaghan, P.W., Jones, A.M., and Probert, S.D., "Thermal Resistance Resulting from Commonly-used Inserts between Stainless Steel, Static, Bearing Surfaces," *Wear*, Vol. 40, No. 3, 1977, pp 339-350.

[15] O'Callaghan, P.W., "Some Aspects of Contact between Solids," PhD Dissertation, University of Wales, Aug. 1971.

[16] Hargadon, J.M., "Thermal Interface Conductance of Thermoelectric Generator Hardware," ASME Paper No. 66-WA/NE-2, 1965.

[17] Fletcher, L.S. and Miller, R.G., "Thermal Conductance of Gasket Materials for Spacecraft Joints," *Thermophysics and Spacecraft Thermal Control, Progress in Astronautics and Aeronautics*, Vol. 35, edited by R. C. Hering, MIT Press, Cambridge, 1974, pp. 335-349.

[18] Yip, F.C., "Effect of Oxide Films on Thermal Contact Resistance," Heat Transfer with Thermal Control Applications, Progress in Astronautics and Aeronautics, Vol. 39, edited by M. M. Yovanovitch, MIT Press, Cambridge, 1975, pp. 45-64.

[19] Roess, L.C., "Theory of Spreading Resistance," appendix to Neills, N.D. and Ryder, E.A., "Thermal Resistance measurements on Joints formed between Stationary Metal Surfaces," Paper presented at Semiannual Meeting, ASME, Heat Transfer Division, 1948, Milwaukee.

[20] O'Callaghan, P.W. and Probert, S.D., "Effects of Transverse Losses on Longitudinal Heat Transport Observations," Measurement and control, Vol. 4, No. 3, 1971, pp. 25-34.

THERMAL CONTACT CONDUCTANCE OF COATED MULTI-LAYERED SHEETS

J. W. Sheffield*
University of Missouri-Rolla, Rolla, Mo.

and

T. N. Veziroglu[+] and A. Williams[≠]
University of Miami, Coral Gables, Fla.

Abstract

Steady state thermal contact conductances of multi-layered electrically insulated sheets are examined theoretically and experimentally. Tests have been conducted for twelve selected combinations of sheet material and surface coatings in environments of vacuum and low pressure helium. Comparisons of previously reported theories with the new planar contact theory show good agreement for small values of the ratio of the dimensionless conductivity number, $K = k_f/k_s$, to the dimensionless gap number, $B = \delta/a$, corresponding to the cases of low fluid gap conductivities, such as those existing under vacuum conditions. The new circular contact theory is used to predict the thermal contact conductance and is compared with experimental results.

Nomenclature

A	= constant
a	= contact element radius or half width
$B=\delta/a$	= gap number
b	= solid contact radius or half width
$C=b/a$	= constriction number

Presented as Paper 80-1468 at the AIAA 15th Thermophysics Conference, Snowmass, Colorado, July 14-16, 1980. Copyright © American Institute of Aeronautics and Astronautics, Inc., 1980. All rights reserved.
 *Assistant Professor, Mechanical and Aerospace Engineering.
 +Professor, Mechanical Engineering.
 ≠Adjunct Professor, Mechanical Engineering, Also Associate Professor, Monash University, Clayton, Victoria, Australia.

$D=L/a$ = thickness number
$f(r)$ = admissible function for temperature along gap
h_f = effective heat-transfer coefficient for interstitial media
J_0 = zeroth order Bessel function of the first kind
J_1 = first order Bessel function of the first kind
$K=k_f/k_s$ = conductivity number; transform kernel
k_c = thermal conductivity of coating
k_f = thermal conductivity of interstitial fluid
k_s = thermal conductivity of solid
L = half thickness of a single sheet
M = Meyer hardness
N = number of sheets in a stack
P_c = apparent contact pressure
T = temperature
T_0 = temperature at midplane
t_c = coating thickness
$U=U_{ca}/k_s$ = thermal conductance number
U_c = thermal contact conductance
x,y = Cartesian coordinates
r,z = cylindrical coordinates
δ = interstitial fluid thickness

Introduction

Stacks of flat, coated, thin sheets of metal, clamped together, are used in most electrical transformers, generators and motors. The steady-state temperatures attained within such an electrical device will usually confine the energy conversion efficiency by limiting the allowable output. These electrical devices have intrinsically low, effective thermal conductivities across the lamination due to the large number of contact interfaces occurring in series, and also due to the electrically insulating coating. Such low thermal conductivities cause an increase in thermal slopes in transferring the heat produced by magnetic and electrical losses within the electrical machinery.

Historically, the heat flow through stacks of lamination has been explored as an adjunct to studies of thermal contact conductance. Imperfect contacts between adjacent faces in stacks of lamination cause thermal contact resistance. The main parameters affecting the resistance are the clamping pressure, the hardness of the metals, the geometry, the surface roughnesses, the thermal properties of the metals in contact and of the interstitial fluid. In the case of stacks laminated with thin sheets of metal having very thin and relatively soft coatings, the true metal-to-metal contacts are destroyed and/or reduced in number.

This paper presents an experimental and theoretical examination of the steady-state thermal contact conductance of multi-layered electrically insulated sheets. The new theory developed herein is used to predict the thermal conductances of stacks of lamination over a wide range of contact pressures and with various interstitial fluid conditions. The validity of the new circular theory was confirmed by recently obtained experimental results. Comparisons of earlier theories with this new theory show satisfactory agreement.

Literature Survey

Earlier studies on contact heat transfer of multi-layered stacks of thin layers have addressed applications ranging from mechanically strong thermal insulators for cryogenic tank supports[1] to those in electrical machinery.[2-7] Thomas and Probert[1] investigated the thermal resistances of multi-layered contacts under static loads as an application of cryogenic thermal insulation. Their conclusions suggest that for the design of insulation supports one should select a material possessing low thermal conductivity, a low number of surface asperities per unit area, and a high value of hardness. In addition, a large number of layers was recommended for the insulator.

Hargadon[2] measured the thermal interface conductance of materials used in the hardware of thermoelectric generators. Special tests were conducted to evaluate the effect of interstitial material on the interface temperature gradient.

Williams[3] introduced an analytical model for the thermal behavior of stacks of enameled sheets. His model relies on the separation of the components of contact conductance attributable to the solid contacts and to the interface fluid. Measurements of the thermal conductivities of stacked laminations were reported by Williams[4] in support of his model for a variety of materials in environments of air, vacuum, and helium. The clamping pressure was varied over a range of 0.340 - 1.03 MPa (50. - 150. psi) to represent conditions typical for turbogenerators. Williams concluded that the smoothness of the lamination played a prominent part in the contact conductance for normal atmospheric conditions. However, as the conductivity of the interstitial fluid increased, the thermal load was carried mainly by the interstitial fluid rather than by the solid contacts.

Jhamnani[5] measured the total thermal resistance of stacks of thin electrical steel laminations. Both grain-oriented and non-oriented electrical steels were tested in air environments and under high vacuum. Although Jhamnani examined stacks containing one, three, five and six discs in lamination, he presented his results as thermal resistance per interface. However, in general, the resistance per contact depends upon the number of layers present. Hence, the results were not generalized. Jhamnani noted a tendency for the thermal resistance per contact to become less dependent upon the number of layers as the number of layers increased, especially for high contact pressure.

Al-Astrabadi et al.[6,8] proposed a simple correlation for resistances of stacks of thin layers under compression. The least-squares straight-line power law fit was made for all available published data and the resulting relationship had a correlation coefficient of 0.95.

Sheffield et al.[7] presented the results of an experimental investigation on thermal contact conductance of multi-layered electrically insulated sheets. The test assemblies were made from pieces of material supplied by manufacturers of electrical machinery so the test measurements are truly representative of current practice, with respect to plate thicknesses, surface finishes, roughness, coatings, etc. The measurements at vacuum conditions confirmed that the major parameter in determining

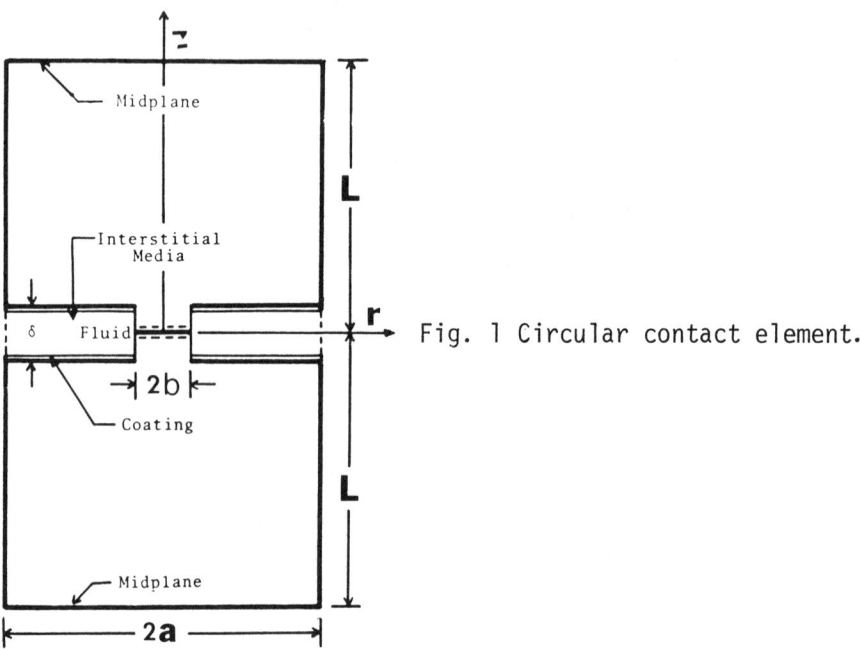

Fig. 1 Circular contact element.

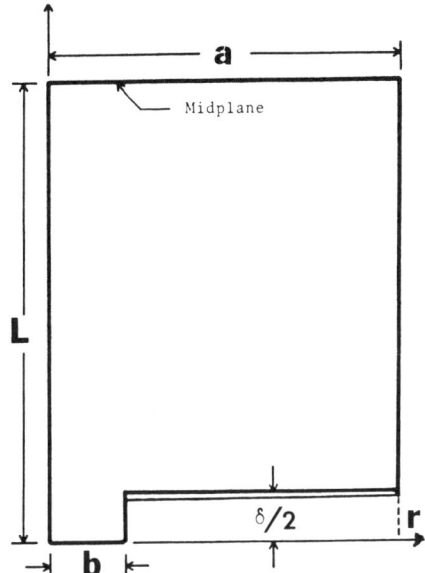

Fig. 2 Quarter element used for evaluation of temperature distribution.

the effective thermal conductivity across the laminations is the thermal contact conductance.

Theory

For the theoretical study the contact interfaces in a stack of lamination are assumed to be composed of a number of similar contact elements. These contact elements have a centered, load bearing contact region surrounded by a non-contact region. For simplicity, the contact elements are assumed to be made up of two solid plates, of thickness 2L each, and having a circular contact in the middle of their surfaces which face each other, with the gap between them filled with a fluid of uniform conductance (see Fig. 1). If the r-axis divides the gap, δ, into two equal gaps of thickness, $\delta/2$, then the system becomes symmetrical with respect to the r-axis from the heat transfer point of view. Similarly, the z-axis divides the contact element width, 2a, into two equal halves. Thus it suffices to consider a one quarter region only, as illustrated in Fig. 2.

The steady-state temperature distribution in the plate must satisfy LaPlace's equation:

$$\frac{1}{r}\frac{\partial}{\partial r}\left(r\frac{\partial T}{\partial r}\right) + \frac{\partial^2 T}{\partial z^2} = 0 \qquad (1)$$

The boundary conditions of the problem can be written as follows:

$$\frac{\partial T}{\partial r}(0,z) = 0 \quad \text{for } 0<z<L \tag{2}$$

$$\frac{\partial T}{\partial r}(a,z) = 0 \quad \text{for } 0<z<L \tag{3}$$

$$T(r,L) = T_0 \quad \text{for } 0<r<a \tag{4}$$

$$T(r,0) = 0 \quad \text{for } 0<r<b \tag{5}$$

$$T(r,0) = f(r) \quad \text{for } b<r<a \tag{6}$$

where T_0 is the uniform temperature at the midplane of the solid plate and $f(r)$ is the unknown temperature profile along the gap. Equation (6) is a new model replacing the conventional convective boundary condition. Thus, the new formulation eliminates the mixed boundary conditions which are more difficult to handle. To solve Eq. (1) subject to boundary conditions given by Eq. (2-6), the temperature is first scaled by T_0 such that

$$\Theta(r,z) = T(r,z) - T_0 \tag{7}$$

An integral transform is used to remove the partial derivative with respect to "r." The integral transformation and inverse transformation are given by

$$\overline{\Theta}(\beta_m,z) = \int_0^a r' K(\beta_m,r') \Theta(r',z) \, dr' \tag{8}$$

$$\overline{\Theta}(r,z) = \sum_{m=0}^{\infty} K(\beta_m,r) \overline{\Theta}(\beta_m,z) \tag{9}$$

where $K(\beta_m,r)$ is the transform kernel and β_m's are the corresponding eigenvalues. The kernel is given by[9]

$$K(\beta_m,r) = \begin{cases} \sqrt{\frac{2}{a}} \frac{J_0(\beta_m r)}{J_0(\beta_m a)} & \text{for } m = 1,2,3,\ldots \\ \sqrt{\frac{2}{a}} & \text{for } m = 0 \end{cases} \tag{10}$$

where β_m's are the positive roots of

$$\beta_m J_1(\beta_m a) = 0 \quad \text{for } m = 0,1,2,\ldots \tag{11}$$

The general solution for this formulation is given by

$$T(r,z) = \frac{T_0 z}{L} + \frac{2}{a^2}\left(1 - \frac{z}{L}\right)\int_b^a rf(r)\,dr$$

$$+ \frac{2}{a^2}\sum_{m=1}^{\infty} \frac{\sinh(\beta_m[L-z])}{\sinh(\beta_m L)} \cdot \frac{J_0(\beta_m r)}{J_0^2(\beta_m a)} \int_b^a rf(r) J_0(\beta_m r)\,dr \qquad (12)$$

Various admissible functions for the unknown temperature, $f(r)$, were assumed. After comparison of the solutions with previously reported theories, a Bessel temperature profile was selected for further analysis. Likewise the analogous planar contact problems were formulated, solved, and evaluated. The unknown constant of the Bessel temperature profile for the circular contact problem was selected as

$$f(r) = A \cdot \left(J_0(C) - J_0\left(\frac{r}{a}\right)\right) \qquad (13)$$

where A is a constant. It should be noted that this temperature distribution satisfies the end condition of $f(r) = 0$ when $r = b$. The constant A is determined by a heat flow balance across the fluid gap as follows:

$$\int_b^a h_f f(r)\Big|_{z=0} r\,dr = \int_b^a k_s \frac{\partial T}{\partial z}\Big|_{z=0} r\,dr \qquad (14)$$

where h_f is the gap conductance and k_s is the solid thermal conductivity. The contact conductance, U_c, was then found to be given by

$$U_c = \frac{h_f}{1-C^2} - \frac{4k_s C \sum_{m=1}^{\infty}\left(\frac{\alpha_m J_1(\beta_m b)}{J_0^2(\beta_m a)} \coth(\beta_m L)\right)}{a^3(1-C^2)[2CJ_1(C) - 2J_1(1) + (1-C^2)J_0(C)]} \qquad (15)$$

where $\alpha_m =$

$$\frac{bJ_0(C)J_1(\beta_m b) - ab\beta_m J_0(\beta_m b)J_1(C) + a^2\beta_m J_0(\beta_m a)J_1(1)}{\beta_m(\beta_m^2 a^2 - 1)} \qquad (16)$$

and C, the constriction number, is given by

$$C = b/a \qquad (17)$$

To obtain a closed form solution, the unknown constant "A" was determined by utilizing Eq. (14) for the limiting case as the constriction number, C, approached zero. The resulting contact conductance was obtained as

$$U_c = \frac{\left(h_f + \frac{k_s}{L}\right)\left(1 - 2J_1(1)\right)}{[(1-C^2)J_0(C) - 2J_1(1) + 2CJ_1(C)]} \quad (18)$$

In practical applications the constriction number tends to have a small value justifying this limiting case.

For comparison, the solution for a planar contact conductance formulation is

$$U_c = \frac{h_f}{(1-C)^2} + \frac{k_s}{L}\left(\frac{1}{(1-C)^2} - 1\right) \quad (19)$$

for the special case of small constriction number, C, and a linear temperature profile for $f(x)$:

$$f(x) = A \cdot (x - b) \quad (20)$$

Experiments

A detailed description of the test apparatus and the experimental procedure used to measure the thermal contact conductance is given in Ref. 7. For each test material, measurements of surface roughness and waviness using a profilometer were made. Thicknesses of the insulating surface coatings were checked using a thickness gauge. In an attempt to find an effective material hardness of the coated electrical steel sheets, a micro-hardness tester was used. Numerous loads were applied so that the effective hardness as a function of penetration depth could be estimated. Reduced loads were used in order to obtain penetration depths less than the surface coating thickness.

Circular discs were punched from the samples. The diameters of the discs were measured using a micrometer after filing the burrs produced during the punching process. Fifty-one discs formed a typical stack of lamination. Three thermocouples were inserted at equal axial spacings. The stack was then wrapped with several layers of thin asbestos cloth making a small sub-assembly for testing.

The sub-assembly was then placed in the thermal contact conductance testing apparatus and the experiments were conducted in a vacuum and low pressure helium using different loads.

Comparison of Present Solutions with Existing Theories

For comparison of the solutions for the thermal conductances of contacts having interstitial fluids, the earlier theories given in Refs. 10, 11, 12, and 13 have been rewritten in terms of the dimensionless numbers. The percentage of deviation of earlier theories (U) from the present theories (U_0), defined as $100 \cdot (U - U_0)/U_0$, has been calculated for several combinations of constriction number, C, and the ratio of the conductivity number, $K = k_f/k_s$, to the dimensionless gap number, $B = \delta/a$. The results are presented in Tables 1 and 2. A study of Table 1 shows that the earlier theories have minimal

Table 1 Comparison of planar contact solutions with existing theories

C	K/B	U_0	% deviation from present solution = $100(U - U_0)/U_0$	
			Ref. 10	Ref. 11
0.1	10^{-4}	0.117	28.2	3.5
	10^{-2}	0.130	23.1	1.9
	1	1.35	-3.7	0.5
	10^2	124.0	-7.6	-18.4
0.3	10^{-4}	0.521	-13.6	-24.3
	10^{-2}	0.541	-9.8	-22.7
	1	2.56	-14.1	-30.6
	10^2	205.0	-44.0	-31.0
0.5	10^{-4}	1.50	-5.33	-37.5
	10^{-2}	1.54	-6.50	-38.2
	1	5.50	-49.9	-57.6
	10^2	401.0	-65.5	-66.3
0.7	10^{-4}	5.06	-25.1	-37.2
	10^{-2}	5.17	-26.3	-38.1
	1	16.20	-63.1	-67.5
	10^2	1120.0	-81.7	-82.0
0.9	10^{-4}	49.5	-32.6	-36.2
	10^{-2}	50.5	-33.9	-37.3
	1	149.0	-73.7	-75.0
	10^2	10000.0	-94.1	-94.3

Table 2 Comparison of circular contact solutions with existing theories

C	K/B	U_0	% deviation from present solution $= 100(U - U_0)/U_0$	
			Ref. 12	Ref. 13
0.1	10^{-4}	0.0107	385.0	167.7
	10^{-2}	0.0208	182.3	62.2
	1	0.103	-----	47.7
	10^2	102.0	10.2	-1.7
0.3	10^{-4}	0.109	66.8	22.7
	10^{-2}	0.121	68.2	19.9
	1	1.33	33.6	-2.9
	10^2	122.0	15.9	-17.6
0.5	10^{-4}	0.408	164.2	125.2
	10^{-2}	0.426	153.6	121.1
	1	2.22	18.6	-11.8
	10^2	182.0	-12.4	-44.3
0.7	10^{-4}	1.51	9.7	133.7
	10^{-2}	1.54	8.1	126.5
	1	5.51	-38.9	-17.9
	10^2	402.0	-56.4	-74.5
0.9	10^{-4}	14.4	-58.8	21.5
	10^{-2}	14.6	-61.0	20.3
	1	44.0	-80.5	-48.9
	10^2	2980.0	-89.8	-82.3

deviations for small constriction numbers. An important note to be made here is that the results from Refs. 11, 12, and 13 are solutions for semi-infinite circular contact elements, i.e., $D = \infty$ and those from Ref. 10 are for finite circular contact elements applicable for laminated stacks. For comparison the current theories were evaluated for $D = 2$ showing satisfactory agreement of the planar contact formulation. The circular contact solutions, however, have significant deviations from the results of Ref. 12 and 13.

The utility of modelling the mixed boundary conditions of the contact plane by an admissible temperature distribution at the fluid gap surface is confirmed by the satisfactory agreement shown in Table 1. The circular contact element formulation will be shown in a later section to have good predictive capabilities for the problem posed in this investigation. Thus, this circular theory was selected as the most appropriate formulation.

Comparison of Theory with Experiments

Figures 3-8 show the measured and predicted values using the circular contact formulation of the thermal contact conductance for stacks of laminations of electrically insulated sheets of electrical steels. They indicate good agreement. The standard AISI (American Iron and Steel Institute) notation for the electric steel is used for referencing the test materials. The experimental value of contact conductance was derived from the temperature drop directly attributable to contacts by subtracting from the overall temperature drop across the test specimen that portion due to the heat flow conducted through the solid components. Each set of experimental values was measured at steady-state thermal conditions for each loading increment. Figures 3-8 refer to an environment of the test specimens of "In Vacuum"; the minimum absolute pressure which could be attained was usually approximately 200 microns of mercury, and thus there is still some conduction heat transfer across the gap region.

Evaluation of the contact conductance from Eq. (18) requires four parameters, C, h_f, k_s, and L. The average half thickness of a single sheet was determined by halving the measured thickness of a sheet. The thermal conductivities of the solids were obtained directly from the manufacturer's specification of the electrical steel cores. The effective coefficient of heat transfer for the interstitial media was defined as

$$h_f = k_f/(\delta/2 + \frac{k_f}{k_c} t_c) \qquad (21)$$

where k_f is the interstitial fluid conductivity, δ is the interstitial fluid thickness, k_c is the thermal conductivity of the coating, and t_c is the coating thickness. The thermal conductivity of the interstitial air was calculated from the following expression given in Ref. 14

$$k_f = 229.4 \, P_{vac} \, \delta/\sqrt{T} \qquad (22)$$

where P_{vac} is the vacuum pressure in lbf/ft^2, T is the fluid temperature in $^\circ R$, and δ is the gap thickness. Reference 15

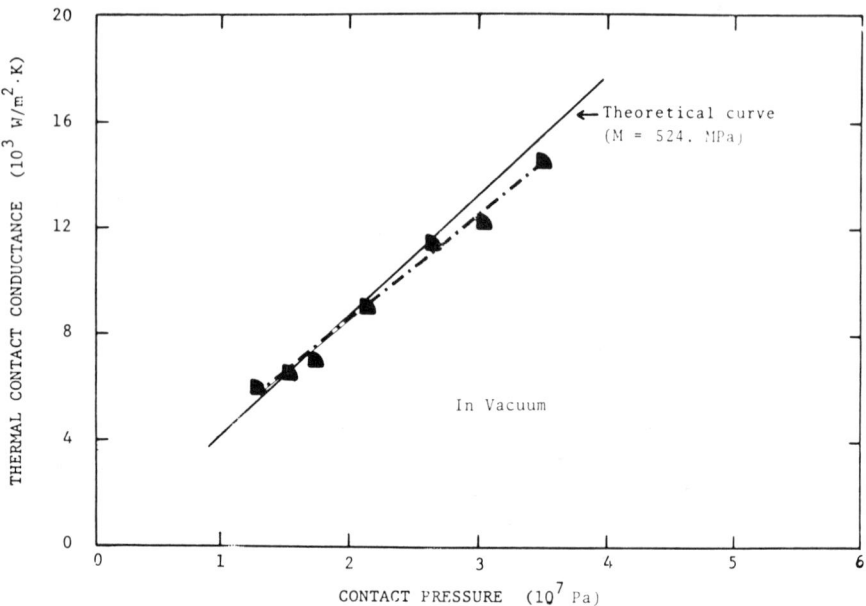

Fig. 3 Contact conductance vs pressure (Westinghouse Electric Company, oriented core/0.35 mm, oxide/5.1 μm).

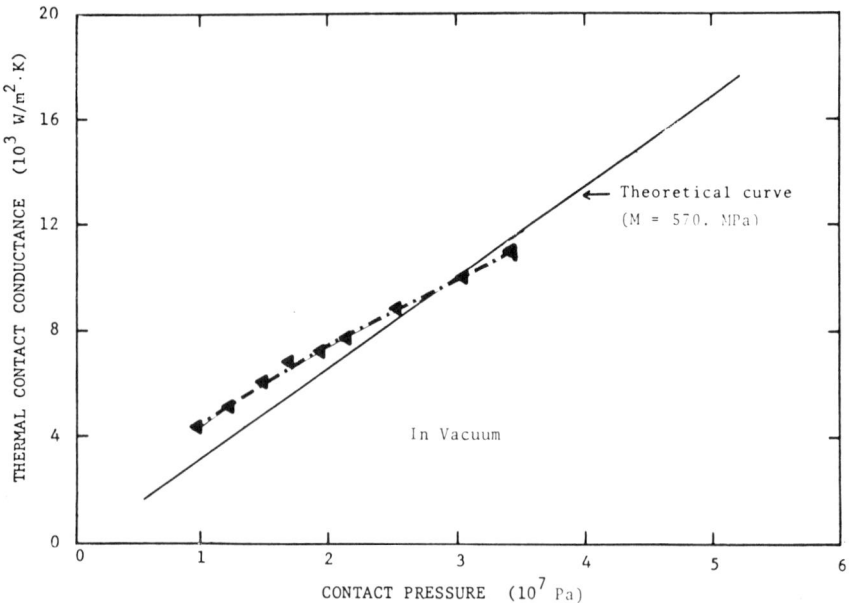

Fig. 4 Contact conductance vs pressure (Westinghouse Electric Company, M-22/0.44 mm, oxide/<1.0 μm).

THERMAL CONTACT CONDUCTANCE OF SHEETS

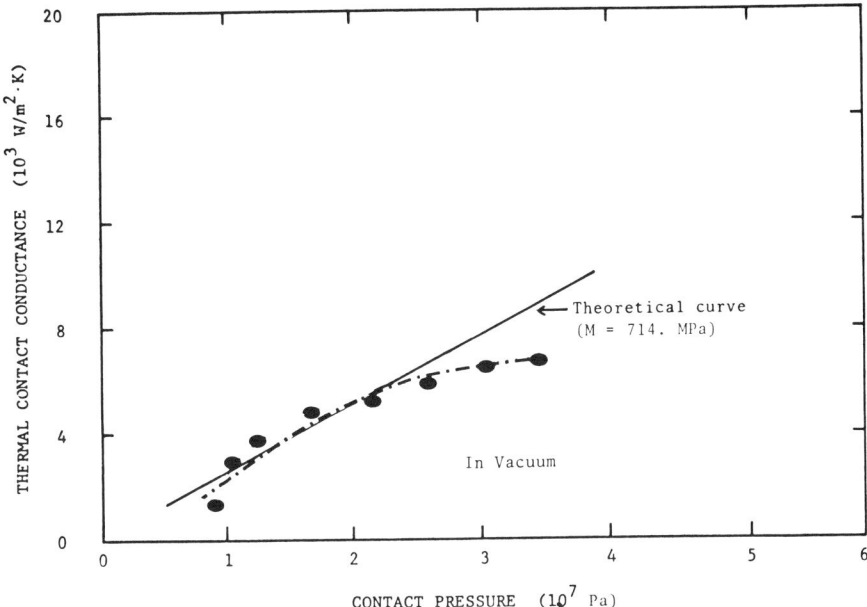

Fig. 5 Contact conductance vs pressure (Westinghouse Electric Company, M-22/0.44 mm, carlite/10.2 μm).

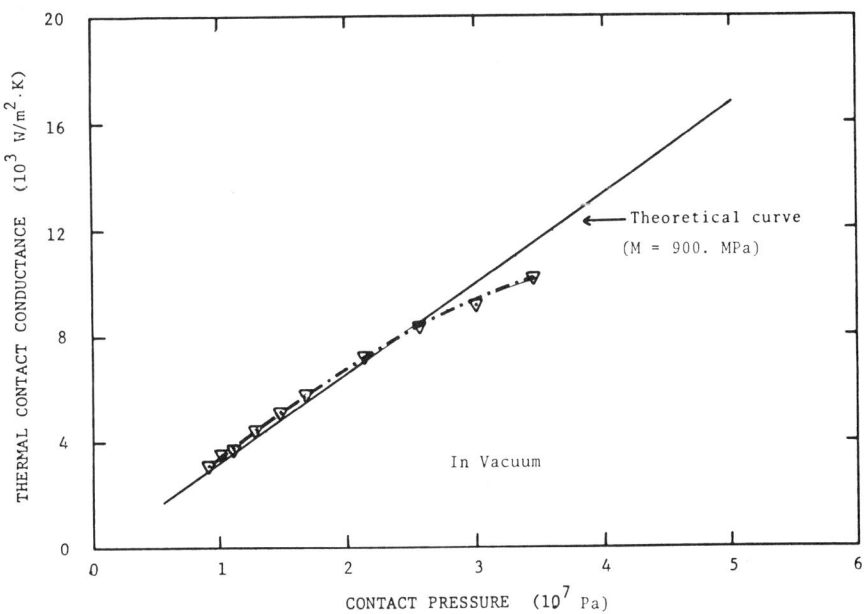

Fig. 6 Contact conductance vs pressure (Beloit Power Systems, M-45/0.61 mm, C-3/10.2 μm).

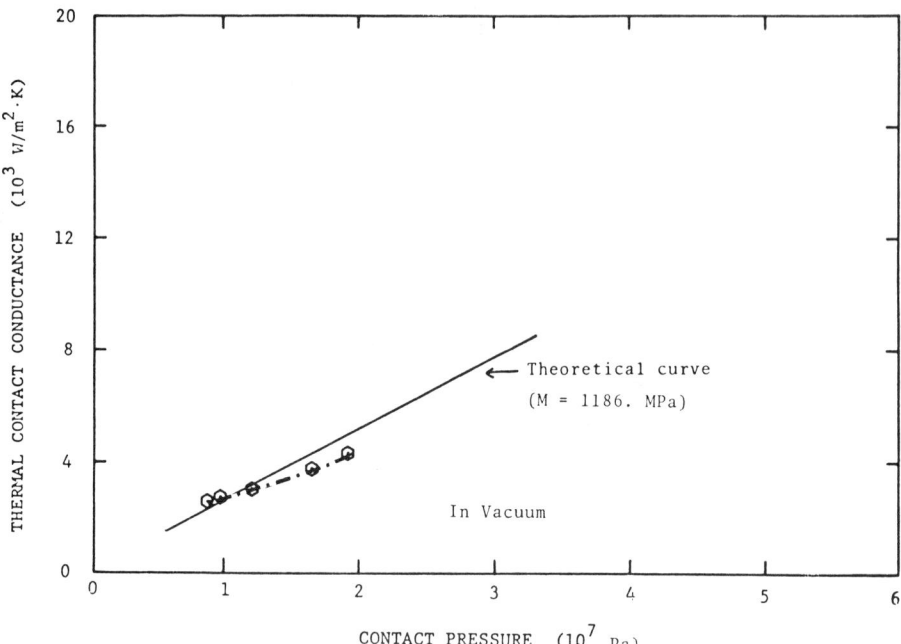

Fig. 7 Contact conductance vs pressure (Lima Electric Company, M-45/0.58 mm, C-5/12.7 μm).

gives a simple relationship for "δ" as a function of surface roughnesses:

$$\delta = 3.56(\delta_1 + \delta_2) \quad \text{for } (\delta_1 + \delta_2) < 7\mu m \tag{23}$$

and

$$\delta = 0.46(\delta_1 + \delta_2) \quad \text{for } (\delta_1 + \delta_2) > 7\mu m \tag{24}$$

where δ_1 and δ_2 are the mean roughness depths of the two contact surfaces. The constriction number is equated to the square root of the ratio of the actual contact area to the apparent contact area, and for increasing pressure is given by the following expression:[15]

$$C = \sqrt{P_c/M} \tag{25}$$

By measuring the hardness at various penetration depths, an effective Meyer hardness, M, was defined as the approximate value obtained at a penetration depth equal to the coating thickness. A detailed description of the determination of the effective Meyer hardness is given by Mentes, et al.[16]

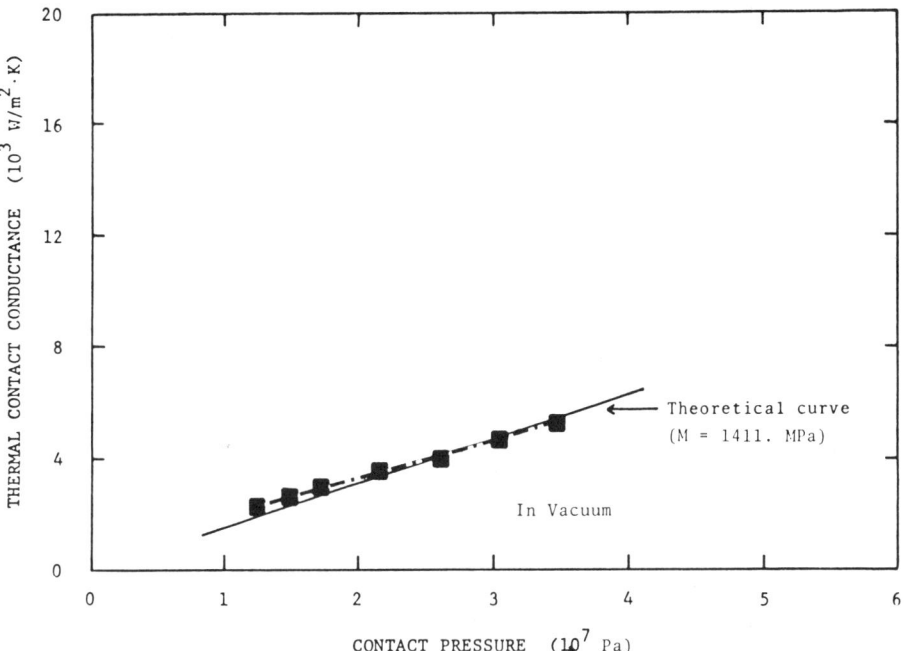

Fig. 8 Contact conductance vs pressure (Westinghouse Electric Company, oriented core/0.35 mm, carlite/10.2 μm).

Conclusions

A theoretical solution has been obtained for the problem of thermal contact conductance of coating multi-layered sheets in lamination by modelling the mixed boundary condition at the contact plane as a prescribed temperature distribution. By a judicious selection of the contact temperature distribution, a closed form solution for circular contacts is suggested for practical applications. Comparisons of earlier theories with this new contact theory show satisfactory agreement. Comparisons with experimental results indicate good predictive capabilities. For improving the prediction, an effective micro hardness is recommended for the coated sheets. In view of the continuing nature of the investigations, several issues are currently being studied and further results are forthcoming.

Acknowledgments

The authors gratefully acknowledge the support of the National Science Foundation, Division of Engineering. We are especially grateful to Dr. Win Aung for his close interest in the project and for his valuable suggestions. We acknowledge the cooperation of those companies supplying samples of test material. Special thanks are given to A. Mentes, R. Samudrala, D. Cunninghan, F. D'Aquino, L. Engel, and S. Engelken for their interest and assistance.

References

[1] Thomas, T. R. and Probert, S. D., "Thermal Resistances of Some Multi-Layer Contacts under Static Loads," International Journal of Heat and Mass Transfer, Vol. 9, No. 7, 1966, pp. 739-754.

[2] Hargadon, J. M., Jr., "Thermal Interface Conductance of Thermoelectric Generator Hardware," ASME Paper No. 66-WA/NE-2, 1966.

[3] Williams, A., "Heat Flow Across Stacks of Steel Laminations," Journal of Mechanical Engineering Science, Vol. 13, No. 3, 1971, pp. 217-233.

[4] Williams, A., "Experiments on the Flow of Heat Across Stacks of Steel Laminations," Journal of Mechanical Engineering Science, Vol. 14, No. 2, 1972, pp. 151-154.

[5] Jhamnani, J., "Thermal and Electrical Properties of Transformer Laminations," M.Sc. Thesis, Cranfield Institute of Technology, School of Mechanical Engineering, 1974.

[6] Al-Astrabadi, F. R., Probert, S. D., O'Callaghan, P. W. and Jones, A. M., "Thermal Resistances on Stacks of Thin Layers under Compression," Journal of Mechanical Engineering Science, Vol. 19, No. 4, 1977, pp. 167-174.

[7] Sheffield, J. W., Veziroglu, T. N. and Williams, A., "An Experimental Investigation of Thermal Contact Conductance of Multi-layered Electrically Insulated Sheets," Heat Transfer, Thermal Control, and Heat Pipes, Vol. 70, Progress in Astronautics and Aeronautics, edited by W. B. Olstad, AIAA, New York, 1980, pp. 130-146.

[8] Al-Astrabadi, F. R., O'Callaghan, P. W., Probert, S. D. and Jones, A. M., "Thermal Contact Conductance Correlation for Stacks of Thin Layers in High Vacuums," Journal of Heat Transfer, Vol. 99, February 1977, pp. 139-142.

[9] Ozisik, M. N., Heat Conduction, John Wiley and Sons, Inc., New York, 1980.

[10] Veziroglu, T. N., William, A., Kakac, S., and Nayak, P., "Prediction and Measurement of the Thermal Conductance of Laminated

Stacks," *International Journal of Heat and Mass Transfer*, Vol. 22, No. 3, 1979, pp. 447-459.

[11] Veziroglu, T. N., Huerta, M. A. and Kakac, S., "Exact Solutions for Thermal Conductances of Planar and Circular Contacts," in *Thermal Conductivity*, Vol. 14, edited by Klemens, P. G. and Chu, T. K., Plenum Publishing Corporation, New York, 1976, pp. 435-448.

[12] Fenech, H. and Rohsenow, W. M., "Prediction of Thermal Conductance of Metallic Surfaces in Contact," *Journal of Heat Transfer*, Vol. 85, February 1963, pp. 15-24.

[13] Laming, L. C., "Thermal Conductance of Machined Metal Contacts," in *International Developments in Heat Transfer*, American Society of Mechanical Engineers, New York, 1963.

[14] *G. E. Heat Transfer and Fluid Flow Data Book*, edited by Roecker, R. H., General Electric Company, Schenectady, New York, 1976.

[15] Veziroglu, T. N., "Correlation of Thermal Contact Conductance Experimental Results," in *Thermophysics of Spacecraft and Planetary Bodies: Radiation Properties of Solids and the Electromagnetic Radiation Environment in Space*, Progress in Astronautics and Aeronautics, Vol. 20, edited by G. B. Heller, Academic Press Inc., New York, 1967, pp. 879-907.

[16] Mentes, A., Sheffield, J. W., Veziroglu, T. N., Williams, A., and Samudrala, R., "Effect of Interface Gases on Contact Conductance," AIAA Paper No. 81-0214, 1981.

Chapter III. Heat Pipes

PERFORMANCE TESTING OF A HYDROGEN HEAT PIPE

J. Alario[*] and R. Kosson[+]

Grumman Aerospace Corporation
Bethpage, N.Y.

Abstract

Test results are presented for a reentrant groove heat pipe with hydrogen working fluid. The heat pipe became operational between 20 and 30 K after a cooldown from 77 K without any difficulty. Steady-state performance data taken over a 19 - 23 K temperature range indicated the following: maximum heat transport capactiy = 5.4 W-m, static wicking height = 1.42 cm, and overall heat pipe conductance = 1.7 W/°C. These data agreed remarkably well with extrapolations made from comparable ammonia test results. The maximum heat transport capacity is 9.5% larger than the extrapolated value, but the static wicking height is the smae. The overall conductance is 29% of the ammonia value, which is close to the ratio of liquid thermal conductivities (24%). Also, recovery from a completely frozen condition was accomplished within 5 min by simply applying an evaporator heat load of 1.8 W.

Introduction

Reentrant groove heat pipe configurations are being developed to improve the performance of extruded axially grooved heat pipes.[1] In particular, the reentrant groove maintains a simple design and provides comparable heat transport capacity, and it also greatly increases the static wicking height over conventional rectangular grooves. This is particularly important with heat pipe systems that use cryogenic fluids, since they characteristically have low wicking height factors (surface tension/liquid density ratio), which can complicate an

---

Presented as Paper 80-022 at the AIAA 18th Aerospace Sciences Meeting, Pasadena, Calif., Jan. 14-15, 1980. Copyright © American Institute of Aeronautics and Astronautics. Inc., 1980. All rights reserved.

*Project Engineer.
+Thermodynamics Technical Specialist.

integrated ground test verification. Also, for cryogenic applications the reentrant groove has the additional advantage of requiring less fluid charge, which can result in lighter weight hardware, owing to lower storage pressures.

To capitalize on these performance advantages with cryogenic fluids, the NASA-Ames Research Center under contract with Grumman developed a reentrant groove extrusion that was suitable for use with hydrogen, and that also could be produced in meaningful quantity.[2] Although the theoretical design was optimized within the limitations of state-of-the-art extrusion technology, acceptable production-type runs (extruded lengths over 100 m) could only be achieved at the expense of a wider nominal groove opening than specified (0.33 vs 0.20 mm). However, dimensional variations of other critical dimensions were within 0.05 mm, which exceeded expectations. The basic 6063-T6 aluminum extrusion (Fig. 1) is 14.6-mm o.d., with a wall thickness of 1.66 mm, and contains 20 axial grooves that surround a central 9.3-mm diam vapor core. Each axial groove is 0.775-mm diam, with a 0.33-mm opening formed by pinched, rounded fin tips. An excess vapor reservoir is provided at the evaporator to minimize the pressure containment hazard during ambient storage.

The analytical aspects of the development effort and a detailed description of the resulting hardware were previously documented,[2] and this paper concludes the description by presenting the results of tests with both ammonia and hydrogen fluids.

Test Evaluation

The heat pipe was first tested in a room temperature environment with varying charges of ammonia working fluid. In addition to obtaining ammonia test data, these tests were used to determine accurately the optimum fluid charge so that the heat pipe could be properly filled with hydrogen. For each measured ammonia charge, the maximum heat transport capacity was determined as a function of adverse tilt.

Once the proper charge had been determined, the heat pipe was filled with research purity hydrogen and tested in a thermal vacuum chamber using a liquid helium cooling system. The hydrogen test program was designed mainly to develop a performance map for the heat pipe, which consisted of tilt/dryout data at nominal operating temperatures.

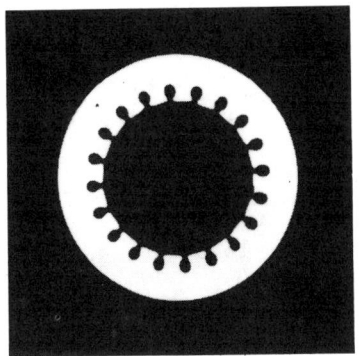
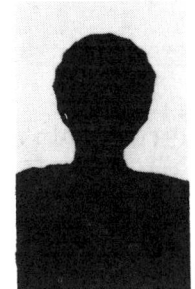

HEAT PIPE TUBE   SINGLE REENTRANT GROOVE

CHARACTERISTICS DIMENSIONS

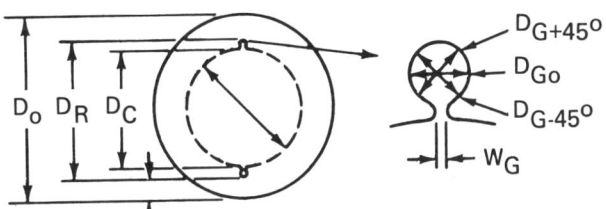

DIMENSION	MEASUREMENT, MM		
	MIN	AVG	MAX
OUTSIDE DIAMETER, D_o	14.58	14.62	14.68
ROOT DIAMETER, D_R	11.33	11.34	11.38
CORE DIAMETER, D_C	9.24	9.30	9.35
MINIMUM GROOVE OPENING, W_G	0.305	0.335	0.355
GROOVE DIAMETERS, D_{G+45}	0.754	0.782	0.805
D_{Go}	0.754	0.769	0.805
$D_{G-45°}$	0.754	0.776	0.805
AVG GROOVE DIA, D_G	–	0.776	–
WALL THICKNESS, T_{WALL}	1.651	1.676	1.679

Fig. 1 Reentrant groove extrusion photomicrograph.

Instrumentation

The location of all instrumentation was the same for both the ammonia and hydrogen test programs, although by necessity the composition of the thermocouple wire was changed. Copper-constantan wire was used for the room temperature tests, and Chromel vs gold -0.07 at % Fe was used for the hydrogen

Table 1 Sensitivity of various thermocouple wires

Thermocouple wire	Sensitivity, µV/K
Chromel-alumel	3.0
Copper-constantan	4.0
Chromel-constantan	6.5
Chromel-gold-iron	16.8

tests because of its superior sensitivity. A comparison with other possible thermocouple materials over the temperature range 10 - 25 K is shown in Table 1.

The instrumentation layout for the heat pipe is shown in Fig. 2. A total of 10 thermocouples, each positioned at the 3 o'clock orientation, were used on the heat pipe proper; four each on both the evaporator and condenser sections, and two on the transport section. Two thermocouples were located on the heat pipe charge tube so that the heat transfer from the reservoir could be measured. Two more thermocouples monitored the reservoir temperature to ensure that it was always slightly warmer than the evaporator. A trace heater was attached to the reservoir so that any liquid trapped in it could be vaporized and returned to the heat pipe. However, since reservoir temperature was never a problem, it was never used during the tests.

In addition to these primary temperature measurements, a limited number of copper-constantan thermocouples were used for backup, although they were never actually needed. One was located on the evaporator (7.5 cm from the end), one on the transport section, and two on the condenser block.

Electrical heat input was supplied to the evaporator by Nichrome ribbon, which was helically wrapped and attached to the pipe by a 1/2-mil layer of double-backed Kapton tape. A second layer of single-backed Kapton tape electrically insulated the heater from the surroundings. The thermocouples at the evaporator were attached directly to the pipe wall, with the bead located between consecutive turns of the heater. All thermocouples were bonded in place with EA 934 adhesive. To minimize stray emf owing to thermal gradients, several turns of thermocouple wire were taken before the wire was routed to the ambient.

PERFORMANCE TESTING OF A HYDROGEN HEAT PIPE   309

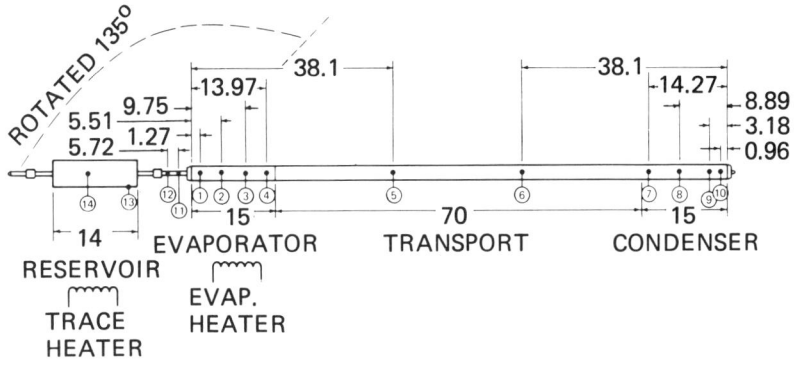

Fig. 2 Heat pipe test instrumentation.

The accuracy of the low-temperature measurements was enhanced by avoiding large temperature gradients between the ends of the thermocouple junctions. Therefore, each leg of the Chromel-gold wire was referenced to an LN_2-cooled cold junction, as indicated in Fig. 3. In this case, the reference junction was a copper tube through which LN_2 was routed before it entered the primary cold wall of the thermal vacuum chamber. The thermocouple junctions were electrically insulated by a layer of Kapton.

A temperature resolution of better than 0.1 K was obtained using a direct-reading, high-resolution precision potentiometer (Leeds and Northrup Type K-3 Universal Guarded Potentiometer, which had a resolution of 0.5 µV per division). This was used to measure the thermocouple emf. Since the Chromel vs gold -0.07 at % Fe has a test range sensitivity of 16 µV/K, the 1.6-µV emf output for each 0.1 K was well within the required measuring capability.

Tests with Ammonia

The primary purpose of this test was to confirm the theoretical (100% fill) charge for the reentrant groove heat pipe so that the optimum hydrogen charge could be accurately determined. In addition, complete performance maps (Q vs tilt) were generated, and repriming behavior from both mechanical and mechanical/thermal dryout was observed.

Test Setup

For the ammonia tests, the heat pipe was configured without the reservoir, and the charge valve remained attached to the heat pipe assembly since the fluid charge was sufficiently large (9 g) that it had little effect on heat pipe performance. The pipe was mounted on a standard laboratory test stand that had the capability of accurate adverse tilt adjustments from level to 7.6 cm (3.0 in.), and up to at least 15 cm (6 in.) when the heat pipe had to be disabled. Testing was done in a room temperature environment, with a cold-water-spray bath used as the condenser heat sink. The remainder of the heat pipe, including valve, was completely surrounded with Armaflex insulation. The copper-constantan thermocouples were continuously monitored with a multipoint Bristol chart recorder, and the steady-state temperature distributions were recorded using a Doric digital voltmeter. Evaporator heat input was regulated by a Variac, and the power was measured using an ammeter and voltmeter.

Results

The maximum heat transport vs adverse tilt performance maps for selected ammonia charges are summarized in Fig. 4.

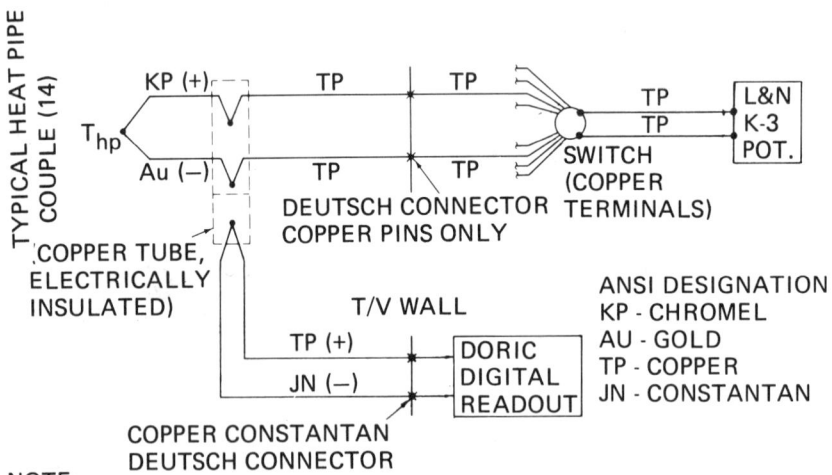

NOTE:
1. READ T_{REF} (TP/JN) → K → μV (FROM KP/Au CALIBRATION) = (μV) REF
2. READ OTHER KP/Au TEMPERATURES = $(\mu V)_T$
3. CORRECT MEASUREMENT FOR REFERENCE: $(\mu V)_{REF} - (\mu V)_T$

Fig. 3 Low-temperature thermocouple schematic.

PERFORMANCE TESTING OF A HYDROGEN HEAT PIPE

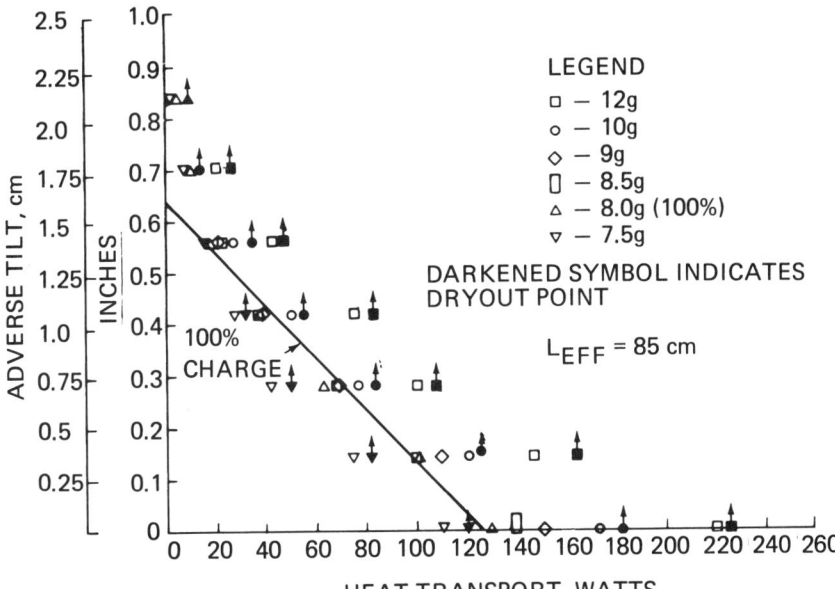

Fig. 4 Performance of NASA-AMES reentrant groove heat pipe with ammonia at 20°C.

The change of slope near zero-tilt values signals a liquid puddle contribution and is indicative of an excess fluid charge. As seen, the 8-g charge resulted in the highest heat transport (130 W) without evidence of a puddle. The lower 7.5-g charge had a concave shape and a lower heat transport, which indicates an undercharged condition. Corresponding pipe-temperature-drop data are given in Fig. 5. They indicate unexpectedly low film heat-transfer coefficients of 2014 and 5362 $W/m^2°C$ for the evaporator and condenser, for an equivalent overall heat pipe conductance of 5.85 W/°C. The postulated reason for these low film coefficients is that the rounded fin shape at the groove entrance creates a relatively thick fluid layer that causes inefficient heat transfer. This was confirmed by subsequent tests with a specially modified groove design. A sample section of tubing was modified to create a groove profile with a convergent entrance, which resulted in a thinner fluid layer and more efficient heat transfer. Comparable test results with ammonia gave improved evaporator and condenser film coefficients of 7900 $W/m^2°C$ and 14000 $W/m^2°C$, respectively. The overall heat pipe conductance increased almost 400% to 20.2 W/°C.

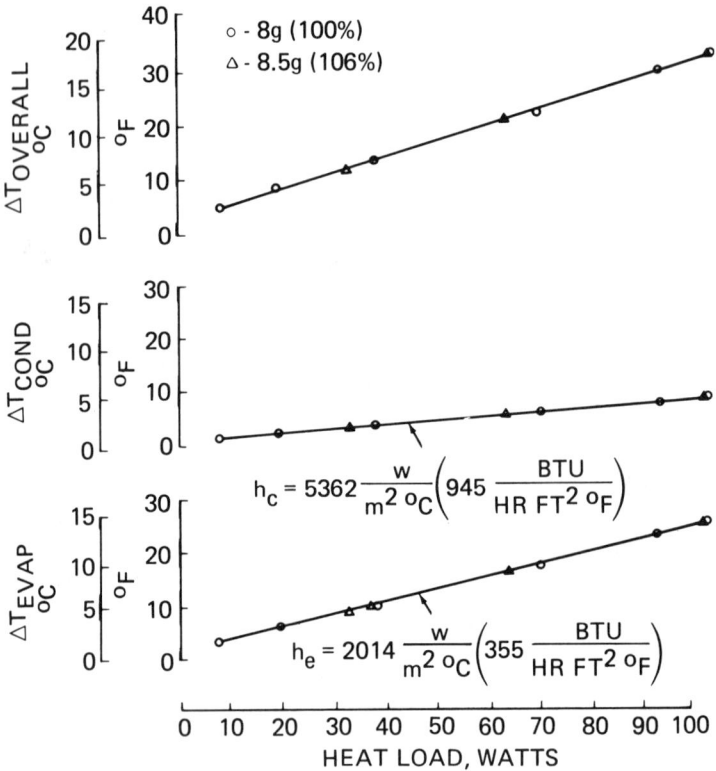

Fig. 5 Reentrant groove extrusion with ammonia, temperature gradients.

Using the original extrusion, tests to monitor recovery from a mechanical/thermal dryout were also run with an 8.5-g ammonia charge. The adverse tilt was first raised to 17 cm. Then 7 W of power were applied for several minutes so that the evaporator temperature exceeded the condenser temperature by 25°C. The heat pipe tilt was then decreased to 1.27 cm (1/2 in.), and the evaporator power was set at a lower value than that previously determined for Q_{max}. Results indicated that the heat pipe became operational at heat loads up to 60% of the Q_{max} value, as evidenced by the convergence of the evaporator and condenser temperatures.

Tests with Hydrogen

This test was designed mainly to develop a performance map for the heat pipe at the nominal 20 K operating tempera-

PERFORMANCE TESTING OF A HYDROGEN HEAT PIPE 313

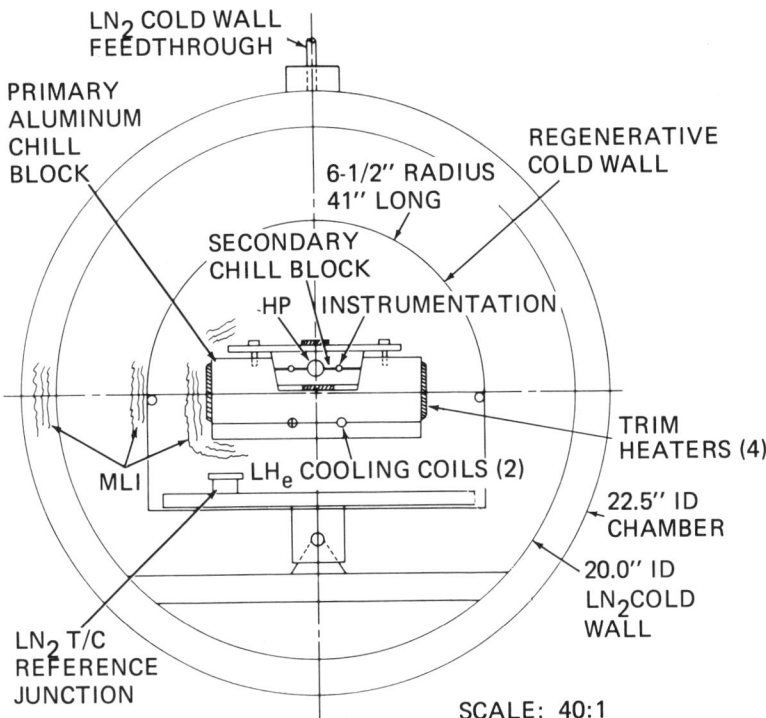

Fig. 6 Low-temperature thermal vacuum chamber test setup (end view).

ture. This consisted of first setting an adverse tilt, then incrementing the evaporator heat input until dryout occurred. Performance at selective off-design temperatures was also measured. The associated temperature profiles were used to determine the overall heat pipe conductance. Recovery of heat pipe operation after a thermal dryout also was monitored. A test point, with the evaporator severely raised to prevent heat pipe operation, was included to determine the effect of heat conduction through the aluminum tube at cryogenic temperatures. A secondary test objective was to examine the transient response of the pipe during cooldown from room temperature, freeze-out, and thaw.

Test Setup

All tests with hydrogen fluid were run in a specially modified end-loaded thermal vacuum chamber (60 cm in diameter by 122 cm long), with condenser cooling provided by a liquid helium cooling system. As shown in Fig. 6, the chamber contained an MLI-backed LN_2 cold wall as the primary means of thermal isolation from the room environment. A regenerative copper cold wall, which was cooled by escaping helium boil-off from the condenser block, served as a secondary thermal barrier. All chamber feedthroughs, except the LN_2 cold wall, are through the removable end wall, which also contains a 7.6-cm diam viewing port. The heat pipe test assembly, including the condenser cooling block, was mounted on a 1.27-cm-thick aluminum plate and isolated from it by low thermal conductance Teflon stand-offs (Fig. 7). The entire test assembly was encased in 30 layers of aluminized Mylar multilayer insulation; the condenser block had a double blanket, or 60 layers.

The chamber was balanced on a knife edge and supported by precision jack stands, so that tilt adjustments of up to 7.6 cm could be made by tilting the entire chamber. A surveyor's transit and metering scales, attached to opposite ends of the pipe assembly and viewed through the window, were used to accurately measure differences in elevation. (Although never used, external metering scales were also mounted to the ends of the chamber as a backup, in case the internal view was obscured.) A comparison and correlation between the two readings was made, using LN_2 coolant in the cold walls and cooling block; they indicated a stable platform.

The condenser section of the heat pipe was firmly squeezed between the two parts of a split, dove-tailed aluminum chill block, which were bolted together. Shallow channels, which were machined into the mating faces of the block in both axial and transverse directions, served as conduits for the thermocouple wires. The temperature of the chill block (and the

Fig. 7 Hydrogen heat pipe final assembly on test stand.

heat pipe) was controlled by an aluminum cooling block to which it was conductively coupled. This primary cooling block contained both liquid helium flow channels and electrical trim heaters. (The latter were never needed.) Additional trim heaters were attached directly to the smaller chill block. These were the only ones actually used to stabilize heat pipe test temperatures.

The desired nominal operating temperature for the heat pipe was established by adjusting the trim heaters on the chill block after a stable helium flow rate had been established. A near-constant heat pipe operating temperature was maintained by decreasing the trim heater input in accordance with the incremental increases in evaporator heat load. All electrical heat inputs were determined by measuring the actual voltage and amperage, which accounted for any variations in electrical resistance.

The helium cooling system (Fig. 8) was supplied by an expendable 500-liter liquid helium Dewar, which was connected to the chamber feedthroughs by vacuum jacketed transfer lines. A stable helium flow rate as high as 12 liters/h was maintained by pressurizing the Dewar to 2.5 psig with gaseous helium, and then adjusting the flow control valve on the Dewar To permit accurate flow measurements, and as a safety precaution to prevent the formation of liquefied air, an inline heater was used to preheat the helium vapor to about 300 K before venting to the atmosphere. The vent line was also fitted with a thermocouple and a calibrated flow meter, so that the boil-off rate could be measured and flow adjustments made to compensate for excessive liquid helium blow-through. During most test points, the actual helium usage nearly matched the theoretical flow rate required by the electrical power dissipations, usually within 5%. This indicates a very efficient cooling system, with minimum heat losses due to either blow-through or extraneous heat leaks.

Tare Heat Leaks

Before liquid helium cooling was actually used, an experimental tare heat leak test was run with liquid nitrogen as the coolant. The block was first cooled to 77 K, then the liquid nitrogen flow was stopped, and the remaining LN_2 residue was purged with nitrogen gas until the block reached 83 K. At that point, a known amount of electrical heat (10.6 W) was input to the primary heaters, and the temperature rise of the block was measured with time. Over a period of 25 min, the block temperature increased from 83.0 to 90.8 K, an average rate of 18.7 K/h. The tare leak can then be estimated by solving the following heat balance:

$$Q_e + Q_L = (MC_p) \, dT/d\tau$$

where

Q_e = electrical heat input = 10.6 W

Q_L = tare heat leak

MC_p = thermal capacitance of the condenser block

$dT/d\tau$ = average temperature rise rate

The condenser block assembly weighs 5.45 Kg (12.0 lb), and the average value of heat capacity for aluminum over this temperature range is 0.39 J/g°C. Thus, the capacitance of the block is 2.12 kJ/°C (1.116 Btu/°F). Solving for the unknown heat leak yields:

$$Q_L = (MC_p) \, dT/d\tau - Q_e$$

$$Q_L = 0.41 \text{ W}$$

The corresponding average temperature difference measured between the condenser block and the mounting plate during the tare test was 50.5 K (90.8°F). Thus, an equivalent heat leak conductance for these temperature levels can be estimated as 0.41/50.5 = 0.0081 W/°C (0.0045 W/°F). The important assumption implicit in this value is that conduction heat leaks predominate over radiation. This was confirmed by calculations showing the radiation leak to be 0.011 W.

The theoretical heat leaks are comprised of:

Conduction through the Teflon stand-offs (six at the condenser block and one at the evaporator).

Radiation through the MLI (60 layers around the condenser block and 30 around the rest of the heat pipe assembly).

Conduction through the instrumentation (thermocouples and heaters).

Conduction from the reservoir through the charge tube and into the end of the evaporator. This leak is greatest at the start of testing when the reservoir is warm, and it decreases to a negligible amount with time.

PERFORMANCE TESTING OF A HYDROGEN HEAT PIPE 317

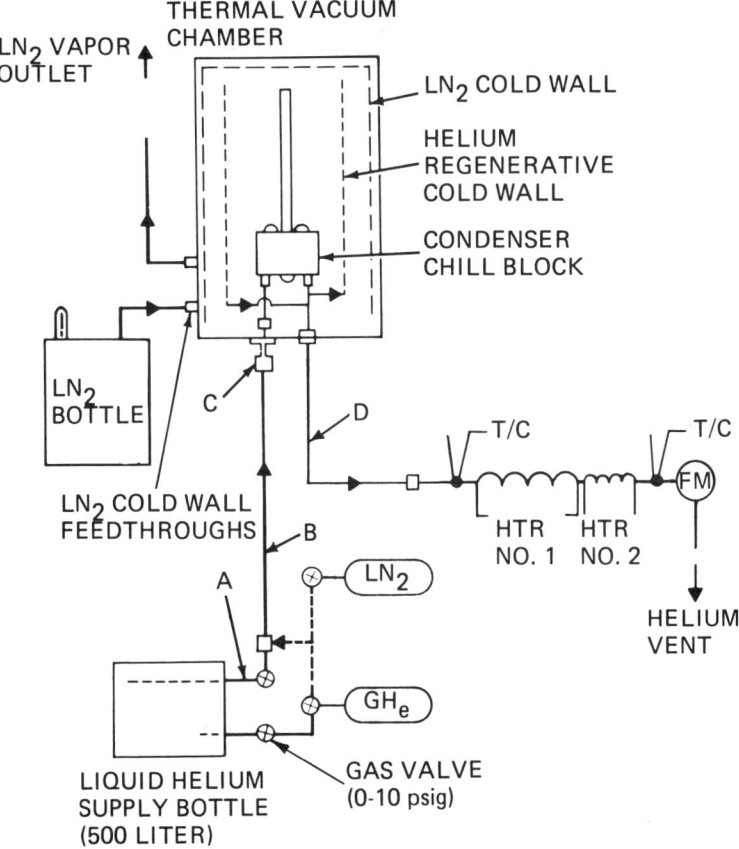

Fig. 8 Low-temperature facility flow schematic.

The estimated heat leaks, accounting for changes in material thermal conductivity with temperature, for both LN_2 and LHe cooling are shown in Table 2.

Tare heat leak test measurements were about 30% below theoretical, which is reasonable considering the small magnitudes (less than 1 W).

Results

The cooldown transient for the hydrogen heat pipe (without an evaporator heat load) shows the spread between selected evaporator, transport, and condenser temperatures at given times (Fig. 9). The helium coolant flow rate averaged about 5 liter/h, with various chill block trim heater inputs used to stabilize the condenser temperature. Heat pipe action started about 2-1/2 h into the cooldown, as evidenced by the convergence of all three temperatures within the expected 15 - 33 K operating range.

The steady-state performance data were taken over a temperature range of 19 - 23 K. They are summarized in Fig. 10 in a plot of net evaporator heat load heat vs pipe tilt. The data (solid line) indicate a static wicking height of 1.42 cm and a maximum transport capacity of 5.4 W-m. Predicted performance extrapolated from the ammonia test data is indicated by the dashed line. The static wicking height agrees with predictions, while the heat transport capacity is only 9.5% larger than anticipated. The errors associated with the mea-

Table 2   Estimated heat leaks

	Calculated heat leaks, W	
	LN_2 cooling	LHe cooling
Mounting plate to HP (conduction)	0.36	0.23
Instrumentation (conduction)	0.21	0.32
MLI (radiation)	0.01	<0.01
Charge tube (conduction)	-	0.60→0.00
Total theoretical	0.58 W	0.55 + Charge tube
Measured at condenser block during tare test:	0.41 W	-

PERFORMANCE TESTING OF A HYDROGEN HEAT PIPE 319

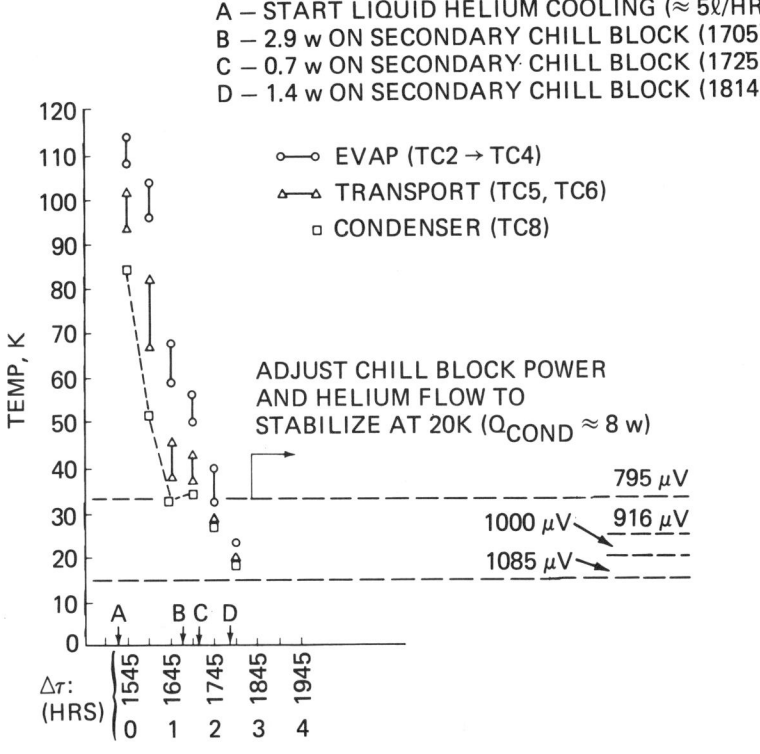

Fig. 9 Cooldown transient.

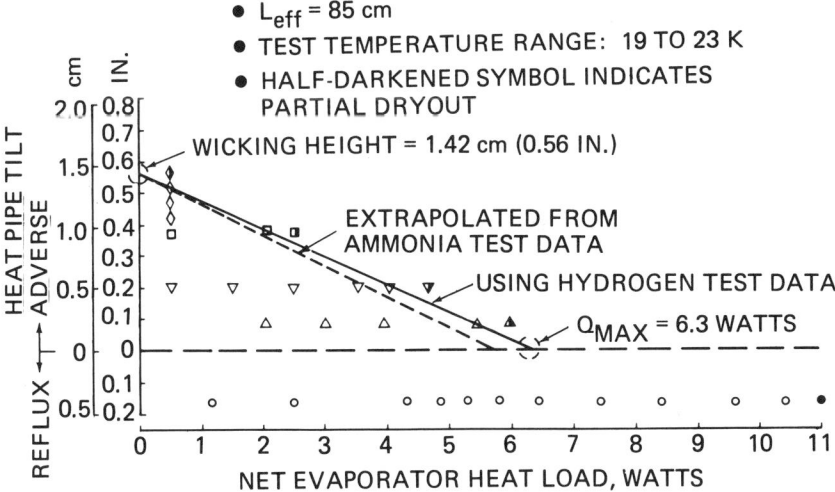

Fig. 10 Hydrogen heat pipe performance data, Q_{max} vs HP tilt.

sured data are less than 0.2% for the electrical heat input, and ±0.063 cm (±0.025 in.) for the tilt measurement.

The corresponding values for the overall average temperature gradient between the evaporator and condenser sections at each heat load are plotted in Fig. 11. The variation is linear and indicates an overall heat pipe conductance of 1.7 W/°C, which is 29% of the ammonia value. It is of interest to note that this is close to the variation in thermal conductivities between the two liquids, which is about 24%.

Detailed temperature profiles at the various tilt positions are presented in Fig. 12. The measurements at the far ends of the heat pipe (1.27 cm at the evaporator and 0.96 cm at the condenser) have been omitted since they were suspect. Post-test examination showed that the evaporator thermocouple had lifted from the heat pipe wall, which would cause it to read higher than the rest of the evaporator. The thermocouple at the end of the condenser may be influenced by variations in clamping pressure, since it typically registered 1 - 2°C warmer than the other condenser measurements.

As seen in Fig. 12, an anomaly exists in the condenser temperature profile since two of the thermocouples are slightly warmer than the adiabatic (transport) temperature. Because of this it was not possible to determine the individual evaporator and condenser heat-transfer film coefficients, which are based on separate temperature differences. A speculated explanation is that this condition resulted from a combination of extraneous heat leaks and poor thermocouple contact. Small heat leaks between the evaporator and condenser sections could

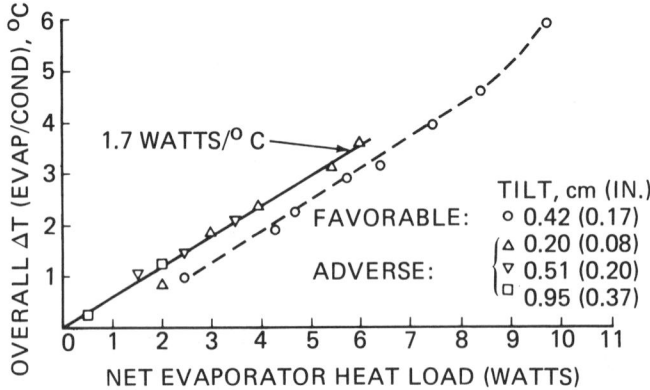

Fig. 11 Hydrogen heat pipe performance, overall temperature gradient.

PERFORMANCE TESTING OF A HYDROGEN HEAT PIPE   321

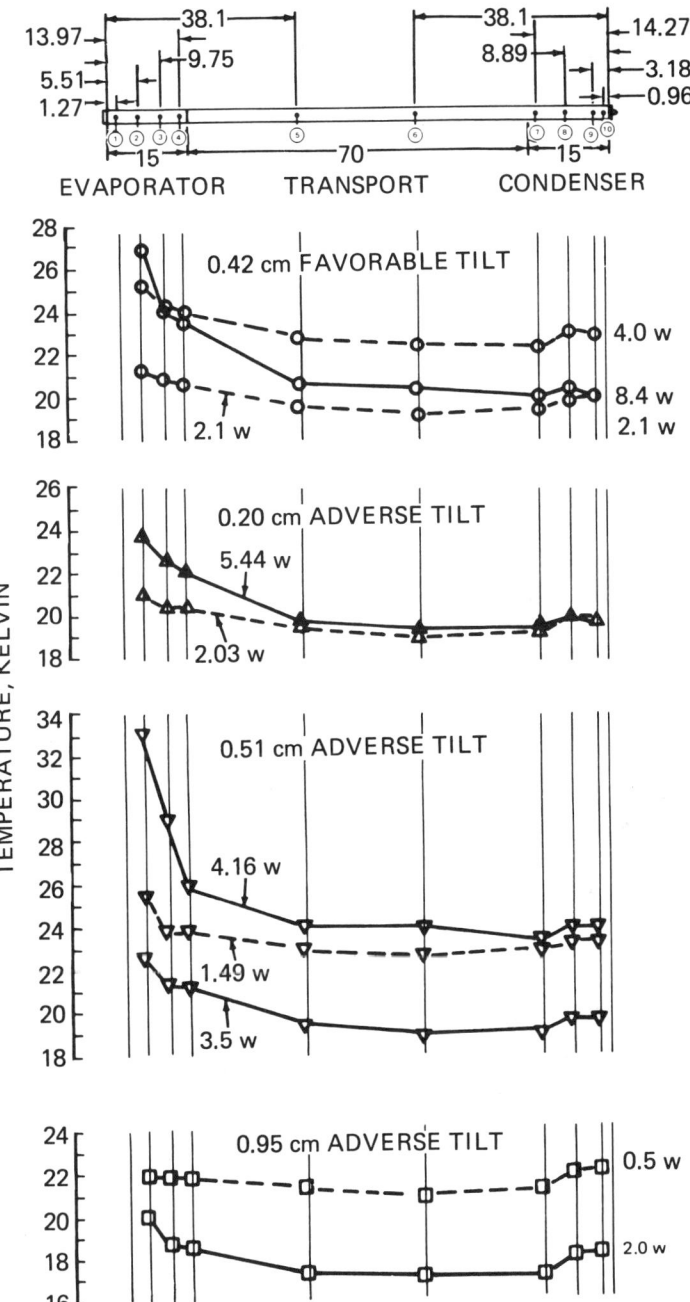

Fig. 12 Hydrogen heat pipe performance, temperature profiles.

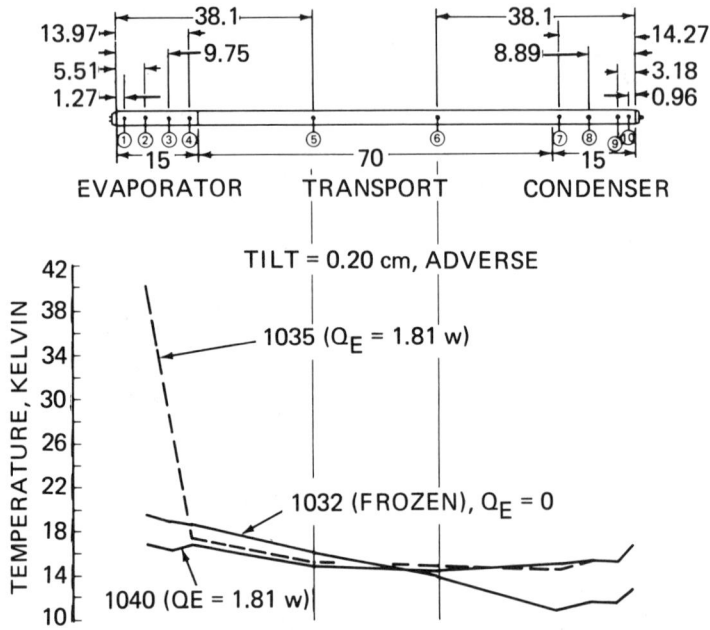

Fig. 13 Heat pipe recovery from frozen condition.

be caused by radiation tunneling between parallel wraps of the multilayer insulation. Also, any slight lifting of the thermocouple bead from the heat pipe would make the junction more sensitive to these radiation tunneling effects.

Recovery of heat pipe operation after a dryout was no problem. Whenever a dryout was witnessed during a steady-state test point, the evaporator heat load was reset to the previous value. The heat pipe always recovered operation within a few minutes.

As a final test, startup of heat pipe operation from a completely frozen condition was also successfully demonstrated. The heat pipe was kept frozen for 3 h by holding the condenser at approximately 12 K with a 0.20 cm adverse tilt. Then a 1.8-W evaporator heat load was applied, and within 5 min the heat pipe was operational at 15 K. The corresponding temperature profiles for the startup sequence are shown in Fig. 13.

Conclusions

Axial groove heat pipes with reentrant groove profiles can be produced in useful quantities.

A slight modification to the groove profile, creating a convergent entrance after the extrusion process would permit heat pipes to be built that are uniquely suited to cryogenic fluids, especially hydrogen. These heat pipes would provide superior overall thermal performance with a significantly lower fluid inventory, which greatly simplifies storage pressure requirements. This can result in lighter weight heat pipes due to thinner walls, or a smaller excess vapor reservoir, which simplifies packaging.

Working with hydrogen fluid is not much different than working with other cryogens, such as methane, which can result in high storage pressures. The important considerations are: materials (avoiding hydrogen embrittlement), storage pressure containment, and low-temperature test requirements. The last item can be expensive, if an expendable liquid helium coolant is used.

References

[1] Harwell, W., Kaufman, W.B., and Tower, L.K., "Reentrant Groove Heat Pipe," AIAA Paper 77-773, June 1977.

[2] Alario, J., Kosson, R., and McCreight C., "A Re-entrant Groove Hydrogen Heat Pipe," AIAA Paper 78-420, July 1978.

DESIGN, DEVELOPMENT, AND TEST OF A 1000 W OSMOTIC HEAT PIPE

A. Basiulis,* G. L. Fleischman,+ and C. P. Minning≠

Hughes Aircraft Company, Torrance, Calif.

Abstract

Osmotic heat pipes can be designed to transport high heat loads. An osmotically pumped heat pipe has been operated continuously for 500 h with a thermal loading of 500 W. A peak thermal transport of 1000 W was demonstrated using water as the working fluid, sucrose as the solute, and tubular cellulose acetate membranes. Concentration polarization effects did not materialize to the degree expected because of self-induced convection currents; leakage of solute through the membranes was too small to measure; and there was essentially no carryover of solute in the vapor. A self-equalizing flow control mechanism was utilized in the experimental heat pipe. However, long life capabilities of osmotic membranes and working fluids are current unknowns.

Introduction

Heat pipe energy transport devices have recently matured to production in temperature control applications. Examples are: thermal control diodes in missile electronics, cylindrical heat pipes which cool shipboard and missile electronics, heat pipe modules for heat recovery, and variable conductance heat pipes for the thermal control of spacecraft components.[1] All heat pipes, however, depend on capillary forces to pump working fluid from the condenser to the evaporator and are limited in their ability to transport thermal

---

Presented as Paper 80-1482 at the AIAA 15th Thermophysics Conference, Snowmass, Colo., July 14-16, 1980. Copyright © American Institute of Aeronautics and Astronautics, Inc., 1980. All rights reserved.
  *Assistant Department Manager, Electron Dynamics Division.
  +Project Manager, Electron Dynamics Division.
  ≠Senior Staff Engineer, Electro-Optical and Data Systems Group.

energy long distances. Osmotic heat pipes[2,3] presently under development for the Air Force Flight Dynamics Laboratory under Contract F33615-77-C-3031 are not limited by capillary pumping. They show a capability of transporting energy long distances and taking over where conventional heat pipes leave off.

Principle of Operation

A conceptual drawing of a typical osmotically pumped energy transport system or osmotic heat pipe is shown in Fig. 1. When two fluids of different concentrations are separated by a semipermeable membrane, as shown in Fig. 1a, the solvent will flow into solution to attain equilibrium. As long as the concentration gradients are high between solvent and solute solution, flow will continue. If the two chambers are connected, as shown in Fig. 1b, the spontaneous pumping action will continue until the concentration of the two solutions reaches a state of equilibrium. At this point, neither solvent nor solute will flow through the membrane. By adding heat to the solution, as shown in Fig. 1c, solvent evaporates and flows to the condenser where it recondenses, giving up heat, and returns to the membrane. Thus, the concentration difference at the membrane is constantly maintained and continuous pumping action is assured.

The thermal energy required to separate pure solvent from the solvent-solute mixture is the same energy, or waste heat, which must be transported from the evaporator region to the condenser.

Design Considerations

Based on a literature survey, research, and laboratory experiments with an osmometer, candidate membrane and working fluid systems were selected.[4,5] Consideration was given to materials compatibility and potential failure mechanisms, which are unique in direct osmosis applications. Failure mechanisms and their effect on heat pipe performance are summarized in Table 1.

Water was selected as the optimum working fluid for the current design. It has the highest mass flow and separation properties in conjunction with current state-of-the-art membranes. Moreover, water has the highest latent heat of vaporization of any of the candidate working fluids. As in conventional heat pipes, a high latent heat of vaporization results in greater heat transport for a given mass flow or pumping rate.

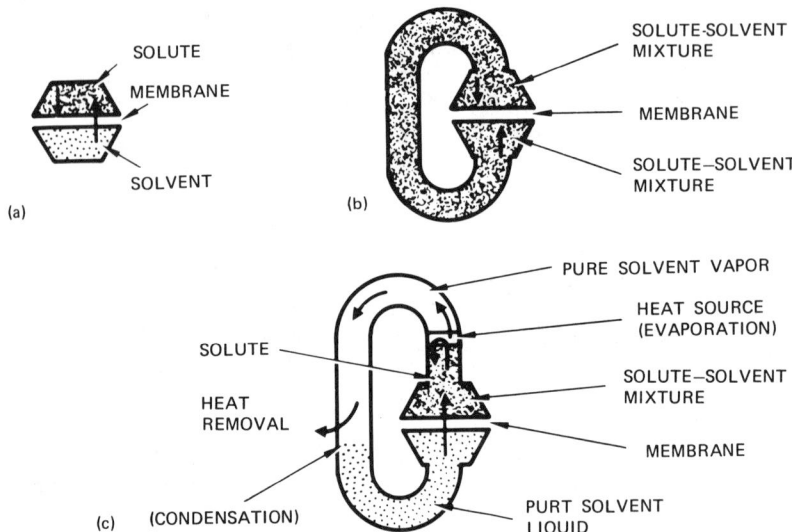

Fig. 1 Principle of operation of an osmotic heat pipe.

Table 1 Potential failure mechanisms

Failure mechanism	Effect on heat pipe performance
Polarization concentration	If concentration of solvent is allowed to accumulate and block the semipermeable membrane, the pumping capacity of the osmotic membrane can go to zero.
Flow controls	In osmotically pumped heat pipes, the liquid pumping rate is independent of the heat input. As a result, at low heat input rates, the evaporator can be flooded causing spillover of solute into the solvent chamber.
Leakage of solute through the membrane	Leakage can reduce the concentration gradient and thus the pumping capacity.
Solute carryover in vapor	The presence of solute in the vapor can cause contamination of the solvent.

Electrolytes such as those formed from salts dissolved in water offer definite advantages as solutes for osmotic pumping. First, ions are rejected by most membranes, better than nonionic solutes. Second, the salts have extremely low vapor pressures, and consequently do not mix with the working fluid vapor. Also, they are highly soluble in water so that high osmotic pumping pressures can be easily obtained.

Organic solutes were selected, however, to avoid corrosion, and possible noncondensable gas generation in a metallic heat pipe. As in a conventional heat pipe, any noncondensable gas would be swept to the condenser surfaces, preventing efficient condensation heat transfer. Sucrose was selected as the solute, then, for the following reasons: it is extremely soluble in water; it is non-volatile; it has good membrane separation properties; and is noncorrosive with envelope material.

Copper and Monel are compatible with water in heat pipes over the design temperature range. Because a non-corrosive solute was selected, the materials compatibility problem reduces to compatibility between the solvent and the envelope material. The extensive data from conventional heat pipe life testing can be used for this purpose.

In order to provide maximum flow per unit volume of the osmotic pump, three different membrane module designs were evaluated.

Hollow Fiber Module

One of the most promising module configurations for osmotic heat pipe applications is the hollow fiber membrane module (Fig. 2). An advantage of this approach, in addition to large surface areas per unit volume, is the elimination of the need for a membrane support material. A hollow fiber module of this type can achieve a surface area of 9,000 ft^2/ft^3 of container volume.

Spiral Wrapped Module

This has been a highly successful design for obtaining large membrane surface areas in compact volumes for reverse osmosis applications. However, the module investigated for use in an osmotic heat pipe (Fig. 3) was specially designed to provide approximately 0.5 in. intermembrane spacing on the solution side. This spacing between wraps was provided to allow adequate fluid circulation and mixing to take place.

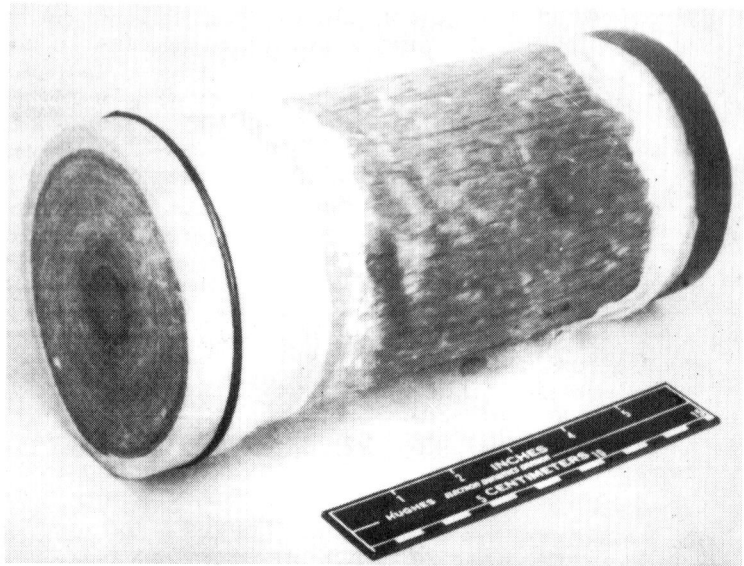

Fig. 2 Hollow fiber module.

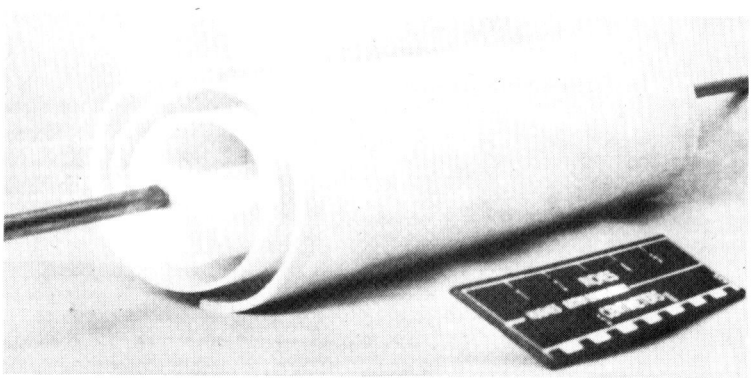

Fig. 3 Spiral wrapped membrane package.

Shell-and-Tube Module

A module consisting of three perforated stainless steel support tubes and a flange (tube sheet) was constructed as depicted in Fig. 4. Tubular cellulose acetate membranes were placed inside the support tubes and sealed at the top and bottom using epoxy. An advantage of these membranes was that the longitudinal seams, which were sealed with epoxy in the spiral wrapped modules, were eliminated.

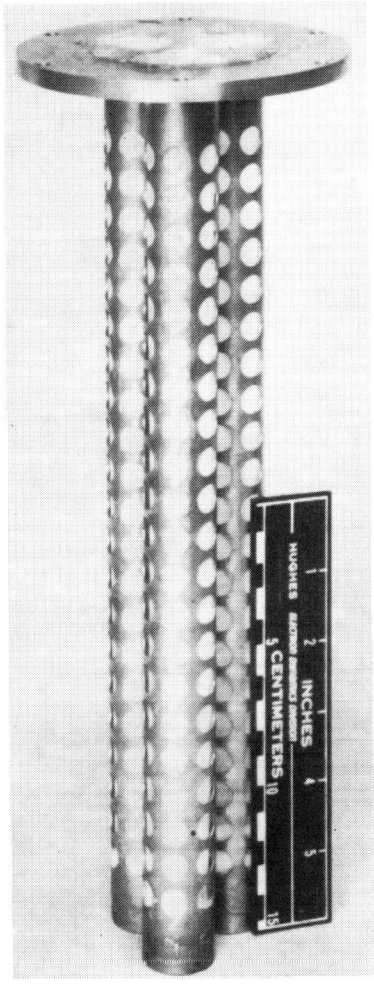

Fig. 4 Tubular membrane module.

Component Evaluation and Testing

Experiments were designed to evaluate heat pipe components, failure mechanisms, and to develop techniques to overcome effects limiting heat pipe performance. A basic test vehicle was designed to evaluate osmotic heat pipe components. It consisted of a working fluid column, membrane module compartment, solvent feed siphon, and evaporator. The test apparatus is shown in Fig. 5.

Osmotic pumping rates were measured by metering solvent feed into the solvent reservoir as illustrated. Since the

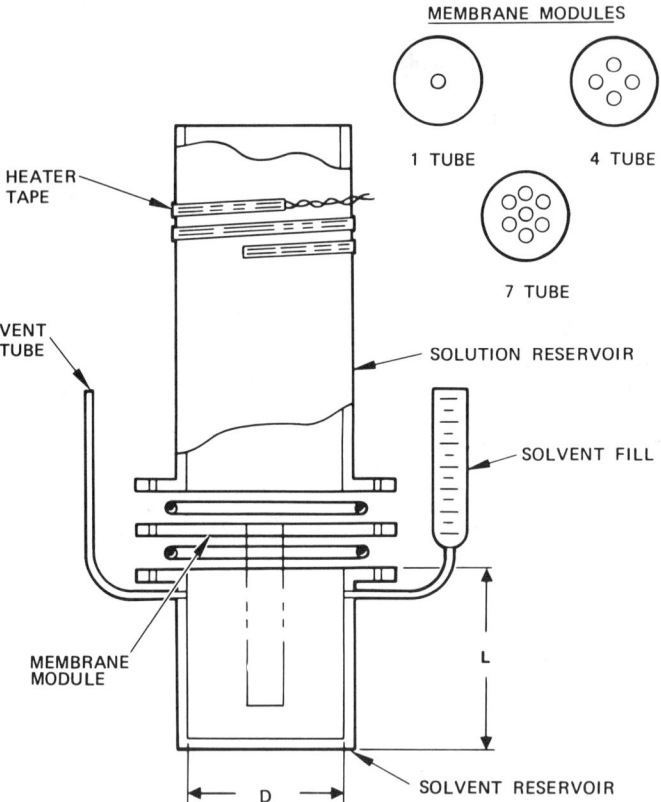

Fig. 5 Component evaluation apparatus.

solvent reservoir was kept full at all times, the rate of decrease of the water level in the glass cylinder was a very accurate measure of the amount of solvent being pumped by the membranes. The solution column was maintained at a constant height by boiling off the excess liquid. This approach maintained an overall constant concentration level in the solution column.

Polarization Concentration Effect

This is a major problem in reverse osmosis. Although it was analytically predicted to be a serious problem, it did not materialize to the degree predicted because of self-induced convection currents.[6] With special flow channel designs, concentration polarization was even further reduced. Figure 6 shows the mixing tube arrangement. In this configuration, concentrated solution returned via the central flow separator tube to the membrane, swept the membrane, and mixed with incoming solvent. Dilute solution then flowed upward through the annulus to the evaporator, where solvent evaporates and the process repeats itself.

Flow Control

Unlike the capillary wick in an ordinary heat pipe, the pumping rate across the membrane in an osmotic heat pipe does not automatically match the heat input rate. Two flow control techniques were developed to shut off osmotic pumping, gas blockage, and solvent depletion. Both techniques were effec-

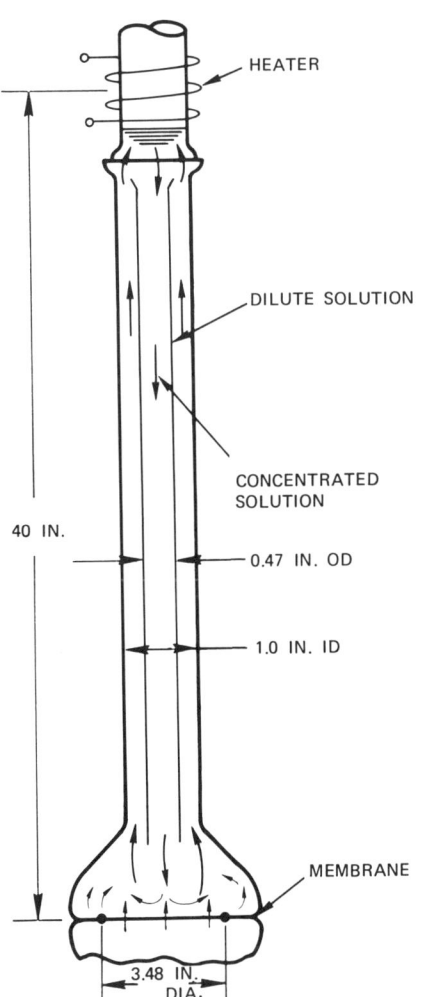

Fig. 6 Thermal syphon experiment.

tive in stopping solvent flow or in modulating the solvent flow.

The gas blockage technique is illustrated in Fig. 7. When the power was turned off, the pressure inside the heat pipe lowered and a large bubble appeared underneath the membrane, as shown. Pumping stopped immediately when the bubble appeared. When the pressure was increased by increasing the heat input, the bubble collapsed and the heat pipe started operating again. Conversely, the heater on the gas reservoir could be actuated to control the presence or absence of the bubble.

An advantage of this technique is that the flow can be stopped without special allowance for storage of the solvent in the evaporator when the heat pipe is turned off. However, it would be difficult to incorporate in the more compact membrane modules. The introduction of gas into the heat pipe could eventually saturate the solvent and enter the vapor space, thus degrading performance.

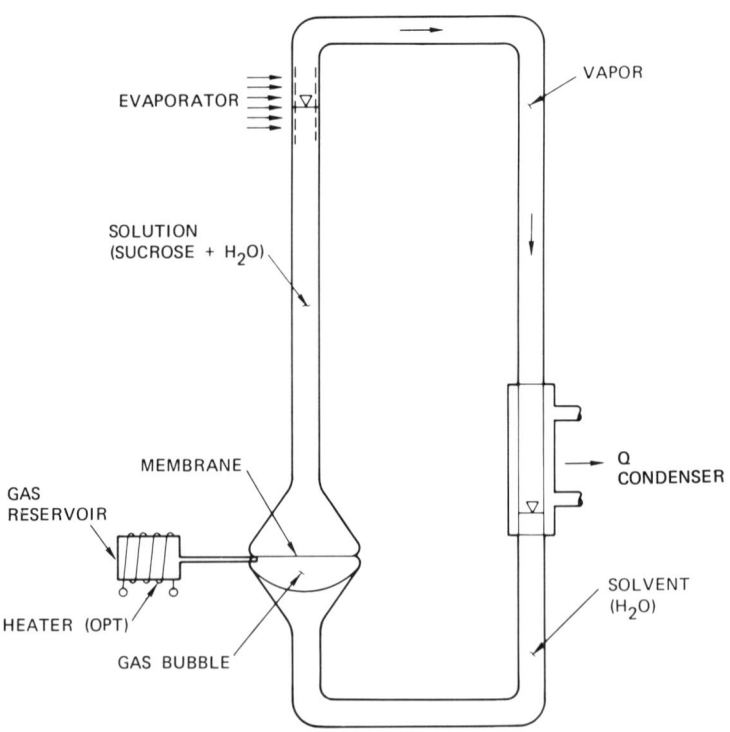

Fig. 7 Gas blockage shut off mechanism for osmotic pump.

Figure 8 illustrates the solvent depletion technique. The addition of a vent tube at the top of the membrane module provides a self-equalizing mechanism for the solvent level. When the heat pipe is turned off, all of the solvent which is needed to fill the membrane module is in the solution reservoir (evaporator). As heat is applied to the evaporator, distillation takes place and the solvent condenses in the condenser. The membrane module begins to fill and osmotic pumping begins through the membrane. If more heat input is applied, the solvent will fill the module, wetting additional membrane area. The maximum power of the heat pipe is reached when the solvent has filled the membrane module (solvent chamber). Note that the solvent level on the condenser side is at all times equal to the height of the solvent column in the membrane module.

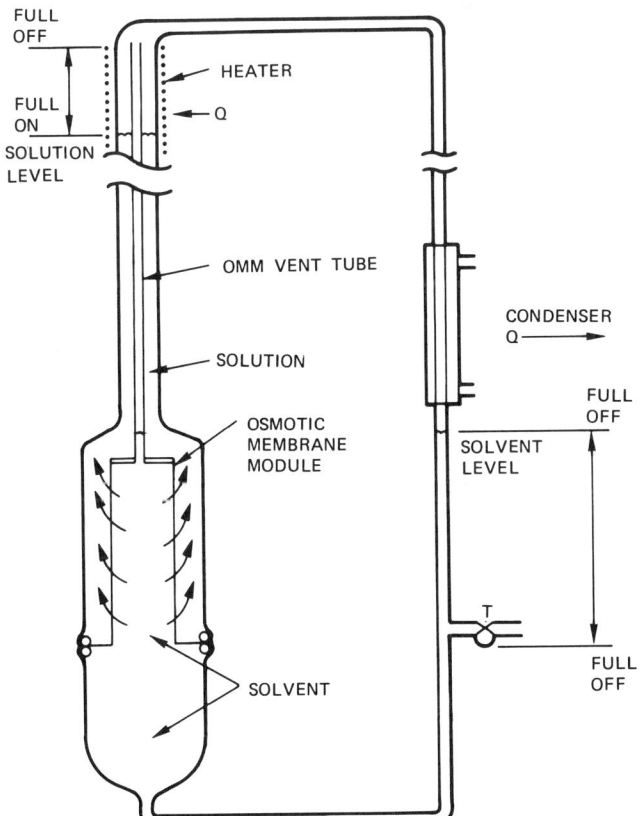

Fig. 8 Solvent depletion technique for shut off and flow control of osmotic pump (condenser column height is balanced by solvent column height in membrane module).

This approach is probably the most practical since the vent tube eliminates any problems with vapor or gas bubbles blocking the membrane area at maximum power conditions. Also, the additional volume required to store solvent in the evaporator is minimized, since only the solvent required to fill the module must be stored.

Leakage Through the Membrane

During the course of the osmotic pumping experiments, samples were taken periodically from the solvent side for chemical analysis. Solutes investigated included polyethylene glycols (molecular weights of 1000 and 6750), magnesium sulfate, and sucrose. The presence of solute was detected in each sample. However, the amounts were so small that quantitative values could not be obtained.

In view of the fact that leakage is a possibility, an analysis was made to evaluate the influence of the ratio of membrane area to solvent compartment volume on long term performance. This analysis was conservative in that the solvent flow was taken to be zero. It was concluded, however, that designs having small solvent volumes such as the spiral wrapped module would be sensitive to leakage of solute through the membrane, or a microscopic leak in the seal. The shell-and-tube module, on the other hand, was relatively insensitive to solute leakage.

Solute Carryover

The carryover experiment was performed to determine how much, if any, solute was being entrained in the vapor and carried over into the solvent compartment during evaporation. Since this is a distillation process, very little contamination due to carryover was anticipated.

The experiment was performed by collecting the condensate in a clean beaker while operating the test apparatus over a period of three days. The results indicated essentially no carryover of solute in the vapor.

Module Performance

Each of the three membrane modules previously described (Figs. 2, 3, and 4) were tested for osmotic pumping performance. The best results were obtained both in reliability and flow rate tests with the shell-and-tube design.

Baseline Design

Based on the results of the experimental work described above, the following goals were established for a baseline design:

1) Heat transport: 500 W.
2) Evaporator temperature 50°C to 100°C.
3) Column height (evaporator above condenser) 1.8 m (6 ft).
4) Passive flow control.
5) Life 1000 h.

Figure 9 shows a cross-section of the baseline design. The heat pipe consists of five major components: 1) osmotic pump module, 2) mixing and flow separator tubes, 3) evaporator, 4) condenser, and 5) heat pipe envelope.

Osmotic Pump Module

The osmotic pump section of the heat pipe includes the solvent reservoir and shell-and-tube membrane module (Fig. 9). A total of 18 perforated stainless steel tubes were used to support each of the cellulose acetate membranes, which were cast in the form of cylindrical tubes. The membranes were 2.45 cm (0.96 in.) in diam by 30.5 cm (12 in.) long, and were sealed to the support tube walls at the top and bottom with RTV. This approach eliminated the need for lengthy seams and small volumes on the solvent side, which could be sensitive to any leakage of solute through the membrane or a microscopic leak in the seal area. In addition to this, the cylindrical membrane geometry could be adapted to the passive start-up and control technique previously discussed. The total membrane surface area was 4225 cm^2 (655 $in.^2$).

The solvent reservoir volume consists of the volume not displaced by the membrane package. Two ports on the side at top and bottom of the reservoir are for vapor venting and solvent entering the reservoir, respectively. Note that this port represents the vent tube previously referred to in the solvent depletion control technique. An O ring seal above and below the membrane module flange sealed the assembly into the heat pipe envelope.

Mixing and Flow Separator Tubes

There are two concentric solution transport tubes in the solution column. The inner tube transports the upward moving

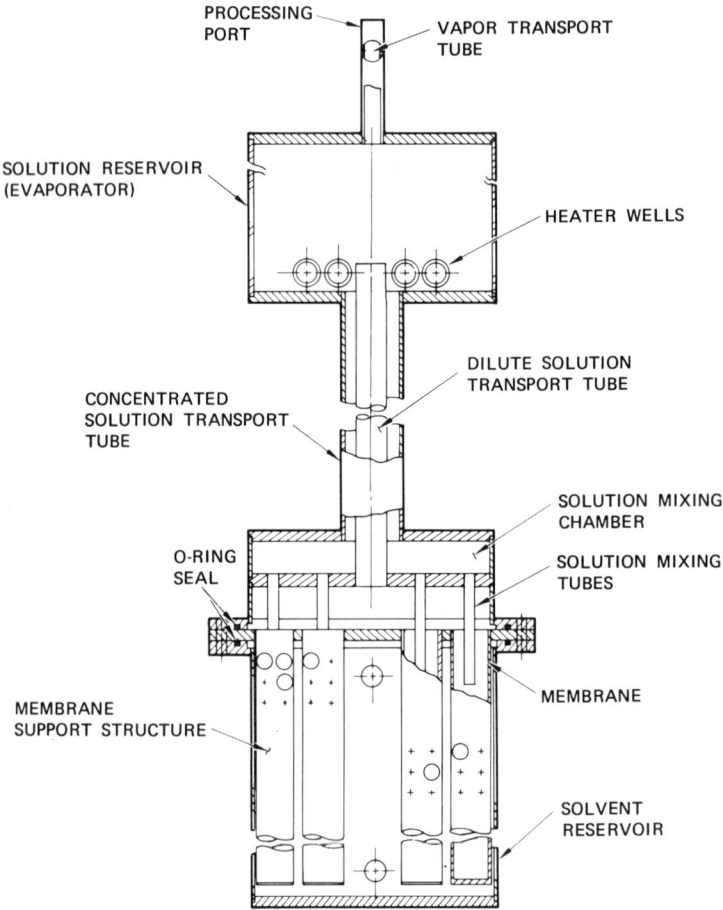

Fig. 9 Baseline osmotic heat pipe cross section.

fresh solvent flowing through the membrane, whereas the outer annulus transports the downward moving solution concentrated by evaporation in the evaporator. The solution mixing chamber and mixing tubes (Fig. 9) direct the concentrated solution down into each membrane cylinder. In this manner, the concentration level is maintained near the membrane surfaces to provide continuous osmotic pumping pressures over long periods of time.

Evaporator and Condenser

The evaporator was designed to be at least equal in volume to the solvent reservoir, as dictated by the solvent depletion control technique. That is, when the heater power is turned off, all of the solvent will be pumped out of the solvent reservoir, and reside in the evaporator (solution reservoir). The evaporator was 15.2 cm (6 in.) in diam and 20.3 cm (8 in.) long. Four heating wells were installed at the bottom of the evaporator. A cartridge heater was placed inside each of the wells.

There were no wicks in the evaporator. Instead, the heater wells were completely immersed in the solution at all times to promote boiling. This eliminated problems related to unwetted surfaces of the heater wells overheating and breaking down the sucrose.

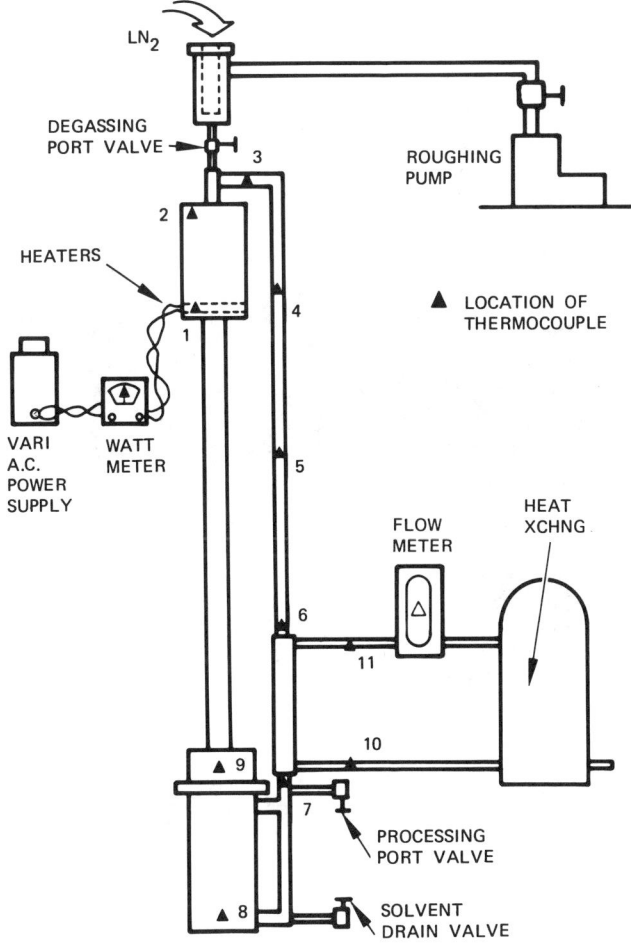

Fig. 10 Baseline osmotic heat pipe test setup.

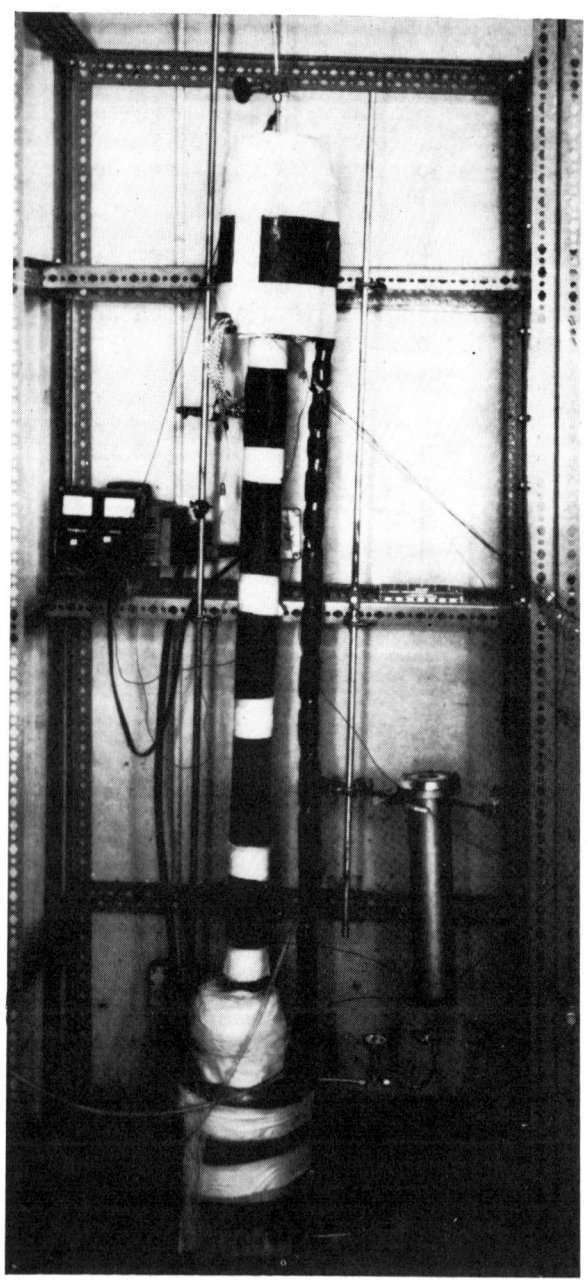

Fig. 11 Baseline osmotic heat pipe with insulation for thermal testing.

Table 2 Maximum power test results

Date	Time	Power, W	\#1	\#2	\#3	\#4	\#5	\#6	\#7	Comments
8/3/79	11:00 a.m.	168	98°F	93°F	92°F	92°F	92°F	93°F	92°F	
	11:30 a.m.		107	102	101	102	101	102	103	
	2:40 p.m.	265	152	142	141	140	141	141	140	
	2:45 p.m.									
	3:15 p.m.		165	157	155	154	153	153	86	
	4:45 p.m.		147	135	135	135	135	134	124	
	5:00 p.m.	0								
8/13/79	9:15 a.m.	160								
	9:30 a.m.	260	129	124	124	124	124	124	124	
	9:45 a.m.	504								
	9:55 a.m.		59°C	56°C	56°C	56°C	56°C	56°C	56°C	Turned power down. Pipe is too hot.
	10:30 a.m.		69	68	67	66	66	67	63	Problems with coolant pump.
	11:00 a.m.	442	75	74	73	72	71	72	60	
	11:01 a.m.									
	11:45 a.m.		85	80	80	81	80	81	52	
	12:05 p.m.	220	83	80	80	80	80	81	53	
	12:10 p.m.									
	12:40 p.m.		83	81	82	80	73	60	49	Turned off to repair coolant pump.
	12:45 p.m.	0								

(Table continued on next page.)

Table 2 (Continued)

Date	Time	Power, W	#1	#2	#3	#4	#5	#6	#7	Comments
	2:50 p.m.	160	61							
	3:00 p.m.		63	60	59	59	59	59	57	Burped non-condensable gas.
	3:30 p.m.		63	60	60	60	60	47	34	
	3:35 p.m.	378		60	60	60	60	60	60	Burped non-condensable gas.
	3:40 p.m.		63	60	60	60	60	60	60	
	3:45 p.m.		64	61	60	60	60	60	60	
	4:00 p.m.		67	62	62	61	61	61	61	
	4:15 p.m.		68	65	65	65	65	65	65	
	4:30 p.m.		69	66	66	66	66	66	66	
	4:50 p.m.	0								
	5:00 p.m.	286								
8/14/79	8:45 a.m.		49	46	46	46	46	46	46	
	9:00 a.m.		52	50	50	50	50	50	50	
	9:20 a.m.		55	52	52	52	52	52	52	
	9:45 a.m.		65	62	62	62	62	62	62	
	10:15 a.m.		72	68	68	68	68	68		
	11:30 a.m.	448								
	11:45 a.m.		70	67	67	67	67	67	67	
	11:50 a.m.		80	75	75	75	75	75	74	
	1:25 p.m.	448								
	1:35 p.m.	639								
	2:05 p.m.		84	80	80	80	80	80	80	

(Table continued on next page.)

Table 2 (Continued)

Date	Time	Power, W	#1	#2	#3	#4	#5	#6	#7	Comments
	2:40 p.m.		89	84	84	84	84	84	–	#7 is bad T.C.
	2:45 p.m.	739								Increased flow rate and started using tap water as coolant.
	4:00 p.m.		93	87	87	87	87	87		
	4:05 p.m.	855								
	4:45 p.m.		63	61	61	61	61	61	–	
	4:50 p.m.	1000								
	5:15 p.m.		65	61	61	61	61	61	–	
	5:30 p.m.		60	56	56	56	56	56	–	
	5:31 p.m.	0								
8/15/79	8:30 a.m.	250								
	10:00 a.m.	1000								
	11:30 a.m.		72	67	67	67	67	67	–	
	1:50 p.m.		75	70	70	70	70	70	–	End maximum power test.
	2:00 p.m.	0								

The condenser was an annular water cooled heat exchanger located on the vapor transport tube just above the membrane module. This aspect of the heat pipe is shown more clearly in Fig. 10.

Heat Pipe Envelope

A material search revealed that neither copper nor Monel were readily available in the large diameter tubes needed for the solution and solvent reservoirs. Therefore, 347 stainless steel was specified for these parts, and Monel was used for the vapor transport tube (Fig. 10) as well as the solution transport tube.

Heat Pipe Testing

After processing, the heat pipe was instrumented for thermal testing as shown in Fig. 10. First, the maximum power or pumping capability was determined by increasing power in 100-200 W increments, allowing the heat pipe to operate at each step for 2 to 4 h. Table 2 is a log of the power and temperature for this test. Performance tests were terminated after the heat pipe operated successfully for 4 h at 1000 W.

After completion of the maximum power test, the heat pipe was set up for life testing. The heat pipe operated continuously with a 500 W heat load for 500 hours. Figure 11 is a photograph of the osmotic heat pipe during life test.

During the test, a slight gas buildup was observed in the condenser. A sample of gas was drawn off into a cylinder, and analyzed using an Inficon IQ 200 residual gas analyzer. No nitrogen was found, indicating that the pipe was leak tight. Neglecting water vapor, which would be expected in any sample drawn from a water heat pipe, the primary constituent of the noncondensable gas was carbon dioxide, with smaller quantities of carbon monoxide, hydrogen, and methane. Hydrogen was probably caused by a slow reaction of water with the stainless steel heat pipe envelope, but the organics present could be due to the oxidation of sucrose.

Conclusions

The basic concepts of osmotic heat pipes are well understood, and components for osmotically pumped systems are available for the temperature range of $25°C$ to $100°C$. Osmotic heat pipes can be designed to transport high heat loads.

Additional development is required to obtain more compact membrane modules for direct osmosis. The hollow fiber module, for example, should not be rejected, based only on the results of the single test reported here. It is conceivable that with proper spacing between fibers, good performance could be obtained.

An osmotic heat pipe made entirely of OFHC copper is recommended for future studies. If the heat pipe contains no dissimilar metals, then certain solutions containing ions could be used without internal corrosion taking place. A nonmetallic heat pipe would be desirable from this point of view.

It is recommended that conventional heat pipes should be used to interface with the heat source and the osmotic heat pipe evaporator. Unlike ordinary resistance type heaters, heat pipes operate at uniform temperatures. Unwetted portions of electrical heaters become overheated resulting in pyrolysis of the solute. This approach would allow wicks to be incorporated into the evaporator for more efficient heat transfer.

Acknowledgments

This paper is based on work performed under U.S. Air Force Contract F33615-77-C-3031. The authors would like to acknowledge C. J. Feldmanis and Dr. L. L. Midolo of the Air Force Flight Dynamics Laboratory for helpful suggestions. Also, R. A. Anderson, for the detail design and testing of the baseline heat pipe, and Dr. J. W. McCutchan of UCLA for providing the tubular membranes and technical advice.

References

[1] Basiulis, A. and Formiller, D. J., "Emerging Heat Pipe Applications," *Third International Heat Pipe Conference*, Palo Alto, Calif., 1978.

[2] "The Passive Pipe in an Active Role," *Vectors*, Hughes Aircraft Co., Volume XII, Spring 1979.

[3] Midolo, Lawrence L., "Heat Transfer Apparatus with Osmotic Pumping," U.S. Patent No. 3,677,337, 18 July 1972.

[4] Minning, C. P., Fleischman, G. L., Giants, T. W., "Development of an Osmotic Heat Pipe," *Third International Heat Pipe Conference*, Palo Alto, Calif., 1978.

[5]Fleischman, G. L. and Anderson, R. A., "An Osmotically Pumped Heat Pipe," *Proceedings of the Fourth Membrane Technology Seminar*, Clemson Univ., Clemson, S. C., March 1979.

[6]Minning, C. P., et al., "Design and Test of a Prototype Osmotic Heat Pipe," *Heat Transfer, Thermal Control, and Heat Pipes*, Progress in Astronautics and Aeronautics, Vol. 70, edited by W. B. Olstad, AIAA, New York, 1980, pp. 307-328.

HEAT PIPE PERFORMANCE WITH GRAVITY ASSIST AND LIQUID OVERFILL

F. C. Prenger, Jr.* and J. E. Kemme+

Los Alamos National Laboratory, Los Alamos, N. Mex.

Abstract

Performance limits for gravity assist heat pipes were investigated with volume of the working fluid and tilt angle as the independent variables. Two limits have been identified, an entrainment limit which results in overheating of the evaporator wall near its midpoint and a pressure drop limit resulting in overheating at the base of the evaporator. The former is a function of the vapor flow characteristics and was observed to be independent of the liquid volume and tilt angle; whereas, the latter is related to the liquid behavior and is strongly related to both the tilt angle and the liquid inventory in the heat pipe.

Nomenclature

A	=	cross-sectional area of the liquid puddle
D	=	heat pipe diameter
F_g	=	gravity force
F_s	=	surface tension force
f	=	Fanning friction factor
G^*	=	dimensionless geometrical parameter (Eq. 13)
g	=	acceleration of gravity
h_{fg}	=	latent heat of vaporization
L	=	length of the heat pipe
q	=	heat pipe power
q^*	=	dimensionless power (Eq. 12)
Re	=	liquid Reynolds number based on puddle depth
u	=	liquid velocity
V	=	volume

Presented as Paper 80-1504 at the AIAA 15th Thermophysics Conference, Snowmass, Colo., July 14-16, 1980. Copyright© American Institute of Aeronautics and Astronautics, Inc., 1980. All rights reserved.

*Staff Member, Advanced Engineering Technology.
+Consultant, Advanced Engineering Technology.

Ve	=	excess liquid volume
δ	=	puddle depth
θ	=	heat pipe elevation angle (horizontal, θ = 0)
ρ	=	liquid density
σ	=	surface tension
φ	=	central angle at any heat pipe cross section
ϕ_0	=	central angle subtended by liquid puddle

Introduction

As terrestrial applications of heat pipes become more numerous the importance of ascertaining performance limits during gravity assist operations increases. The use of simple wick structures or textured internal surfaces in the gravity assist heat pipe enables interaction between the counterflowing liquid and vapor. The wick or textured internal surface serves to distribute the liquid circumferentially however, this same distribution system provides a high impedance, axial flow path for the liquid as it moves, with gravity assist, from the condenser to the evaporator. Heat pipe designers have used the technique of overfill to provide excess working fluid which then forms a free flow liquid passage outside the wick or textured surface. The excess liquid creates a parallel, low impedance flow path and the heat pipe capacity is greatly increased. Substantial improvements in heat pipe performance have been reported at working fluid volumes approaching one-third of the total heat pipe volume.[1]

There appear to be at least three mechanisms which limit heat pipe capacity when excess liquid is present: boiling at the evaporator wall in the pool region,[2,3] azimuthal dryout resulting from pressure gradient limitations,[4] and entrainment of the liquid by the counterflowing vapor.[5] It is

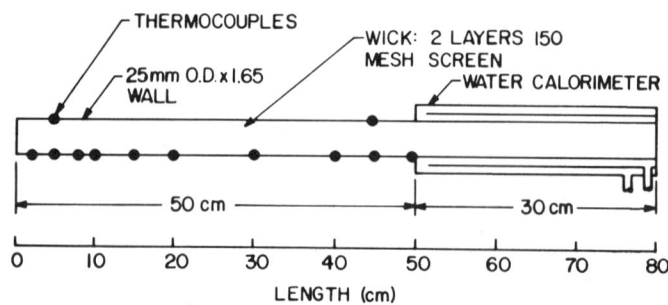

Fig. 1 Heat pipe schematic showing thermocouple locations.

also possible that these mechanisms may interact forming a complex set of conditions which interrupt liquid flow to the evaporator resulting in evaporator dryout.

The authors have long been interested in formulating a physical model describing the fluid dynamics of the gravity assist heat pipe with liquid overfill. In this paper we present the results of experiments with gravity assist heat pipes. We have correlated the results of these experiments with a liquid flow model of the heat pipe.

Experiments

A schematic of the heat pipe tested is shown in Fig. 1. The outer wall is 25.4 mm in diam and 1.65 mm thick. The evaporator is 0.5 m long and contains 12 thermocouples, ten of which are located along the bottom of the heat pipe. The condenser is 0.3 m long and attached directly to it is a two-pass, water-cooling jacket. The heat pipe wick is composed of two layers of 150 mesh screen with a wire diameter of 0.066 mm. All materials are stainless steel. Heat is supplied by induction heating of the evaporator wall, and the power transported by the heat pipe is determined using the flowrate and temperature rise of the condenser cooling water.

The heat pipe was mounted with provision to vary the tilt angle between 0 and 90 deg. Additional working fluid was added between runs at ambient temperature when the heat pipe internal pressure was below atmospheric pressure. Performance limits were obtained by increasing the input power until overheating of the evaporator wall was observed. Two distinct patterns of dryout occurred. At low liquid inventories the bottom of the evaporator would overheat which was an indication of insufficient circulation of the working fluid. This limit could be increased by adding more working fluid or by changing the tilt angle. The second overheating pattern occurred near the midpoint of the evaporator wall and at much higher liquid inventories. It is believed that this limit results from entrainment of the returning liquid by the vapor stream causing dryout along the evaporator wall. Although there is still sufficient working fluid to form a liquid pool in the evaporator, the vapor-liquid interaction limits the circulation of working fluid in this case.

Three working fluids were investigated at various fill volumes and at various tilt angles. Two of the fluids were organic (toluene and methanol), and the third was inorganic (water). The highest performance was obtained with water, whereas the easiest startup was achieved with methanol.

Toluene had the highest operating temperature and extended the temperature range of the test data.

Figure 2 shows the performance data for methanol as a function of tilt angle and working fluid inventory. The volume of the heat pipe exclusive of the wick is 292 cm^3. The wick holds an additional 9.5 cm^3; therefore, liquid volumes of 106, 155, and 203 cm^3 represent approximately one-third, one-half, and two-thirds fill, respectively. The data indicate improved performance with increasing liquid inventory and a strong maximum at 45 deg tilt. At a liquid volume of 13 cm^3 the maximum performance remains constant but the performance at both low and high tilt angles continues to increase with additional working fluid. At 38 cm^3 of working fluid the limit becomes independent of tilt angle but continues to be a weak function of liquid volume. Evaporator dryout occurs at its midpoint and the limit is vapor dependent. A similar behavior is observed for water and toluene, Figs. 3 and 4. For water, however, the entrainment limit begins to decrease at the higher fill levels due to the lower vapor density and hence higher velocity required at a given power level.

The data for all working fluids display a maximum near a 45 deg tilt angle at low liquid inventories. The increase

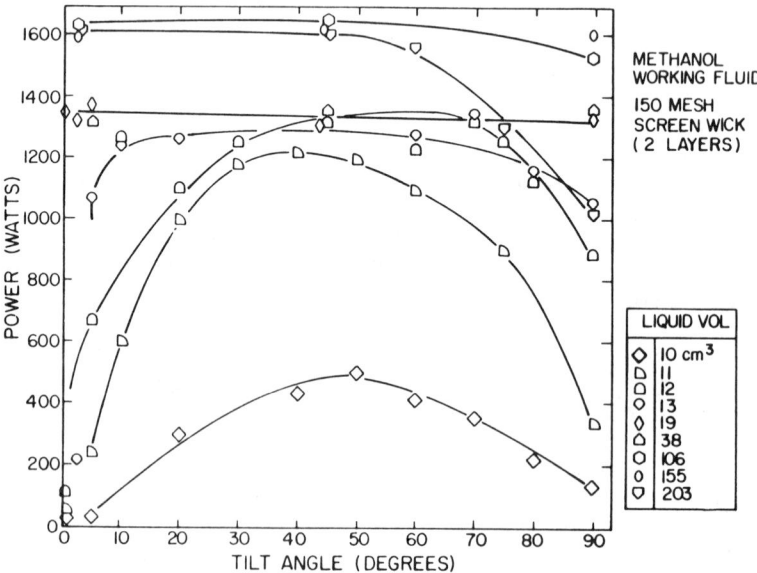

Fig. 2 Heat pipe performance for methanol working fluid.

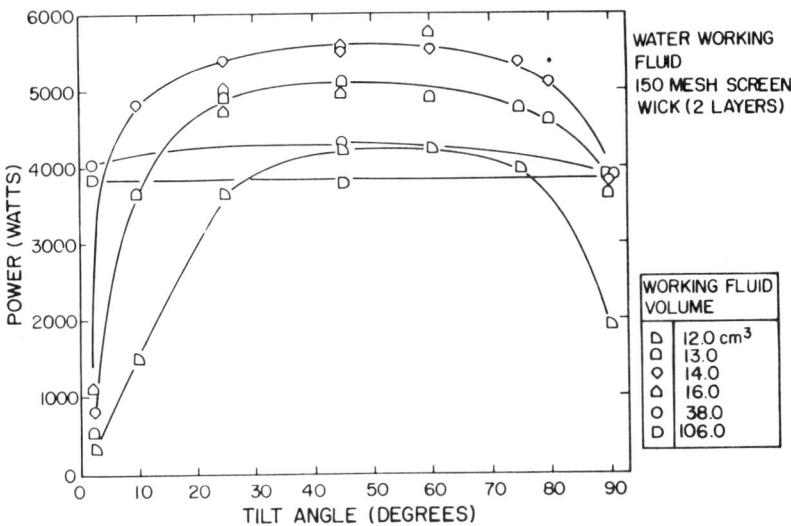

Fig. 3 Heat pipe performance for water working fluid.

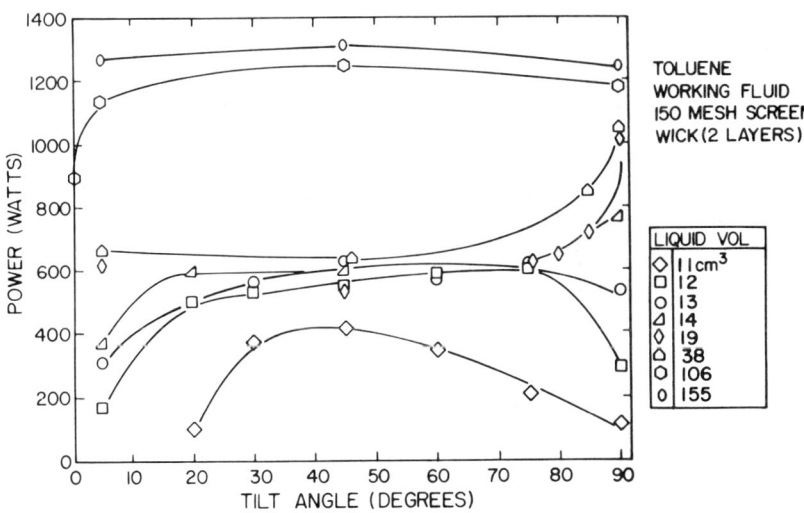

Fig. 4 Heat pipe performance for toluene working fluid.

in performance up to 45 deg tilt is expected since the influence of gravity on the liquid return is increasing. However, the decrease in performance at higher tilt angles is unexpected and may be caused by a change in the liquid flow pattern. At low tilt angles the excess liquid may return in a puddle or trough, whereas at higher tilt angles the liquid may distribute itself around the inside of the wick and fall as a thin film. The accompanying large increase in surface area allows for a stronger frictional interaction of the liquid with both the wick and the counterflowing vapor. This would reduce the liquid flow and also the performance. The tilt angle at which this transition occurs may also be a function of the wick structure.

Some of the data for toluene at low fill levels exhibit contrary behavior. The performance limit increases instead of decreasing at the higher tilt angles. The toulene behavior is similar to both water and methanol at the very low and high liquid volumes but at volumes of 14, 19, and 38 cm^3 this opposite trend occurs. Additional tests with the toluene may be required to understand this phenomenon.

Liquid Limit Model

At low vapor velocities the process whereby the circulation of the heat pipe working fluid is interrupted is strongly dependent on the liquid flow characteristics. In particular since the performance limits reach a maximum near 45 deg tilt angles, there must be a second mechanism retarding the liquid flow which offsets the increasing hydrostatic force as the tilt angle is further increased. Any successful model of the liquid flow inside the heat pipe must include this effect.

The formation of a free flow puddle or trough of liquid is augmented by gravity when the heat pipe is horizontal. Opposing the gravity force and causing a spreading of the liquid around the heat pipe circumference is the capillary or surface tension force. A force balance is shown in Fig. 5. To predict the heat pipe performance it is necessary to relate the depth of the puddle to the tilt angle and liquid volume.

A Bond number can be defined for this configuration as follows

$$B_o = \frac{\text{Gravity force}}{\text{Surface tension force}} = \frac{F_g}{F_s} \qquad (1)$$

The surface tension force is

$$F_s = \sigma L \qquad (2)$$

HEAT PIPE PERFORMANCE

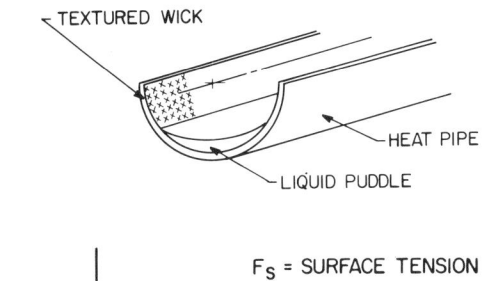

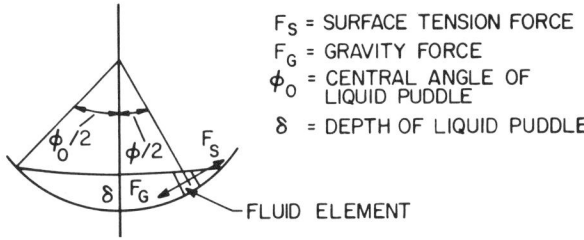

Fig. 5 Excess liquid model of gravity assist heat pipe.

and the gravity force is

$$F_g = \rho g \int_0^{V(\phi_0)} \sin \frac{\phi}{2} \, dV(\frac{\phi}{2}) \qquad (3)$$

where the limit of integration ϕ_0 is the central angle subtended by the liquid puddle. The volume element is

$$dV(\frac{\phi}{2}) = \frac{LD}{2} \delta(\phi) d(\frac{\phi}{2}) \qquad (4)$$

Assuming the surface of the liquid puddle is flat, the depth is related to the central angle by

$$\delta(\phi) = \frac{D}{2} \left[1 - \frac{\cos \phi_0/2}{\cos \phi/2} \right] \qquad (5)$$

Substituting Eq. (5) into Eg. (3) and integrating gives

$$F_g = \frac{\rho g D^2 L}{4} \left[1 - \cos \frac{\phi_0}{2} (1 - \text{Log} \cos \frac{\phi_0}{2}) \right] \qquad (6)$$

In Eq. (6) the term

$$1 - \text{Log} \cos \frac{\phi_0}{2} \doteq 1$$

for small values of ϕ_0. For ϕ_0 less than 60 deg the error is less than 6%, however, the error increases rapidly with increasing ϕ_0, reaching 59% for ϕ_0 = 150 deg. This corresponds to a liquid puddle filling one-third of the cross section of the tube. Simplifying Eq. (6) and noting that

$$\frac{2\delta_0}{D} = 1 - \cos\frac{\phi_0}{2} \tag{7}$$

gives

$$F_g = \rho g D L \delta_0 / 2 \tag{8}$$

Using Eqs. (2) and (8) the Bond number for $\theta = 0$ (no tilt) becomes

$$B_{o_{\theta=0}} = \frac{\rho g D \delta_0}{2\sigma} \tag{9}$$

When the gravity and surface tension forces are of the same order the Bond number is unity and Eq. (9) can be solved for δ_0 giving

$$\delta_{o_{\theta=0}} = \frac{2\sigma}{\rho g D} \tag{10}$$

Equation (10) can be generalized for other tilt angles by introducing the cosine of the tilt angle or

$$\delta_o = \frac{2\sigma \cos\theta}{\rho g D} \tag{11}$$

The limiting performance of the heat pipe at low vapor velocities is reached when the pressure drop of the returning liquid equals the hydrostatic head, then

$$\rho L g \sin\theta = 2 f \rho u^2 \frac{L}{\delta_o} \tag{12}$$

Equation (11) can be rewritten in terms of the heat pipe power as

$$q = \rho h_{fg} A \left(\frac{\delta_o g \sin\theta}{2f}\right)^{1/2} \tag{13}$$

Substituting the expression for δ_o, Eq. (11) into Eq. (13) and noting that

$$V_e = AL$$

gives

$$\frac{q}{\left[\rho h_{fg}^2 D^3 \sigma\right]^{1/2}} = \frac{1}{\sqrt{f}} \left(\frac{V_e}{D^2 L}\right) (\sin\theta \cos\theta)^{1/2} \tag{14}$$

Equation (14) is expressed in dimensionless form. The left side contains the heat pipe power and fluid properties while the right side contains the excess liquid volume, tilt angle, and heat pipe dimensions. The trignometric term in Eq. (14) has a value of zero at tilt angles of 0 and 90 deg. Although the heat pipes were not tested at tilt angles less than 2.5 deg, the heat pipe performance at both low tilt angles and in the vertical position were finite. Therefore, two phase angles θ_1 and θ_2 are introduced into the trignometric term of Eq. (14) to account for these experimental observations. The values for the phase angles were found to be

$$\theta_1 = 2.4 \text{ deg}$$

$$\theta_2 = 10 \text{ deg}$$

Eq. (14) can now be written as

$$\frac{q}{\left[\rho h_{fg}^2 D^3 \sigma\right]^{1/2}} = \frac{1}{\sqrt{f}} \left(\frac{V_e}{D^2 L}\right) [\sin(\theta-2.4)\cos(\theta-10)]^{1/2} \quad (15)$$

By defining the following dimensionless groups

$$q^* = \frac{q}{\left[\rho h_{fg} D^3 \sigma\right]^{1/2}} \quad (16)$$

and

$$G^* = \frac{V_e}{D^2 L} [\sin(\theta-\theta_1)\cos(\theta-\theta_2)]^{1/2} \quad (17)$$

a correlation of the form

$$q^* = C(G^*)^n \quad (18)$$

is obtained. The test data are shown in Fig. 6 in terms of q^* and G^* and are correlated by the equation

$$q^* = 22.5 \ G^* \quad (19)$$

The slope of the line in Fig. 6 is 1.0 which is in agreement with the model. A constant puddle depth along the heat pipe axis was assumed in the derivation. Inclusion of a variable puddle depth would change the form of Eq. (3) requiring a double integration. Also the inclusion of a vapor pressure drop term may improve the accuracy, although for low powers the vapor velocities are small.

Some of the data are influenced by liquid-vapor interaction, notably the data for water at

$$G^* > 1.0 \times 10^{-2} \qquad (20)$$

It appears in this case that the liquid limit is no longer controlling and that entrainment may be occurring. This implies that the liquid limit model here presented may not be valid for values of $G^* > 1.0 \times 10^{-2}$. Other heat pipes may exhibit entrainment limits at lower G^* values. Unfortunately more data are needed to verify this aspect.

The form of the correlation, Eq. (14), implies a constant friction factor. By considering the liquid flow to be lami-

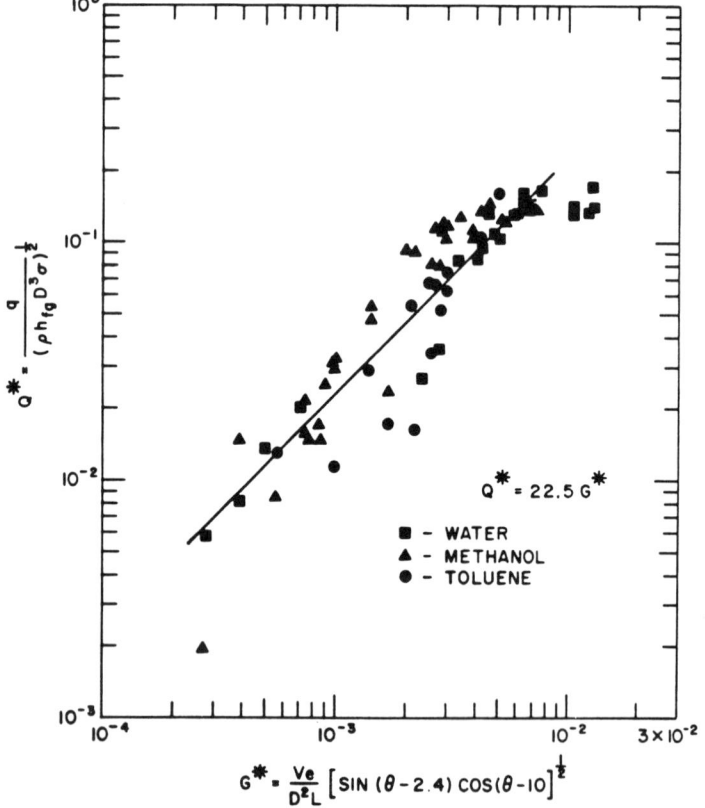

Fig. 6 Data correlation for liquid limit.

nar, a variable friction factor such as

$$f = 16/Re \qquad (21)$$

could be included. This would introduce the liquid viscosity into the definition of q* which is a logical addition and should be considered in future work.

Since the same heat pipe was used for all tests the effect of wick geometry on the correlation was not evaluated. Further testing using different liquid distribution systems such as knurled or threaded surfaces is planned. These heat pipes will also have different physical dimensions and will be used to evaluate the appropriateness of the correlating parameters.

Conclusions

1) Performance limits of gravity assist heat pipes using the working fluids water, methanol, and toulene were experimentally investigated.

2) Two evaporator dryout mechanisms were observed. One occurred at the beginning of the evaporator and was observed at low tilt angles and small excess liquid volumes. These limits are successfully described by a liquid pressure drop model. The other mechanism produced dryout near the evaporator exit and was associated with high power and large excess liquid inventories. This limit was independent of tilt angle and liquid inventory and is believed to be associated with entrainment. Additional work in this area is planned.

3) A correlation of the test data was made using a liquid pressure drop model with a constant friction factor. The results are given by Eq. (19) in terms of a dimensionless power and a dimensionless geometric parameter.

Acknowledgments

The authors express their gratitude to the Office of Basic Energy Sciences of the U.S. Department of Energy for financial support of this work. We also thank Michael Elder for building the heat pipes and for providing technical support during the testing phase.

References

[1] G. M. Grover, private communication, Nov. 1979.

[2] H. N. Chi and A. Abaht, "Performance of Gravity-Assisted Copper-Water Heat Pipes with Liquid Overfill," Third International Heat Pipe Conference, Palo Alto, Calif., May 22-24, 1978.

[3] H. Nguyen-Chi and M. Groll, "The Boiling Limit of Gravity-Assisted Copper-Water Heat Pipes with Screen Wick Structure and Liquid Overfill," AIAA 14th Thermophysics Conference, Orlando, Florida, June 4-6, 1979.

[4] C. A. Busse and J. E. Kemme, "The Dry-Out Limits of Gravity-Assist Heat Pipes with Capillary Flow," Third International Heat Pipe Conference, Palo Alto, Calif., May 22-24, 1978.

[5] C. L. Tien and K. S. Chung, "Entrainment Limits in Heat Pipes," Third International Heat Pipe Conference, Palo Alto, Calif., May 22-24, 1978.

THE HEAT PIPE THERMAL CANISTER

W. Harwell*
Grumman Aerospace Corporation, Bethpage, N.Y.

and

S. Ollendorf[+]
NASA Goddard Space Flight Center, Greenbelt, Md.

Abstract

The thermal canister represents a new approach to instrument thermal control for Shuttle experiments which require tight temperature control. The canister substitutes a known, benign thermal environment for the variable and uncertain environment of space, the space environment being neutralized by a system of feedback controlled variable conductance heat pipes. A proto-flight unit has been fabricated and this paper describes the acceptance thermal vacuum test results of the thermal canister experiment which will fly on an early Space Shuttle flight.

Nomenclature

$-Z_{LV}$ = -Z axis in local vertical
$-X_{SI}$ = -X axis solar inertial
$+Z_{SI}$ = +Z axis solar inertial
PTC = Passive Thermal Control (shuttle constraint)

Introduction

In 1975, the thermal branch at NASA Goddard Space Flight Center began to examine the technology base required to sup-

---

Presented as Paper 80-1461 at the AIAA 15th Thermophysics Conference, Snowmass, Colo., July 14-16, 1980. Copyright © American Institute of Aeronautics and Astronautics, Inc., 1980. All rights reserved.
*Senior Engineer, Thermal Systems Department.
+Head, Spacecraft Component Development and Analysis Section.

port their small scientific instrument development program, these instruments being ultimately destined to fly on the Space Shuttle. As part of their program, they funded a number of studies to examine the thermal canister approach to simplifying thermal design of instruments. The outcome of these studies was proof-of-concept hardware, which was tested in 1977 (see Ref. 1). This hardware showed that an active feedback control system with variable conductance heat pipes as the active element could give the tight, uniform temperature environment required by sophisticated instruments.

This resulted in NASA funding a flight demonstration unit to be flown on an early Space Shuttle mission (see Ref. 2). This unit was designed and fabricated by Grumman Aerospace Corporation and has passed a rigorous series of tests at GSFC. These tests included vibration, launch phase simulation, blowdown, acoustics and a preliminary thermal vacuum test. Additional electronics were added and the flight configuration was subjected to a second vibration test, electromagnetic/radio frequency interference tests and an acceptance thermal vacuum test. This paper describes the thermal canister experiment to be flown on the Orbital Flight Test Vehicle No. 4 (OFT-4) as part of the Office of Space Sciences Payload (OSS-1) and discusses the acceptance thermal vacuum test results.

Design

Control Concept

The Thermal Canister Experiment (TCE) consists of a rectangular box, the aluminum side skins being isothermalized by a system of fixed conductance heat pipes (FCHPs) which collect the thermal energy dissipated by enclosed electronic boxes and conducts this heat load to a system of variable conductance heat pipes (VCHPs) which ultimately reject the heat to space.

VCHPs operate by virtue of a noncondensible gas which is normally contained in a gas reservoir at one end of the pipe. The gas expands out of the reservoir into the condenser as the active pipe temperature decreases. The noncondensible gas in the condenser blankets the condenser surface, effectively stopping heat pipe action in that part of the pipe. The length of condenser rendered inactive depends upon the temperature distribution along the pipe. Such a device can be designed to operate in the passive mode over approximately a 30°C temperature range, the actual operating point depending upon the evaporator heat input and the equivalent external environment. Tighter temperature control can be obtained by controlling the reservoir temperature using active feedback control (see Ref. 3). In the TCE approach, as shown schematically in Fig. 1, the temperature sensor of the control loop is located on the canister wall (or by uplink command on an experiment node). The controller is a proportional controller with a 1°C proportional control band and the controlled element is an electric heater bonded to the VCHP reservoir. As the temperature at the controlled point falls, the heater is turned on, thus forcing noncondensible gas out of the reservoir. When the heater is turned off, the reservoir temperature falls and the gas contracts into the reservoir. In an active feedback controlled VCHP control system, the temperature excursion at the control point depends upon the controller characteristics and reservoir parameters. In the TCE the range has been set at less than ±1°C.

Design

The TCE, as shown in Figs. 2 and 3, is a 3 m x 1 m x 1 m rectangular box, the four sides being bolted to corner longerons.

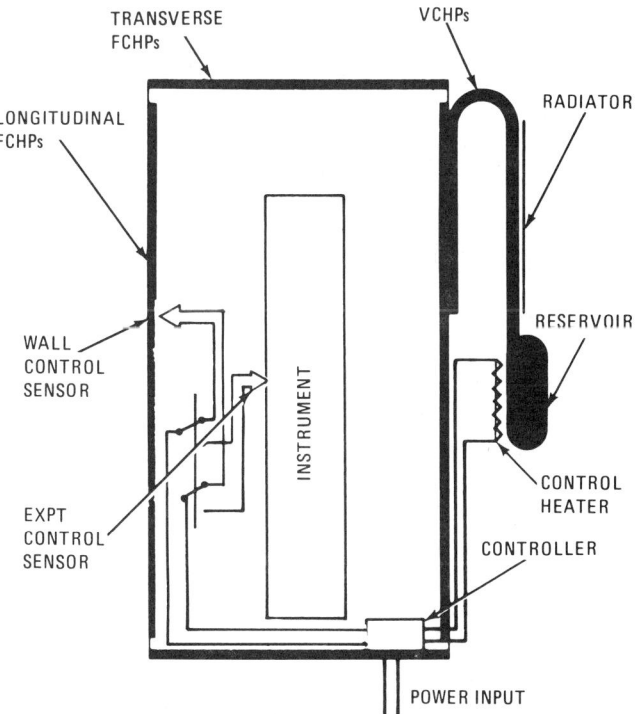

Fig. 1 Control concept

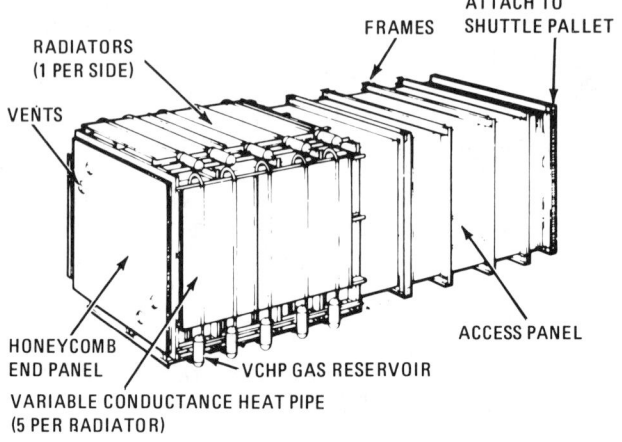

Fig. 2 Canister assembly

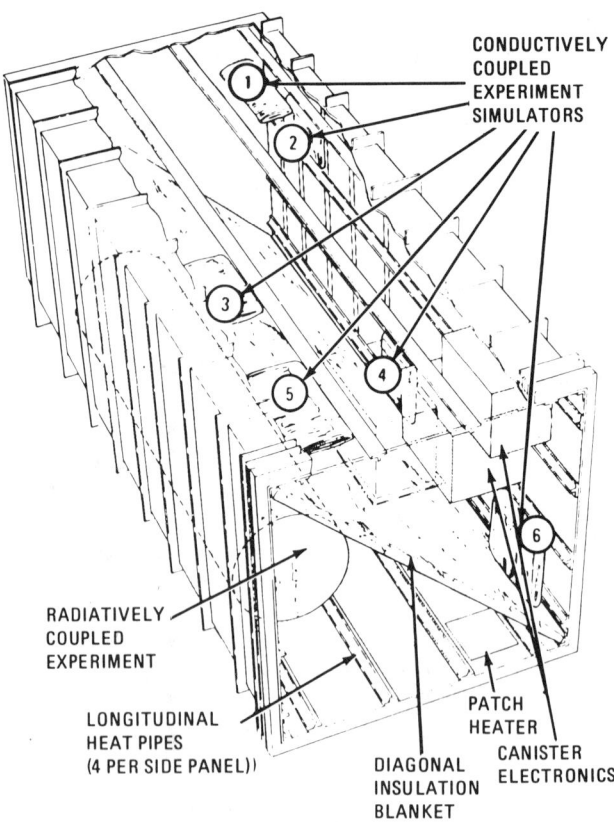

Fig. 3 OFT4 canister experiment configuration

Each side has four 3 m-long FCHPs and two 1 m-long FCHPs bolted to it to provide both longitudinal and transverse isothermalization. The enclosure is completed with two 9.5 mm-thick honeycomb panels bolted to the end frames. A limited degree of thermal isolation between the side skins is achieved by having fiberglass shims between the side skins and longerons.

Heat rejection from the TCE is through four silver-backed Teflon-coated radiators mounted to the top section of the canister. Each radiator consists of five radiator segments, each segment being coupled to the canister through a VCHP. The segments are isolated from each other by fiberglass shims and mechanically supported away from the canister walls by fiberglass channel supports. Multilayer insulation blankets are positioned between the canister wall and the radiators and cover all external surfaces, except for the radiators and small windows left in the insulation around each reservoir. This window is covered with silver-backed Teflon and provides a known coupling between the reservoir and its surroundings. All multi-layer blankets consist of fifteen layers of double aluminized Kapton with Dacron spacers. An outer layer of Beta cloth is used to provide a diffuse external coating. The canister assembly is shown in Fig. 2.

The TCE is end mounted, through four tension bolts, to a titanium adaptor plate which in turn is mounted by twelve bolts to the OSS-1 pallet.

All heat pipes on the TCE, as shown in Fig. 4, were fabricated from axial grooved aluminum extrusion with an integral flange (see Ref. 4). The charge fluid is ammonia and the noncondensible gas in the VCHPs is nitrogen.

<u>Control System</u>

Active feedback control is achieved by using the output of a sensor at the control point as the input signal to a hardwired proportional controller, which in turn, controls a small heater bonded to each reservoir. Four separate controllers are installed, each controlling the five reservoir heaters on a radiator assembly. The controllers have a 1°C proportional control band, proportionality being achieved by duty cycling the heaters. The set point of each controller can be selected by uplink command with sixteen selectable control temperatures in the range -5 to 30°C (i.e., approximately 2°C increments). Additional electronics are included to allow selection of control sensor which may be positioned either on a side wall or on an experiment node. The controllers are wired in pairs to provide backup capability in the event that one fails.

The TCE has an on-board microprocessor which has been programmed as a smart controller. The basic control algorithm is proportional band control, but refinements include the following: 1) each reservoir heater is separately controllable, rather than in banks of five; 2) upper and lower temperature thermostat routines to prevent temperature excursions by the reservoir out of the control range; 3) comparing the desired control temperature to real time computed min/max controllable range based upon measured internal dissipation and external environment and adjusting the control temperature correspondingly; 4) projecting from a transient to steady state condition to prevent over/undershoot. Additional details of the control system are given in Ref. 5.

Experiment Details

Experiment simulators have been built into the TCE to represent radiatively coupled and conductively coupled components. The radiatively coupled simulator is a 2 m x 0.38 m D cylinder representing a telescope tube. The cylinder is painted black and has five electric heaters bonded to it. These heaters are separately controllable by uplink command. The conductively tied simulators are six aluminum plates, each weighing 3 to 8 lbs, with different conductive couplings to

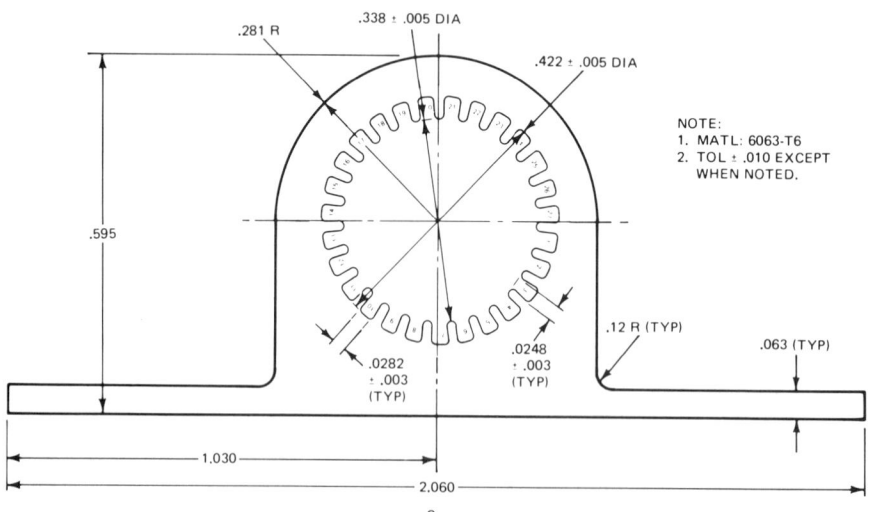

Fig. 4 Heat pipe extrusion

the canister wall and each with an electric heater of 10 or 40 W. The heaters are also separately controllable by uplink command. Although all four side walls are separately controlled, the arrangement of the experiment for OSS-1 maintains only two temperature regions - the radiative enclosure and the conductive enclosure. The experiment configuration is shown in Fig. 3.

Flight Electronics

The TCE has its own instrumentation and electronics to support the flight. Instrumentation consists of 43 thermistors on the side panels and experiment simulators, one platinum resistance thermometer on each of the 20 reservoirs, and one platinum resistance thermometer on each of the 20 radiator segments. There are also housekeeping measurements of voltage and current. The electronics consist of power control and signal conditioning units, a PCM, microprocessor, and a tape recorder. The microprocessor serves three functions:

1) It has a programmed timeline controlling the simulation heaters and changing set point temperature to a predetermined mission timeline. This timeline can be changed by uplink command.

2) It formats the data to give three parallel data streams which are recorded on various tape recorders.

3) It has a built-in algorithm which allows it to replace the proportional controllers and act as a "smart" controller.

Power for the TC will be drawn from the 28V Shuttle bus.

Test Configuration

The TCE has heat pipes running in three directions. It was therefore impossible to put the canister into any one orientation which would allow all heat pipes to be tested simultaneously in the capillary flow mode. Axial grooved FCHPs have been flown on a number of spacecraft - the Orbiting Astronomical Observatory - Copernicus (OAO-C), Applications Technology Satellite-F (ATS-F), International Ultraviolet Explorer (IUE), Solar Maximum Mission (SMM) - and their operation is well understood. Therefore, for the TCE acceptance thermal vacuum (ATV) test, it was decided that operation of the FCHPs in the boiler mode would be acceptable, provided all VCHPs were operating in the heat pipe mode. For the ATV test, the canister was set up vertically in the GSFC 10 ft x 15 ft chamber. In this configuration, the VCHPs have a built-in tilt of 3 mm with the evaporator elevated above the condenser while the FCHPs are either vertical (longitudinal pipes) or horizontal (transverse pipes). The experiment simulators and flight electronics are mainly situated towards the lower end of the canister. Therefore, by applying heat to the lower end of the canister and rejecting from the upper end, the vertical FCHPs would start operating in the reflux mode (liquid return by gravity down the pipe walls).

The incident external environment was simulated by electric heaters bonded to the underside of the radiators, reservoirs, flux sensors and to the outer layer of the multi-layer insulation blankets. These heater circuits were all controllable outside the thermal vacuum (TV) chamber. The chamber wall was held at liquid nitrogen temperature.

In addition to the flight instrumentation, approximately 250 thermocouples were distributed about the canister to monitor its performance during ATV test.

Test Series

The ATV test was performed in three phases. Familiarization or hands-on operating experience was gained during Phase I. In Phase II, eight commands available with a back-up command system were verified. Phase III was a typical mission profile. Only Phase III test data are discussed in this paper.

All testing was controlled by a programmed timeline burned into the microprocessor. Real time adjustments were made via simulated vehicle telemetry to incorporate experience gained during the running of the test.

Discussion of Results

The object of the TCE flight experiment was to take advantage of a typical OFT-4 mission to demonstrate temperature control at a number of set point temperatures with control sensors on either the canister wall or an experiment node, with both constant and variable experiment dissipation and with control options of hardwired proportional control mode, microprocessor algorithm control or passive control. The design OFT-4 mission has four orbital conditions ranging from the very cold $-Z_{LV}$ to the hot $+Z_{SI}$ (see Table 1). The limited energy budget available to support the TCE flight experiment required low temperature set point control to be demonstrated in cold environment conditions and high temperature control in hot environments.

A total of 19 control points were demonstrated during Phase III of the ATV test. These control points are summarized in Table 1. At each of the test points where the wall temperature controlled, the recorded variation in temperature at the control points were less than ±0.75°C, compared to the design value of ±1°C. With experiment node control and duty cycling heaters, the variation in temperature at the control point was ±4°C on the radiative tied cylinder and ±5°C on the conductive plates. The design goal for experiment control was ±4°C.

Some typical test points are discussed below.

Table 1 Summary of control points

Test Point No.	Flux Cond ①	Control Mode ②	Control Temperature °C				Power	
			Radiative		Conductive		Rad Watts	Cond Watts
			Wall	Expt	Wall	Expt		
1	PTC	PC	14	–	14	–	37	113 ③
3	PTC	MP	14	–	14	–	120	121
5	$-Z_{LV}$	PC	–	14	–	14	95	173
6	$-Z_{LV}$	PC	–	16	–	14	95	173
7	PTC	PC	–	16	–	14	93	113
8	$-Z_{LV}$	PC	9	–	9	–	113	124
10	PTC	PC	4	–	4	–	70	113
11	$-X_{SI}$	PC	14	–	14	–	79	253/114
12	$-X_{SI}$	PC	–	14	–	11	78/0	157/115
13	$-X_{SI}$	PC	–5	–	–5	–	71	155/114
14	$-X_{SI}$	MP	0	–	0	–	98	126/86
15	$-X_{SI}$	PASSIVE	–	–	–	–	71/0	86
18	$-X_{SI}$	MP	9	–	–1	–	103	165
21	$-X_{SI}$	PC	–1	–	9	–	103/0	195/115
22	$-X_{SI}$	PC	–1	–	16	–	103	193
23	$+Z_{SI}$	MP	10	–	20	–	38	86
24	$+Z_{SI}$	PC	11	–	21	–	38	114
26	$+Z_{SI}$	PC	–	25	21	–	63	114
27	$+Z_{SI}$	PC	–	25	23	–	64/0	114

①   PTC   Passive Thermal Control   ②   PC   Hardwired Proportional Controller
     $-Z_{LV}$ –Z Axis In Local Vertical   MP   Microprocessor Algorithm Control
     $-X_{SI}$ –X Axis Solar Inertial   Passive   Controllers Disabled
     $+Z_{SI}$ +Z Axis Solar Inertial   ③   Includes Electronics

Panel Wall Control

Typical temperature histories with panel wall control are shown in Fig. 5. The figure shows the transient response of the side walls local to the control points as the canister temperature increased from a 4°C control point (test point 10) to a 14°C control point (test point 11) with hardwires proportional control mode. The figure also shows the temperature of the blocked section of four of the radiator elements responding to the XSI absorbed environment. The blocked section of the radiator corresponds to the section of the VCHP blanketed by noncondensible gas and the radiator in this area will operate as a reradiating surface insulated on the back side. Finally, in Fig. 5, the temperature histories of two of

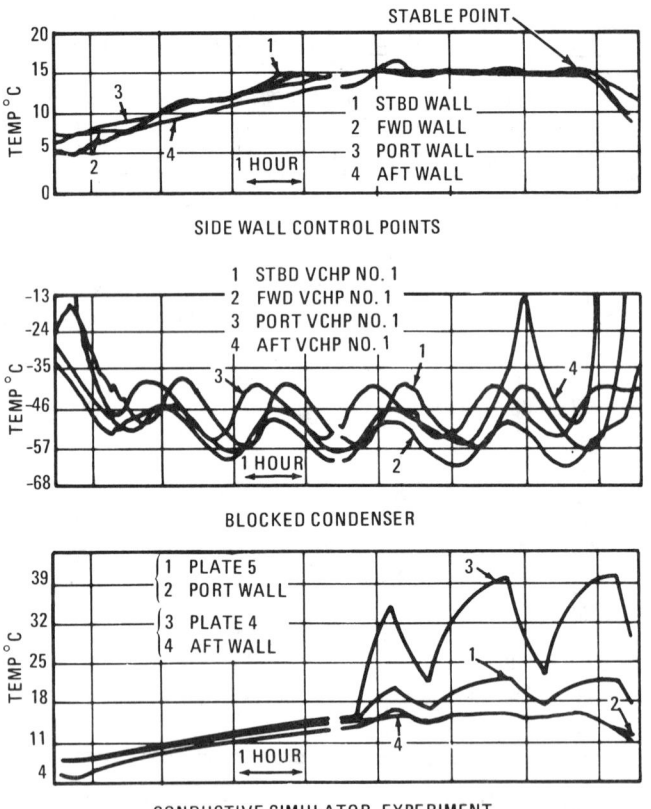

Fig. 5 Temperature history approaching stable point - control sensors on wall at 14°C set point, proportional control (TEST POINT #11).

the conductive plate simulators are compared to their corresponding wall control point variations. During the last 3½ hr leading up to the 14°C stable point, these two heaters were being duty cycled 60 min ON (40 W), 30 min OFF. The difference in response of Plates 4 and 5 is due to their different masses, which were 6 lbs and 8 lbs, respectively. While the radiators were cycling ±10.5°C and the Plate 5 experiment simulator ±9.3°C, the side walls were cycling ±0.6°C. The maximum spatial temperature variation recorded on the starboard and forward side panels were each 2°C, with a simulated radiatively tied experiment input of 79 W. The corresponding variation in the port and aft panels was 5°C and 4.3°C, respectively, with heat input varying from 253 W (60 min) to 113 W (30 min). Local hot spots of approximately 12°C above control point temperatures were recorded in the port and aft panels local to Plates 5 and 6 when these heaters were enabled. During the ATV test this was thought to be due to nucleate boiling in the liquid pool filling the bottom of each longitudinal FCHP. The test conditions have since been simulated on a 3 m-long FCHP operating in the boiler mode, but the anomaly did not reappear. The reasons for the local hot spots remains unexplained at present and it will be interesting to determine if they reoccur during zero-gravity flight.

The temperature histories for flight thermistors local to the hardwired proportional controller wall control sensors at two microprocessor control points are summarized in Fig. 6. Both control points were with $-Z_{LV}$ absorbed environment fluxes, a very cold condition. Control was set for 4°C with 107 W in the radiatively tied experiment and 117 W in the conductively tied experiment. The wider variation in wall temperatures shown in Fig. 6, compared to Fig. 5, is due to the way the algorithm operates. The algorithm uses a number of control sensors on each panel compared to the one sensor used by each hardwired controller; the algorithm uses average panel temper-

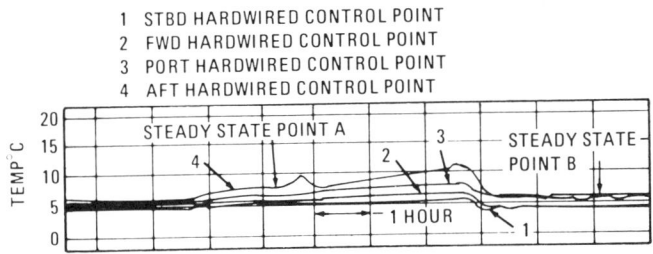

Fig. 6 Temperature history, microprocessor control mode at 4°C with control sensors on wall (Phase I testing).

atures rather than single point temperatures; the algorithm attempts to minimize gradients within each panel. The recorded spatial temperature variations in the side panels at steady state point A (see Fig. 6) were 1.3, 1.7, 2.1 and 2.3°C in the starboard, forward, port, and aft panels, respectively.

At Point A, the dissipations were increased to 138 W in the radiatively tied experiment and 142 W in the conductively tied experiment. Although the condensers of the VCHPs were not fully open the side panel temperatures slowly increased. This was traced to the reservoir minimum temperature "thermostat" subroutine in the algorithm being set too high, thus preventing the reservoir from cooling down to allow the VCHPs to open up. The thermostat subroutine was modified by uplink command, the VCHPs opened up and the canister snapped into control at Point B (see Fig. 6).

Zone Control

Temperature histories with two zone temperature control are shown in Fig. 7. The data shown corresponds to 38 W in the radiative tied experiment and 114 W in the conductive tied experiment (test point 24). At the stable condition, the panel wall temperatures varied by less than ±0.6°C and ±0.75°C in the radiative and conductive enclosures, respectively.

Experiment Node Control

Experiment node control is shown in Fig. 8. In this mode, the radiative cylinder was being controlled to 14°C and

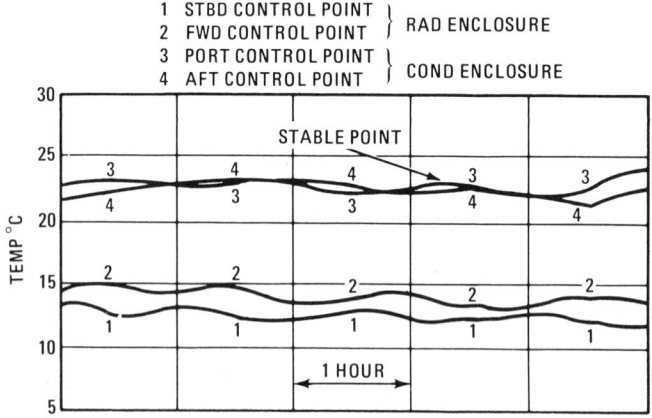

Fig. 7 Temperature history, side wall temperature control, radiative enclosure at 11°C conductive enclosure at 21°C with hardwired controller (Test point #24).

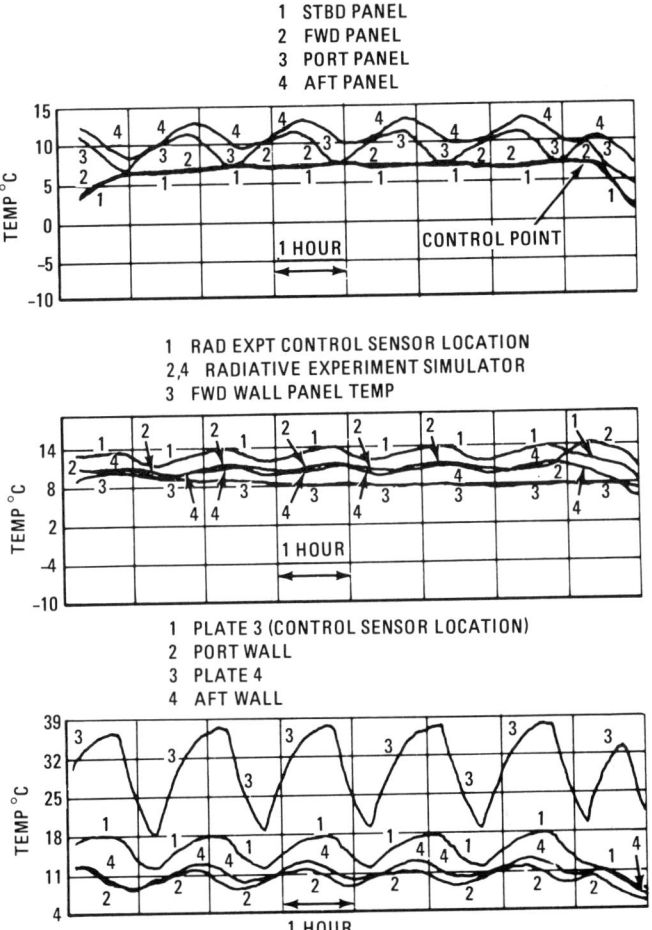

Fig. 8 Temperature history, experiment node control, rad at 14°C cond at 11°C hardwired proportional control (Test point #12)

conductive Plate 3 to 11°C. The flight timeline burned into the microprocessor calls for the temperature of two radiative nodes and two conductive nodes to be used by the proportional controllers. Early in Phase I testing it became obvious that the two controllers in an enclosure were fighting each other to try to maintain control. This was particularly true with conductive experiment simulators (Plates 3 and 4), as their different conductive couplings (0.88 and 1.53 W/°C) and heater dissipations (10 W and 40 W) were forcing cross-talk between the two controllers in an enclosure. (This would not have been a problem if the two sides had been conductively and radiatively decoupled from each other.) In Phase III, the timeline was changed by uplink command to put the controllers into the redundant mode; that is, operating with only one active controller per enclosure and only one control sensor per experiment. The data shown in Fig. 8 correspond to duty cycling the radiative experiment 78 W (60 min) to 0 W (30 min) and the conductive experiment 157 W (60 min) to 115 W (30 min) with $-X_{SI}$ absorbed fluxes (test point 12). The temperature of the radiative cylinder local to the control point cycled ±2°C, while the adjacent panels cycled ±0.5°C. The radiative cylinder control sensor was near a 40 W heater which was being cycled On/Off. Conductive Plate 3 cycled ±3°C, while the adjacent panel cycled ±2°C. Conductive Plate 3 had a 10 W heater which was being cycled On/Off. The differences in temperature excursions between the cylinder and plate was due to the mass capacitance effect and the hard conductive coupling, compared to the soft radiative coupling. This test point indicates the difficulty of controlling an experiment node which has low mass combined with internal dissipation.

Combined Enclosure/Experiment Control

Simultaneous control of the radiatively tied experiment and the conductively tied enclosure are shown in Fig. 9. At this point in the test the radiative experiment was cycling 64 W (60 min) to 0 W (30 min), while the conductive experiment was dissipating a total of 114 W with only Plate 1 heater enabled. The external flux condition being simulated was the 100% suntime $+Z_{SI}$ case. The conductive simulation plate ran at constant temperature, while the side panels cycled approximately ±0.5°C in phase with the external environment. Part of the variation in side panel temperature would also be due to hunting by the hardwired controller. The radiative cylinder, on the other hand, cycled ±4°C and the side panels ±2°C in phase with the cycling heaters.

Active Reservoir Control

The concept of active control with VCHPs is that as the temperature at the control point falls, the heater on the reservoir should rapidly drive up the reservoir temperature, forcing noncondensible gas into the condenser section and hence closing down the pipe. The reservoir heater must therefore be sized to compensate for uncontrolled heat leaks, thermal capacity of the reservoir and the controlled heat leak through the reservoir "window". The thermal capacity of each reservoir is approximately 0.1 Wh/°C and the uncontrolled heat

HEAT PIPE THERMAL CANISTER

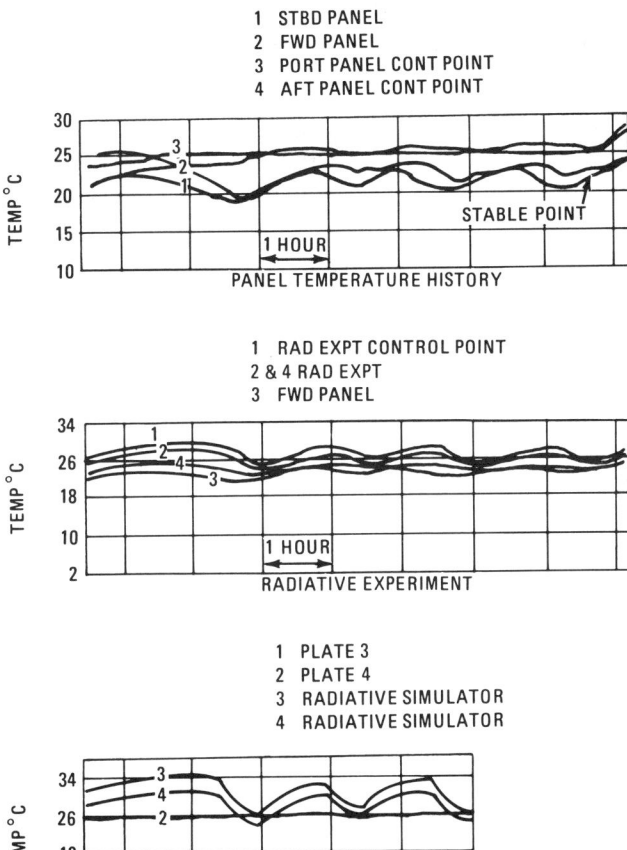

Fig. 9 Temperature history, radiative experiment at 25°C, conductive enclosure at 23°C, proportional control mode (Test point #27).

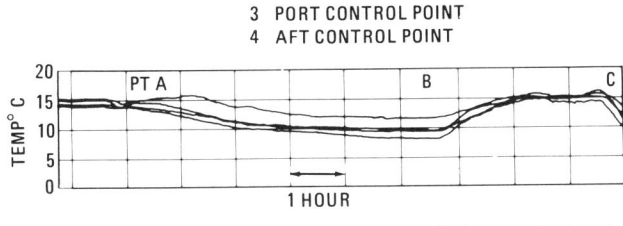

Fig. 10 Temperature history (Phase I testing)

leak approximately 0.25 W. The 5 W of heater power available to each reservoir should therefore be able to rapidly drive the reservoir temperature up to a steady state condition of approximately 60°C, at which point the VCHP would be completely closed down and losing heat only by conduction along the pipe walls (approximately 2.5 W per pipe in the worst case). The total heat leak from the canister would then be approximately 50 W through the VCHPs, 45 W through the multi-layer insulation blankets and 20 to 30 W through the canister feet. A total heat input of approximately 120 W should therefore be sufficient to hold the control point with the reservoir heaters duty cycling below 100%. At several points in the ATV test, however, the reservoir heaters were not able to hold the canister temperature, although they were fully on and the VCHPs would not shut down. This is shown in Figs. 10 and 11. At point B, Fig. 10, the canister temperature was drifting down, although the total internal dissipation was 350 W in a $-Z_{LV}$ external environment. As shown in Fig. 11, some VCHPs were partially open and the reservoir heaters would not close the pipe down. Using the measure data a heat balance on the reservoirs shows:

Heat In

Absorbed environment	0.5 W
Through canister supports	0.04 W
Control heater	5.0 W
Total	5.54 W

Heat Out

Through MLI	0.06 W
Along low "k" section	0.11 W
Window loss	2.14 W
Total	2.31 W

This leaves 3.23 W heat leak unaccounted for. The most likely mechanism to explain this heat loss is a natural convection cell set up between the low "k" section and the radiator.

At point B, the simulated environment was changed to the PTC mode, the dissipation reduced to 240 W, the canister temperature rapidly increased and snapped into control. In this hotter environment the reservoir/radiator temperature differ-

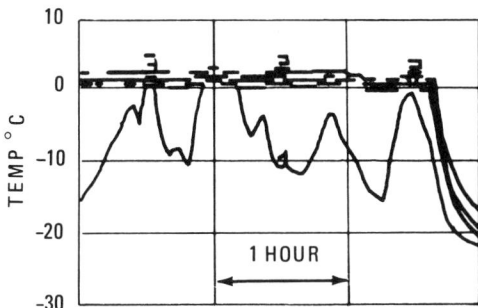

Fig. 11 Radiator temperature distribution corresponding to point B, fig. 10

ence causing the natural circulation would be much lower than in the $-Z_{LV}$ environment and hence, the 5 W of power would enable the VCHPs to close down. In zero-gravity space this natural convection cell would not be set up and the 5 W of heater power per reservoir should be sufficient to ensure hard control points in all environments, provided the minimum heat leak requirements are met.

Conclusions

A thermal canister has been acceptance tested for flight on an early Space Shuttle mission. All major test objectives were met in thermal vacuum testing of a flight configured thermal canister experiment. These objectives included:

1) Demonstration of ±1°C control of wall panels with both hardwired proportional control and microprocessor algorithm control over the temperature range -5°C to 14°C.

2) Demonstration of zone control with one enclosure at 9°C and the other at 1°C in one test and then 11°C and 21°C in a second test.

3) Demonstration of experiment node control.

4) Demonstration of simultaneous control of radiatively tied node in one enclosure and panel wall in the other enclosure.

5) Demonstration of passive control.

Active control in some very cold environments was difficult to achieve except at high interval dissipations. This problem has been explained on the basis of a convection cell in the low conductance/blocked radiator sectors. This cell would not be set up in zero-gravity and control should be achievable in flight at considerably lower dissipations.

Acknowledgment

Performed under contract NAS 5-24321, issued by the Goddard Space Flight Center of the National Aeronautical and Space Administration.

References

[1] W. Harwell, R. Haslett & S. Ollendorf, "Instrument Canister Thermal Control", AIAA paper No. 77-761, 12th Thermophysics Conference, Albuquerque, N.M.

[2] R. McIntosh & S. Ollendorf, "A Thermal Canister Experiment for the Space Shuttle", Paper No. 78-456, 3rd International Conference on Heat Pipes, Palo Alto, Calif.

[3] W. Bienert, P.J. Brennan & J.P. Kirkpatrick, "Feedback Controlled Variable Conductance Heat Pipes", AIAA Paper No. 71-421, 6th Thermophysics Conference, Tullahoma, Ala.

[4] W. Harwell & R. McIntosh, "The Thermal Canister Approach to Instrument Thermal Design", Paper No. 80-ENAs - 23, 10th Intersociety Conference on Environmental Systems, San Diego, Calif.

[5] "Heat Pipe Thermal Canister Experiment", GAC Final Report on Contract NAS 5-24321.

VAPOR CHAMBERS FOR AN ATMOSPHERIC CLOUD PHYSICS LABORATORY

G. L. Fleischman[+]
Hughes Aircraft Company, Torrance, Calif.

and
T. R. Scollon, Jr.[+]
General Electric Company, Valley Forge, Pa.

and
J. D. Loose[≠]
NASA George C. Marshall Space Flight Center
Huntsville, Ala.

Abstract

The methanol/stainless steel vapor chambers (flat plate heat pipes) discussed in this paper were developed for use in spaceborne atmospheric cloud chambers. This application imposed stringent thermal and mechanical requirements on the design. Flatness, low thermal mass, vibration, and structural integrity requirements were achieved in addition to precision temperature uniformity and thermal transport. Heat transfer coefficients on the order of 0.34 to 0.40 $W/cm^2-°C$ were measured. The vapor chambers are capable of transporting 170 W-cm per cm of width in either the axial or side-to-side direction.

Introduction

The vapor chambers described herein were designed, built, and tested by Hughes Aircraft Company to specifications applicable for their use in an orbiting atmospheric cloud physics

---

Presented as Paper 80-1462 at the AIAA 15th Thermophysics Conference, Snowmass, Colo., July 14-16, 1980. This paper is declared a work of the U.S. Government and therefore is in the public domain.
 *Project Manager, Electron Dynamics Division.
 [+]Thermal Engineer, Space Division.
 [≠]Thermal Systems Engineer, Systems Engineering Division.

research facility. Such a facility was designed by the Space Division of the General Electric Company under contract to the George C. Marshall Space Flight Center of the National Aeronautics & Space Administration. The low gravity environment of a space facility lets scientists advantageously study cloud microphysics for long observation periods without gravity induced disturbances -- an advantage not available in terrestrial facilities.[1]

The vapor chambers form inner walls of experiment volumes and provide high temperature uniformity surfaces[2,3] needed for precise temperature and humidity conditioning of experiment air samples. Figure 1 shows four vapor chamber configurations. These are used in pairs to form a continuous flow diffusion chamber (CFD), a static diffusion liquid chamber (SDL), and two sections of a saturator (SAT-S and SAT-R). For identification, each configuration herein is given the name of the corresponding experiment volume.

This paper defines key technical requirements and describes the vapor chamber development program from design generation through fabrication, assembly, and testing.

Design Requirements

Key vapor chamber technical requirements are listed in Table 1. The chamber surface which serves as the experiment volume wall is designated as the critical surface. The critical surface is divided into zones defined by isothermality requirements. Figure 2 shows the zone boundaries, and worst case surface heat flux profiles. Note that two different profiles exist for the CFD configuration, designated by CFD-H and CFD-C. By analyzing maximum flux variations across these zones, a minimum vapor-to-external surface heat transfer conductance can be derived for each vapor chamber. Consequently, surface-to-vapor conductance is also a key parameter in verification of temperature uniformity.

Side-to-side heat transport capacity is required for adequate working fluid distribution between the critical surface and a row of thermoelectric modules (TEMs) mounted along one edge of the chamber. The TEMs control temperature level. They also act with a linear heater mounted along the opposite edge, which can be activated to sweep noncondensable gas to harmless areas under the TEMs. This feature was provided in the event gas appears during the ten year vapor chamber life. Critical surface heat flux profiles (Fig. 2) also impose an implicit lengthwise heat transport capability.

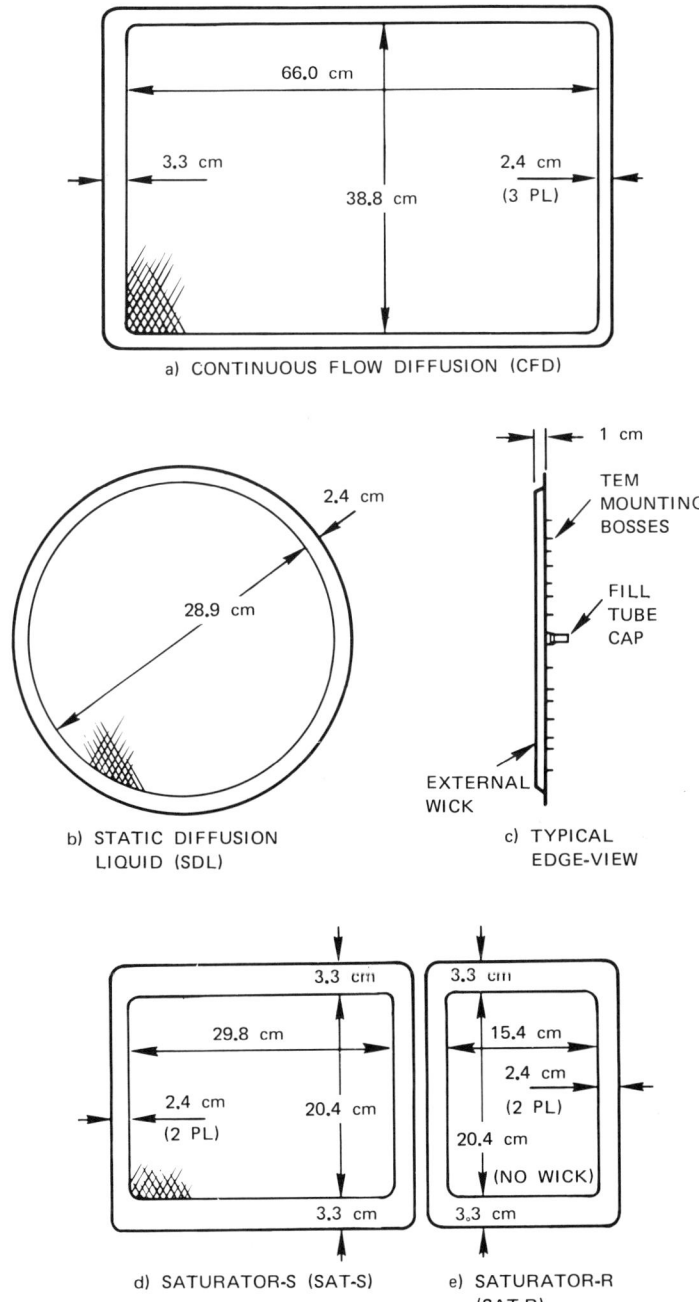

Fig. 1 Vapor chamber configurations for atmospheric cloud physics laboratory.

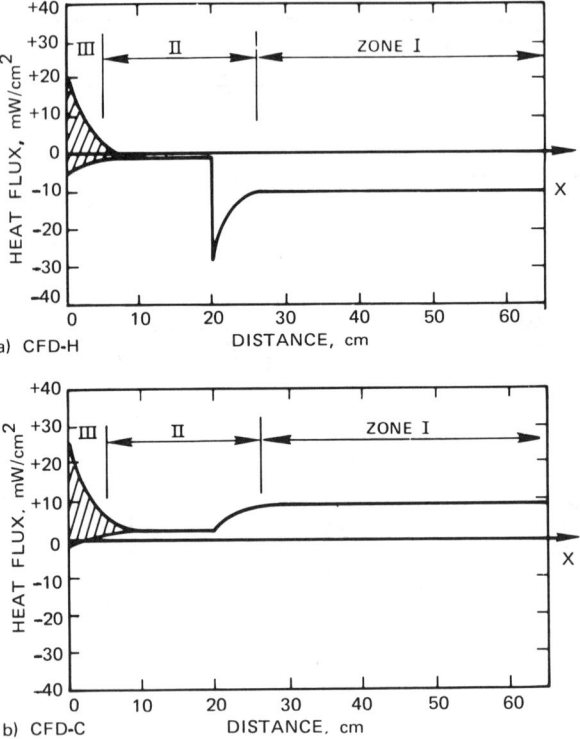
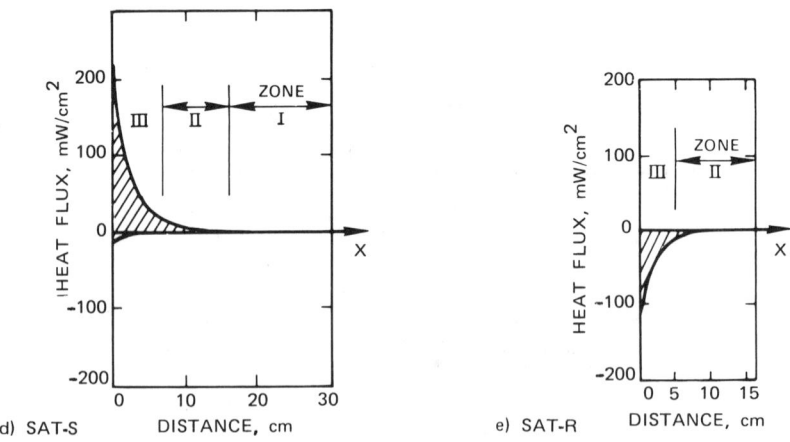

Fig. 2 Critical surface heat flux profiles (+ is into chamber).

VAPOR CHAMBERS

Table 1 Key vapor chamber technical requirements

Isothermality:	Within:
- Zone I	±0.01°C
- Zone II to I	±0.1°C
- Zone III to I	±1.0°C
Temperature Range	
Operating	-25°C to +30°C (+60°C SAT-R)
Survival	-40°C to +65°C
Side-To-Side Heat Transport	
CFD	72 W
SDL	~69 W
SAT-S	49 W
SAT-R	24 W
Weight	<3.5 g/cm^2 of critical surface
Leak Rate	Less than 5×10^{-9} scc (helium)/s
Structural	Proof at $2.5 \times \Delta P$ max
	Burst at $4.0 \times \Delta P$ max
Deflections Plus Flatness	±0.003 in. TEM interface surface
	±0.005 in. other mating surfaces
	±0.010 in. critical surface
	±0.020 in. all other areas
Life	10 years (operate 2 months/year)
Dynamic Environments	Shuttle launch, landing

Critical surface flatness/deflections were specified to maintain stringent tolerances on the experimental volume spacing between the pair of vapor chamber critical surfaces. Moreover, the vapor chamber must provide thermal sensor wells for mounting precision temperature sensors near the critical surface without perturbing its flatness. Structural, dynamic, and safety requirements were derived from requirements for manned spacecraft.

Design and Analysis

Methanol was selected as the working fluid for its favorable thermal characteristics and its low vapor pressure in the operating temperature range. A stainless steel, all welded construction provided working fluid compatibility and structural strength.

Figure 3 illustrates the basic vapor chamber design. Isothermality requirements dictated high thermal conductance of the wall/wick system. Thin stainless steel sheet metal (0.048 in. thick, 18 gage) was selected as the envelope material. The critical surface wick was a single layer of diffusion bonded 150 mesh screen. The interior of the vapor chamber was bridged by small diameter (0.125 in. diam) studs covered with a 120 mesh sock wick. These studs were placed on 1 in. x 1 in. centers, and provided structural strength, as well as liquid coupling between the critical and noncritical surface wicks. The noncritical surface (TEM surface) had coarse mesh screen wicks to provide both liquid transport and reservoir capacity for working fluid expansion. Sensor wells for mounting precision temperature sensors (thermistors) were provided by drilling through selected studs from the backside to the critical surface (Fig. 3).

Stud and Critical Surface Thermal Analysis

The studs were thermally modeled to analyze their potential disturbance to critical surface isothermality. One model with thermistor installed, and results of a specific computer run are depicted in Fig. 4. The heat transfer coefficients shown were based on Hughes test data for the critical surface, and calculated from equivalent liquid thicknesses for the other surfaces. Several cases were run varying the stud diameter with and without the thermistor. If the stud was too small, a temperature depression was created over the stud. If too large, a local temperature rise was created. An optimum diam. of 0.125 in. (3.2 mm), with a 0.067 in. i.d. (1.7 mm) to match the thermistor, had little disturbance effect. The results in Fig. 4 indicate that the disturbance is less than 0.01°C for this case with 170 mW/cm^2 input to the critical surface, a flux much greater than typical Zone I flux levels (Fig. 2).

The ability to maintain isothermality requirements between zones was checked by calculating the minimum critical surface conductance needed to keep the temperature difference in the compared zones within the corresponding temperature differential limit, ΔT. The most demanding heat flux differential, ΔQ, and ΔT combination exists between Zones I and II of the CFD-H. From Fig. 2, the average Zone I flux is -10 mW/cm^2. The maximum deviation, ΔQ, from this in Zone II is 20 mW/cm^2(-10 mW/cm^2 - (-30 mW/cm^2)). The thermal resistance between these two locations (two wall/wick sections and negligible vapor flow thermal resistance) is 2 $\Delta T/\Delta Q$ or twice that of one section. In this worst case the minimum one section conductance required, $\Delta Q/\Delta T$, is (20 mW/cm^2/0.1 or 200 mW/cm^2-°C. Adjusting for the wall thickness results in a required minimum heat transfer coefficient, h, of 240 mW/cm^2-°C or 430 Btu/h-ft^2-°F. This is well below the predicted design value of 581 Btu/h-ft^2-°F. Table 2 lists the minimum h requirements derived for each vapor chamber configuration.

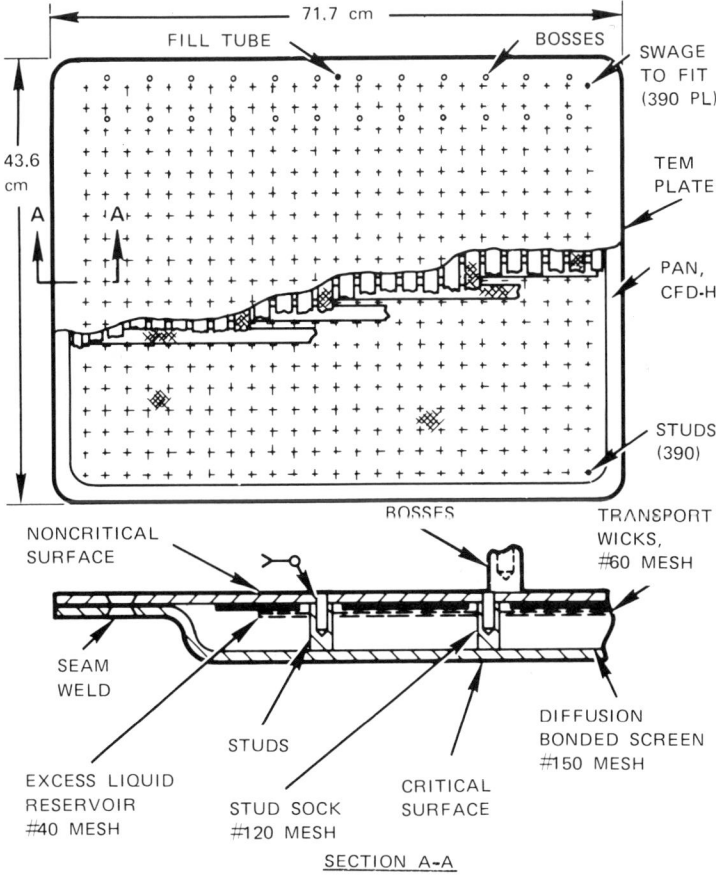

Fig. 3 Basic vapor chamber design details.

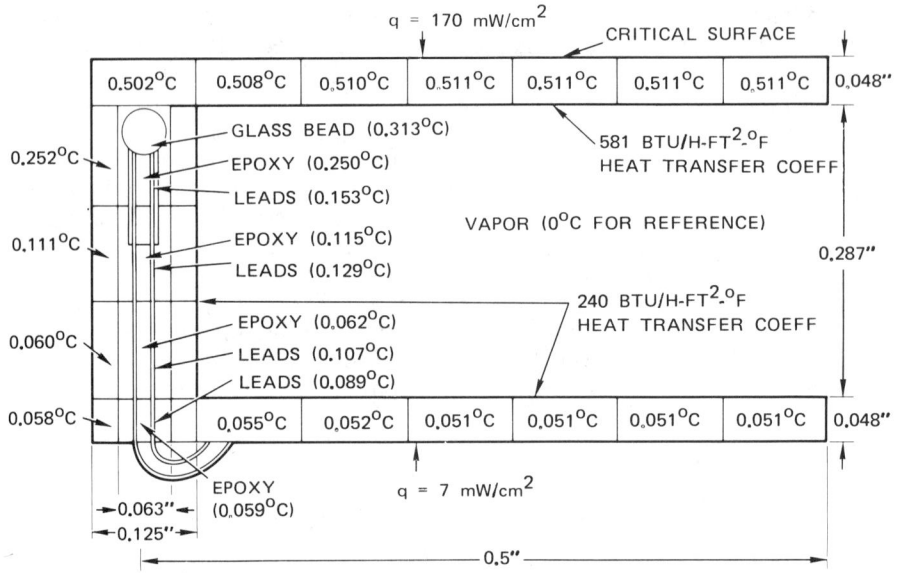

Fig. 4 Stud thermal model and analysis case.

Wick Analysis

In the preliminary design, only the side-to-side, sock transfer, and critical surface wick elements were included in the analysis. The CFD and SDL engineering models were built to this design. However, anomalies which occurred during the isothermality testing of these units led to a more extensive wick analysis, taking into account the two-dimensional nature of the vapor chambers.

In this analysis, the wick equations were programmed for rapid computerized investigation of: 1) all vapor chamber configurations; 2) wick parameters, such as number of layers, mesh size, etc.; 3) axial as well as side-to-side transport limits; and 4) body force and temperature effects. An electrical network analogy to liquid flow was used. Conductors, G, were set up in the following form:

$$G = KA/\nu L$$

where ν = liquid kinematic viscosity; K = wick permeability; A = wick cross-sectional area; and L = flow length.

The two-dimensional nature of the problem was handled by standard series/parallel conventions. To predict maximum

Table 2 Minimum h for interzone isothermality

Unit	Zones Compared	Max ΔQ (mW/cm^2-$^\circ$C)	Limit ΔT ($^\circ$C)	h min (BTU/h-ft^2-$^\circ$F)
CFD-H	1-2	20	0.1	430
	1-3	60[a]	1.0	110
CFD-C	1-2	9	0.1	170
	1-3	48[a]	1.0	90
SDL	1-2	0	0.1	0[b]
SAT-S	1-2	20	0.1	430
	1-3	190[a]	1.0	400
SAT-R	2-2	9	0.1	170
	2-3	180[a]	1.0	380
DESIGN PREDICTION:				581

[a] Includes 30 mW/cm^2 additional heat leak loading not shown in Fig. 2.

[b] h therefore not dictated by interzone comparison.

transport, mass flow (heat input) was varied until the local pressure difference across the liquid-vapor interface was equal to the maximum capillary head at the most highly stressed point in the wick system. Due to different mesh sizes, this point did not always occur at the end of the liquid flow path (as generally occurs in a "classical" heat pipe).

Results showed that the initial wick system was limited in its axial transport capacity, particularly with respect to the high leading edge fluxes (Fig. 2). However, the addition of axial wick strips yielded large safety factors on axial transport, and also enhanced side-to-side capability.

Improved liquid reservoirs of coarse 40 mesh screen were also incorporated at this time to compensate for liquid volume

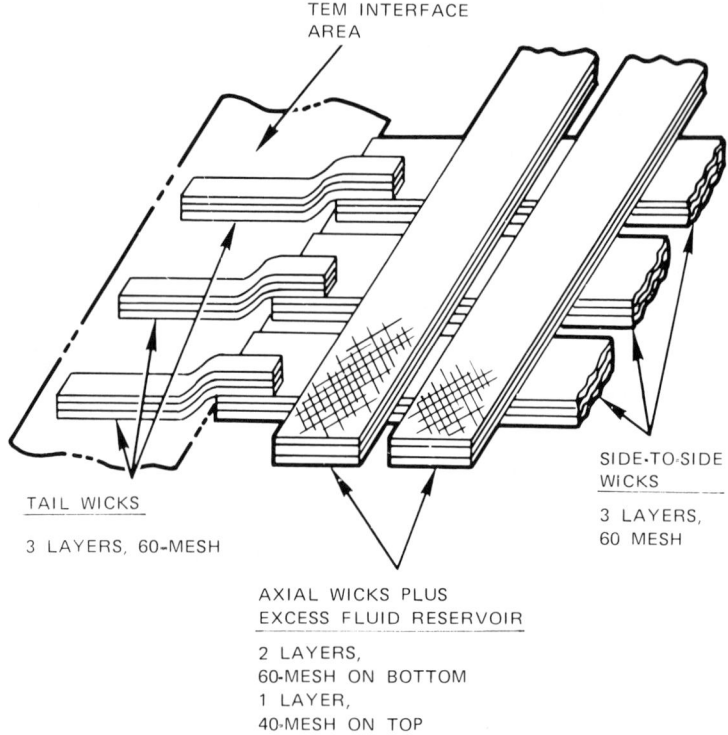

Fig. 5 Transport/reservoir wick system (noncritical surface). Note: Stud sock wicks between wick strips not shown.

temperature variations and decrease performance sensitivity to fluid inventory.

The final wick system is depicted in Fig. 5. Analytical results correlated well with the CFD and SDL engineering models, which were built without the axial wicks, and the SAT-S unit, which included the improved wicking. This verified the model for prediction of all units with the axial wick/reservoir addition.

Stress Analysis

As noted in Table 1, the maximum specified survival temperature was 65°C. The vapor pressure of methanol is 1.0 atm at this temperature. Therefore, the maximum pressure difference, ΔP, across the vapor chamber walls could be 1.0 atm if the external environment was a vacuum, as in depressurization of the Spacelab. Alternatively, at the minimum survival temperature of -40°C the vapor pressure is

0.002 atm. The pressure difference would also be 1.0 atm, in this case, if the unit was at minimum temperature with atmospheric external pressure.

The maximum deflection,[4] δ_{max}, for this geometry is given by the following expression:

$$\delta_{max} = 0.063 \, \Delta P \, a^4/E \, t^3$$

where a = spacing of studs; t = thickness of material; and E = modulus of elasticity.

It was found that the maximum deflection of the vapor chamber surfaces would be slightly less than 0.001 in. over the operating temperature range. Also, the worst case from a stress point of view is the necked down region of each stud. A safety factor of 2.1 was calculated for the proof pressure, based on yield stress, and a safety factor of 2.6 was calculated for the burst pressure level, based on ultimate stress.

Dynamic Analysis

The natural frequency of the vapor chamber construction was analyzed by assuming an equivalent flat plate. The validity of these calculations was later verified using a standard vibration analysis computer program. Results showed that the natural frequencies are very high relative to the imposed sinusoidal frequency for this structure. A value of 502 Hz was obtained for the SAT-S configuration, whereas the maximum imposed frequency was only 50 Hz. It was concluded that the design is a rigid structure, and produces no significant relative amplification factors.

A safety factor of 5.3 was calculated for the worst case random vibration acceleration of 25g. This was based on failure of a stud weld in shear.

Mechanical Design and Fabrication

The construction of flat plate heat pipes is, in general, more complex than conventional tubular heat pipes. Moreover, the design of a flat plate heat pipe to operate in zero-gravity with weight and flatness restrictions represented a significant challenge in the fabrication of heat pipe hardware. Stainless steel was selected as the envelope material because it has good weldability, machinability, and forming properties. Its stiffness-to-weight ratio is good in comparison with other materials, especially in the annealed condi-

Fig. 6 Sintered metal wick (400X).

tion. An all-welded construction was used for strength, leak tight integrity, and compatibility considerations. Figure 3 illustrates the vapor chamber construction.

As previously described, the critical surface wick was a single layer of 150 mesh stainless steel screen diffusion bonded (sintered) to the 18 gage envelope material. Typical results are shown in Fig. 6. Uniform sintering was verified by laboratory wick tests and extensive visual examination. The sintering technique resulted in selected sheets having 100% bonding over the required area. After sintering, the sheet metal was deep drawn to form a shallow pan. Cross sections after drawing showed little, if any, effect in the working areas; however, there was some wick compression along the edges.

Studs were projection welded to the critical surface, using weld fixtures drilled on a numerically controlled tape machine. Holes in the noncritical surface were drilled using the same tape to match the stud locations (Fig. 7).

Sock wicks were then spot welded to the studs. The critical surface end of the wick was cut square and butted against the sintered wick. The opposite end was formed into tabs which extended beyond the stud and pressed against the noncritical surface when assembled. Tabs were avoided at the critical surface to minimize wick nonuniformities and, thus, temperature disturbances. Transport wicks were spot welded to the noncritical surface.

In closure, the opposite ends of the studs were welded in place using a tape controlled electron beam welder. The vapor chamber perimeter was sealed using a water cooled seam welder to prevent distortions.

Flatness

Two problems were encountered in achieving the required flatness. First, diffusion bonding the wick significantly warped the sheet metal. Although subsequent deep drawing stretched and smoothed the metal, small residual distortions remained after welding the studs and seam welding. Second, the electron beam weld beads on the noncritical surface prevented a smooth interface for the TEMs. Both problems were solved by electroform depositing a thin layer (0.01 in., minimum) of nickel on these surfaces and finishing to the required flatness.

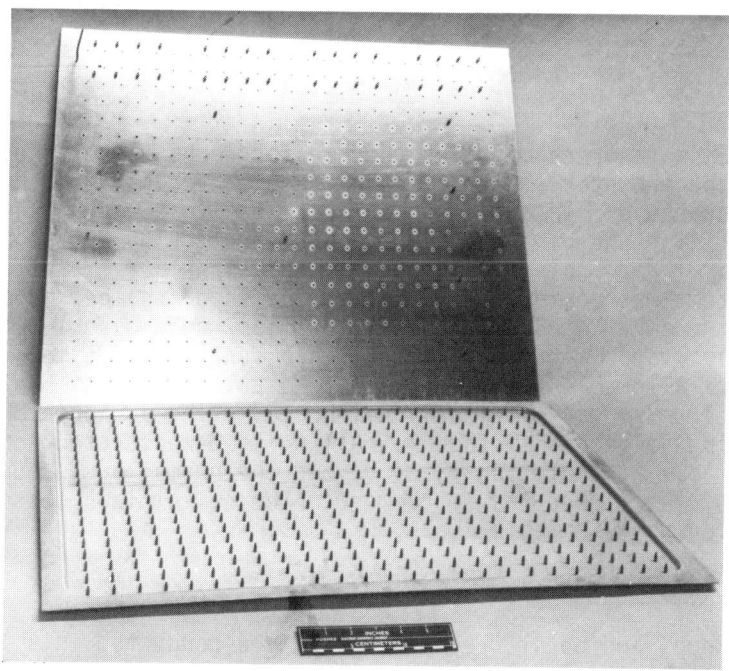

Fig. 7 CFD studs installed and TEM plate drilled.

Working Fluid Processing

Because the critical surface can act either as an evaporater or condenser in the cloud physics application, extreme care was required in processing. The vapor chambers must be essentially gas-free prior to pinchoff.

The working fluid was filtered through a molecular sieve for water removal, and then transferred by double distillation into a graduated burette prior to filling the unit. The first charge was refluxed at 50°C for 4 h. This charge was flushed from the unit and the process repeated. An 8 h reflux operation at 50°C was imposed on the second fill. Finally, the fluid was degassed by chilling the unit with dry ice and venting to vacuum.

A hot pinchoff technique was used to seal the fill tube. Closures using this technique were pressure tested. The minimum failure pressure was 7250 psig. Final verification of fill was obtained by weighing the units to ±1 g.

Test Activity

The development and qualification testing consisted of mechanical as well as thermal performance verification.

Leak Rate

Each vapor chamber was checked for leaks using a helium mass spectrometer (leak detector) before processing with working fluid. In addition, a final leak test was performed after fill tube pinchoff using an Inficon IQ-200 Mass Analyzer (residual gas analyzer) tuned to the mass number of methanol. This test was performed by placing the vapor chamber in a vacuum chamber at 1×10^{-6} Torr, or less, and measuring the methanol vapor accumulation rate as a function of time. The maximum allowable rate, equivalent to a leak rate of 5×10^{-9} scc helium/s, was previously calibrated using a precision metering valve with methanol. All units demonstrated no measurable accumulation rates beyond the normal background rate of the closed vacuum chamber. The helium leak tests revealed no detectable leaks.

Pressure Integrity

Proof and burst pressure tests were conducted in a 1 atm environment, with only positive internal pressure since the vapor chamber structure is most sensitive to that mode.

Internal vapor pressure was elevated to provide scaled up pressure differentials listed in Table 1 by heating the unit to the necessary fluid temperature. All vapor chambers were proof pressure tested. The SAT-S pressure was also increased to the burst pressure level. No ruptures or permanent deformations were observed in any units.

Flatness/Deflection

Deflection of the surfaces was found to be less than 0.001 in. over the operating pressure range. Deflections were measured using a dial indicator attached to a height gage. Flatness was measured by using a machinist's scale of sufficient length to span the entire surface. The scale was placed on edge in contact with the surface, and flatness was verified by inserting precision shim stock in the gaps underneath the straight edge.

After the electroformed nickel was finished to the desired smoothness, the flatness was, in general, within ±0.020 in. (±0.50 mm), and ±0.010 in. (±0.25 mm) in certain specified areas. No change in flatness was observed after exposing the unit to 1 atm inward pressure differential, and 2.5 atm outward.

Dynamic Test

General Electric dynamically tested an engineering model to verify structural integrity and assure vibration caused no functional degradation. The CFD was tested because it was the largest and, therefore, most vulnerable configuration. TEM mass simulators and functional thermistors were attached to simulate the complete configuration.

Prior to vibration, the CFD underwent extensive configuration and performance checks for comparing against postvibration results. Checkouts included: 1) an accurate determination of dimensions, weight, and surface flatness; 2) x-ray inspections which clearly showed all wick elements; 3) heat transport tests; 4) checks for noncondensable gas; and 5) heat transfer coefficient checks.

The CFD was supported around its perimeter and vibrated in each of its three principle axes. Response was monitored with five accelerometers. Low level sine modal surveys revealed a first resonance of 280 Hz normal to the critical surface with a dynamic amplification factor of 24 at the center. Sine vibration tests to 5.4 g and random vibration tests to 6.45 g RMS were then imposed. The unit center experienced 33 g RMS during the random series.

All previbration checks were then repeated. No damage occurred and performance matched well with the prior data. The vapor chamber was thus dynamically qualified for a space flight environment.

Life Test

A stainless steel/methanol heat pipe has been operated at Hughes for 20,000 h at 115°C with only a slight increase in temperature drop down the heat pipe.[5]

Based on corrosion and oxidation theory, Anderson[7] developed a model which predicts a passivating film growth rate in a methanol/stainless steel system. Initially the film growth and thus gas evolution will occur at a parabolic rate. Over long periods of time, uniform gas generation with a linear time dependance occurs. With this approach the life of the unit was estimated to be on the order of 10^6 h at the design temperature. The life is determined by the length of time to fill a volume allocated to noncondensable gas storage along one edge where the TEM modules are located. One SAT-S unit has been designated as a life test unit to be tested at a nominal temperature of 65°C over a period of three months in order to qualify the design.

Thermal Performance

Heat transport and isothermality capability over the operational temperature range were measured in performance testing. Isothermality was inferred by determining the critical surface heat transfer coefficient, h, for both inward and outward imposed heat fluxes. The contingency gas sweeping technique was also tested.

Vapor chamber temperatures were monitored with calibrated fine wire (0.005 in. diam), chromel-constantan, thermocouples spot welded directly to the surface. Thermocouples were referenced to a Kaye Instruments electronically controlled ice point. Output was measured with a Fluke 8502A precision digital multimeter, with an accuracy of 0.005% plus 5 digits. A Lewis Engineering Co. thermocouple switch was used to rapidly access successive thermocouples. Overall steady state temperature uncertainty was judged by various checks to be within ±0.05°C.

Temperature level was controlled by circulating liquid through a narrow linear channel which simulated the mounting of TEMs. A 1 in. wide linear heater, mounted on the edge opposite the linear coolant channel, was activated for heat transport tests, the gas sweeping test, and some heat transfer coefficient tests conducted at +30°C. Vapor chambers were horizontally mounted, edge supported from a frame structure on thermal isolation pads, and enclosed in foam insulation. Voltmeters, ammeters, and thermocouples were used to monitor and control heater power and circulating coolant temperatures, respectively.

Heat Transport. Side-to-side heat transport capacity permits vapor chamber temperature control by the TEMs. Three thermal load conditions define the integrated transport requirements listed in Table 1. First, the critical surface loads (Fig. 2) must be transported to and from the TEMs. Second, the linear heater load, if gas sweeping is required, must be transported. Third, the TEMs must change the critical surface temperature at specified rates. Table 1 specifies the equivalent watts between the linear heater and TEM row (or linear channel).

Transport tests were conducted at -25°C and 30°C with the critical surface facing upward and with the linear heater elevated 0.25 in. (6.4 mm) above the linear channel (adverse tilt). Though this tilt gives slightly conservative results, it assured that stud wicks were fully tested. Figure 8 shows the CFD test configuration.

The SDL, SAT-S, and CFD were each tested at single preset linear heater powers. In addition, progressive tests of the CFD at +30°C were conducted by both Hughes and General Electric in which linear heater power was increased gradually. Two dryout conditions were arbitrarily defined: 1) initiation occurs in progressive tests when the temperature differential, ΔT, between the linear heater and the thermally unloaded portion of the critical surface takes its first positive irreversible step; 2) dryout occurs when this ΔT equals 1°C. Consistent stable operation was, however, possible even beyond the arbitrarily defined dryout condition.

Results are summarized in Table 3. Only the SAT-S had, at this time, axial wick elements. These augment not only axial but also side-to-side heat transport. Both the CFD (at -25°C) and the SAT-S (at -25°C and +30°C) exhibited stable, less than 1°C ΔT, operation at their preset linear heater powers. Their capacities exceeded their requirements.

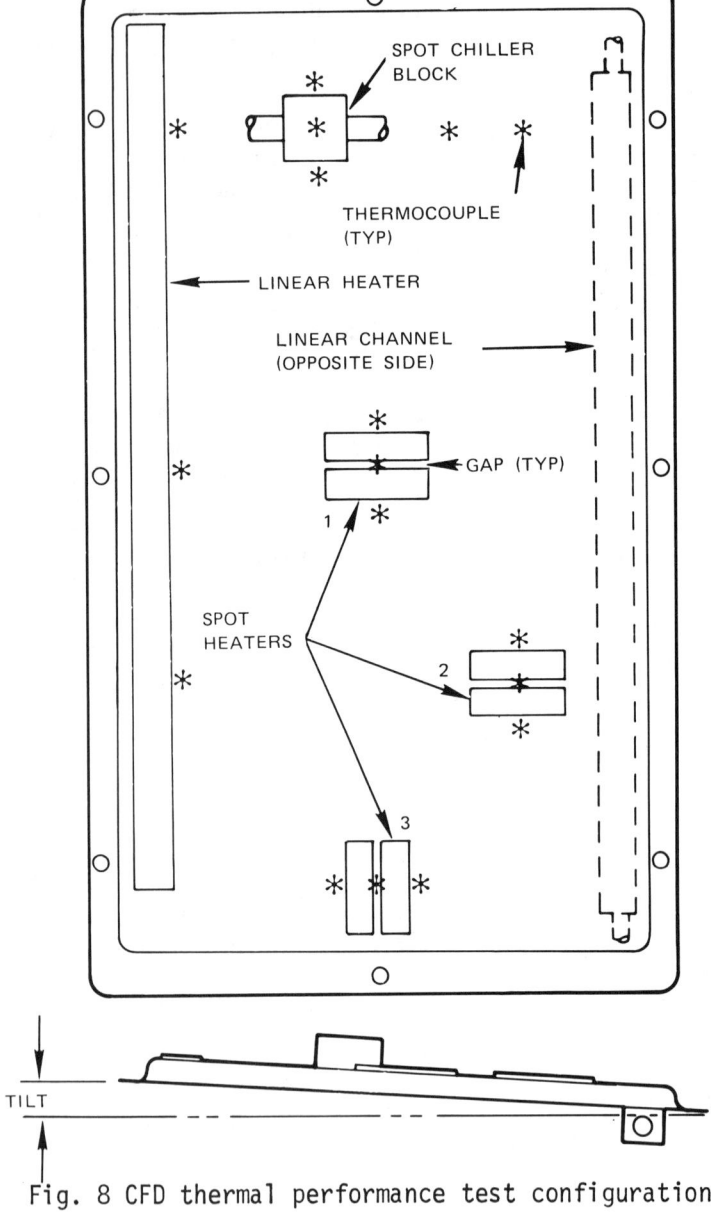

Fig. 8 CFD thermal performance test configuration.

The SDL incurred significant ΔTs at 80 W indicating transport capacity, without augmentation from axial wicks, was less than 80 W but not necessarily less than the 69 W required. The progressive CFD tests showed substantial CFD capacity margin.

Table 3 Side-to-side heat transport capacity

Unit	Temperature, °C	Req'd W	Preset heater	Test results, W Progressive test		Predicted performance, W[b]	
				Initial $\Delta T \to +$	D-0 $\Delta T = 1°C$	Initial $\Delta T \to +$	D-0 $\Delta T = 1°C$
CFD	-25	72	>100[a]	-	-	157	220
	+30	72	-	141[a]	212[a]	259	370
				145[a,c]	199[a,c]		
SDL	-25	~69	<80[a]	-	-	67	100
	+30	~69	<80[a]	-	-	109	150
SAT-S	-25	49	>50	-	-	98	140
	+30	49	>50	-	-	160	230
SAT-R	-25	24		NOT TESTED		47	70
	+30	24				76	110

[a] Unit without axial wicks.
[b] Unit with axial wicks.
[c] General Electric test results.

These data were used with the analytical wick models to predict the capacity for all units with axial wicks (two right-hand columns of Table 3).

With axial wicks, all units are predicted to substantially exceed their transport requirements; one exception is the SDL at -25°C where the capacity is marginal. In consideration of the factors that determined the SDL requirement, the marginal capacity at -25°C in effect only limits the rate at which its critical surface temperature can be changed. The rate available, however, exceeeds the temperature driving capability of the TEMs at this low temperature making TEMs, not transport, the SDL operational restriction. Side-to-side heat transport capacities were thus shown to be adequate for the application.

<u>Isothermality (Heat Transfer Coefficient)</u>. As previously discussed, a quantifiable approach to verifying isothermality performance is to verify the critical surface wick heat transfer coefficient, h. Table 2 shows isothermality requirements of all units will be met if h exceeds 430 Btu/h-ft^2-°F. Tests were constructed to measure h against this derived requirement.

Heat loads were locally imposed into the critical surface (evaporation) by taped down (film) "spot heaters." These were typically 1 in. x 1 1/2 in. rectangular. Heat loads were locally extracted (condensation) using a "spot chiller." The spot chiller consisted of a 1 in. x 1 in. thin Teflon spacer sandwiched between the surface and an impressed copper pad whose temperature was controlled by circulating liquid. Chiller heat loads were calculated from the spacer's dimensions and thermal properties, the estimated (contact) conductances at the interfaces, and the controlled and measured temperature difference across the spacer. Power was measured to the spot heaters.

In the simplest approach, h is determined from the heat load imposed, Q, the spot load area, A, and the measured temperature differential, ΔT, between the loaded area and an unloaded (adiabatic) point on the critical surface; i.e., $h = Q/A\Delta T$. Two drawbacks, however, negate this simplistic approach. First, physically mounting a temperature sensor under the spot load spoils the interface. Second, accurate measurement cannot be obtained from a sensor mounted adjacent to the load. High h performance creates steep surface temperature gradients at spot load edges. Measured ΔTs are highly sensitive to the distance from the load and, in all cases, are less than the ΔT under the load, leading to incorrectly high h values.

These difficulties were avoided by splitting the spot loads down the center with a narrow gap, mounting the sensor in the center as shown in Fig. 8, and thermally modeling the resultant configuration. Gap width and h variations were analyzed parametrically. Results correlated h directly with the spot load dimensions, the measured heat load, and the measured ΔT. Figure 9 shows one surface temperature profile from the analysis. Edge gradients are dramatic. Surprisingly, reduced but significant gradients exist within the gap. However, the measured ΔT is now increased, less sensitive to position, and conservative (high) if the sensor is not at the exact gap center.

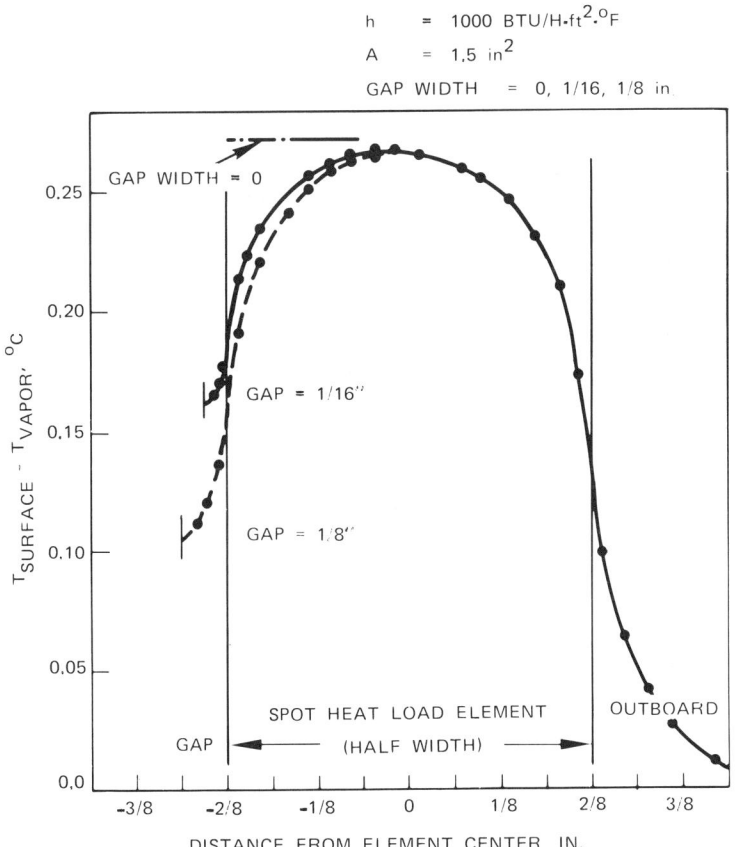

Fig. 9 Surface temperature profile.

In coefficient testing, fill, orientation, chamber temperature, and heat flux were varied. Only the SAT-S had the axial wick improvement. Fill was varied at 90%, 100%, and 110% of each unit's calculated design fill to determine fill sensitivity and verify the design fill adequacy. The critical surface was horizontal, facing either up or down. Tests were conducted at $-25^\circ C$ and $+30^\circ C$ to cover the operational temperature range.

Test fluxes were elevated above operational levels to yield large ΔTs relative to the $\pm 0.05^\circ C$ uncertainty in their measurement. Test fluxes varied from 100 to 600 mW/cm^2, whereas operational fluxes (Fig. 2) are typically between 0 to 30 mW/cm^2. This upward scaling is permissible as long as the internal wicking transport capability is not exceeeded and the heat transfer mode across the critical surface wick (conduction) is maintained. Previous investigations[7,8] using similar wicks and working fluids demonstrated fluxes over 600 mW/cm^2 could be imposed before the mode changed from conduction to nucleate boiling with a corresponding change in h.

The CFD was tested first and exhibited occasional, systematic occurrences of anomalously low values of evaporator (heater) h. These occurred only during $+30^\circ C$ test runs and predominantly only at heaters 2 and 3 (Fig. 8). Figure 10 (with constant h lines imposed) shows this behavior for heater 2. At first unexplainable, this behavior was eventually coupled with operation of the linear channel which was operative during $-25^\circ C$ runs, to maintain the low temperature, but inactive in the $+30^\circ C$ runs. The linear channel, when operating provided a favorably located condensation area, or liquid source, from which liquid was easily transported through the side-to-side transport wicks to the heaters. When not operative, the only liquid source was the spot chiller. Since the liquid flow path between the chiller and heaters was tortuous, it was suspect. The low heater h's at $+30^\circ C$ thus resulted not from a temperature related wick inadequacy but instead from a breakdown of internal liquid transport; i.e., the necessary chiller to heater axial transport capacity was not available.

This hypotheses was corroborated by two subsequent $30^\circ C$ tests. In the first, the linear channel was activated and 30 W were applied to the linear heater. Heater 2 and 3 h's improved dramatically from approximately 100 previously to values exceeding 600 $Btu/h\text{-}ft^2\text{-}^\circ F$. Heater 1 improved from 430 to 650 $Btu/h\text{-}ft^2\text{-}^\circ F$. In the second $+30^\circ C$ test, with the

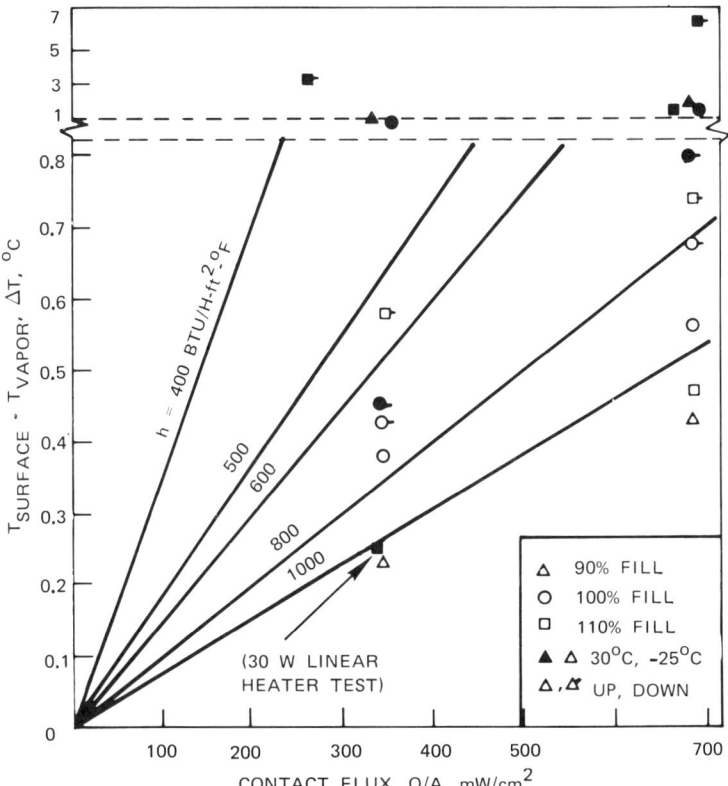

Fig. 10 CFD heat transfer coefficient performance (spot heater #2).

linear channel and heater inactive, a small cooling load was placed on the critical surface at a point on the side in line with the side-to-side transport wick leading to heater 3 (Fig. 8). Heater 3 h improved again, from 100 initially to 650 Btu/h-ft-°F. This second test confirmed that operating the linear heater and channel in the first test had not introduced some unknown enhancement effect.

The SAT-S, which had the internal axial wicking improvements, showed no effect from operating the linear channel. Similarly, no coupled effect of fill and orientation was indicated. These observations confirmed that the previous problems were due to internal wick inadequacies which were now corrected.

Results for one spot heater (heater 1) are shown in Fig. 11. Again, h varies with temperature scatter but generally lies between 500 and 800 Btu/h-ft^2-°F. A slight improvement at higher fluxes is indicated. The h at two other spot heaters varied from 350 to 600 Btu/h-ft^2 and 340 to 510 Btu/h-ft^2-°F, respectively. These ranges are lower than observed on the CFD and SDL but not unexpectedly. In the SAT-S, an air gap heating technique was substituted for the bonded film heaters and a simple radiation/conduction model was used to compute the air gap heat transfer. The model could not, however, quantitatively include the effects of shims and thermocouple leads within the narrow air gap. Both effects artificially give low h results, as low as 2/3 of actual values, and bonded heater results must, therefore, be considered more reliable. An important aspect of SAT-S tests, however, was the observed insensitivity to fill, orientation, and linear channel operation.

SAT-S spot chiller h's ranged between 330 and 500 Btu/h-ft^2-°F with a mean value of 400 Btu/h-ft^2-°F.

When data points known to be affected by axial transport breakdown or overfill are excluded, the total data base of the three units consists of 86 evaporator h data point and 23 condensation h data points. The mean evaporator h is 591 Btu/h-ft^2-°F. The mean increases to 716 Btu/h-ft^2-°F if the conservative SAT-S results are excluded. The mean condensa-

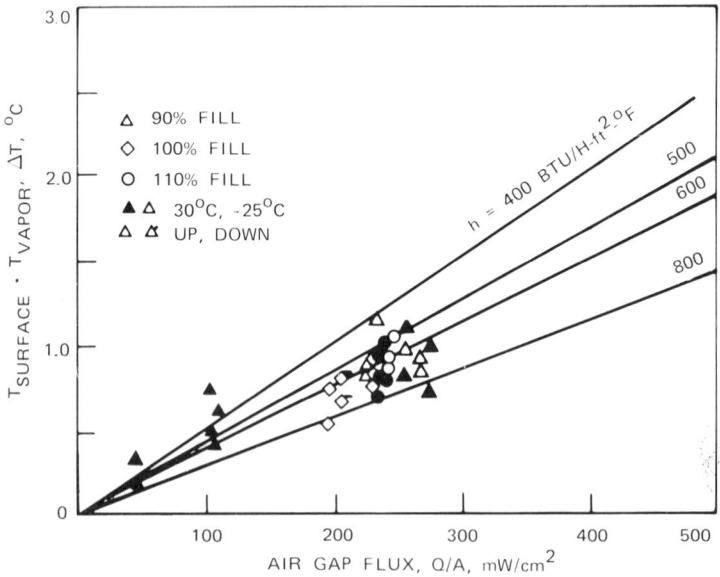

Fig. 11 SAT-S heat transfer coefficient performance (heater #1).

tion h is 449 Btu/h-ft^2-°F. The condensation h data points are believed to be slightly low since the spot chiller conductance was not calibrated but instead calculated from intentionally pessimistic assumptions for the spacer and interface conductances. More realistic assumptions would increase condensation h values by 25%.

More accurate refinement of the observed h values would require more elaborate and calibrated heat loading methods and more sophisticated measurement techniques than employed here. Such a refinement, however, is not a prerequisite to verifying vapor chamber design acceptability. The test approach demonstrated that h is greater than 430 Btu/h-ft^2-°F, insensitive to the test variables, and therefore sufficient to maintain the Table 1 isothermality limits.

Gas Sweeping. A small charge of dry nitrogen gas was intentionally introduced into the CFD after h testing to verify the contingency gas sweeping technique. With the linear channel controlling the temperature to 30°C and with elevated spot heater and chiller fluxes, critical surface temperature gradients were measured first without and then with the linear heater activated. The overall critical surface gradient was reduced from 5.8°C, initially, to 0.55°C with 30 W applied to the linear heater. Temperatures indicated a gas pocket shifted from a broad area surrounding the spot chiller to a small area under the TEMs. Note that the TEMs are located along one edge of the unit in a noncritical zone. This edge is not exposed to the experimental volume, and any temperature gradients around the TEMs due to the presence of noncondensable gas would not affect isothermality of the critical surface. The spot chiller h increased dramatically from 13 to 570 Btu/h-ft^2-°C, clearly demonstrating gas sweeping effectiveness. The linear heater can reestablish isothermality if called upon.

Conclusions

It is concluded that the vapor chambers described have been developed and qualified to provide isothermal walls for a spaceborne atmospheric cloud physics facility. The integration of several advanced manufacturing techniques resulted in successfully meeting all design requirements.

Basically, the vapor chambers are stainless steel, welded sheet metal containers with conventional metallic mesh wicking and methanol as the working fluid. Specifications are summarized below:

Size:	1 cm thick; up to 44 cm wide x 72 cm length
Weight:	2.8 g/cm^2 without external water wick 3.0 g/cm^2 with external water wick
Heat transport:	Approximately 170 W-cm per cm of chamber width
Heat transfer:	>500 Btu/h-ft^2-°F (>0.28 W/cm^2-°C)
Isothermality:	±0.01°C with cloud chamber-type thermal loading (Fig. 2)

Acknowledgments

This work was sponsored by the NASA Marshall Space Flight Center, Alabama, under Contract NAS 8-32668.

References

[1] Moses, J. L., Fogal, G. L., and Scollon, T. R., Jr., "Atmospheric Cloud Physics Laboratory Thermal Control," ASME Paper No. 78-ENAS-9, July 1978, San Diego.

[2] Fleischman, G. L., Marcus, B. D., et. al., "Flat-Plate (Vapor Chamber) Heat Pipes," AIAA Paper No. 75-728, May 1975, Denver.

[3] Basiulis, A. and Formiller, D. J., "Emerging Heat Pipe Applications," _Proceedings, 3rd International Heat Pipe Conference_, May 1978, pp. 59-62, Palo Alto.

[4] Den Hartog, J. P., _Advanced Strength of Materials_, McGraw-Hill Book Co., New York, 1952, p. 134.

[5] Basiulis, A. and Prager, R. C., "Compatibility and Reliability of Heat Pipe Materials," _Radiative Transfer and Thermal Control_, Progress in Astronautics and Aeronautics, Vol. 49, edited by A. M. Smith, AIAA, New York, 1976, pp. 515-529.

[6] Anderson, W. T., Edwards, D. K., Eninger, J. E., Marcus, B. D., "Variable Conductance Heat Pipe Technology: Final Research Report," NASA Report No. CR-114750, March 1974.

[7] Marto, P. J., Mosteller, W. L., "Effect of Nucleate Boiling on the Operation of Low Temperature Heat Pipes," ASME Paper No. 69-HT-24, Aug. 1969, Minneapolis.

[8] Abhat, A. and Seban, R. A., "Boiling and Evaporation from Heat Pipe Wicks with Water and Acetone," ASME Journal of Heat Transfer, Vol. 96, Aug. 1974, pp. 331-337.

A PROTOTYPE HEAT PIPE RADIATOR FOR THE GERMAN DIRECT BROADCASTING TV SATELLITE

R. Schlitt* and R. Meyer[+]
ERNO Raumfahrttechnik GmbH, Bremen, FRG

Abstract

The paper describes development, fabrication and test of a high performance heat pipe radiator to cool high dissipating repeater units for future communications satellites. According to design guidelines established during a general radiator study a configuration with 11 parallel and two crosswise heat pipes, bonded between face sheets of a honeycomb sandwich, has been selected. The heat pipe is an axially grooved aluminium profile with ammonia as working fluid. The cross section of the pipe is rectangular to fit the 30 mm sandwich height which was required for structural reasons. Before building the radiator a special test program was performed to investigate the one-sided heat input into the heat pipe for this type of radiator interface. Results show that the radiator efficiency may be substantially improved by the use of saddle blocks bonded to the heat pipe, which actually represents a three-sided heat input. The completed radiator was thermally tested in a liquid nitrogen shrouded vacuum chamber to simulate deep space environment. Test results correlate very well with calculated temperatures.

Introduction

After the successful launch of the first European communications satellites (SYMPHONIE and OTS) a new generation of telecommunications systems is currently being studied which allows a direct communications link between the satellite

---
Presented as Paper 80-1460 at the AIAA 15th Thermophysics Conference, Snowmass, Colo., July 14-16, 1980. Copyright © American Institute of Aeronautics and Astronautics, Inc., 1980. All rights reserved.
*Presently responsible for thermal control in the EUROSATELLITE project group for TV-SAT/TDF-1 at MBB, Munich, Federal Republic of Germany.
+ Thermal Specialist, Thermodynamic Section.

HEAT PIPE RADIATOR FOR TV SATELLITE

and an individual user's home. Such systems are characterized by high radio frequency power and consequently high thermal dissipation within the payload.

A typical example is the German direct broadcasting television satellite (TV-SAT), which is now under development. The payload module of the preoperational version with three (out of five) communications channels operating is schematically shown in Fig. 1. The fully operational system will have five out of ten working channels. The high dissipating units of the payload, i.e., the traveling wave tubes (TWT) and the electrical power conditioners (EPC), are mounted on the side walls of the satellite structure to permit thermal control by radiation into space. Not considering the free radiating TWT collectors, each pair of TWT and EPC conducts about 200 W into the side walls. From there the heat must be spread along the panel surface with up to a total of 600 W to be radiated into space when three channels operate on one spacecraft side.

Heat pipes has been used in the past to spread heat of high concentrated sources over a radiator panel. The best known example is certainly the thermal design of the ATS-F spacecraft,[1] where heat pipes are integrated into the north/south payload panels. The work presented here is aimed to improve this technology to include applications with high heat flux densities in the equipment/heat pipe interfaces and to develop heat pipe profiles which can be integrated into large panels with thermally nonoptimized honeycomb thickness. Furthermore the high number and the associated

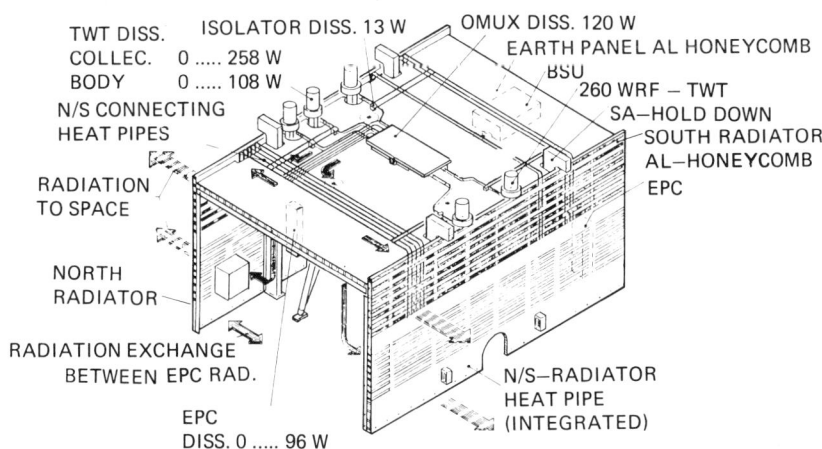

Fig. 1 Layout of the German TV-SAT, payload module.

weight of the heat pipes involved (a total of about 70,000 will be integrated in the payload module of Fig. 1) made it necessary to conduct a general study to optimize the heat pipe layout in terms of maximum thermal performance with minimum weight, fabrication and test complexity.

General design guidelines for heat pipe radiators have been established at ERNO in recent years [2,3] and can be summarized as follows:

1) For a dissipation of about 200 W or higher heat pipe radiators exhibit substantial weight savings against solid panel radiators.

2) Since spacecraft sidewalls are typically structured as sandwich panels, heat pipes should be integrated between the panel facesheets to reduce temperature gradients through the sandwich and to facilitate surface coating and equipment mounting. Heat pipes for this purpose have typically rectangular or square cross sections to obtain large bonding areas between facesheets and heat pipes.

3) In order to perform individual heat pipe tests the selected heat pipe/radiator bonding technique should allow integration of the charged heat pipes.

4) For the heat pipe layout the first choice should be the parallel arrangement to ease radiator and satellite system tests.

These guidelines were applied for fabricating and testing the prototype radiator described in this report. The same philosophy has been adapted for the design of the TV-SAT payload radiators, as seen in Fig. 1.

Although not within the scope of this paper it should be mentioned that the thermal design of the payload module (Fig.1) represents a heat pipe layout with severe restriction on ground testing. Heat pipes are located in the three orthogonal planes of the spacecraft and ground testing was found to be only possible with the horizontal Earth panel facing down which will provide reflux operation modes for the U-shaped north/south connecting heat pipes.[4]

Heat Pipe Integration

According to the design guidelines mentioned earlier a layout is selected with heat pipes integrated between the panel facesheets. The integration of heat pipes into a

HEAT PIPE RADIATOR FOR TV SATELLITE

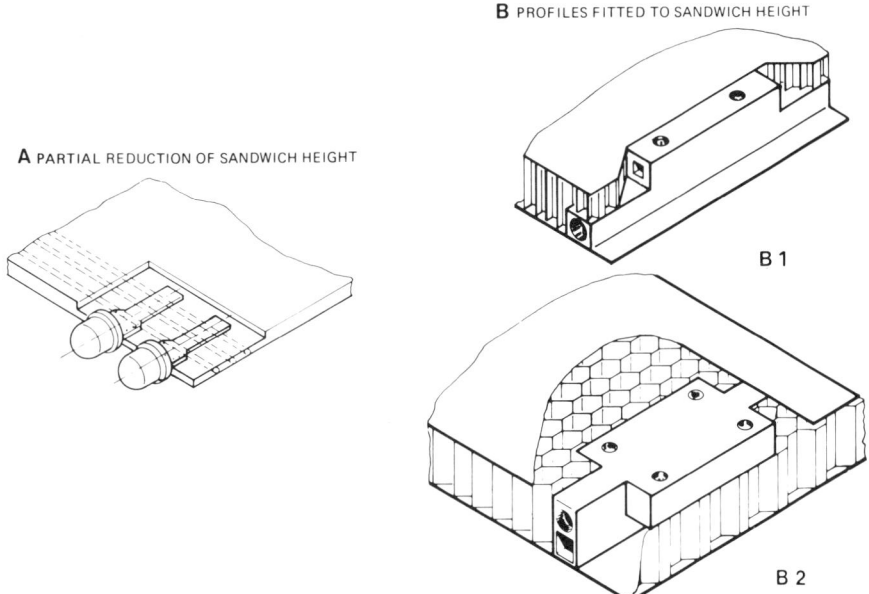

Fig. 2 Heat pipe radiator design concepts.

honeycomb structure must consider the structural requirement of the panel which represents a load carrying part of the spacecraft structure. A sandwich plate for this purpose has ideally a large honeycomb height and thin facesheets. This is opposite to thermal requirements which call for thin sandwich structures with thick facesheets in order to minimize the thermal resistance between equipment and radiating surface. The structural behavior of a sandwich plate with integrated parallel heat pipes should be more favorable than a simple honeycomb structure, but the stiffness is only improved in direction of the heat pipes. In order to obtain an equivalent stiffness normal to the heat pipes external stringers on the panel must be used, which, of course, could be disadvantageous with respect to the layout of the spacecraft system.

Because of missing system parameters in the beginning of this project and in order to obtain a radiator design which is generally applicable, the stiffening effect of the heat pipes were not considered in this program. With this constraint the sandwich height of load carrying panels for high power communication satellites is typically 30 to 40 mm, whereas a heat pipe diameter optimized for this purpose is about 15 mm. In order to bridge this sandwich height by heat pipes the concepts shown in Fig. 2 may be considered.

In concept A heat pipes with thermally optimized dimensions are bonded to the outside (radiating) facesheet and the sandwich is partially reduced in height in order to mount the dissipating equipment directly on the heat pipes. This concept represents the best thermal solution and is optimal concerning radiator weight, however, the stiffness of such a sandwich plate is reduced and the arrangement obviously assumes that structural requirements are less severe.

In concept B the integrity of the sandwich is maintained and heat pipe profiles are provided which have outside dimensions to fit the required sandwich height. Two heat pipe profiles which may be considered for this purpose are illustrated under B1 and B2 in Fig. 2. In both versions a rectangular bridging profile has been added to a square heat pipe with thermally optimal dimensions to obtain a rectangular cross section. In concept B2 the actual heat pipe is located adjacent to the inside facesheet and close to the dissipating equipment. The heat is directly conducted into the heat pipe and from there via the bridging profile to the outside facesheet. Since the heat flux is low, the sidewalls of the bridging profile are thin, which results in a lightweight heat pipe profile.

Alternatively, the actual heat pipes may be located close to the outside facesheet (concept B1) with the heat dissipation of the equipment conducted through the bridging profile into the heat pipe. Here, the relative high heat flux requires a heavy bridging profile, which however may be milled away in areas where no equipment is mounted. The selection of one of these concepts will be influenced by the equipment layout of the radiator. Generally, a radiator layout with little high dissipating equipment may favor concept B1, since areas with the heavy bridging profiles are small. On the other hand radiator layouts with large and/or redundant equipment would increase the heat pipe weight of concept B1 and concept B2 may be the more optimal solution.

Other factors such as mounting of additional equipment in a late stage of a project, fabrication complexity and thermal performance must also be considered.

Especially the thermal performance of heat pipes integrated into sandwich structures has a strong impact on the radiator design, since it establishes the temperature gradient between equipment and heat pipe and consequently the overall radiator efficiency. Due to the importance of this parameter is was decided to perform a test program in order to determine experimentally the heat conductance coefficient of a typical radiator heat pipe interface. This program is described in the following paragraph.

Performance Investigation of Heat Pipes Integrated into Sandwich Structures

Test Program

Heat conduction into a heat pipe may be expressed by the so-called film coefficient which defines the heat flow between the outside heat pipe wall and the vapor of the working fluid, related to the unit area of the internal heat pipe surface and assuming uniform radial heat flux.

This coefficient has been experimentally determined for axial grooved heat pipes and several working fluids.[5] For ammonia this value is 0.7 W/cm² K or related to the evaporator length 2 W/cm K (assuming an usual i.d. of 0.9 cm).

However, uniform heat input into the evaporator is not existent in the heat pipe-radiator arrangements described in Fig. 2, and consequently, the film coefficient will be much lower.

The reduced heat conduction into heat pipes with one-sided input has long been known. First experimental investigations were carried out during the ATS program.[1]

According to this work a conductance value of only 0.18 W/cm K was measured for a heat pipe with square cross section bonded between facesheets. In order to improve this value, so-called saddle blocks were used in the ATS program which enlarge both the bonding area to the facesheets and the heat input surface of the heat pipe and result in a conductance value of 0.54 W/cm K. Theoretical investigations of one-sided heat input of heat pipes also showed a strong decrease in thermal performance, when compared to uniform radial heat flux.[6]

Since both the experimental and theoretical studies of Refs. 1 and 6 are not generally applicable, it was decided to test specific interface configurations for this type of radiator. The work was performed in cooperation with the University of Braunschweig[7] and included an extensive test program for three different heat pipe-radiator

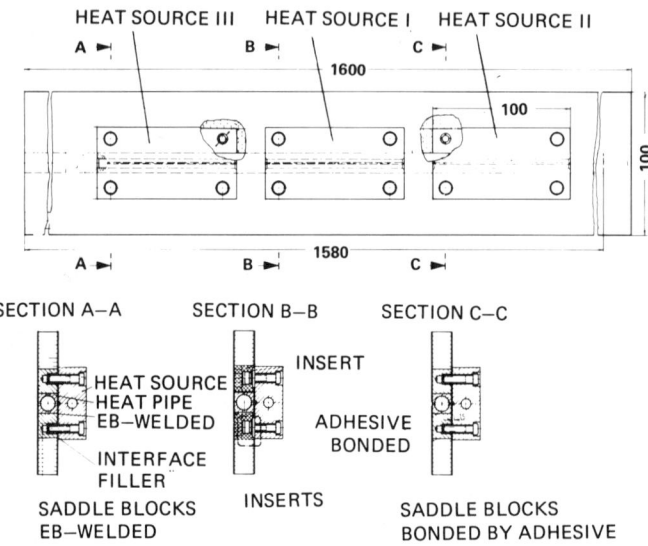

Fig. 3 Layout of test radiator.

interface configurations. The test item is shown in Fig. 3 and consists of a 1600 x 100 mm sandwich radiator with one integrated axially grooved heat pipe of square cross section.

In the center part of the radiator 3 heat sources with equal dimensions are mounted above the heat pipe. The sources consist of aluminium blocks heated by a cartridge heater and their interface areas to the radiator correspond to a mounting area section of a typical traveling wave tube.

Thermal tests were carried out with three attachment methods for the heat sources:

1) Heat source I is mounted over inserts to the radiator.

2) Heat sources II and III are bolted to saddle blocks which in turn are bonded to the heat pipe by an epoxy adhesive in configuration II and EB welded to the heat pipe in configuration III.

Between heat sources and radiator three different interface fillers were used: SIGRAFLEX, CHOTHERM 1661 and RTV 11 with Teflon foil.

HEAT PIPE RADIATOR FOR TV SATELLITE

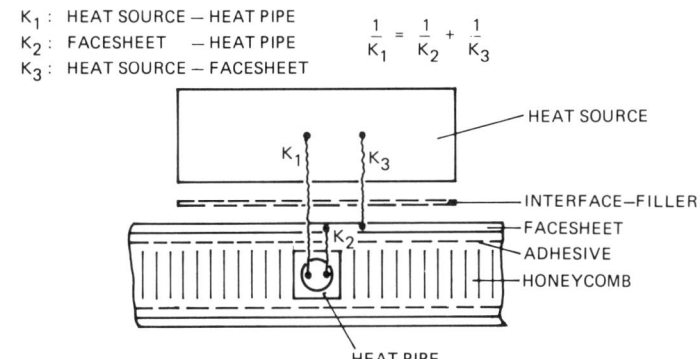

Fig. 4 Definition of conductance values.

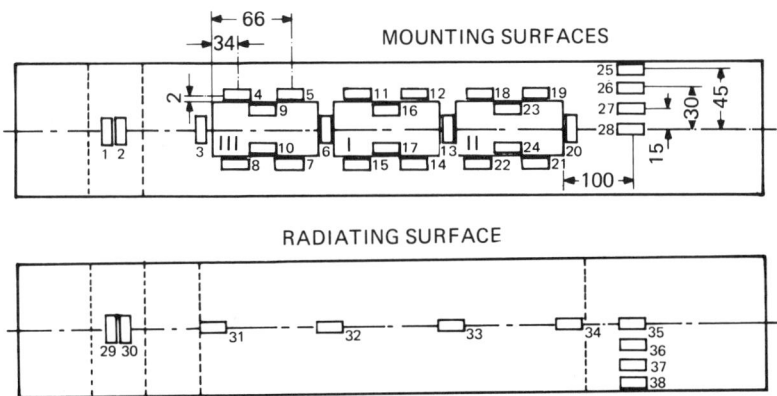

Fig. 5 Thermocouple location.

The test objective was threefold and included the determination of

1) Thermal conductance values for the heat source-heat pipe interface as defined in Fig. 4.

2) The performance of different interface fillers.

3) The influence of the 1 g environment on the heat conductance into the heat pipe.

For the purpose of this test program the radiator was equipped with 38 thermocouples as shown in Fig. 5. During the test the mounting side of the radiator including heat sources was insulated by multilayer insulation forcing the heat to be conducted to the opposite radiator side, which was prepared for radiation heat transfer with black mat paint. Tests were carried out in a cold (LN_2) shrouded vacuum chamber to simulate space environment.

Thermal Conductance of the Heat Source-Heat Pipe Interface with Different Interface Fillers

For these tests the radiator was placed horizontally in the vacuum chamber with the heat sources facing down.

The heat sources (one at a time) were powered in steps with 30, 60, and 100 W and temperatures were measured after each step at steady state conditions. Tests were repeated with three different interface fillers, as mentioned above.

From the measured temperatures the conductance values k_1 and k_2 were calculated as follows:

$$k_1 = \frac{Q}{L_{HS}(T_{HS} - T_{HP})} \qquad k_2 = \frac{Q}{L_{HS}(T_{FS} - T_{HP})}$$

with Q = heat input to the heat source; L_{HS} = length of heat source along the heat pipe; T_{HS} = average temperature of heat source, for example (see Fig. 5): $T_{HS} = (T_9 + T_{10})/2$ for heat source III; T_{FS} = average temperature of facesheet, for example: $T_{FS} = (T_3 + T_4 + T_5 + T_6 + T_7 + T_8)/6$ for heat source III; and T_{HP} = temperature of heat pipe vapor, considered to be the temperature measured on the heat pipe in a multilayer insulated radiator section (T_1, T_2).

For the investigated mounting configurations and interface fillers the conductance values are plotted in Fig. 6.

These results may be summarized as follows:

1) Heat conductance in a heat pipe with one-sided heat input (k_2 with insert mounting) can be drastically improved by the use of saddle blocks which, in fact, represents a situation with three-sided heat input. The improvement depends on the quality of saddle block bonding. The conductance is roughly a factor 2 times higher with adhesive bonding and 3 times higher with a welded junction.

2) The overall conductance between heat source and heat pipe (k_1) is smaller than the heat pipe film coefficient (k_2) due to the additional resistance of the interface filler. SIGRAFLEX filler, a graphite foil of 0.2 mm thickness

yields the highest conductance value of 0.75 W/cm K for welded saddle blocks. The conductance for insert mounting and SIGRAFLEX is only 0.4 W/cm K.

3) Due to the higher conductance values the use of saddle blocks may substantially reduce radiator weight and/or the number of heat pipes. This technique should therefore be considered in optimizing high performance heat pipe radiators.

Influence of 1-g Environment

As mentioned earlier the conductance tests were carried out with the heat sources facing down. This, obviously, is the preferred test orientation in a 1-g environment, since excess liquid in the heat pipe may form a puddle at the heat input side, which would prevent a premature dryout of grooves in that region. However, in satellite system tests a configuration may well exist when heat sources are placed on top of a heat pipe, which, in turn, could re-

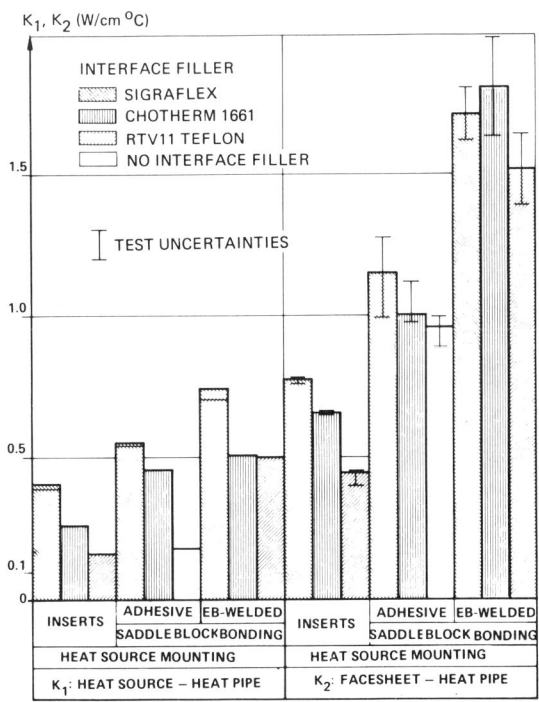

Fig. 6 Conductance of heat source - heat pipe interface with different mounting configurations and interface fillers.

sult in a partial dryout of grooves and a decrease in conductance values.

In order to investigate the 1-g influence, some of the tests described earlier were repeated with the heat sources configured to be on top of the heat pipes.

The results, given in Fig. 7, show that

1) Conductance values are 3% to 5% lower at 30 W heat input and 6% to 10% lower at 100 W heat input

2) The decrease in conductance is especially pronounced for insert mounting (10% vs 6% for welded saddle blocks, both with 100 W heat input). This would suggest that a partial dryout exists in both cases and that heat is conducted to the wetted section only through the heat pipe wall in case of insert mounting and through the saddle block/heat pipewall combination with low thermal resistance in the other case.

3) Although a linear dependence of the decrease in conductance with power input has been drawn in Fig. 7, this must be verified in more detailed testing. A linear dependence would suggest a continuous depletion of more grooves with larger heat input.

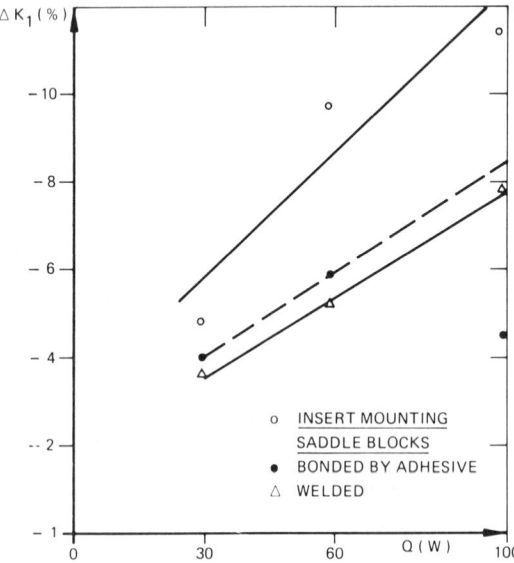

Fig. 7 Decrease in thermal interface conductance when heat source is facing up.

Table 1 Evaluation of design concepts

Concept	thermal	Structure weight	Radiator flexibility		Fabrication
A	very good	poor	good	good to poor	poor
B1	poor	good	good to poor	poor	good
B2	good	good	good	good	good to poor

4) The test results show that with respect to thermal performance in a 1-g environment saddle block configuration should be preferred.

Radiator Design

In order to arrive at an optimal radiator design the design concepts shown in Fig. 2 were evaluated with respect to thermal and structural performance, minimum weight, radiator flexibility and fabrication complexity. The results are summarized in Table 1. The evaluation has been performed according to the following criteria.

Thermal Performance

A) Saddle blocks are possible, however decreased capability to radiate from panel backside into the satellite.

B1) No saddle blocks are possible, decreased capability to radiate from panel backside into the satellite.

B2) Saddle blocks and radiation from panel backside are possible, however, higher temperature drop between facesheets as compared to concept A.

Structural Performance and Weight

A) Based on the heat pipe layout of Fig. 8 and a radiator area optimized for one redundant repeater channel a relative low heat pipe weight of 1.7 kg has been estimated for this version. However, additional structure weight to restore plate stiffness will invalidate this weight advantage.

B1) For this version a high heat pipe weight of 3.6 kg has been estimated due to the large mounting areas of the redundant equipment layout.

B2) The heat pipe weight for this concept was calculated to 2.5 kg and is independent from equipment layout.

Radiator Flexibility

A and B1) Relocating or mounting of additional radiator equipment requires redesign of the radiator (only in some cases for A).

B2) Equipment can be mounted anywhere on the heat pipes.

Fabrication Complexity

A) Nonconstant radiator height complicates fabrication.

B1) Nonconstant heat pipes cross section complicates fabrication of honeycomb heat pipe interfaces.

B2) Honeycomb milling only required in areas with saddle blocks.

According to this evaluation the design concept B2 has been selected for the development of a prototype radiator.

Development and Test of the Prototype Radiator

Radiator Layout

The radiator layout is based on accommodating a TWT and EPC of a redundant repeater channel as developed by AEG-Telefunken.

The thermal model of the TWT is shown in Fig. 9.

The equipment consists of a free radiating collector and of the TWT-body, which is attached to the readiator. Thermal dissipation and temperature limits are shown in Table. 2.

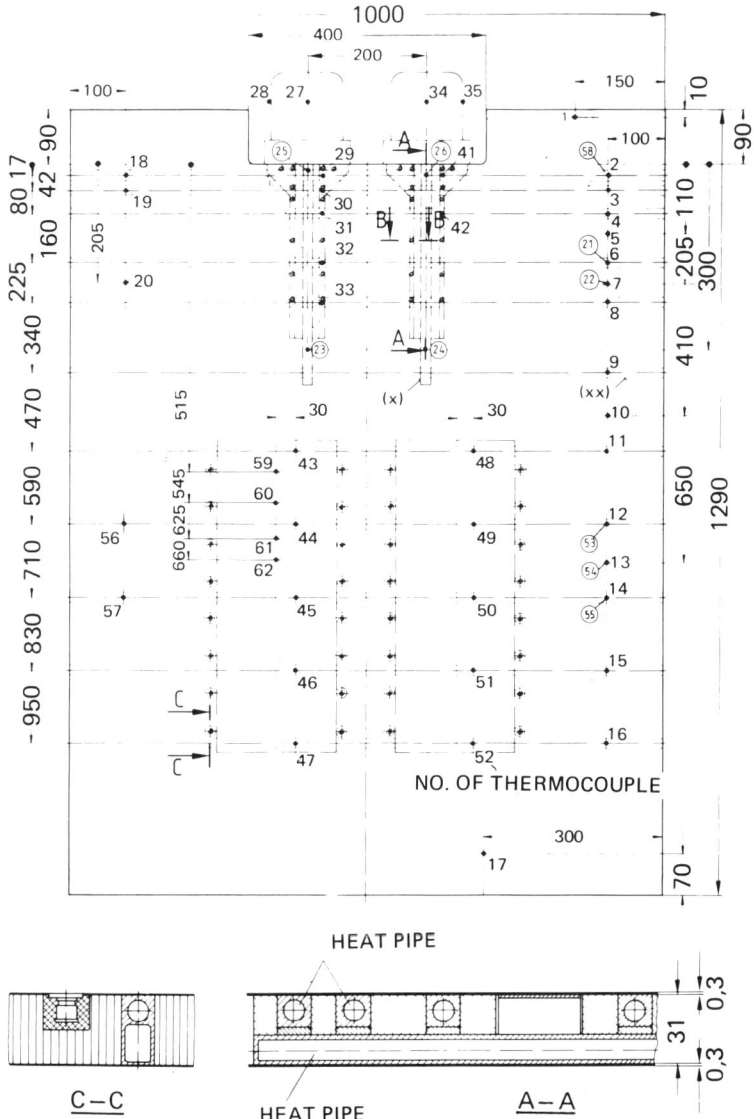

Fig. 8 Heat pipe-radiator layout (numbers indicate thermocouple location for test program).

The heat distribution within the TWT is such that about 80% of the heat dissipation is concentrated in the first 60 mm of the TWT body close to the collector. The TWT baseplate has been made wider in this area to decrease the heat flux density (see Fig. 9). Due to this uneven heat distri-

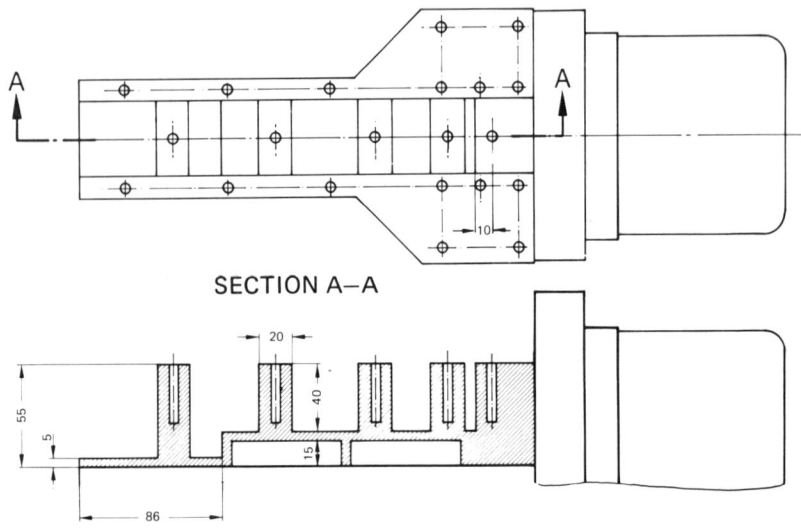

Fig. 9 TWT-TL 12260 thermal model.

Table 2 Thermal dissipation and temperature limits

	Dissipation	Temperature limits (°C)	Contact area (mm²)
TWT collector	251 Watt	0 to 350	free radiating
TWT body	108 Watt	0 to 70	270 x 70 (110)
EPC	110 Watt	-10 to 45	400 x 200

bution a radiator with parallel heat pipes would have a steep temperature gradient and, consequently a low efficiency since heat conduction of the radiator normal to the heat pipes is minimal. It was therefore decided to incorporate additional short heat pipes parallel to the TWT axis. These heat pipes would help to equalize the radiator temperature and, consequently, maximize the radiator efficiency.

Due to the more even heat distribution in the EPC baseplate a crosswise heat pipe layout was not necessary in this radiator section.

The selected radiator design as shown in Fig. 8 represents a typical cut-out of the 1.3 m high and 2.4 m wide

TV-SAT payload module panel (see Fig. 1). The width of this development panel is actually too large for the equipment dissipation given in the Table 2, but was selected in order to study the radiator fabrication process with a representative panel size. The necessary radiating surface was optimized by analysis and finally established by covering part of the surface with multilayer insulation (MLI) as shown in Fig. 10.

Due to the different operating temperature ranges of TWT and EPC the two radiator areas for these equipment have been separated by an "unused" panel section with low thermal conductivity.

The EPC section contains five parallel heat pipes and the TWT section six parallel and underneath each TWT one short heat pipe. All heat pipes are integrated between the panel facesheets to obtain flat mounting surfaces for thermal coating and equipment attachment.

For the heat pipes an axially grooved profile with ammonia as working fluid and AlMgSi 0.5 as wall material has been selected for this project. As mentioned earlier

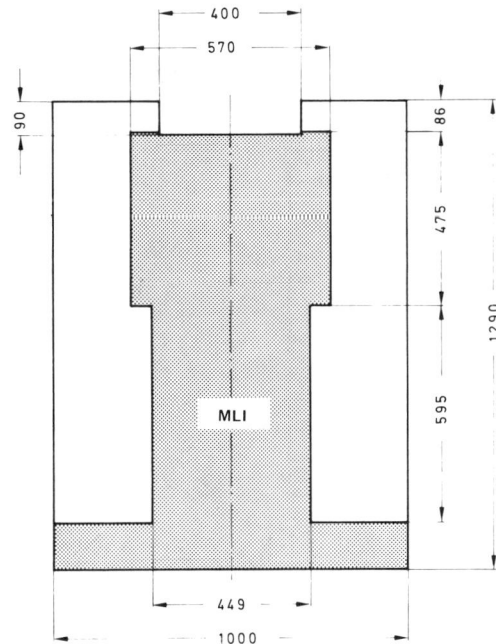

Fig. 10 Multilayer insulation coverage of radiator frontside.

the profile has a rectangular cross section to fit the sandwich height.

This profile, as shown in Fig. 11, has been derived from an existing spacequalified heat pipe with square cross section and 24 grooves of 0.8 mm width. The typical transport capability of the heat pipe is also shown in Fig. 11.

The design of the radiator was supported by a mathematical model in order to determine the MLI coverage of the radiating surface and to predict the thermal performance.

Since the heat pipes are almost isothermal along their length, it was decided to extend the finite elements over

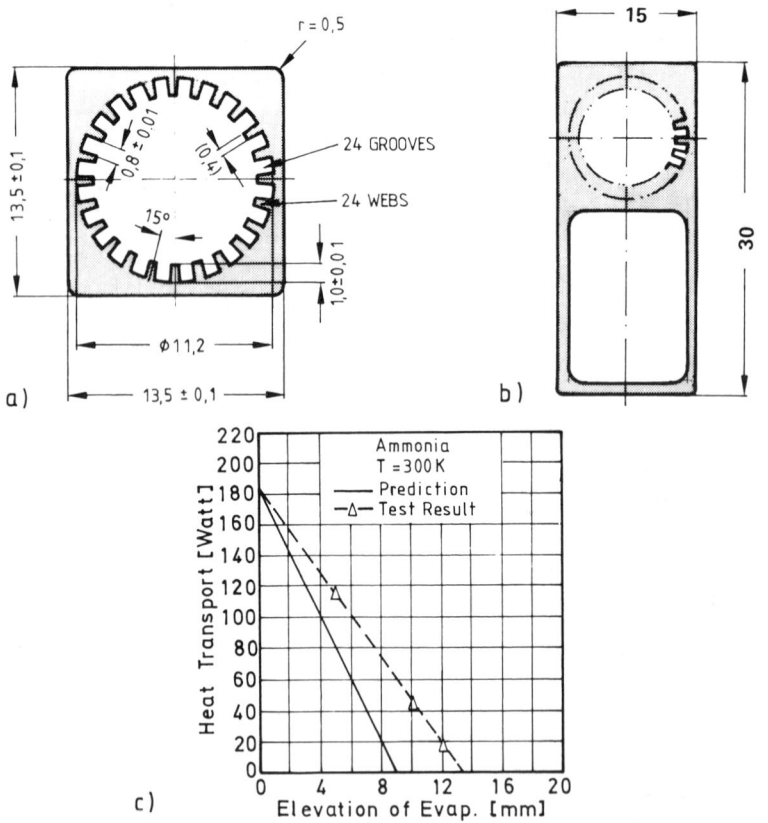

Fig. 11 Heat pipe, a) existing spacequalified profile with square cross section, b) radiator heat pipe with rectangular cross section and identical groove dimensions, c) typical transport capability.

the full width of the radiator and have a fine element subdivision normal to the heat pipes. The model consists of 232 finite elements. The large number of elements was selected in order to investigate in detail the heat fluxes associated with the equipment-radiator and radiator-heat pipe interfaces. The exact knowledge of these interfaces would enable smaller node models for similar configurations in the future by maintaining the same confidence in the calculation.

In the mathematical model a finite element model with 13 nodes was included for the TWT. This model was delivered by the tube manufacturer and structured according to internal heat path distribution and thermal resistances. The correlation of calculated data and test results is given later in this paper.

Fabrication

Heat pipe fabrication includes machining of the parts, welding of the pipe assembly, cleaning, charging and final pinch off. All pipes were indivdually acceptance tested before delivery for integration into the radiator.

Fabrication of the radiator plate involves the conventional technique of bonding of honeycomb sandwich, except that 13 heat pipes and the associated saddle blocks are added for bonding between the facesheets. The problem associated with bonding of filled heat pipes is the high curing temperature of the adhesive which in turn produces a high internal heat pipe pressure. The standard adhesive used at ERNO for structure bonding cures at 120 °C and produces an internal heat pipe pressure of about 100 bar. Heat pipes are designed and tested to obtain a safety margin of 1.5 against bursting.

Several fabrication steps are illustrated in Figs.12-14.

The network of heat pipes in the TWT radiation area is shown in Fig. 12. Note, that saddle blocks have been used in heat pipe corssing areas in order to improve heat conduction between the pipes. Although saddle block bonding by an adhesive has been found to be somewhat less effective than welding (see Performance Investigation of Heat Pipes Integrated into Sandwich Structures), the former technique has been applied for this prototype radiator because it represents the more cost effective approach. Figure 13 shows the entire heat pipe system as it is bonded to the lower facesheet (equipment side). In the next step strips of

Fig. 12 Radiator before bonding of saddle blocks in heat pipe crossing areas.

honeycomb material are inserted between the heat pipes and bonded to facesheets and pipes (Fig. 14).

Thermal Test

The test objective was to verify the interface conductance values between equipment radiator and radiator/heat pipe with this rather complex radiator design. For this purpose thermal tests were carried out in a liquid nitrogen shrouded vacuum chamber in order to simulate cold space environment.

The test object was prepared as follows:

1) Radiating surface painted black and partly insulated by MLI according to analysis (Fig. 10).

2) Mounting surface of the radiator equipped with thermal dummies for TWT and EPC. Entire surface insulated by MLI.

3) A total of 68 thermocouples was distributed on the radiator according to Fig. 8.

The radiator as installed in the test chamber is shown in Fig. 15.

Tests were conducted by heating the thermal dummies to preset values and recording temperatures after steady-state condition was established. Several test steps were performed, with different heater power and different equipment combinations operating.

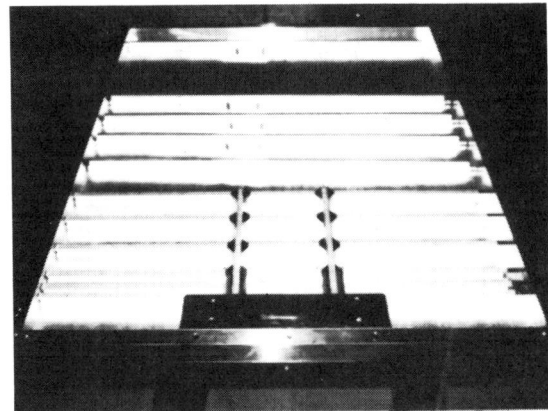

Fig. 13 Radiator after bonding of heat pipes to lower face sheet.

Fig. 14 Radiator after bonding of honeycomb and before bonding of second facesheet.

Test Results and Data Correlation

According to temperatures measured at both ends of the heat pipes, all pipes are isothermal. It was therefore correct to extend the finite elementes for analysis over the full heat pipe length.

Figure 16 shows a typical temperature distribution of the radiator in a direction normal to the heat pipes. Note that heat pipe temperatures are higher than neighboring radiator parts which leads to the typical uneven temperature distribution of a heat pipe radiator.

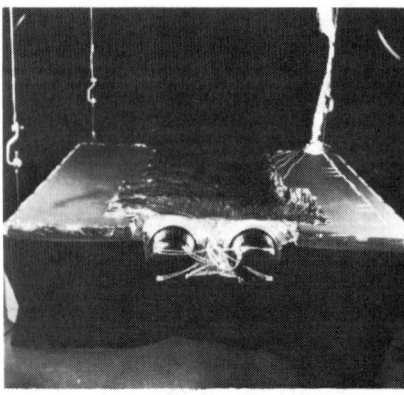

Fig. 15 Radiator installed in test chamber.

The heat load distribution over the radiator is also shown in Fig. 16. The strongly uneven distribution along the TWT body is here of special interest. Three heaters adjacent to the collector head are each producing 23 W whereas the remaining two heaters are only powered to 6 W. The body part which is connected to the collector receives an additional 52 W across this interface which amounts to a heat load of 75 W within the first 3 cm of the TWT body. Despite this uneven heat load distribution the associated temperature gradient of the radiator and TWT body is far less pronounced (Fig. 16). This is due to the short crossing heat pipe underneath each TWT.

From test results the performance characteristic of the TWT-radiator can be summarized as follows:

Radiator efficiency: $\eta = 94\ \%$
Max. gradient of radiator $\Delta T = 11\ K$
Max. gradient of TWT-body $\Delta T = 3.5\ K$

Temperature difference of
TWT/radiator interface at $\Delta T = 4\ K$
max. heat flux location

Typical interface conductance
TWT/heat pipe $k = 11.5\ W/K$.

The radiator efficiency is defined as measured radiator heat load divided by the heat load which would be radiated at the maximum measured radiator temperature.

According to measured temperatures of the EPC, which were 20 to 50 °C higher than predicted, the interface conductance between equipment and radiator was much lower

HEAT PIPE RADIATOR FOR TV SATELLITE

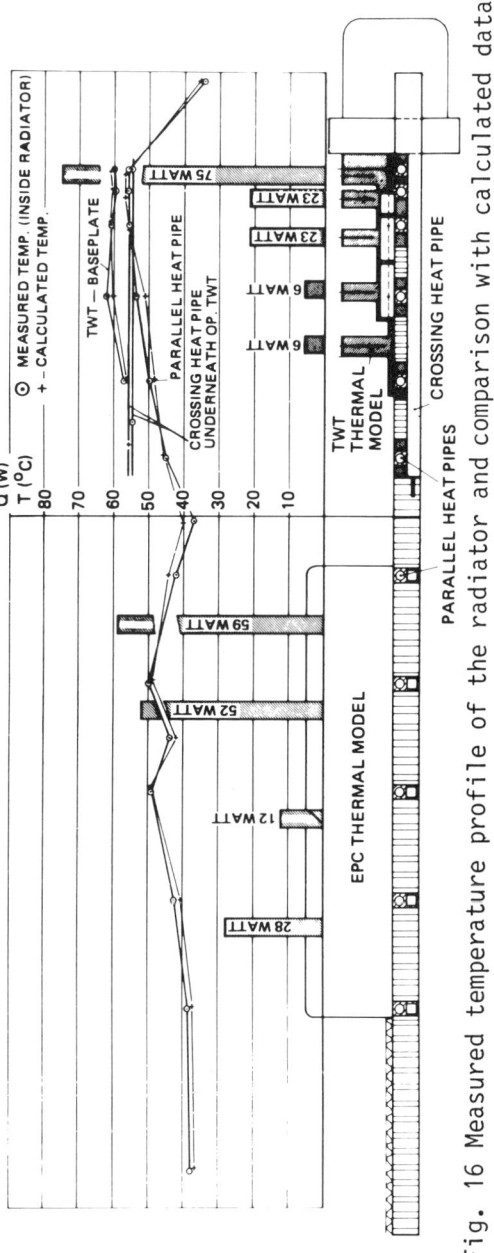

Fig. 16 Measured temperature profile of the radiator and comparison with calculated data.

than expected. The reason for this behavior was assumed to be the large mounting area of the box (500 mm x 200 m) which was bolted to the radiator only at the circumference. This attachment method leads to deformation of the equipment base plate and low contact pressure in the center of the plate where the heat sources are concentrated. This discrepancy between test and prediction was not studied in detail because a new base plate design was proposed by the electronic supplier during the course of the project. In this design the attachment method is improved by distributing additional mounting holes over the entire base plate and close to heat sources. There is good reason to assume that the temperature discrepancies are not repeated with this configuration.

The characteristics of the EPC radiator can be summarized as follows:

Radiator efficiency $\eta = 89\ \%$
Max. gradient of radiator $\Delta T = 11\ K$
Max. gradient of EPC base plate $\Delta T = 35\ K$

Conclusion

The technology to integrate heat pipes in large sandwich structures has been demonstrated with the fabrication and test of a prototype radiator for a direct broadcasting television satellite, now under development in Germany.

The developed technique provides excellent radiator performance due to the high heat conduction from dissipating equipment into the heat pipes and because of the practically isothermal heat distribution over the radiator area.

A crosswise heat pipe layout has been found to be very effective for the thermal control of equipment with highly uneven heat source distribution.

The results of this program phase indicate that the developed technology is well suited to optimize the thermal control of future high power spacecraft projects.

Acknowledgment

The major part of the work was performed under contract to BMFT(BPT), Bonn, Porz, Contract No.: 01 YM 038-AK/KS-WRT 5050.

References

[1] Berger, M. E. and Kelly, W. H., "Application of Heat Pipes to the ATS F Spacecraft" ASME Paper No. 73-ENAs-46, July 1973, San Diego.

[2] Schlitt, R., "Layout of Heat Pipe Radiators for High Dissipating Payloads of Future Communication Satellites" (in German), DGLR Nr. 78-174, DGLR Jahrbuch 1978.

[3] Schlitt, R. "A Heat Pipe Radiator for Shunt Electronics: Development and Test Results", Proceedings of Spacecraft Thermal & Environmental Control Systems Symposium, 1978, also ESA Sp-139, Nov. 1978.

[4] Schlitt, R., Laux, U., and Meyer, R., "Problems Associated with Thermal Testing of Large Heat Pipe Systems for Space Application," to be presented at the 4th International Heat Pipe Conference, London, England, 7-10th Sept. 1981.

[5] Schlitt, K. R., Kirkpatrick, J. P. and Brennan, P. J., "Parameter Performance of Extruded Axial Grooved Heat Pipes from 100 to 300 K," AIAA Paper No.74-724, July 1974, Boston.

[6] Kamotani, Y., "Effects of One-Sided Heat Input and Removal on Axially Grooved Heat Pipe Performance," AIAA Paper No. 77-191, Jan. 1977.

[7] Brandt, R., "Investigation of One-sided Heat Input in Heat Pipes of Spacecraft-radiators", Jan. 1979, unpublished report of the University of Braunschweig, Germany.

Chapter IV. Thermal Control

DEVELOPMENT OF A 5 W 70 K PASSIVE RADIATOR

J.P. Wright*
Rockwell International Corporation, Downey, Calif.

Abstract

Future space based, infrared, Earth surveillance systems may one day be cooled by large heat pipe radiators operating at 70 K or even as low as 40 K. The use of passive radiators could eliminate the need for mechanical refrigerators and the attendant problems of power consumption and long life reliability. A large two-stage heat pipe radiator was developed for testing on the ground to verify the thermal performance and structural integrity of large, multiwatt, passive cryogenic radiators. The Radiator has a projected base area of 8.1 m^2 and is designed to reject a 5 W heat load to space at 70 K. The radiator design includes three oxygen heat pipes which transport the 5 W heat load from a simulated detector focal plane and distribute it over the second-stage radiator surface. The radiator is also designed to withstand the launch environment imposed by a Shuttle orbiter or a Titan III launch. Design tradeoffs are discussed, and the final radiator design configuration is described. Thermal vacuum test results and thermal model predictions are presented. Agreement is very good. Subsequent acoustic testing of the cryogenic radiator test unit and thermal vacuum tests with a 5.1 m^2 third stage added are also discussed.

Nomenclature

A = area
C_1, C_2 = coefficients of insulation thermal conductivity, Eq. (3)

Presented as Paper 80-1512 at the AIAA 15th Thermophysics Conference, Snowmass Colo., July 14-16, 1980. Copyright © American Institute of Aeronautics and Astronautics, Inc., 1980. All rights reserved.
*Supervisor, Aerothermal Systems Group.

H	= wicking height factor
k	= thermal conductivity
N_ℓ	= liquid transport factor
R	= radius
T	= temperature
W	= width
Y	= height
α	= angle between equatorial and ecliptic planes
β	= angle between Earth-sun vector and orbit plane
γ	= angle between sun shield and plane of radiator
$\varepsilon, \bar{\varepsilon}$	= emittance
η	= fin efficiency
θ	= time
ν	= kinematic viscosity
σ	= Stefan-Boltzman radiation constant
ϕ	= angle from Earth-vehicle vector to Earth horizon

Subscripts

c	= cold boundary
ER	= end of run
h	= hot boundary
ℓ	= liquid
MLI	= multilayer insulation
R	= radiator
S	= helium shroud
SS	= steady state
TP	= triple point
v	= vapor

Introduction

Passive radiators are an attractive means of cooling space infrared sensors and other equipment requiring operation at cryogenic temperatures. The passive radiator has several potential advantages over closed cycle mechanical refrigerators or solid cryogen coolers. They include long life potential, intrinsic reliability, and lack of electrical power consumption. Cryogenic heat pipes and multiple radiator stages make it possible to achieve significant levels of cooling at temperatures of 70 K or below. Recent tests on experimental two- and three-stage radiators have demonstrated heat rejection capacities of 0.7 W at 70 K and 0.03 W at 40 K, with a minimum (no load) temperature of 33.5 K.[1] System tradeoffs show that, for small heat loads, passive radiators have a lower overall system weight compared to mechanical refrigerator systems at temperatures above 67 K.[2] However, for long life applications (greater than two years), passive radiators may be desirable down to temperatures as low as 35 K.

The next logical step toward orbital testing of a large multiwatt radiator system is to demonstrate extending this radiator technology to a large, structurally sound, flightweight system and its performance on the ground in a simulated space environment. This paper discusses the current status and results to date of an Air Force program to develop a large, flightweight, multistage, heat pipe radiator system.

Multistage Radiator Concept

The general theory and operating principles for the multistage radiator concept are documented in the literature.[3,4] The multistage radiator concept consists of two or more stages that radiate to a cold environment. Each stage is thermally isolated from the other by multilayer insulation and low-conductivity support posts to minimize heat conduction. The principle of performance is based on each intermediate radiator stage intercepting the parasitic heat leakage through the insulation below and radiating it to space. Thus, each successive stage "sees" a colder and colder boundary temperature. This allows each successive stage to attain lower and lower temperatures. The intermediate stages can also reject heat at higher temperatures to cool other elements of a system such as optics, baffles, and electronics. Heat pipes transport heat from the source to the radiator and distribute it over the radiator surface to increase overall fin efficiency.

Radiator Design Summary

The cryogenic radiator test unit (CRTU) is a two-stage heat pipe radiator. The overall configuration of the CRTU is shown in Fig. 1. It consists of two radiator stages, insulation blankets, sun and side shields, heat pipes, a structural support pan, and structural mounting supports. The first stage is mounted off the structural support pan. Low thermal conductance support posts are used. The second stage is similarly supported off the first stage. Forty layer blankets of multilayer insulation (MLI) are located between the pan and the first stage and second stage. The radiator stages are made from 1.27 cm aluminum honeycomb with 0.025 cm facesheets. The first-stage radiator is rectangular measuring 2.1 m x 3.86 m. Three sides are bent upward to form the sun and side shields. The sun shield is 38 cm high and is angled up 60 deg from the plane of the radiator. The side shields are angled up 90 deg from the plane of the radiator and are triangular (see Fig. 1). The second-stage MLI blanket covers all but 0.9 m² of the first stage and 0.4 m² of the sun shield. The exposed areas are coated with a high emissivity black paint (ε_R = 0.9).

The second-stage radiator is rectangular and measures 1.83 m x 3.76 m. The exposed surface area is 6.88 m² and is also coated with high emissivity black paint. Heat is distributed over the radiator surface by a 2.4 m distributor heat pipe, which is bonded along the centerline of the radiator. Heat is transported from the heat source to the radiator via two transport heat pipes (for redundancy) that are mechanically and thermally coupled to the radiator heat pipe by a thermal clamp block. The transport heat pipes are L shaped and measure 1 m x 0.25 m and are at a right angle to the plane of the radiator. Heat is input to the transport heat pipes by cartridge-type heaters. The heat pipes are insulated with MLI and an aluminum thermal shroud.

The structural support pan is a riveted aluminum sheet metal structure that supports the radiator and simulates the spacecraft interface. It is stiffened with zee section longerons and intercostals to provide a rigid structure to which the radiator stages can be attached.

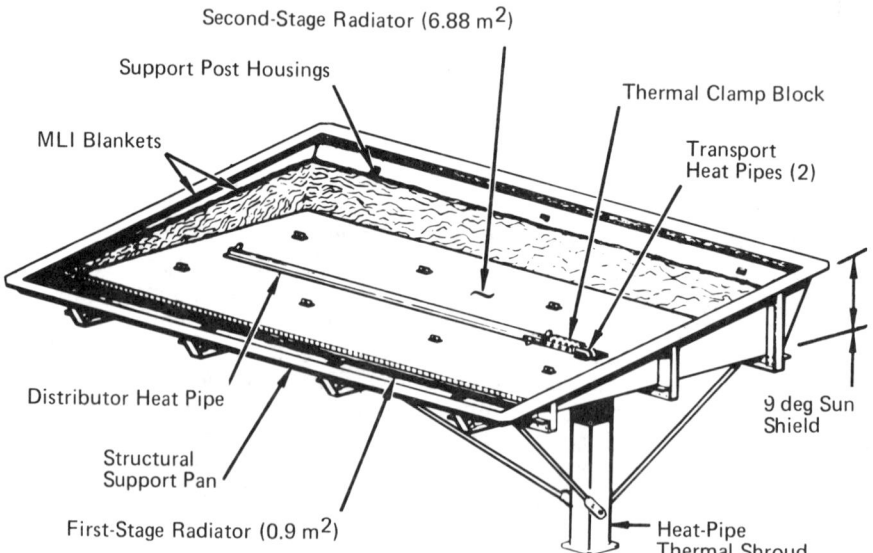

Fig. 1. Radiator configuration.

DEVELOPMENT OF A 5 W 70 K PASSIVE RADIATOR

Radiator Design Requirements

The design requirements for the CRTU are:

1. Heat rejection capacity of 5 W at an average radiator temperature of 70 ± 2 K.

2. Temperature drop from the radiator to a heat source located 1 m from the radiator to be less than 3 K.

3. Shielding capable of preventing direct solar impingement on the radiator in a geosynchronous orbit.

4. Ability to withstand the dynamic launch environment imposed by the Shuttle orbiter or a Titan launch vehicle.

5. Compatibility with a 2.74 m diam design envelope (Titan III booster).

6. Design operational life expectancy of at least 5 yrs.

In addition, the CRTU was to be designed to be tested in a space simulation (thermal vacuum) chamber. The CRTU was also required to be designed for later addition of a third-stage radiator.

Design Tradeoff Analyses

A number of design tradeoff analyses were performed to determine the optimum configuration of the radiator, shields, and heat pipes. Results of these analyses are discussed below.

Radiator Size and Geometry

Based on the analytical results of Wilson and Wright,[3] a two-stage radiator is optimum (minimum weight) at 70 K. The heat rejection capacity and optimum stage area ratio for a two-stage radiator is a function of the effective emittance of the insulation between the stages and the warm boundary temperature. Figure 2 shows the optimum area ratio and heat rejection capacity of a two-stage radiator as a function of the effective emittance of the insulation blankets. Based on previous test experience with a smaller two-stage radiator,[1] a value of 0.005 to 0.01 was felt to be an achievable value for blanket emittance. For a blanket emittance of 0.0075, the optimum ratio of the second-stage area to the total projected area is 0.87; the corresponding heat rejection capacity for an isothermal radiator with an emittance of 0.9 is 0.93 W/m^2.

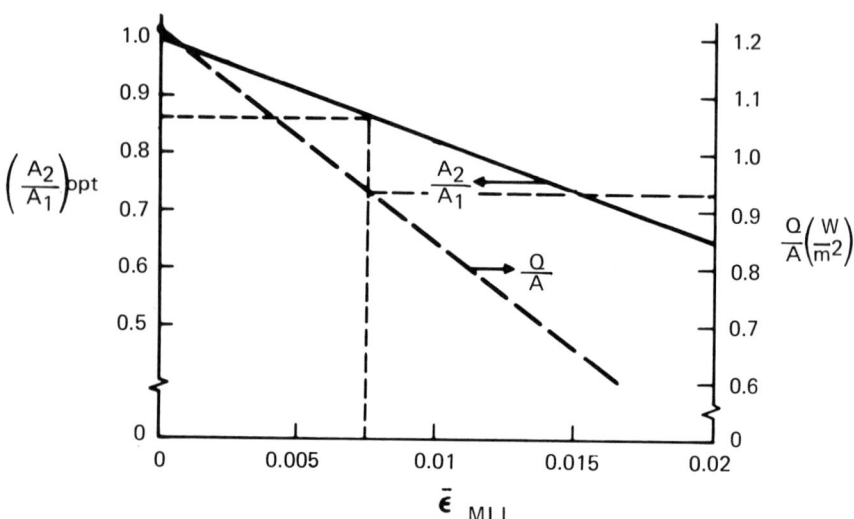

Fig. 2. Optimum stage area ratio vs. insulation emittance.

Assuming a fin efficiency of 0.8, the required radiator area is 6.7 m². A design value of 8.1 m² was selected to provide a 20% design margin.

Radiator Shielding

Shielding requirements for the CRTU are for a spacecraft in Earth orbit at geosynchronous altitude (35,800 km). The primary axis of the spacecraft is assumed to be in the direction of the velocity vector. In the generalized configuration shown in Fig. 3, the orbit plane is inclined to the equatorial plane at an angle i. The plane of the radiator is tilted above the orbit plane such that the radiator looks above the Earth's horizon. For six months of the year, the Earth-sun vector is below the orbit plane; for the other six months it is above the orbit plane. The radiator can be shielded from any direct solar incidence by erecting a shield equal to the angle ϕ between the Earth-vehicle vector and the vehicle-horizon vector. When the sun moves above the orbit plane, the spacecraft is yawed 180 deg such that the radiator faces south instead of north. This yaw maneuver must be repeated at six month intervals. An oblique view of the radiator orbit and shielding geometry is shown in Fig. 4.

At geosynchronous altitude, the required shield angle, ϕ, is 8.5 deg. A 9 deg shield angle was selected to provide some design margin for attitude control variations.

DEVELOPMENT OF A 5 W 70 K PASSIVE RADIATOR 435

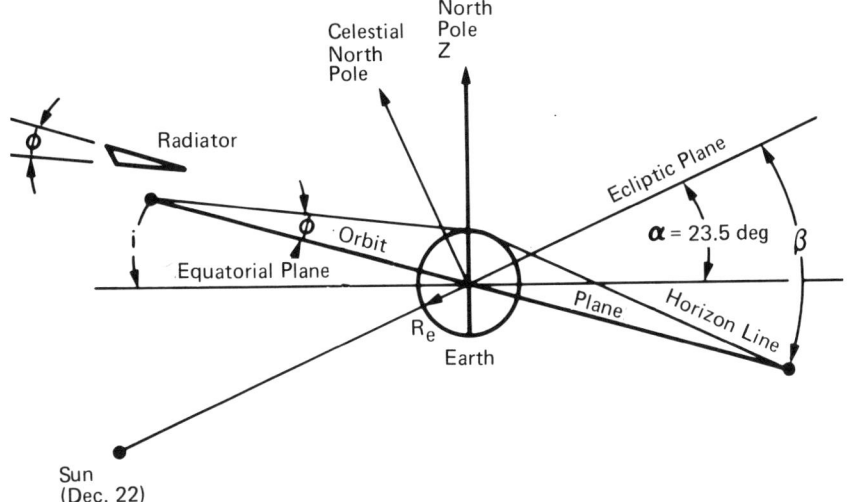

Fig. 3. Radiator shielding geometry - geosynchronous orbit.

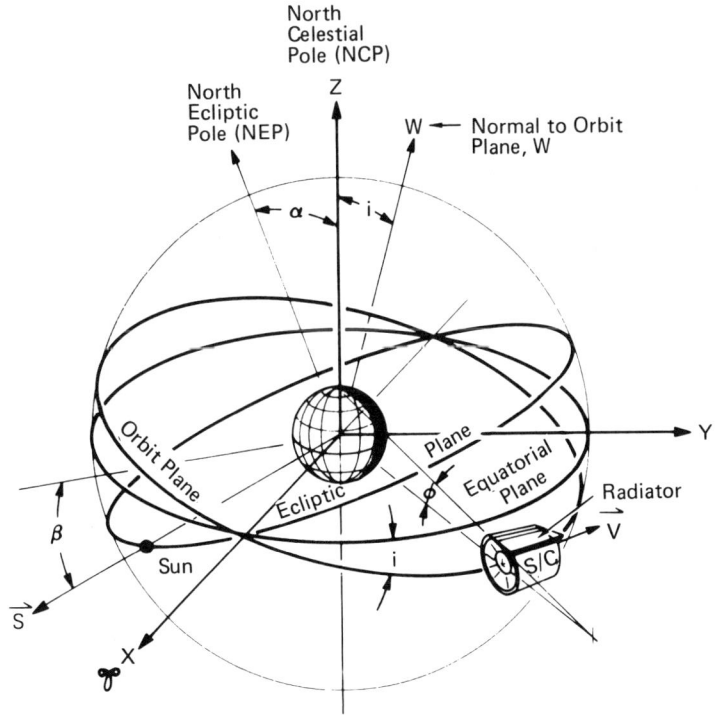

Fig. 4. Orbital and radiator shielding geometry.

The shield geometry is also constrained by the requirement for compatibility with a Titan booster payload shroud. Figure 5 shows the radiator and shield geometry within a 2.74 m diam envelope (maximum dynamic envelope). The maximum radiator width, W_R, is a function of the shield tilt angle, γ. It is desirable to tilt the shield back (γ less than 90 deg) to minimize the view factor of the shield to the second stage of the radiator. On the other hand, as γ becomes very small, the allowable radiator width, W_R, and the distance from the spacecraft centerline to the base of the radiator, Y, diminish rapidly. A shield tilt angle of 60 deg was selected as a good compromise. This corresponds to a radiator width of 2.1 m and a Y dimension of 0.48 m.

Radiator Fin Efficiency

For a large radiator such as the CRTU, heat pipes are required to distribute the heat load over the radiator surface. The radiator fin efficiency is a function of the fin material and thickness, the radiator and sink temperature, and the heat pipe spacing. Figure 6 shows the radiator fin efficiency at 70 K as a function of fin thickness and the number of distributor heat pipes. Based on structural considerations, it was desirable to make the radiator out of aluminum honeycomb sandwich for stiffness. A standard 0.127 cm (1/2 in.) panel with 0.025 cm facesheets has an equivalent conduction thickness of 0.069 cm. This gives an efficiency of 0.8 with one heat pipe, 0.91 with two heat pipes, and 0.95 with three heat pipes. In terms of maximizing the performance-to-weight ratio, two heat pipes appear to be the optimum number. Based on structural considerations, however, it was desired to

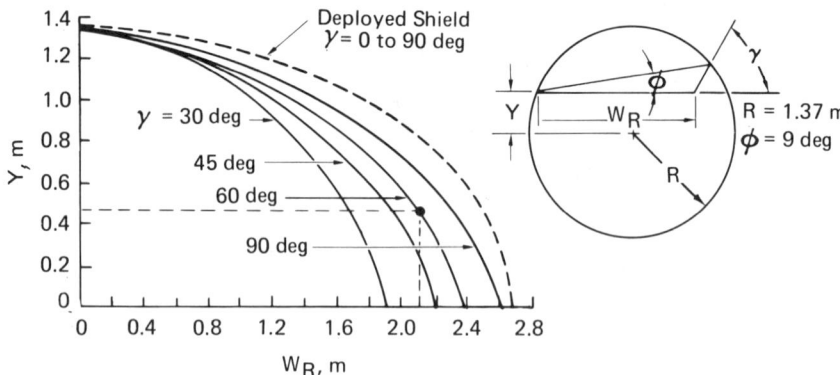

Fig. 5. Radiator design envelope compatibility.

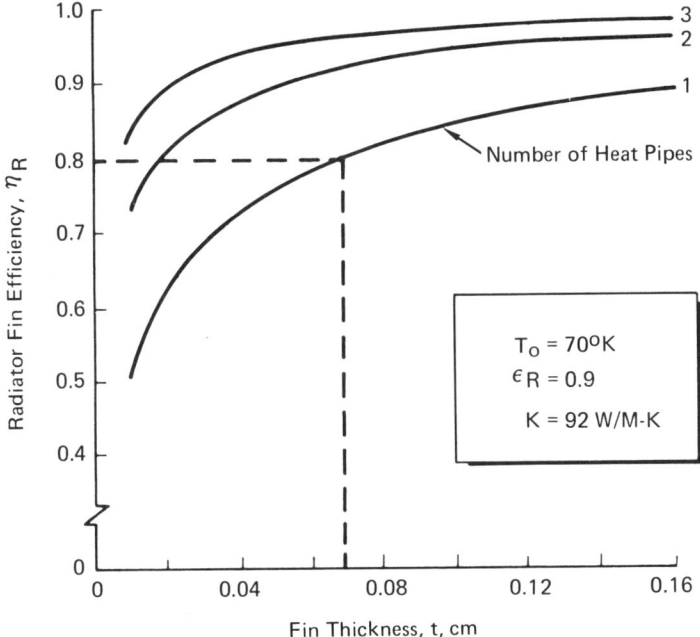

Fig. 6. Radiator efficiency vs. fin thickness.

minimize the mass of the second-stage radiator to minimize the static and dynamic loads that must be carried by the structural supports. As it turned out, the structural supports were stronger than predicted, and two heat pipes would probably be optimum for this size radiator.

Structural Supports

In addition to withstanding the structural loads imposed by a Shuttle or Titan launch, the structural supports' primary design considerations are thermal shrinkage of the radiator and parasitic heat leakage through the supports. Maximum axial and lateral acceleration load factors of 13 g and 8 g respectively were used for determining the design (static) loads. Calculation of dynamic loads was based on typical vibration input levels for the Titan IIIC and the Shuttle orbiter vehicles.

Because of the size of the radiator, up to 2 cm of lateral deflection of the radiator stages due to thermal contraction must be accommodated by the structural supports. As a result, the selected design approach was to support the radiator stages along two sides with rod end type posts

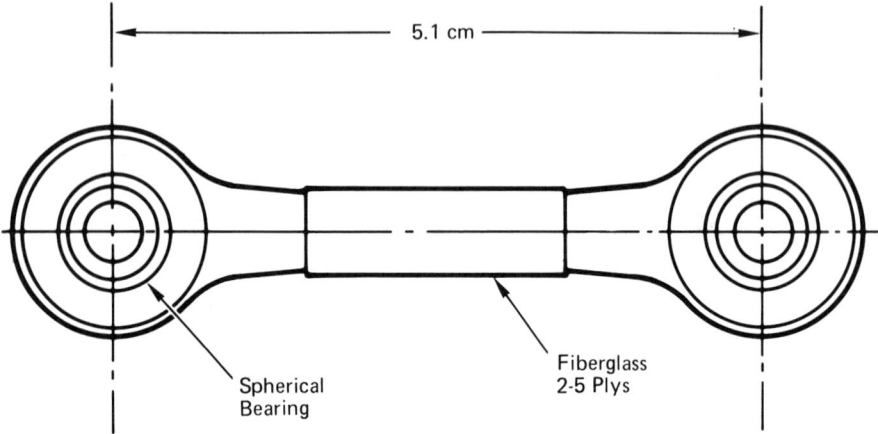

Fig. 7. Structural support post configuration.

to allow shrinkage in two dimensions. Rod end standoff posts were also used in the dimension perpendicular to the plane of the radiator. A post spacing of approximately 1 m was selected as a compromise between minimizing the structural load on each post and minimizing the total number of penetrations through the insulation. A typical post configuration is shown in Fig. 7. The posts are made of fiberglass with spherical rod end bearings. The side support posts have a tensile and compressive strength in excess of 2200 N and a thermal conductance of 5.1×10^{-5} W/K. The standoff posts have a maximum tensile strength of 1500 N and a compressive strength in excess of 2200 N, with a thermal conductance of 1.5×10^{-4} W/K.

Heat Pipe Design

The major design considerations for the heat pipes are transport, operational range, pressure containment, 1-g sensitivity, and complexity. The major design variables are the working fluid and the wick design. At 70 K, the candidate fluids are nitrogen and oxygen. Figure 8 shows the relative liquid transport factors and the pressure containment requirements of these fluids. As can be seen, oxygen has a liquid transport capability of 1.8 x that of nitrogen at 70 K. On the other hand, the required pressure containment for oxygen is about 20% higher than that for nitrogen. Table 1 lists some other properties of oxygen and nitrogen at 70 K. Oxygen has a higher static wicking height factor than nitrogen, which improves 1-g testability. Oxygen also has a lower vapor-to-liquid viscosity ratio than nitrogen, which results in lower vapor losses. Its triple-point temperature is significantly lower than nitrogen; hence, its operating temperature range is much lower. The lower operating temperature range is desirable since it improves the startup characteristics of the radiator from a cold condition and allows the radiator to be tested at lower temperatures. For these reasons, oxygen was selected as the working fluid despite the higher pressure-containment requirement.

The ATS groove design[5] was selected as the baseline wick design. This was based on a number of considerations. First the ATS groove design has been proved to be a reliable configuration, and the predicted transport at 70 K exceeds the 5 W transport requirement. The design safety factor for heat transport was 1.28 and was considered adequate to meet the program goals. Figure 9 shows the predicted transport capacity of the heat pipes as a function of temperature. Test data at zero tilt are shown also for comparison.

Producing a grooved extrusion with a wall thick enough to contain the high internal pressure was also a consideration. In general, thick walled extrusions are harder to produce since the flow of aluminum into the thin groove lands is

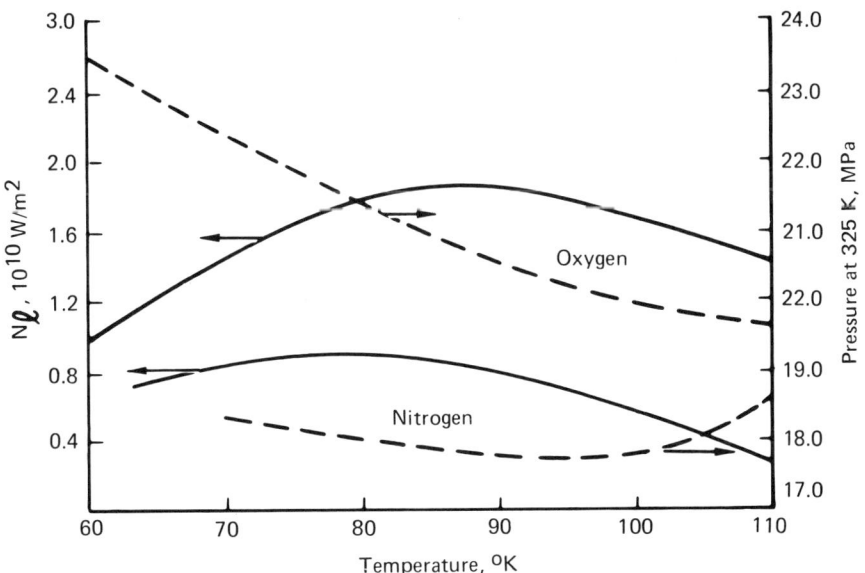

Fig. 8. Candidate heat pipe fluid design properties.

Table 1. Properties of Oxygen and Nitrogen at 70°K

Wicking height factor, H 10^{-8} m	Vapor/liquid viscosity, ν_v/ν_ℓ	Triple point, T_{TP} °K
Oxygen		
152	10	54.3
Nitrogen		
125	58	63.1

more difficult. However, the simple design of the ATS grooves minimizes this problem and reduces the development risk. An alternative to a thick wall extrusion is use of a superheated reservoir. This reduces the pressure in the pipe but adds considerable complexity to the radiator design since three reservoirs would have to be added. Additional complexity is introduced since the reservoir temperature must be controlled to maintain the required liquid inventory. The final extrusion geometry is shown in Fig. 10. The square cross section and flanges simplify the thermal coupling of the heat pipes.

Final Radiator Configuration

Figure 11 shows the final configuration of the CRTU after assembly. The CRTU is shown mounted on a ground transport dolly. Total weight of the CRTU is 145 kg including heaters, wiring, and instrumentation. Of the total, the structural support pan, which was not weight optimized, accounts for approximately 62 kg. In a flight application, the radiator may be attached directly to the spacecraft structure, in which case a support pan would not be required.

Thermal Test Program

A thermal vacuum test was conducted to determine the operating characteristics of the CRTU in a simulated space environment. Test objectives were:

1) To demonstrate the CRTU's ability to transport 5 W of power from a simulated detector focal plane and distribute it effectively over the radiator surface at 70 ± 2 K.

DEVELOPMENT OF A 5 W 70 K PASSIVE RADIATOR

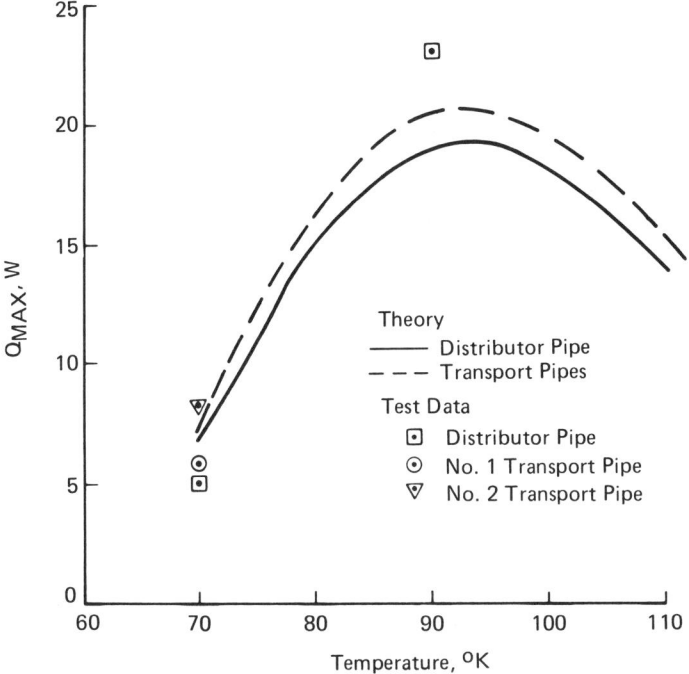

Fig. 9. Heat pipe transport vs. temperature.

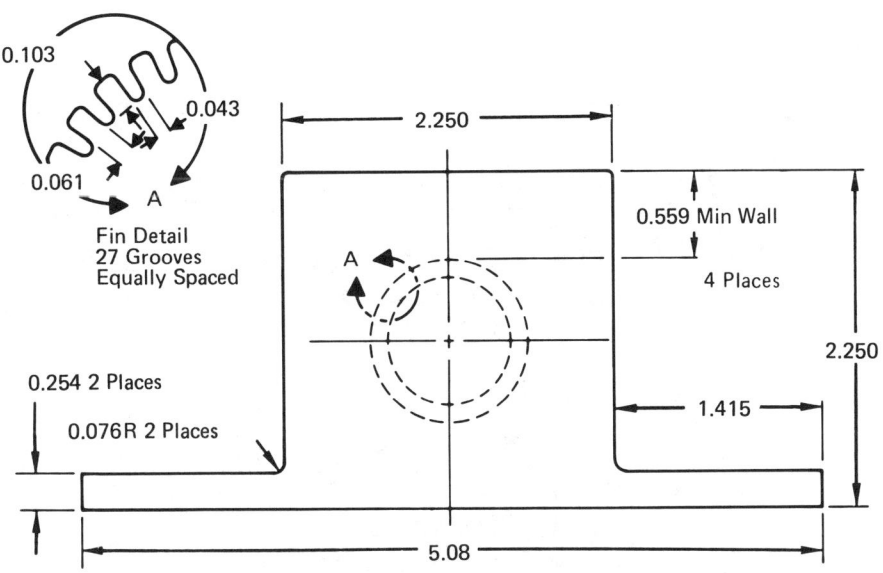

Fig. 10. Heat pipe extrusion geometry.

2) To determine radiator performance for various heat loads to the second-stage radiator.

3) To determine the effect of adding up to 20 W of power to the first stage on the second-stage temperature.

4) To determine the sensitivity of second-stage temperature to the sun shield temperature.

5) To assess the thermal performance of the insulation blankets and structural support posts.

Test Setup

The thermal vacuum test was conducted in the Mark I space chamber at the Arnold Engineering Development Center (AEDC). The test setup is shown in Fig. 12. The CRTU is mounted with its handling dolly on the floor structure of the Mark I chamber. The radiator stages and the distributor heat pipe are in the horizontal plane. Four screwjacks at the base of the dolly are used to level the heat pipe. The evaporator leg of the transport heat pipes is vertical (reflux mode).

Above the radiator is a gaseous helium cooled shroud that simulates deep space. The shroud is finned and painted with a high emissivity black paint ($\varepsilon_s \geq 0.95$). The helium shroud is maintained at 20 $\pm$ 5 K during the test. An MLI closeout blanket is located around the periphery between the

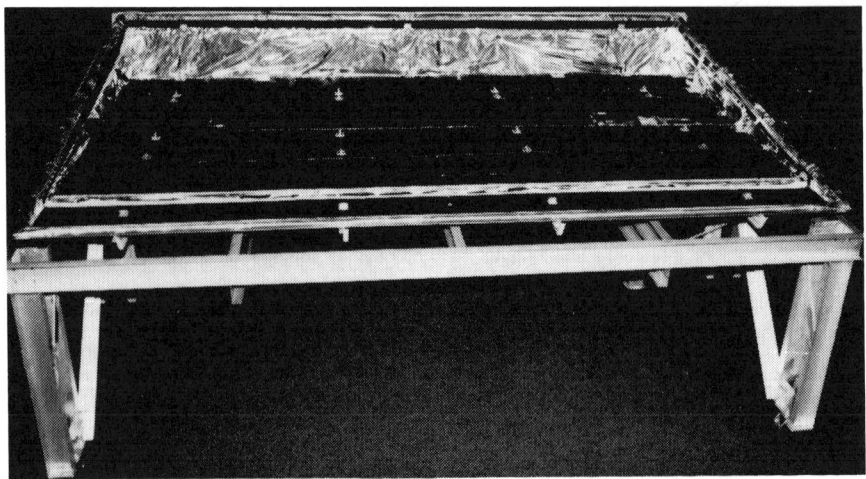

Fig. 11. Final radiator configuration.

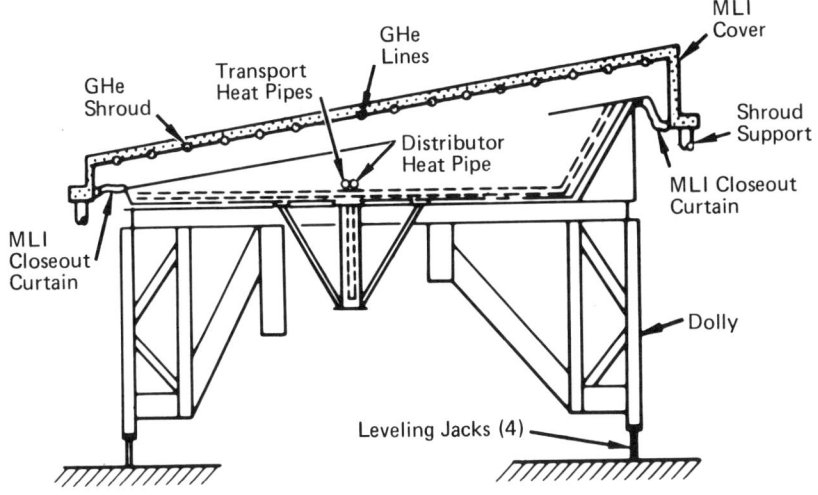

Fig. 12. Thermal vacuum test setup.

CRTU and the helium shroud to prevent any stray thermal radiation from the chamber of structure from impinging on the radiator surfaces.

Instrumentation

Test instrumentation on the CRTU consists of 50 thermocouples and 6 silicon diode temperature sensors. The thermocouples are ANSI Type E (chromel constantan) and are attached to various locations on the first- and second-stage radiators, the heat pipes, the structural support pan, and the sun shield. The thermocouples are connected through the chamber to the AEDC data system. The overall measurement accuracy of the system is ± 25 µV, which corresponds to a maximum error of ± 1 K. The thermocouple wire was calibrated at five temperature points between 77 K and 300 K. The calibration data are used in the computer data system to generate output temperature data.

The silicon diode temperature sensors are used to provide accurate temperature data at critical locations on the CRTU such as the heat pipes. The diodes are calibrated at 95 points in the range of 4 K to 300 K and have an overall end-to-end accuracy of less than 0.1 K.

Table 2. Thermal Vacuum Test Matrix

Test run	1st-stage load, W	2nd-stage load, W	Pan/side shield temp., °K	Sun shield temp., °K	No. of days
1	0	5	290	290	3
2	0	0	290	290	2
3	20	5	290	290	2
4	0	7	290	290	2
5	0	3	290	290	2
6	20	10	290	290	2
7	0	5	290	325	2
8	0	5	290	290	?

Heaters

Heat input from the simulated detector focal plane is supplied by a standard cartridge type heater (two heaters are used for redundancy). Heater power is determined by measuring both the current and voltage drop across the heater at the connector interface to the radiator. Heater power is measured to an accuracy of less than 0.02 W.

Heater power to the first stage is supplied through a row of Kapton film heaters bonded to the bottom of the first-stage radiator just underneath the exposed edge. Similar heaters are used on the structural support pan and on the back of the sun shield to maintain them at the desired temperature during the test.

Test Conditions

Eight test conditions were run. They are summarized in Table 2. Under all test conditions, the vacuum chamber pressure was less than 1×10^{-5} Torr. The structural support pan was maintained at an average temperature of 290 ± 5 K. Each test condition was held for the number of days indicated in Table 2 to ensure near steady state conditions at the end of each test run.

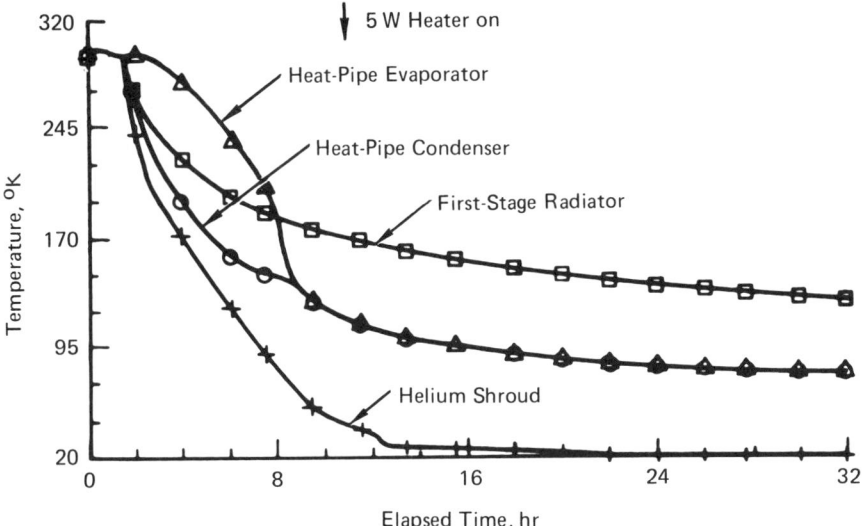

Fig. 13. Temperature histories - Test Run No. 1.

Test Results

Figure 13 shows the cooling down of the helium shroud and radiator temperatures during Test Run 1. Note that the evaporator of the transport heat pipe lags behind the condenser until the condenser reaches approximately 145 K. After that the heat pipe cools isothermally. At 145 K, the oxygen in the heat pipe starts to enter the two-phase regime where the heat pipe becomes operational. The heater power was increased to 5 W approximately 11 h after the beginning of the cooldown. At the end of Test Run 1, the radiator had cooled to 70.80 K and was still cooling at the rate of 0.06 K/h. The evaporator region of the transport heat pipes was at 73.2 K to 74.9 K. Most of the temperature drop was in the evaporator because of the column of excess liquid that occurs with the heat pipe vertical (reflux mode). The first-stage temperature was 106.1 K at the end of the run and was cooling at the rate of 0.3 K/h.

Steady state temperatures were projected from the end-of-run data. The projection uses the relation

$$T_{SS} = T_{ER} + \frac{C \left(\frac{dT}{d\theta}\right)}{\frac{dq_E}{dT}} \qquad (1)$$

Table 3. Thermal Vacuum Test Results

Test run	First-stage temperature		Second-stage temperature	
	End of run	Steady state[a]	End of run	Steady state[a]
1	106.09	105.67	70.80	70.48
2	103.32	103.10	45.24	45.31
3	147.71	147.75	71.78	72.70
4	107.31	106.54	76.53	77.05
5	104.12	103.99	62.84	62.13
6	147.77	147.77	84.51	87.77
7[b]	104.81	104.84	70.52	70.12
8	105.94	104.97	70.97	70.33

[a]Projected using Eqs. (1) and (2).
[b]3250°K sun shield.

where $dT/d\theta$ is the temperature rate of change at the end of the test run.

The term in the denominator is the derivative of the emissive power of the radiator stage with respect to its temperature and is given by

$$dq_E/dT = 4\sigma \ \epsilon_R \ \epsilon_S \ A_R \ T_R^3 \qquad (2)$$

Using Eqs. (1) and (2), the projected steady state temperatures for the first- and second-stage radiators are 105.7 K and 70.5 K, respectively. End-of-run and projected steady state test results for the other test conditions are summarized in Table 3.

Thermal Math Model

A thermal math model (TMM) consisting of 21 nodes was used to simulate CRTU performance and to correlate test results. The thermal analyses program SINDA was used to generate the data.[6] The extremely low conductances between stages as well as the low emissive power of the radiator stages at the operating temperatures permitted use of large

DEVELOPMENT OF A 5 W 70 K PASSIVE RADIATOR

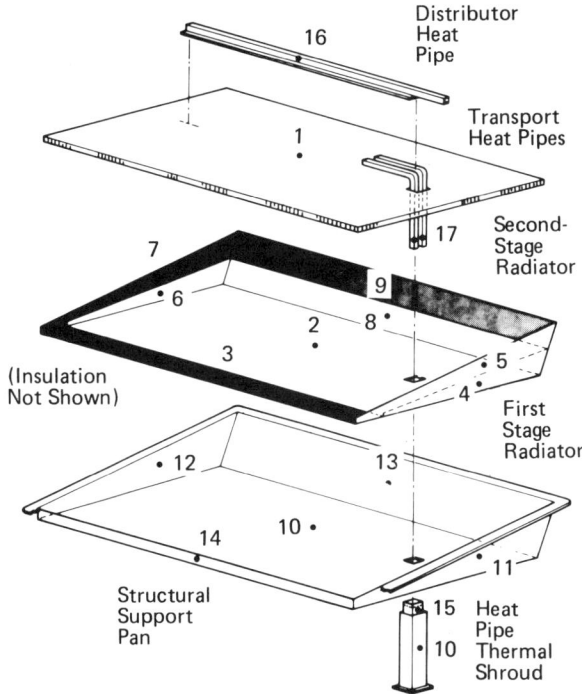

Fig. 14. CRTU thermal math model nodal breakdown.

surface area nodes. The nodal breakdown is shown in Fig. 14. There are 15 diffusion nodes representing the major structural elements of the radiator and three zero capacitance nodes. The zero capacitance nodes simulate the outer layers of the exposed insulation (MLI) surfaces. Three boundary nodes are used to represent the space sink (helium shroud), the spacecraft boundary (structural pan), and the sun shield.

The heat pipe performance is simulated by empirical data determined from testing. The data are input in an array of heat pipe transport capability as a function of evaporator temperature. The outer heat pipe shroud (node 10) is held at the spacecraft boundary temperature. The node for the inner heat pipe shroud (node 15) is thermally shunted to the first stage.

Empirical data for the insulation performance were incorporated into the thermal model. The insulation conductivity was expressed as a first- and fourth-order function of temperature:

$$k_{MLI} = C_1 (T_h + T_c) + C_2 (T_h^4 - T_c^4)/(T_h - T_c) \quad (3)$$

where C_1 and C_2 are constants empirically determined from test data relating to the MLI system type, the number of layers and the layer density.

For the 40 layer insulation blankets, the values of C_1 and C_2 are 2.4×10^{-10} W/K and 4.2×10^{-13} W/K,[4] respectively. For analysis and correlation of test results, two sets of values were used for C_1 and C_2. In the first case, the above values were used; in the second case, the C_1 and C_2 constants were degraded by a factor of 2. Additional conductors were added to account for heat leaks through structural posts, heater and instrumentation wires, and MLI grounding straps.

Correlation of Test Results

Figure 15 shows a correlation between test data and predicted temperatures for the second-stage radiator for the cases involving a heat load to the second stage only with a constant sun shield temperature (Test Runs 1, 2, 4, 5, and 8). The temperatures correspond to the fin root temperature of the radiator skin adjacent to the distributor heat pipe (node 1), and were adjusted to steady state levels by use of Eqs. (1) and (2).

Three sets of predicted temperatures are shown. The original test predictions were based on an insulation degradation factor of 2.0 and a fin efficiency of 0.8. As shown in Fig. 15, the TMM overpredicts the stage temperatures by about 1 K for the 7 W case and 7.5 K for the no load case. A second case was run in which the fin efficiency of the second-stage radiator was computed as a function of the fin root temperature. As shown in Fig. 15, this change somewhat improves the correlation at both ends of the test range. In the third case, the efficiency was allowed to vary with temperature, and no insulation degradation was assumed. In this case, the predicted temperature is 75.3 K compared to a steady state test value of 77 K for the 7 W case, 69.6 K vs. 70.5 K at 5 W, and 46.2 K vs. 45.3 K at zero heat load. This case gives the best correlation at the low end but underpredicts by 1.7 K for the 7 W case.

While none of the cases predict within 1 deg for the entire range of heat load, the agreement is generally very good and appears to be fairly well bracketed by the various TMM cases. Discrepancies are attributed to heat losses through insulation and supports or stray radiation from the pan and first-stage radiator, which cannot accurately be predicted. The form of Eq. (3) or the relative values of C_1 and C_2 over a wide temperature range may also be suspect.

The correlation between test data and predicted temperatures for the remainder of the cases is summarized in Table 4. Predicted temperatures are for a variable fin efficiency and an insulation degradation of 2.0. Again the TMM tends to overpredict the second-stage temperatures.

Conclusions

Results of this development program demonstrate that multiwatt heat loads can be cooled to 70 K or below with passive radiators. The combination of cryogenic heat pipes with multistage radiator technology provides a means of cooling large infrared sensors aboard Earth orbiting satellites. The radiator can be shielded from the sun with a relatively small (9 deg) fixed shield. The heat pipes allow the radiator to be located away from the focal plane for optimum shielding.

Test data for the second stage radiator temperature are within 1-5 K of predictions. With 5 W of input power, the second stage cooled to 70.8 K with a projected steady state temperature of 70.5 K, compared to a design specification of 70 $\pm$ 2 K. The addition of 20 W to the first-stage radiator (Test Run 3) only increased the second-stage temperature by about 2.2 K. This demonstrates that multiple cooling levels, such as for optics, baffles, or electronics, as well as the focal plane, can be provided by use of multistage radiators. The test data also shown that the second-stage temperature is very insensitive to the temperature of the sun shield. Increasing the sun shield from 290 K to 325 K had a negligible effect on either the second-stage or the first-stage temperatures.

Post Thermal Test Activities

After the first-phase thermal test, the CRTU went through an acoustic vibration test that simulated launch conditions in the Shuttle orbiter payload bay. Induced accelerations were lower than predicted and there was no damage to any radiator performance. Static tests on the support structure elements, which were developed under this

Table 4. Test Correlation for Runs 3, 6, and 7

Test run	Stage	Predicted temperatures, K Steady state	Test temperatures, K End of run	Test temperatures, K Steady state[a]
3	Second	75.3	71.8	72.7
	First	163.5	147.7	147.8
6	Second	86.6	84.5	84.8
	First	165.9	147.8	147.8
7	Second	71.9	70.5	70.1
	First	112.8	104.8	104.8

[a]Projected temperatures using Eqs. (1) and (2).

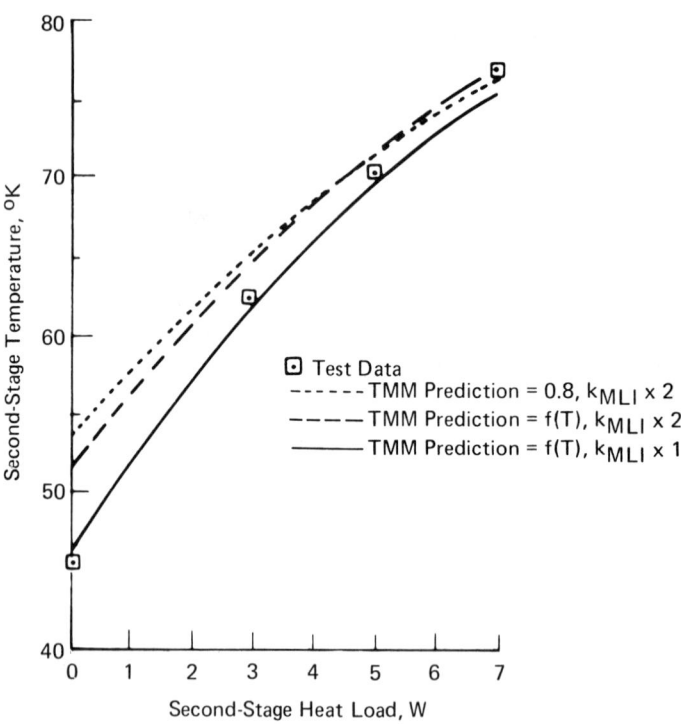

Fig. 15. Thermal vacuum test correlation.

program, showed them to be at least 18% stronger than the design ultimate loads.

A 5.1 m^2 third-stage radiator was added to the CRTU. With the third stage, the radiator system demonstrated a heat rejection capacity of 0.24 W at 41.2 K and reached a minimum (no load) temperature of 36.8 K. Results of this development and test progam are summarized in the final report.[7]

Acknowledgments

The author wishes to acknowledge the contribution of D.E. Wilson, A.M. Lehtinen, G.W. Gurr, and C.D. Rosen of Rockwell International Corporation for their support in the design and fabrication of the CRTU. The contributions of W. Warmbrod, project engineer for the thermal vacuum test at ADEC, and J. Ackerson and W. Haskin of AFWAL, are also gratefully acknowledged.

References

[1] Wilson, D.E. and Wright, J.P, "Development and Testing of Two and Three Stage Heat Pipe Radiators," AIAA Paper No. 79-1060, June 1979. Orlando, Fla.

[2] Haskin, W.L. and Dexter, P.F., "Ranges of Application for Cryogenic Radiators and Refrigerators on Space Satellites," AIAA Paper No. 79-0179, Jan. 1979. New Orleans, La.

[3] Wilson, D.E. and Wright, J.P., "The multistage Heat Pipe Radiator-Air Advancement in Passive Cooling Technology," AIAA Paper No. 77-760, June 1977. Albuquerque, N.M.

[4] Wright, J.P. and Wilson, D.E., "Development of Thermal Control Methods for Specialized Components and Scientific Instruments at Very Low Temperatures," Rockwell International Space Systems Group, SD 76-SA-0230, Nov. 1976.

[5] Brennan, P.J. and Kroliczek, E.J., "Axially Grooved Heat Pipes-1976," AIAA Paper No. 77-747, June 1977. Albuquerque, N.M.

[6] Smith, J.P.,"SINDA Users Manual," TRW Corporation, 14690-H001-R0-00, April 1971.

[7] Lehtinen, A.M.; Wilson, D.E.; and Wright, J.P., "Cryogenic Test Unit, Space Systems Group, Rockwell International, AFWAL-TR-80-3124, Dec. 1980.

TRANSIENT RESPONSE OF THERMAL LOUVERS WITH BIMETALLIC ACTUATORS

Han Hwangbo[*]
MRJ, Inc., Fairfax, Va.
and
W. H. Kelly[+]
Fairchild Space and Electronics Co.,
Germantown, Md.

Abstract

Thermal louvers that use a bimetallic spiral coil as the control actuator exhibit significant time lag and hysteresis in response to temperature changes of the louver base plate. A detailed analysis has been made to show that the hystersis can be attributed to the interaction between the coil and supporting structures. The thermal capacitance between the coil and base plate, and the thermal conductance from the louver blade to the coil are chosen as design parameters. Analytical results indicate that under certain combinations of environment and design conditions, the bimetallic coil delays the response to temperature changes in the base plate which results in a time lag of opening and closing of the louver blades. A thermal vacuum test is performed to determine the time lag of the actuator coil in response to the base plate temperature. Test data are utilized to verify the baseline thermal model which is then parametrically varied to provide guidance for actuator design improvement and understanding louver transient performance.

Introduction

Thermal louvers actuated by bimetallic coils have been used as thermal control devices on many spacecraft (e.g., Pegasus, Nimbus, OAO, ATS-6, and Voyager). Their advantage

---

Presented as Paper 80-1539 at the AIAA 15th Thermophysics Conference, Snowmass, Colo., July 14-16, 1980. Copyright © American Institute of Aeronautics and Astronautics, Inc., 1980. All rights reserved.
*Member of Technical Staff.
+Member of Technical Staff, Comsat Laboratory .

over passive systems lies in providing an adjusting effective emittance which compensates for coatings degradation, internal power fluctuation, seasonal flux changes, and other mission varying properties.

The heat rejection capability of louvers in various orbital environments has been the subject of extensive analytical and experimental studies.[1-7] The louvers' performance is characterized by relating the radiating energy at a given temperature to the blades opening angle. To this end, thermal vacuum testing of louver assemblies are conducted with and without solar simulation to establish the matrix of effective emittance and solar absorptance as functions of blade angle and flux incidence angle. The data from these tests form a basic input to thermal modeling of spacecraft.

Testing is generally performed under steady-state conditions with no special provisions made for the time lag or hysterisis effects of the actuator when responding to transient changes in the external environment or base plate temperature. In actual spacecraft thermal subsystems, the massive components are usually mounted directly on louvered base plates. Hence, the transient louver response times are, in most cases, of secondary importance due to the inherently large thermal time constant of the louver base plate/component assembly. However, if the louver is mounted on the compartment frame some distance away from the equipment mounting surface (as in Bay 2, shown in Fig. 1), or if the component has a large transient power pulse and low thermal mass, a delay in

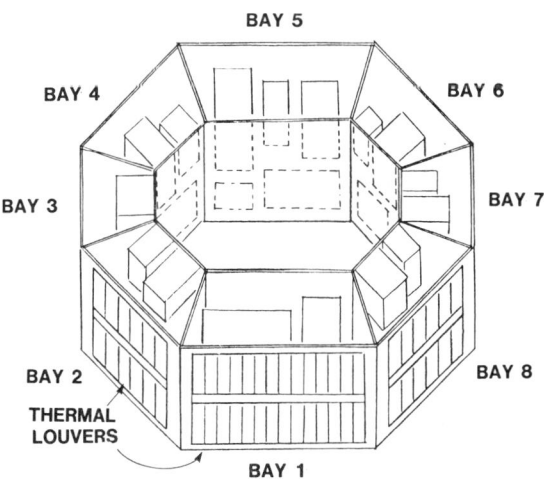

Fig. 1 Spacecraft equipments module with thermal louvers.

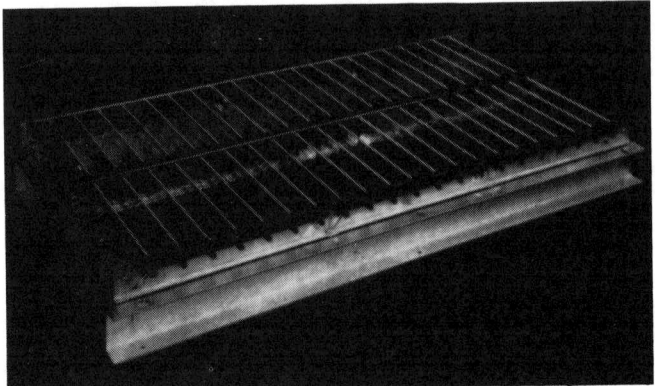

Fig. 2 FSEC/GSFC MMS thermal louver.

the louver actuator response could lead to a temperature increase above qualification limits. The development of lightweight, high-power density spacecraft components and the attractiveness of accessible equipment compartment has emphasized the need for more detailed knowledge of the transient behavior of bimetallic actuated louvers.

This paper describes the development of an analytical thermal model of a louver system which may be used to predict transient louver performance. The results of a thermal vacuum transient response test are utilized to validate the model which is then used to evaluate the effects of the actuator thermal capacitance and the radiative and conductive coupling between actuator and base plate on transient behavior. An example of bimetallic coil selection is given to illustrate how the actuator temperature response may be improved.

Louver Description

The type of louver assembly considered in this paper consists of an aluminum frame, multiple blades, bimetallic actuator coils and an actuator housing cover. Fig. 2 shows this design as is currently being used on the FSEC/GSFC MMS program. Each blade is constructed from two aluminum sheets with a rectangular shaft bonded at each end. The shafts from an opposing pair of blades are inserted into a spool which is attached to the inner spiral end of the bimetallic actuator coil. The outer end of the coil is fixed to the adjustment screw which is used to set the blade calibration positions. Fig. 3 depicts a cross-sectional view of the actuator housing and bimetallic coil. The bimetallic is wound around the spool so that the strip of material with the larger coefficient of expansion forms the outer surface of the coil. This construction causes the coil to contract as its temperature increases and apply a torque to the louver blade which rotates toward the open position. The aluminum actuator housing is conductively coupled to the base plate and provides a heat sink to the actuator. The actuator is enclosed by a housing which is covered by a multiple layered insulation to protect the bimetallic from the external environment.

Bimetallic Actuator Coil Analyses

The angular deflection D_A of the bimetallic coil, shown in Fig. 4, corresponding to a temperature change dT is given by

$$dD_A = K_D(Lm/t)dT \tag{1}$$

where K_D=coil deflection constant; L=active length of element, t=thickness of element; m=specific deflection; and dT=temperature change of element.

The specific deflection is defined as the ratio of the actual deflection, corresponding to a particular temperature change, to the free deflection for the same temperature change. The specific deflection varies from 0, for the case of complete restraint, to 1 for free deflection. The angular deflection of the coil due to a force P acting at the free end of the coil can be related to a mechanical spring rate by the

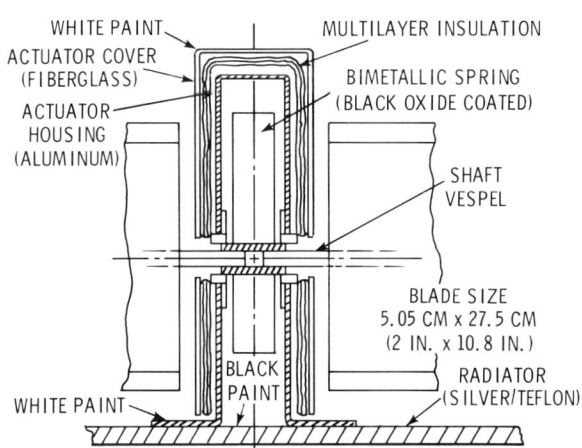

Fig. 3 Thermal louver actuator.

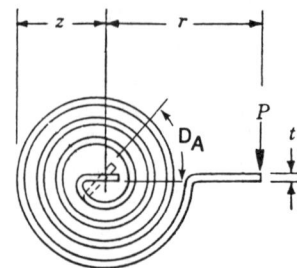

Fig. 4 Bimetallic coil geometry.

formula
$$D_A = rLP/K_p bt^3 \qquad (2)$$

where K_p=strip torque constant; b=width of element; and r= radius to point of load application.

The thermal torque generated by the coil due to temperature change can be translated into an effective force dP by combining Eq. (1) and Eq. (2):

$$dP = K_D K_p (bt^2/r)(1-m)dT \qquad (3)$$

The volume of the bimetallic element can be expressed by combining Eq. (1) and Eq. (3) to give

$$V = bLt = c/m(1-m) \qquad (4)$$

where

$$C \equiv (r/K_D^2 K_p)(dD_A/dT)(dP/dT) \qquad (5)$$

C is nearly a constant for the range of temperature changes and thermal torques considered. Differentiating Eq. (4) with respect to m and setting dV/dm equal to zero, one can show that the specific deflection m=0.5 gives the minimum bimetal volume.[7] Previous design experience indicated, however, that a specific deflection between 0.5 and 0.75 is desirable to insure sufficient torque to overcome bearing friction. The weight of the bimetallic is shown in Fig. 5 as a function of louver open to close temperature range. As the bimetallic actuation range decreases, the weight increases significantly which influences both overall louver weight and the response time of the louver/base plate assembly.

The energy balance equation for a coil in the louver assembly may be written as

$$\rho C_p V(dT_c/d\theta) = FA\sigma(T_c^4 - T_h^4) + K(T_c - T_{bl}) \qquad (6)$$

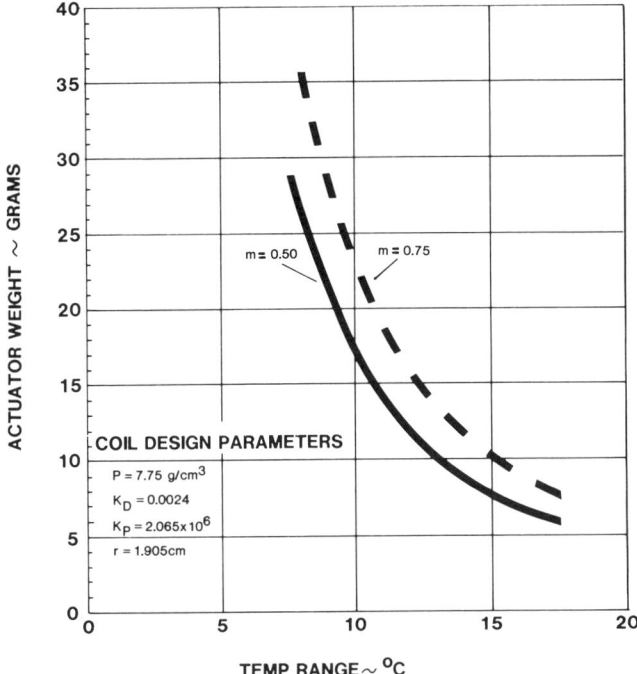

Fig. 5 Coil weight vs temperature actuation range.

where ρ=density of element; C_p=specific heat of element; F= script-F of element to heat sink; A=surface area of element; T_h=temperature of heat sink; T_c=temperature of coil; T_{bl}= louver blade temperature; K=conductance between element and louver shaft; θ=time; and σ=Stefan-Boltzmann constant. It is seen from Eq. (6) that the parameters which influence coil temperature history are the thermal mass of the actuator, $\rho V C_p$, the script-F from actuator to heat sink, and the conductance K between the actuator and the louver blades. Substituting Eq. (4) for the minimum bimetal element volume into Eq. (6), the time rate change of angular deflection is found from the relation

$$(4\rho C_p/K^2_D K_p)(dP/dT_c)(dD_A/d\theta) = \qquad (7)$$
$$F A \sigma (T_c^4 - T_h^4) + K(T_c - T_{bl})$$

Louver Model

Fig. 6 shows a 24 node thermal math model for a section of the louver assembly. The model was developed to evaluate the effect of varying the actuator coil design parameters. The nodal lumped mass representation for the louver is described in Table 1.

The effective emittance values as function of base plate temperature are available from steady-state test data and are used in the radiative couplings from louver base plate nodes to space heat sink. Transient runs were made using SINDA computer program in order to assess the correlation between the 24 node model and the louver transient test data.

Louver Test

A thermal louver measuring 46cm x 58cm was mounted to an aluminum honeycomb panel (0.06cm facesheets) with a hemispherical emittance of 0.8. Kapton film heaters were bonded to the rear surface of the base plate for temperature control. A multi layer insulation blanket was installed on the backside of the panel and around the sides of the louver assembly to reduce radiation losses from all surfaces except the louver radiating area. The emittance of the multilayer insulation blanket was determined from a previous test to be less than 0.015.

The louver effective emittance is defined from the following energy balance on the louver unit in the vacuum chamber:

$$Q_H = \sigma F_e A_b (T_{bp}^4 - T_{ch}^4) + Q_L \qquad (8)$$

Table 1 Nodal description of louver model

Node	Description
1	Bimetallic coil
2	Coil heat sink or base plate
3 to 7	Actuator housing
8 to 10	Outer cover of the housing
11	Actuator spool
12	Louver base plate
13 to 15	Louver blades
16	Louver side frame
17	Louver base plate
18 to 20	Louver blades
21	Louver side frame
22 to 24	Space heat sink

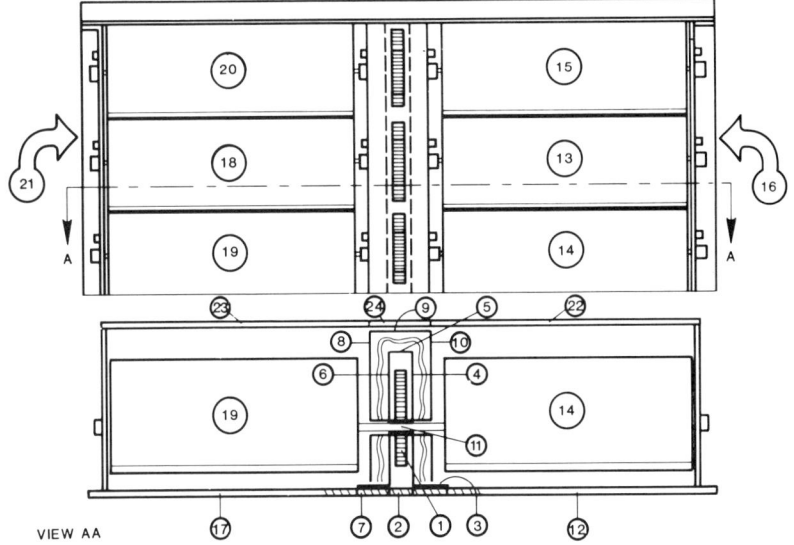

Fig. 6 Louvers thermal model.

where Q_H=heater power; T_{ch}=chamber wall temperature; T_{bp}= average base plate temperature; Q_L=system heat leak due to conduction and insulation loss; F_e=effective emittance; and A_b=base plate area. A series of thermal vacuum test points were obtained to establish the effective emittance of the louver as a function of blade angle and base plate temperature. The test results, using the definition in Eq. (8), are shown in Fig. 7.

The louvers were calibrated to close fully at 13°C and at 30°C in the steady-state environment chamber.

A transient test was conducted to determine the time lag of louver blades opening and closing when the heat input to the honeycomb panel is suddenly increased. The base plate was initially maintained at 13°C by a steady-state power dissipation of 11.7W. A step increase of power to 75W was applied for 15 min followed by a reduction back to 11.7W. The louver blades experienced a rotation from the fully closed to fully open position and back again.

Comparison of Test with Predictions

The analytical louver model with measured properties (bi-metallic mass, base plate emittance, etc.) was used to predict

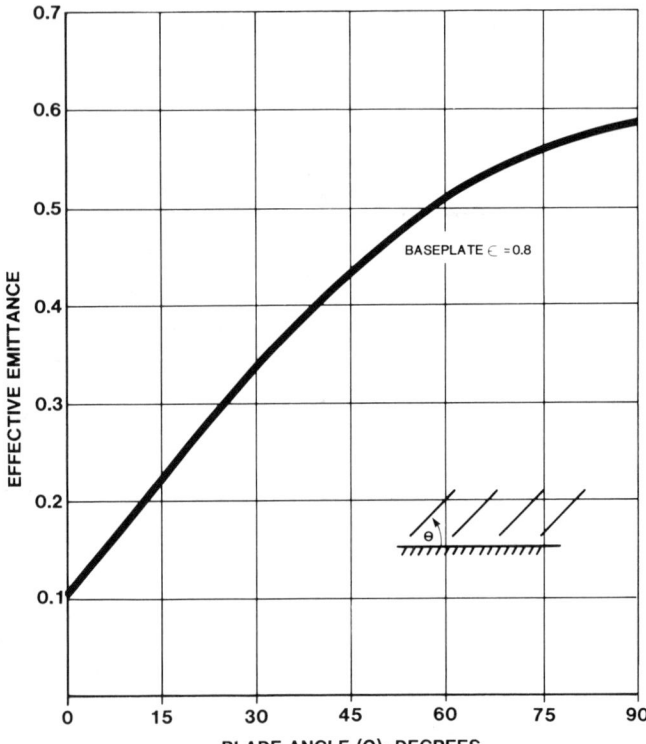

Fig. 7 Louvers effective emittance

the transient performance during test by simulating a stepped rise and reduction in the power input to the base plate nodes. The predicted base plate temperatures are compared with test data in Fig. 8. The temperature profile of the base plate indicates that the louver blades open fully at 6 min after power increase and are stabilized in the fully open position after 8 min. The "overshoot" in base plate temperature is caused by the time lag of the bimetallic coil in responding to the increase in heater power. Within 16 min after reduction in heater power, the louver is once again stabilized in the fully closed position.

Deviations from the baseline values of the louver thermal conductances were incorporated in the model until good correlation between analysis and test data was obtained. (The uncertainty in louver model conductances lies in the contact resistance between the blade shaft/spool and spool/bimetallic coil and in the radiation interchange between the bimetallic coil and the actuator housing.) The predicted base plate

TRANSIENT RESPONSE OF THERMAL LOUVERS

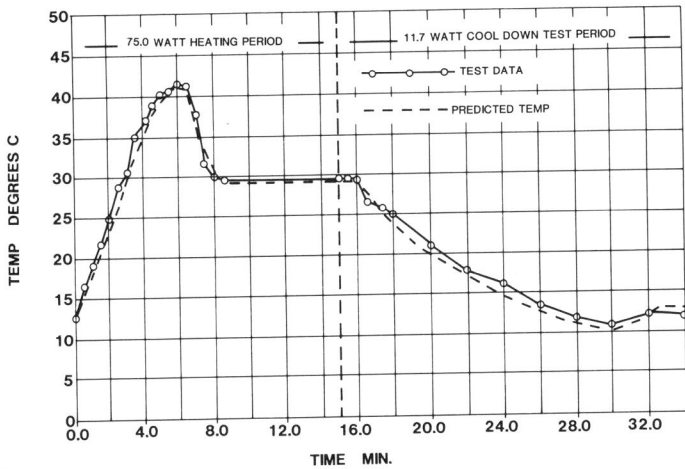

Fig. 8 Louvers base plate temperature response.

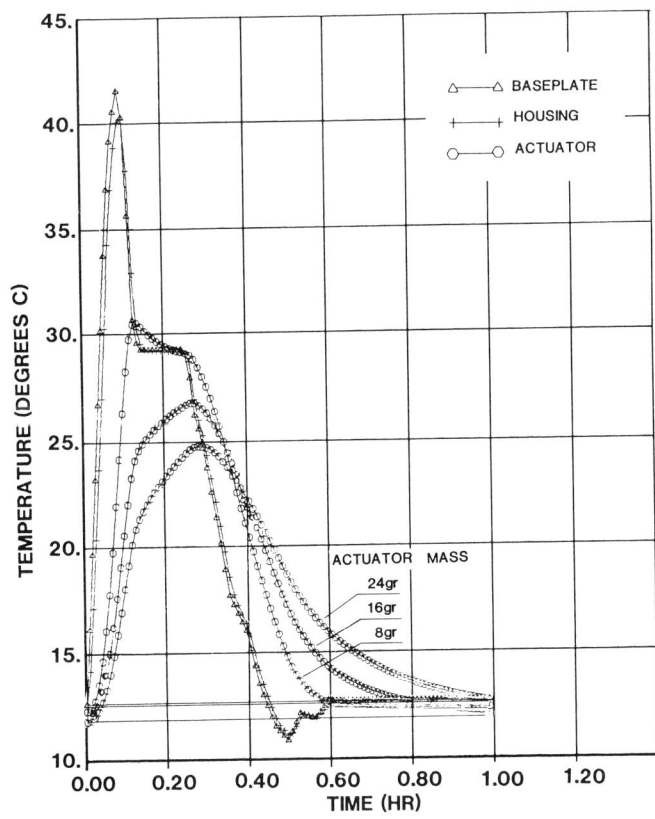

Fig. 9 Actuator mass effect on transient response.

temperatures from the corrected thermal math model, as shown in Fig. 8, show good agreement over the entire range of louver blade positions.

Results of Analysis

After validation by the results of the transient thermal vacuum test, the math model was used to study the effects of the actuator thermal mass, coil/housing script-F, and coil/louver shaft conductance on the response. The base plate emittance and the open/close set points were fixed at the values determined in previous steady-state tests.

Fig. 9 shows the predicted response of various size actuators to louver base plate and actuator housing temperature (the 8g bimetallic coil weight is the test louver actual

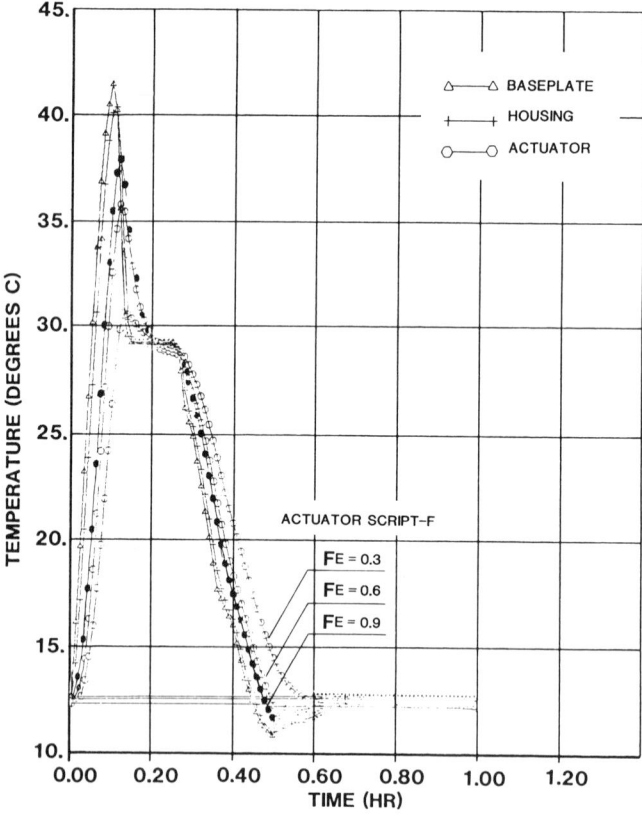

Fig. 10 Actuator radiation coupling effect on transient response.

weight). As expected, the response of the heavier bimetallic coil to base plate temperature is slower than that for the lighter coil. In a similar manner, Fig. 10 presents the actuator response as a function of the radiative coupling between the bimetallic coil and the actuator housing base plate. The louver actuator response is enhanced by providing a large radiative coupling. It should be pointed out that the actual louver response is a result of the combined effects of radiative coupling and bimetallic mass. As the bimetallic mass is increased to obtain a narrower temperature range for blade rotation from open to close, the surface area of the coil also increases, thus leading to a sufficient increase in the total radiative coupling to maintain a constant time response over a reasonable range of bimetallic sizes.

The effect of the value of conductance between the louver blade shaft and the actuator coil is shown in Fig. 11. It is

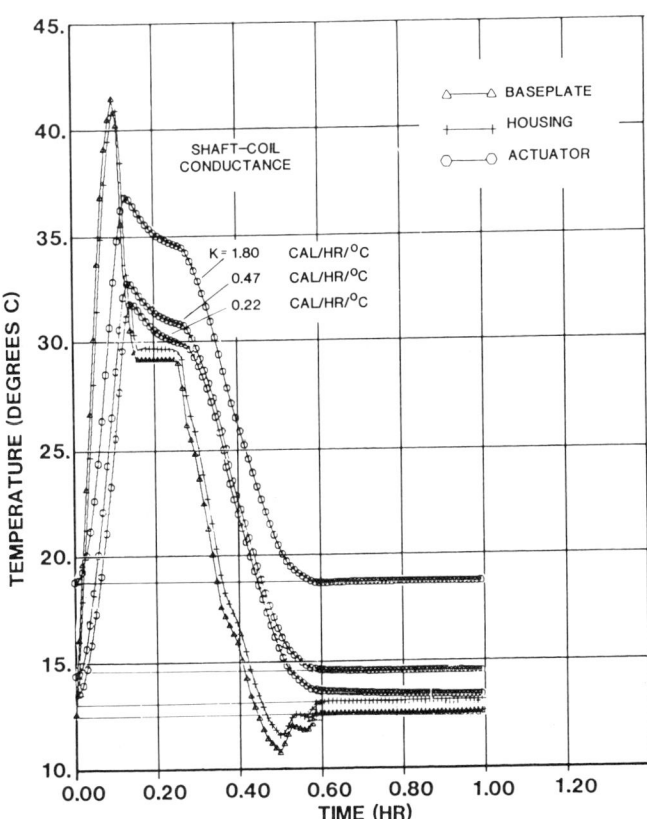

Fig. 11 Blade conductance effect on transient response.

desirable to minimize the conductance as much as possible to prevent the external environment from influencing the actuation response. The curves presented in Fig. 11 assume a blade temperature of 177°C. It should be noted that in addition to producing a time response lag in louver actuation, a large thermal conductance between the louver blade and actuator also causes a temperature "lag" in steady-state louver actuation tracking.

Discussion

The transient response of louver blades can be calculated as a function of the actuator temperature history by utilizing data similar to those shown in Fig. 9-11 together with the angular deflection equations. In determining the size of the bimetallic, provisions must be made to insure overcoming the frictional forces in the louver blade bearings in addition to providing the proper angular span for a specified temperature. Otherwise, the actuator would not track the base plate temperature and a lag in the angular blade position would be experienced over the entire range of operation.

As an illustration of the interdependency of the actuation range and bimetallic coil properties, the blade rotation and developed torque from a bimetallic coil 0.55cm x 0.025cm x 75cm are calculated as a function of specific deflection over a temperature range of 17°C (set point temperature of closing at 13°C and fully open at 30°C). The physical properties of the bimetal coil are given as follows:

Density of bimetallic	$7.75 g/cm^3$
Specific heat	$0.12 cal/g-K$
Emissivity	0.9
Radius of load application	1.9 cm
Coil deflection constant, K_D	0.0024
Coil torque constant, K_p	2.065×10^6

The blade rotation angles and torques are calculated from Eq. (1) and (3) shown in Fig. 12 as a function of the actuator temperature. The selection of the coil specific deflection, M=0.75, produces a torque of 7.07g-cm while operating over the temperature range of 13°C to 30°C. As the coil temperature exceeds 30°C, a significant increase of torque develops in the coil. A maximum torque of 28.3g-cm will develop in the coil

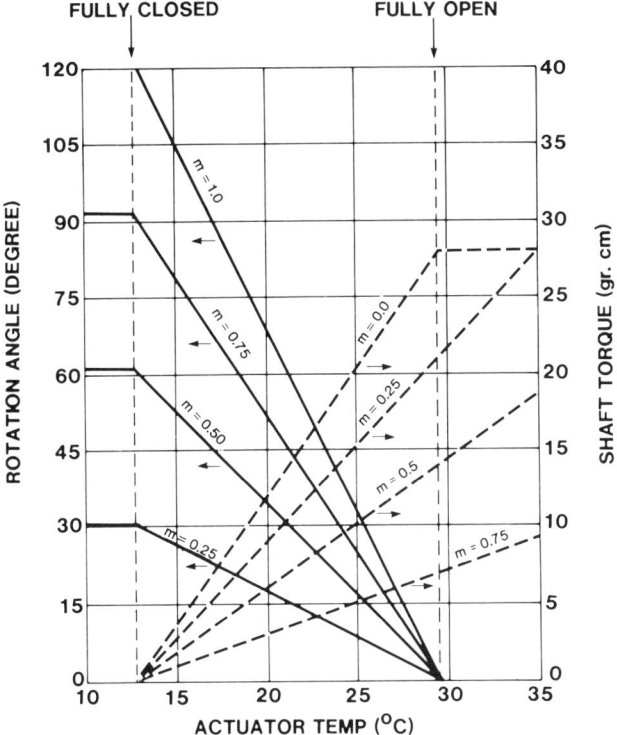

Fig. 12 Shaft rotation and torque vs actuator temperature.

when both ends of the coil are restrained or when the louver blades are fully open.

Conclusion

The transient performances of a louver assembly is analyzed by varying the design parameters for the bimetallic actuator. The analysis shows that the response of the actuator to its heat sink (base plate) temperature can be improved through proper selection of bimetallic size and geometry and by increasing the internal radiative coupling to the heat sink while decoupling the actuator from the louver blades. Exact information on the time response of the actuator is desirable, particularly when narrow temperature limits are required for a small self-supporting package in space environment. A transient response test to determine the louver actuator lag is recommended for inclusion in the basic louver characterization tests normally performed in a space qualification program.

References

[1] Russell, L. and Linton, R., "Experimental Studies of the Pegasus Thermal Control Louver System," NASA FSEC, AIAA Paper No. 67-308.

[2] Parmer, J.F. and Buskirk, D.L., "Thermal Control Characteristics of Interior Louver Panels," ASME Paper No. 67-HT-64.

[3] Michalek, T.J., Stipandic, E.A., and Coyle, M.J., "Analytical and Experimental Studies of an All Specular Thermal Control Louver System in a Solar Vacuum Environment," NASA GSFC, AIAA Paper No. 72-268.

[4] H. Hwangbo, J.H. Hunter, and W.H. Kelly, "Analytical Modeling of Spacecraft with Heat Pipes and Louvers," <u>Thermophysics and Spacecraft Thermal Control, Progress in Astronautics and Aeronautics</u>, Vol. 35, edited by R. G. Hering, AIAA, New York, 1974.

[5] Kelly, W.H. and Hewitt, Dennis R, "Thermal Design of a Communications and Data Handling Module," Ninth Intersociety Conference on Environmental Systems, July 1979, San Francisco.

[6] Furukawa, Masao, "Analytical Studies on Design Optimization of Movable Louvers for Space Use," <u>Journal of Spacecraft</u>, Vol. 16, No. 6, 1979.

[7] "Bimetal Engineering," Chase Thermostatic Bimetals.

α_s/ε_H MEASUREMENTS OF THERMAL CONTROL COATINGS ON THE P78-2 (SCATHA) SPACECRAFT

D. F. Hall[*] and A. A. Fote[†]

The Aerospace Corporation, El Segundo, Calif.

Abstract

The ML12 experiment on board the P78-2 spacecraft includes 16 calorimetrically mounted thermal control coating (TCC) samples. The solar absorptances, α_s, of these samples are deduced from on orbit measurements of temperatures and prelaunch measurements of thermal emittances, ε_H, and residual heat leaks. P78-2 was inserted into a 27,600 × 43,300 km orbit on 2 February 1979 after 3 days in a transfer orbit. In the subsequent year, the α_s values for black paints, polished metals, and a fused quartz mirror [optical solar reflector, (OSR)] were approximately constant, but the α_s values of the In_2O_3-coated OSR, silvered FEP, aluminized Kapton, quartz fabric, and conductive paint samples all increased significantly. The flight data are presented along with a discussion of measurement errors.

Introduction

This paper presents the analysis of data acquired over the first year on orbit of a thermal control coating (TCC) experiment designated ML12. The ML12 experiment[1] is on board the United States Air Force (USAF) Space Test Program P78-2 [spacecraft charging at high altitudes (SCATHA)] spacecraft.[2] The P78-2 was launched on 30 January 1979 into a 176 × 43,278 km transfer orbit. On 2 February, it was injected into a

---

Presented as Paper No. 80-1530 at the AAIA 15 Thermophysical Conference, Snowmass, Colo., July 14-16. Copyright © American Institute of Aeronautics and Astronautics, Inc., 1980.

*Research Scientist, Chemistry and Physics Laboratory.
†Member of the Technical Staff, Chemistry and Physics Laboratory.

27,600 × 43,300 km, 7.9 deg inclination final orbit. Thus, it passes through geosynchronous altitude (35,786 km) twice per day and is always within 23% of that altitude. The vehicle (Fig. 1) is a spin-stabilized right cylinder approximately 1.75 m in both length and diameter.

The ML12 apparatus consists in part of two trays of calorimetrically mounted thermal control coating samples that are exposed to the space environment. In this part of the experiment, changes in the optical properties of these 16 samples due to the effects of the space environment and of contamination are observed. The flight trays, one of which is shown in Fig. 2, were installed on P78-2 in mid-November, 1978 at the Eastern Test Range and, therefore, were not exposed to possible contamination during the thermal vacuum testing of the vehicle. (With the exception of the Grafoil, paint, aluminum, and tape mounted fabric samples, the flight samples were not exposed to the thermal vacuum testing of the trays either. Furthermore, before delivery all smooth samples had been cleaned with a special 10% ethanol/90% trichloroethane

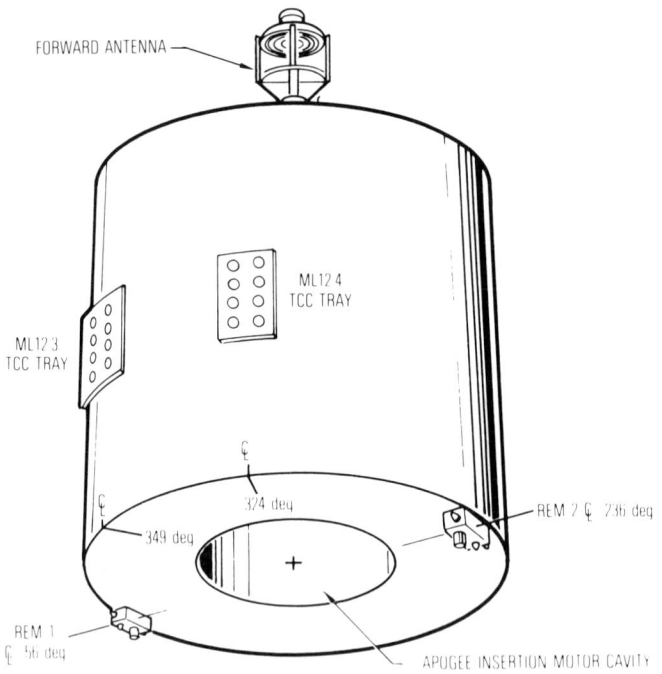

Fig. 1 P78-2 Spacecraft (booms not shown) with locations of TCC trays, hydrazine thrusters, and apogee insertion motor cavity.

THERMAL CONTROL COATINGS ON P78-2 SPACECRAFT

Fig. 2   Flight TCC tray.

solution on cleaning cloths purchased from Applied Research Labs.) The dust covers, shown removed in Fig. 2, remained on the trays almost continuously until a few hours before launch. They were removed through a special door in the space vehicle fairing.

The P78-2 program cleanliness requirements are summarized in Table 1. They are fairly typical of recent Air Force space vehicle programs, although past programs have had both more and less severe formal requirements. However, it is likely that P78-2 was cleaner than most spacecraft because the need for vehicle cleanliness was stressed from the inception of the program through launch.

Thruster operations and appendage deployments represent transient sources of contamination that potentially could have reached and been adsorbed on TCC samples. As shown in Fig. 1, there are two clusters of four hydrazine thrusters, known as rocket engine modules (REMs), located at the lower periphery of the vehicle. The tangential thrusters were used to change the angular velocity of the vehicle, whereas the thrusters parallel with the axis of rotation are used in a pulsed mode to adjust the angle that the vehicle axis forms with the sun line. These thrusters are discussed elsewhere.[11]

Also shown in Fig. 1 is the cavity that contained the solid propellant apogee insertion motor (AIM). The AIM injected the vehicle into final orbit and was ejected from the vehicle 1-2/3 days later (on Julian day 35 of 1979). No observable contamination was traced to the firing of the AIM.[12]

Table 1. Summary of P78-2 cleanliness requirements

Restricted Item	Restriction
Materials of Construction	≤ 0.1% volatile condensable materials (VCM); ≤ 1% total weight loss;[3,4] space qualified.
Surface Contamination	Components: "Free of all visible contamination."[5] Vehicle witness plate before installation of launch vehicle fairing: Level 500A.[6]
Environment	Assembly and test areas: Class 100,000.[4,7,8]
	Eastern Test Range Spin Test Facility and gantry[9] areas: HEPA filtration of inlet air to remove 99.9% of particles over 0.3 μm.
	Launch vehicle fairing.[10]

Five folded booms carrying sensors of other experiments were deployed during the early weeks on orbit. Also deployed were some instrument covers and two 50 m unfurlable tubular antennas. No correlation has been noted between these transient events and α_s values.

Instrument Design

The instrument consists of three interconnected units designated ML12-3, ML12-4, and ML12-5. These units are a modification and augmentation of hardware originally designed and fabricated at TRW Systems.[13] The ML12-3 and ML12-4 are the sample trays that carry eight 1.25 in. diam samples each. The ML12-5 carries the electronic circuitry required to monitor the TCC coupons. The location of the trays on the P78-2 vehicle is shown in Fig. 1. The spin axis of P78-2 is held approximately normal to the sun line.

On the trays, each test sample is mounted on an Al disk (Fig. 3) by means of diluted Eccobond EC57C conducting epoxy (except for the fabric/tape sample, which did not require additional adhesive). Thermal isolation is accomplished by having the disk supported by three 0.57 in. long, 14 mil o.d.

stainless steel tubes, which are thermally insulated from the
base of a cup by fiberglass sleeving. The length and diameter
of the instrumentation leads to the disk were also chosen to
minimize conduction. The volume enclosed by the disk and cup
walls is filled with a combination of multilayer and open cell
polyurethane insulation. The underside of each Al disk carries
two heaters in series and three thermistors. The heaters,
which are actually strain gauges, were chosen because of their
low temperature coefficient of resistance. The heaters are
included for preflight calibration and for thermal desorption
cleaning of six of the samples on orbit. (Hardware constraints
precluded making all samples heatable on orbit.) The three
thermistors span low, medium, and high temperature ranges. On
one-half of the samples, one of the three thermistors is monitored on the basis of the expected temperature range. On the
other samples, two of the thermistors are monitored. In addition, there are two thermistors located on each of the sample
trays to measure the temperatures of the supporting cups.

The electronics package, ML12-5, contains ohmmeter circuits for measuring the 14 thermistors and 2 calibration
resistors associated with each tray. Two stepping circuits
simultaneously scan these signal sources on the two trays,
dwelling for 1 s at each source. The ohmmeter circuits
convert the resistances into analog 0.0-5.1 V signals, which
are digitized by an 8 bit digital-to-analog converter in the
spacecraft telemetry system prior to recording by one of the
vehicle tape recorders. Finally, ML12-5 generates, on command, a programmed heating sequence for six of the samples.
This sequence consists of four heating-cooling cycles, each of

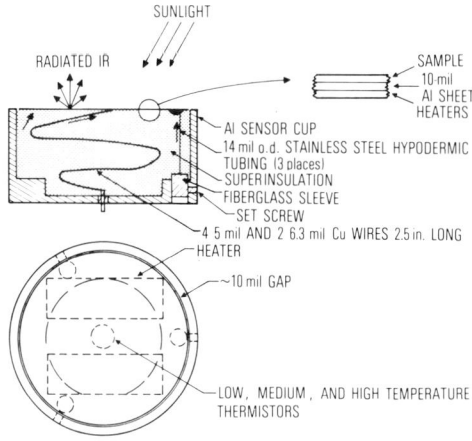

Fig. 3   Arrangement for mounting TCC sample disks in cup.

which achieves a higher equilibrium temperature than the preceding cycle. The cooling cycle permits the determination of any change in α_s, as a result of desorption of contaminants, during the heating. This capability has not yet been exercised on orbit.

Sample Selection

The sixteen samples chosen for testing are given in Table 2. These samples were chosen for a number of reasons:

1) To provide a wide range of initial α_s and ε_H values in order to facilitate contamination detection in case a given contaminant has the same α_s/ε_H as some of the substrates.

2) To include materials that are expected to be relatively space stable in the absence of contamination, such as the polished Al samples, the Au sample, and the OSR sample.

3) To provide for some redundancy to permit checks for consistency. Thus, two black and two polished Al samples, one on each tray, were included.

4) To permit the comparison of similar materials. Thus, both the OSR and the OSR coated with a conductive In_2O_3 layer were included. Similarly, an In_2O_3-coated and an uncoated aluminized Kapton were chosen. Two thicknesses of silvered Teflon, 2 mil and 5 mil, were used. Finally, two samples of Astroquartz fabric, one with an FEP/Al backing and one without such a backing, are being tested.

5) To test a new material. To our knowledge, the In_2O_3/Kapton/Al sample has not been flown before.

6) To include samples identical to those studied in the satellite surface potential monitor package on board P78-2, which is designed to measure the amount of charging of thermal control coating samples.[2] These samples are the aluminized Kapton, the OSRs, the Au sample, the Au-coated Kapton, an Astroquartz fabric, and the 5 mil silvered Teflon.

Data Reduction

The temperature of a sample is determined with the use of Eqs. (1) and (2).

$$C_v(T)\,(dT/dt) = F\left(T_c^4 - T^4\right) + C\,(T_c - T) + \varepsilon_H A\sigma\left(T_o^4 - T^4\right) + \alpha_s S\phi(t) + P \quad (1)$$

Table 2. Flight samples

ML12-3 Tray	ML12-4 Tray
Grafoil (Union Carbide, GTA Grade)	10 nm Au on 5 mil Kapton/Al (Sheldahl Lot 105788, Coated by SRI)
Black Paint (3M401C10)	Astroquartz Fabric (J. P. Stevens 581)/FEP/Al
OSR (OCLI SI-100 Mirror)	In_2O_3/OSR (OCLI SI-100 Mirror)
Vacuum Deposited Au (optically opaque) on Al	22 nm In_2O_3 on 2 mil Kapton/Al (prepared by General Electric)
Diamond Polished 2024 T3 Al	Diamond Polished 2024 T3 Al
FEP (2 mil)/Ag (Sheldahl)	Black Paint (3M401C10)
FEP (5 mil)/Ag (Sheldahl)	Astroquartz Fabric (J. P. Stevens 581)/Tape (Sheldahl 405900)
Yellow Paint (NASA-Goddard No. NS43G)	Kapton (5 mil)/Al (Sheldahl)

$$\phi(t) = AE \sin \omega t \text{ for } 0 \leq \omega t \leq \pi \quad (2)$$
$$\phi(t) = 0 \text{ for } \pi \leq \omega t \leq 2\pi$$

where T = sample temperature; T_c = tray temperature; T_o = temperature of surroundings; F,C = thermal coupling constants between the sample and tray; $\varepsilon_H A\sigma$ = hemispherical emittance times sample area times Stefan-Boltzmann constant; $C_v(T)$ = heat capacity of samples; $\phi(t)$ = radiant power falling on sample; ω = angular velocity of spacecraft; E = solar irradiance; α_s = solar absorptance; S = factor between 0 and 1 to account for shading resulting from spacecraft booms; P = power applied by means of sample heater; and t = time.

Albedo and earth emission are expected to be negligible at near synchronous orbit and were not included in Eq. (1). It is necessary, however, to use the appropriate solar irradiance value E for the day of the data evaluated since this value varies approximately 7% between perihelion and aphelion.

Under normal space conditions, $P = T_o = 0$. The solution of this equation is then given by

$$T = T_{ind} + \Delta T \qquad (3)$$

where T_{ind} is the solution of the time-independent equation

$$F(T_c^4 - T_{ind}^4) + C(T_c - T_{ind}) - \varepsilon_H A\sigma(T_o^4 - T_{ind}^4) + \alpha_s SAE/\pi = 0 \qquad (4)$$

and ΔT is time dependent and given by

$$\Delta T = -\gamma + \left[\gamma\beta/(1 - e^{-\beta})\right]e^{-(\beta\theta/\pi)}$$
$$+ \gamma\beta\left[(\beta/\pi)\sin\theta - \cos\theta\right] \quad 0 \leq \theta \leq \pi$$

$$\Delta T = -\gamma + \left[\gamma\beta/(1 - e^{-\beta})\right]e^{-(\beta\theta/\pi)}e^\beta, \quad \pi \leq \theta \leq 2\pi \qquad (5)$$

where

$$\theta = \omega t$$
$$\beta = \pi\left[C + 4T_{ind}^3(F + \varepsilon_H \sigma A)^{-1}\right](C_v\omega)^{-1}$$
$$\gamma = SEA\,\alpha_s/\beta C_v \omega$$

For the data shown under "Experimental Results," α_s was calculated from T_c and T by the following procedure: 1) Take $\Delta T = 0$ and $T_{ind} = T$ (i.e., the measured instantaneous value). 2) Calculate α_s from the time independent equation [Eq. (4)]. 3) Calculate ΔT with above value for α_s. 4) Calculate $T_{ind} = T - \Delta T$ with above value for ΔT. 5) Return to step 2. ε_H is explicitly assumed constant in these calculations.

Preflight Calibrations and Calculations

The use of Eqs. (4) and (5) to calculate α_s requires a calibration of each of the sample and tray thermistors; a determination of the thermal coupling constants, C, F, and $\varepsilon_H A$; a measurement of the heat capacity C_v of each sample; and a calculation of the shading factors S as a function of spacecraft orientation with respect to the sun.

The 24 sample and 4 tray thermistors were calibrated against precision Cu constantan thermocouples (two used per tray) by placing the trays in a commercial temperature-regulated oven/refrigerator and by recording resistances and temperatures over the range of 203-388 K. The data were fitted to the equation

$$T = \sum_{i=0}^{4} C_i (\ln R)^i \qquad (6)$$

by a least squares routine. The standard errors of the fits varied between 0.05 and 0.27 K; all but three of the 28 values were less than 0.13 K. In comparison, the resolution of the spacecraft telemetry system yields an uncertainty of ±0.5 K.

The determinations of C, F, $\varepsilon_H A$, and C_v were performed in an ion pumped vacuum chamber containing a liquid nitrogen cooled shroud surrounding the trays. The shroud was painted black to ensure proper radiative coupling between it and the samples. The trays were held in a temperature-controlled aluminum holder and kept near room temperature to duplicate the conditions expected on orbit. Under these conditions, $\phi(t) = 0$, and $T_o = 77$ K [Eq. (1)].

The thermal coupling constants, C, F, and $\varepsilon_H A$, were determined by applying a measured amount of current to the sample heaters and by measuring the equilibrium temperatures T and T_c. Five sets of data were generated at five different power levels that were used to obtain C, F, and εA from Eq. (1). A least squares technique was used, with P and T as the variables and T_c as a known constant. The values of ε_H, as determined by this procedure, are given in Table 3.

The error in the value of α_s measured on orbit and due to the uncertainties in this calibration, $\Delta\alpha_{s,c}$, can be estimated as follows: Normally under space conditions, P – 0, and $\phi(t) \neq 0$ in Eq. (1), whereas, during calibration, $P \neq 0$ and $\phi(t) = 0$. Thus, P plays the same role during calibration that $\alpha_s \phi(t)$ plays on orbit. In space, $\overline{\phi(t)} = 0.342$ W. Therefore,

$$\Delta\alpha_{s,c}/\alpha_s \simeq (\overline{\Delta P^2})^{1/2}/0.342\, \alpha_s \qquad (7)$$

and

$$\Delta\alpha_{s,c} \simeq (\overline{\Delta P^2})^{1/2}/0.342 \qquad (8)$$

where $\overline{\Delta P^2}$ is the average value of the square of the uncertainty ΔP in the least squares fit. The values of $\Delta\alpha_{s,c}$ calcu-

Table 3. Preflight determination of conductive heat leaks, C; radiative heat leaks, F; normal emittance, ε'_N; hemispherical emittance, ε_H; solar absorptance, α'_s; and uncertainties in the orbital determination of α_s due to calibration uncertainties, $\Delta\alpha_{s,C}$; and telemetry quantization, $\Delta\alpha_{s,Q}$. Also shown are initial orbital values for solar absorptance, $\alpha_{s,o}$, for comparison with α'_s.

	C, W/K	F, W/K^4	ε'_N	ε_H	α'_s	$\alpha_{s,o}$	$\Delta\alpha_{s,Q}$	$\Delta\alpha_{s,C}$
					ML12-3			
Grafoil	6.58E-4	9.05E-12	0.35	0.34	0.66	0.655	0.013	0.001
Black Paint	9.81E-4	4.72E-12	0.90	0.87	0.97	0.905	0.009	0.002
OSR[a]	8.73E-4	6.01E-12	0.80	0.81	0.08	0.075	0.010	≤0.0005
Au on Al[a]	7.18E-4	6.88E-12	0.02	0.03	0.21	0.190	0.010	0.001
Polished Al	7.25E-4	6.78E-12	0.03	0.05	0.14	0.145	0.009	0.001
FEP (5 mil)/Ag[a]	9.83E-4	3.70E-12	0.80	0.80	0.11	0.121	0.009	0.001
FEP (2 mil)/Ag[a]	1.01E-3	3.96E-12	0.68	0.68	0.06	0.075	0.009	0.001
GSFC Yellow	9.90E-4	2.51E-12	0.91	0.87	0.31	0.320	0.010	0.001
					ML12-4			
Au/Kapton/Al[a]	8.99E-4	5.84E-12	0.12	0.42	0.53	0.525	0.011	≤0.0005
Quartz Fabric/FEP/Al	1.06E-3	2.23E-13	0.86	0.68	0.20	0.190	0.008	0.003
In$_2$O$_3$/OSR	7.93E-3	6.36E-13	0.78	0.76	0.09	0.075	0.014	≤0.0005
In$_2$O$_3$/Kapton/Al	7.10E-4	9.40E-12	0.78	0.71	0.40	0.385	0.016	0.001
Polished Al	8.22E-4	5.55E-12	0.03	0.04	0.14	0.140	0.014	0.001
Black Paint	7.85E-4	5.84E-12	0.90	0.92	0.97	0.960	0.010	0.002
Quartz Fabric/Tape	9.15E-4	2.64E-12	0.85	0.60	0.19	0.185	0.012	0.016
Kapton (5 mil)/Al	7.42E-4	7.53E-12	0.86	0.81	0.48	0.510	0.010	0.001

[a]Sample replaced following thermal vacuum test.

lated in this way are given in Table 3. These errors are found to be small compared to the random errors given below. The larger error for one of the fabric samples on tray 4 is the result of a thermistor having become defective on that sample during calibration. Scheduling limitations did not permit replacement and recalibration of this thermistor. However, the error is comparable to the random error introduced because of telemetry resolution. Furthermore, the temperature of this sample on orbit is monitored by an undamaged backup thermistor. There may be, of course, sources of systematic errors in addition to that addressed here.

To determine C_v, heater power was applied, and the data were recorded in the form of temperature vs time using 30 s intervals. Values of dT/dt for use in Eq. (2) were calculated by fitting a parabola to seven consecutive data points (t,T) by means of a least squares routine. The derivative of the parabola at the center data point was taken as the value of dT/dt there. With dT/dt determined and with T, T_c, and P measured, Eq. (1) was solved to provide C_v as a function of temperature. The resultant heat capacities of the samples are usually between 1.0 and 2.0 J/K. The heat capacities of the 1.25 in. diam 10 mil thick supporting Al disk alone should be 0.45 J/K. Thus, the experimentally determined values of C_v for the sample assemblies are quite reasonable.

The shading factors, S, are a function of the angle ρ between the spacecraft spin axis and the sun, primarily because of one of the booms. Maximum shading occurs for the Grafoil and yellow paint samples at ρ = 90 deg and represents a 10% drop in radiation. Values of S were calculated by Systems, Sciences, and Software for 80 < ρ < 100 deg in 1 deg steps for each of the 16 samples.[14] The samples are not shaded for ρ outside this range.

Random Error Analysis

An analysis of the degree of random error to be expected in calculating α_s has been performed. For each sample, the various input parameters that are used in the equations were varied by amounts equal to the resolution inherent in the 8-bit telemetry processing, and the percentage of change in each calculated value of α_s was noted. The parameters of importance are: 1) the data from the first two channels that are used in converting telemetry voltage to resistance of the thermistors; 2) the tray temperature; 3) the sample temperatures; and 4) the shading factor. The last source of error depends upon the spacecraft orientation and arises from both the 1 deg resolution of the table of shading values versus ρ and the uncertainty of the angle ρ at a given time. Usually, it has been possible to eliminate the contribution of error from boom shading by selecting data obtained when the orientation is such that the shadows miss the sample trays. At those times, the random errors are those denoted $\Delta\alpha_{s,0}$ in Table 3, and the largest contributor to random error is the resolution with which the tray temperature is transmitted. This contribution can be reduced by searching for times when the tray temperature changes telemetry values, but it is a time consuming process, and the data presented here are not from such special times.

Experimental Results

Prior to mounting the flight samples on the calorimeters, values of α_s and normal emittance ε_N were determined optically at the Air Force Wright Aeronautical Laboratory. The solar absorptance was calculated from sample spectral reflectance measured with a Beckman DK2 spectrophotometer fitted with a Gier Dunkle integrating sphere[15] and a 25 equal energy band approximation to the Thekaekara solar spectrum.[16] The ε_N values were obtained with a Gier Dunkle DB100 inspection instrument.[15,17] These values are labeled α_s' and ε_N' in Table 3. The accuracies of the ε_N' and α_s' measurements are approximately ±0.02, and ±0.01, respectively, except in the case of the α_s' measurements of the polished aluminum samples. In these instances, instrumental problems degraded the α_s' accuracy to $\sim$ ±0.05. Note that there are some discrepancies between ε_N' and the values ε_H that were measured calorimetrically during calibration. These discrepancies can be attributed to a number of causes:

1) The calorimetric method yielded hemispherical emittance, ε_H, while the optical method yielded normal emittance, ε_N. For low ε materials, $\varepsilon_H/\varepsilon_N$ may reach $\sim$1.3; for high ε materials, $\varepsilon_H/\varepsilon_N$ is as low as $\sim$ 0.95.[18]

2) The calorimetric method requires heat flow through the sample. If sample conductance is low, as in the case of the two quartz fabric samples, the effective emittance will be lower than that deduced from reflectance measurements. Previous measurements[19] of quartz fabrics also revealed that $\varepsilon_H < \varepsilon_N$, although the differences were not as large as those in Table 3.

3) Less than perfect bonding between the samples and the supporting disks would also lead to a reduced thermal conductance and thus affect the calorimetric determination of ε_H.

4) Samples with high surface roughness often are found to have a smaller value for $\varepsilon_H/\varepsilon_N$ than is usual. The two quartz fabric sample surfaces are quite rough.

5) Partial loss of the Au overcoat probably occurred on the Au on Kapton sample. Visual inspection revealed that this sample was not uniform in appearance. Schedule constraints did not allow for the replacement and recalibration of this sample.

The early orbital values of $\alpha_{s,o}$ can be compared with the prelaunch value α_s' in Table 3. When measurement errors are considered, $\alpha_{s,o}$ and α_s' agree for all but one of the black paints and the 5 mil aluminized Kapton sample. Of these, only Kapton exhibited an apparent increase in α_s from laboratory to orbit, and as much as ~0.02 of the 0.03 increase could have come from the recognized sources of measurement error previously discussed. Since all 16 samples were exposed to the same environments from the time the calorimeters were mounted on the trays, it is concluded that any contamination acquired in tray calibrations, testing, and launch was not sufficient to affect any of the early orbital α_s values measurably. The Kapton increase is likely the result of unidentified and probably unique measurement errors or of contamination specific to that sample (such as that acquired from improper cleaning or handling). The apparent decrease in the black paint sample, α_s, on ML12-3 is likely the result of a problem in the flight instrumentation, since the 0.90 value is lower than expected and not confirmed by the sample of the same paint on ML12-4.

Plots of sample α_s vs time on orbit are shown in Figs. 4-10. These values were calculated with the procedure discussed following Eq. (5) using quick-look data provided by the USAF

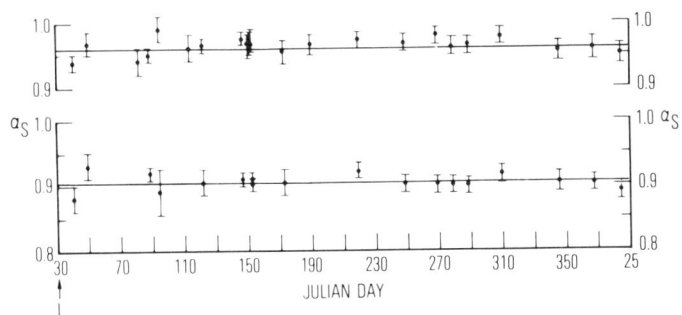

Fig. 4   Solar absorptances of black paint samples from day of launch (day 30).

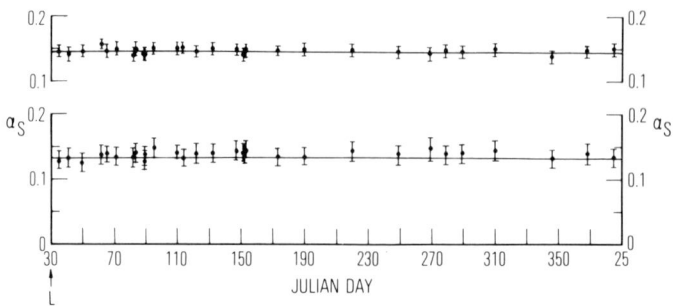

Fig. 5 Solar absorptances of polished Al samples from day of launch (day 30).

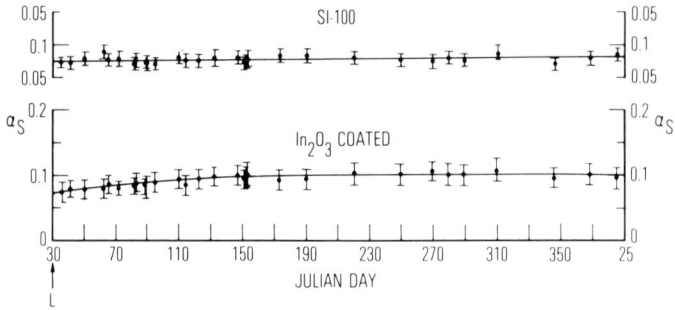

Fig. 6 Solar absorptances of standard and In_2O_3-coated fused-silica mirror samples from day of launch (day 30).

Satellite Control Facility. More data, and perhaps reductions in some error bar magnitude, will be forthcoming, and at that time formal curve fitting procedures will be employed. For the present, curves have been drawn through the points as follows: Whenever possible, a line of zero slope was hand fit to the data. Failing that, a line of nonzero slope was tried. Failing that, a simple curve was faired through the points.

Figures 4-10 reveal that the solar absorptances of the black paints, polished metals, and the uncoated OSR were approximately constant. However, the solar absorptances of the In_2O_3-coated OSR, silvered-FEP, aluminized Kapton, quartz fabric, and conductive paint samples all exhibit significant increases in α_s.

Summary and Discussion

This paper is an interim rather than a final report because the analysis of collected data and data collection are continuing. Therefore, the conclusions drawn are tentative.

THERMAL CONTROL COATINGS ON P78-2 SPACECRAFT

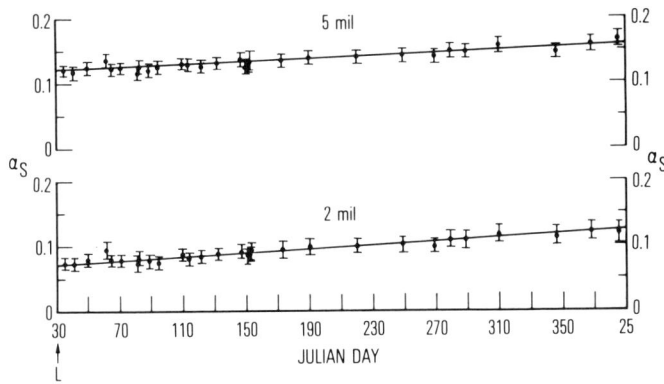

Fig. 7 Solar absorptances of 2 and 5 mil thick silvered-FEP Teflon samples from day of launch (day 30).

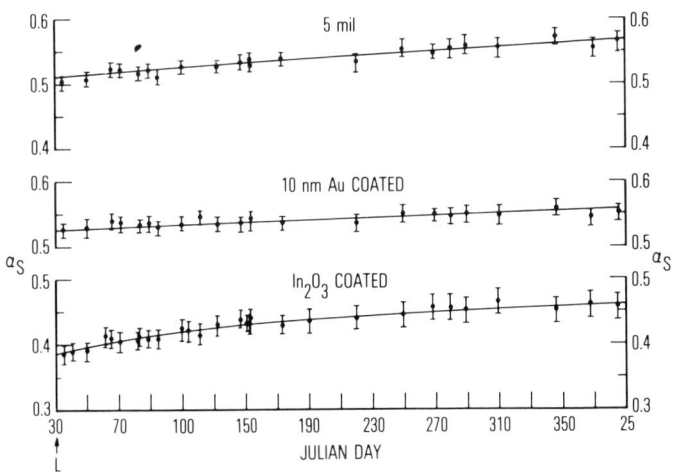

Fig. 8 Solar absorptances of plain, Au coated, and In_2O_3-coated aluminized Kapton samples from day of launch (day 30).

Since the time of launch, analyses have been performed on quick-look data provided by the USAF Satellite Control Facility. A detailed evaluation of the random errors introduced into the calculations of the solar absorptances from the raw data was also performed. These errors are caused primarily by the resolution of the telemetry system and by uncertainty in the amount of shading of the samples resulting from spacecraft booms. Reduction of the uncertainty resulting from telemetry resolution may be possible by averaging large data sets contained on magnetic tape and/or by searching for those times when the tray temperatures change telemetry values.

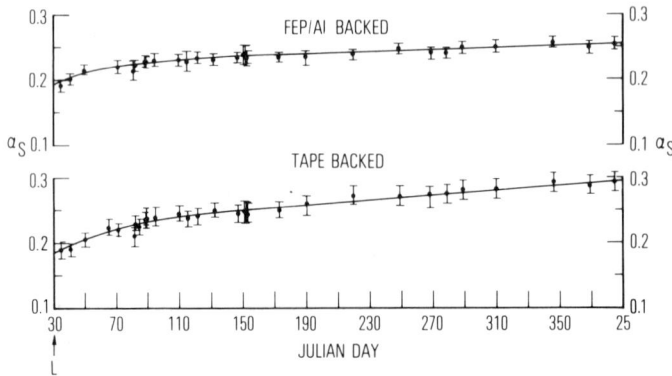

Fig. 9 Solar absorptances of Astroquartz fabric samples backed with silvered-FEP Teflon and double adhesive tape, respectively, from day of launch (day 30).

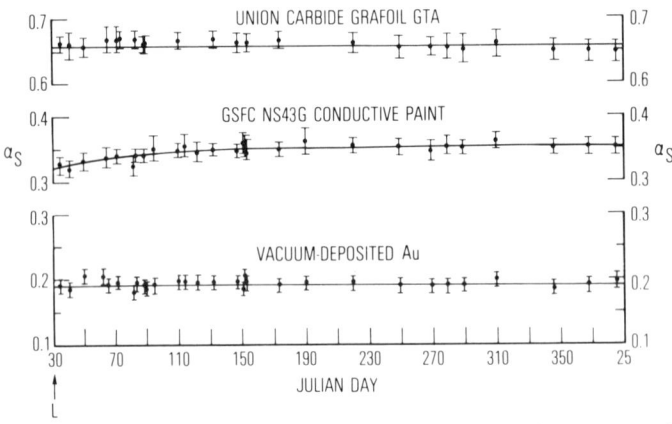

Fig. 10 Solar absorptances of Grafoil, conductive yellow paint, and Au samples from day of launch (day 30).

For most samples, the early orbital values of α_s are in good agreement with prelaunch values obtained from optical reflectance measurements. Apparently, there was little contamination acquired by the samples during launch and prelaunch activities.

Temporal data have been presented for the 16 TCC samples over 12 months into the flight. Because the satellite spins, π hours are required to accumulate one equivalent sun hour. Of course, sample property changes resulting from vacuum and energetic particle bombardment effects are not "slowed" in this way. However, contamination related changes may be slowed because of sunlight effects on contaminant adhesion[20] and optical properties.[21]

The solar absorptances of the black paints, Grafoil, polished metals, and an uncoated fused-quartz mirror all remained approximately constant over the year on orbit. This suggests that these samples are reasonably space stable and that contamination effects on these samples over this time period were minor. (The apparent bleaching of the black paint samples reported earlier[22] was the result of an erroneous assumption of constant solar irradiance in the data analysis.)

A nearly transparent In_2O_3 coating was carried on two of the samples; namely, a fused silica mirror and an aluminized Kapton sample. This coating is electrically conductive, and when it is properly connected to the spacecraft frame, it should prevent these dielectrics from acquiring a large electrostatic charge during the magnetic substorms that occasionally occur at synchronous altitude. (No attempt was made to ground the coating on these samples in this experiment. Therefore, the surfaces may charge and affect the kinetic energy of arriving charged particles just as happens with uncoated samples.)

Comparison of the α_s vs time curves of the conductively coated samples with their uncoated counterparts suggests that both coated samples degraded more rapidly during the first 4 months on orbit than did the uncoated samples.

Other samples exhibiting higher rates of degradation during the first 4 months than during the subsequent 8 months are the 2 quartz fabric and the conductive paint coatings. This functional dependence of α_s on time is frequently observed in both flight and terrestrial experiments. In particular, the fabric performance is quite similar to similar samples flown at much lower altitude on the P72-1 vehicle.[23]

Fabrics are believed to be particularly susceptible to optical property changes induced by contamination. This is because they have a strong affinity for contamination and are hard to clean after installation. The affinity comes from a high surface area to volume ratio, a result of the silicon fibers used in the construction of the fabric. In addition the high-temperature bake normally used to remove the sizing from fabric intended for space use produces highly chemically active fiber surfaces.

Note that the rate of change of the conductive paint α_s approaches zero after only 4 months. Most paints continue to degrade for a longer period.

As with all thermal control coating experiments and applications, one would like to know what fraction of the degradation was the result of contamination and the source of that contamination. The instruments that comprise the rest of the ML12 experiment are quartz crystal microbalances,[1,20] and it is hoped that analysis of the data from them will help answer these questions, at least for P78-2 thermal control coating materials.

Acknowledgments

This work was largely supported by the United States Air Force Wright Aeronautical Laboratory (AFWAL/MLBE) under Space Division Contract (SD) F04701-80-C-0081. D. Prince of AFWAL helped select, characterize, and supply most of the thermal control coating samples. Special thanks are due D. Boucher, D. M. Clark, J. R. Hribar, and S. W. Ritter of The Aerospace Corporation for assistance with data analysis, J. Brown of TRW systems for helpful discussions of the data presented here, and the MCC-F Mission Control Team of the Air Force Satellite Test Center for flight operations.

References

[1] Hall, D. F., Borson, E. N., Winn, R. A., and Lehn, W. L., "Experiment to Measure Enhancement of Spacecraft Contamination by Spacecraft Charging," NASA SP-379, National Technical Information Service, Springfield, Va., 1975, pp 89-107.

[2] Stevens, J. R. and Vampola, A. L., eds., "Description of the Space Test Program P78-2 Spacecraft and Payloads", SAMSO-TR-78-24, National Technical Information Service, Springfield, Va., 1978.

[3] "Vacuum Stability Requirements of Polymeric Materials for Spacecraft Applications," NASA-SP-R-0022, Rev. A, 9 Sept. 1974.

[4] Clippinger, D., "P78-2 Parts, Materials, and Processes Selection and Control Requirements--Materials Selection List," Doc. No. 85800010205, Rev. G, Martin Marietta Corp., Denver, Colo., 16 June 1978.

[5] Clippinger, D., "P78-2 Product Cleanliness Levels, Clean Area Control, and Contamination Control Requirements," Doc.

No. 85800010505, Rev. A, Martin Marietta Corp., Denver, Colo., Oct. 1976.

[6]"Military Standard Product Cleanliness Levels and Contamination Control Program," MIL-STD-1246A, Department of Defense, Washington, D.C., 18 Aug. 1967.

[7]"Federal Standard Clean Room and Work Station Requirements, Controlled Environment," Fed. Std. No. 209B, General Services Administration, Washington, D.C., 24 Apr. 1973.

[8]"Clean Area Control," Doc. No. STP 72102, Rev. G., Martin Marietta Corp., Denver, Colo., 5 Nov. 1974.

[9]"Space Test Program Flight P78-2 Space Vehicle/Launch Vehicle ICD," CDRL A050, USAF Contract F04701-76-C-0116, Martin Marietta Aerospace, Denver, Colo., 8 Apr. 1977

[10]"Delta Spacecraft Design Restraints," Doc. No. DAC 61687B, McDonnell Douglas Corp., Huntington Beach, Calif., May 1976.

[11]Berliner, E., "Design Considerations and Operational Performance of P78-2 (SCATHA) Program Reaction Control System," Paper No. AIAA-80-1295, AIAA/SAE/ASME 16th Joint Propulsion Conference, Hartford, Conn., 30 June 80.

[12]Fote, A. A. and Hall, D. F.,"Contamination Measurements During the Firing of the Solid Propellant Apogee Insertion Motor on the P78-2 (SCATHA) Spacecraft," Proceedings of the Society of Photo-Optical Instrumentation Engineers Technical Symposium East 181, Vol. 287, 1981.

[13]Luedke, E. E. and Kelley, L. R., "Development of Flight Units for Thermal Control Coatings Experiment," AFML-TR-72-233, AFML/MBE, Wright-Patterson Air Force Base, Ohio, 1972.

[14]Steen, P. G., "SCATHA Experiment Shadowing Study," SSS-R-78-3658, Systems, Science, and Software, La Jolla, Calif., 1978.

[15]Pettit, R. R., "Evaluation of Portable Optical Property Measurement Equipment for Solar Selective Surfaces," Journal Engineering for Power, Vol. 100, 1978, p. 489.

[16]Thekaekara, M., "The Solar Constant and the Solar Spectrum Measured from a Research Aircraft," NASA Tech. Rept. R-351.

[17]Nelson, K. E., Luedke, E. E., and Bevans, J. T., "A Device for the Rapid Measurement of Total Emittance," Journal of Spacecraft and Rockets, Vol. 3, 758, 1966, p 758.

[18] Jakob, M., Heat Transfer, Vol. 1, Wiley, N.Y., 1949, p. 52.

[19] Eagles, A. E., "Fabric Coatings for Satellite Temperature Control, Vol. II: Design Handbook," AFML-TR-77-65, Vol. II, AFML/MBE, Wright-Patterson Air Force Base, Ohio, 1976.

[20] Hall, D. F., "Flight Experiment to Measure Contamination Enhancement by Spacecraft Charging," Paper No. 216-15, Optics in Adverse Environments, SPIE, Vol. 216, Bellingham, Wash., 1980, pp 131-137.

[21] Fleischauer, P. D., and Tolentino, L., "The Far Ultraviolet Photolysis of Polymethyl-phenylsiloxane Films on Quartz Substrates," NASA SP-336, U. S. Government Printing Office, Washington, D.C., 1973, pp 645-650.

[22] Hall, D. F. and Fote, A. A., "Preliminary Flight Results from P78-2 (SCATHA) Spacecraft Contamination Experiment," ESA SP-145, Proceedings of an ESA Symposium on Spacecraft Materials, December 1979, pp 81-90.

[23] Prince, D. E., "ML-101 Thermal Control Coating Spacecraft Experiment," AFML-TR-75-17, AFML/MBE, Wright-Patterson AFB, Ohio Aug. 1975.

FREE CONVECTION IN ENCLOSURES EXPOSED TO COMPRESSIVE HEATING

R. P. Bobco*

Hughes Aircraft Company, Los Angeles, Calif.

Abstract

An experimental study of heat transfer in a vertical annulus and a three-dimensional gap was used to establish the influence of compressive heating on the convective process in enclosures. In the annulus, the inner cylinder was heated electrically and the outer cylinder was passive; in the gap, a heated square plate was opposite an equal adiabatic surface. Test runs were made using helium gas with compressive rates of 6, 15, and 30 psi/min. Temperature and pressure histories were reduced to film coefficients based on nodal modeling of the test geometries. The data were correlated in terms of free convection parameters, that is, Nusselt and Rayleigh moduli. The heat-transfer correlations show virtually no influence of compression rate and only a slight dependence on geometry. The correlations will be applied to the design of a vented Galileo mission descent module parachuting into the Jupiter atmosphere.

Nomenclature

A	= area
C	= nodal heat capacity, (mass) x (specific heat)
c_p, c_v	= specific heats at constant pressure and volume, respectively
$\mathscr{F}$	= radiation interchange factor
g	= force-to-mass conversion factor
H	= hA, nodal convective coefficient
$\tilde{H}$	= gas enthalpy

Presented as Paper 80-1536 at the AIAA 15th Thermophysics Conference, Snowmass, Colo., July 14-16, 1980. Copyright© American Institute of Aeronautics and Astronautics, Inc., 1980. All rights reserved.

*Senior Scientist, Thermophysics Department, Space and Communications Group.

$h, \bar{h}$	=	single surface and annular mean value film coefficients, respectively
J	=	mechanical to thermal energy conversion factor
K	=	kA/L, nodal thermal conductance
$K_{\dot{p}}$	=	compressive heating conductance
k	=	thermal conductivity
L	=	hydraulic radius [Eqs. (11) and (15)]
ℓ	=	height of annular volume
M	=	mass of gas in nodal volume
$\dot{M}$	=	rate of mass addition to nodal volume
Nu	=	hL/k or $\bar{h}L/k$, Nusselt modulus
p	=	gas pressure
Q	=	rate of energy added to nodal volume
R	=	radius
$\tilde{R}$	=	$\sigma A \mathscr{F}$, nodal radiation coefficient
Ra*	=	$g\beta\rho^2 L^3 c_p \Delta T^*/\mu k$, Rayleigh modulus based on compressive and convection heating of nodal volume
T	=	absolute temperature
V	=	volume
β	=	$1/T$, gas volume expansivity
γ	=	c_p/c_v, isentropic exponent
μ	=	gas dynamic viscosity
ρ	=	density

Subscripts

$(\ )_b$	=	balsa wood
$(\ )_e$	=	equipment node
$(\ )_g$	=	gas node
$(\ )_h$	=	heated surface node
$(\ )_s$	=	heat sink node
$(\ )_v$	=	pressure vessel surface node
$(\ )_w$	=	passive surface node
$(\ )_\infty$	=	gas entering nodal gas volume

Introduction

The Galileo mission to the planet Jupiter (launch date 1984) will include a thin walled vented descent module which will fall, subsonically, into the atmosphere to gather scientific data. During descent, the hydrogen/helium atmosphere will enter equipment bays, creating an internal convective environment for science and electronics packages and structural components. Fig. 1 is a schematic representation of the Galileo descent module showing the arrangement of equipment, structure, and vents. Fig. 2 shows representative pressure and temperature histories for a descent module parachuting into a 78% hydrogen, 22% helium atmosphere.[1] The mission will carry the descent module to a pressure of at least 13 bars, where the atmospheric temperature will be in the range 335 - 408 K (variation for "cool" to "warm" model atmospheres).

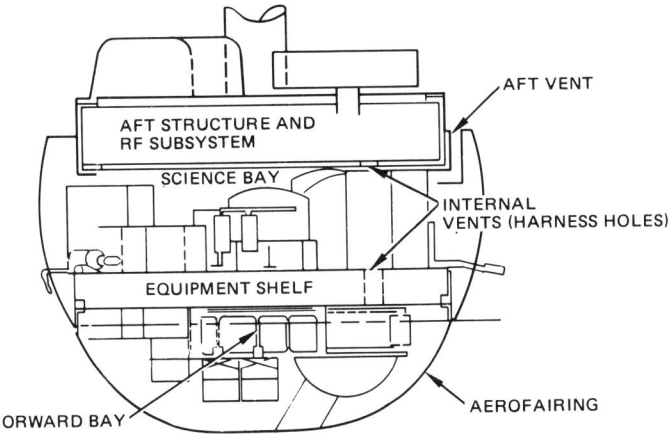

Fig. 1 Galileo descent module thermal design features.

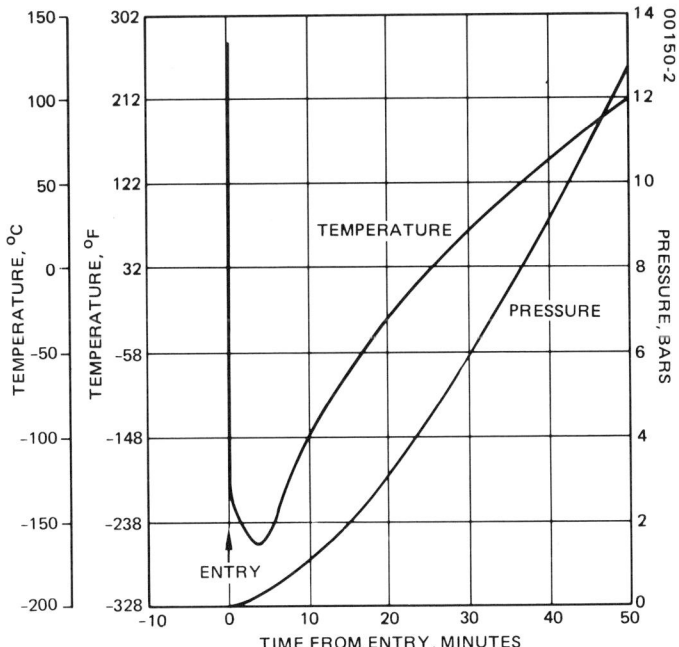

Fig. 2 Typical descent module trajectory: atmospheric temperature and pressure histories, nominal model atmosphere (78% H_2, 22% He).

Predicting heat-transfer coefficients in equipment bays containing electronic "black boxes" has been as much an art as a science for many years. In the Galileo application the heat-transfer problem is especially difficult because the process of atmospheric ingestion does compression work on the internal fluid and increases the fluid temperature as if in response to a volumetric heat source. The influence of rapid compression on convective heat transfer in simple containers was investigated by Means and Ulrich,[2] but the results were not appropriate to the Galileo descent module with its complex internal geometry and low compression rate (0.41 bar/min). The combination of geometric complexity and fluid dynamic uncertainty (i.e., forced and/or natural convection) required an experimental study to evaluate the nature and magnitude of convective heat transfer under conditions approximating those expected in the Galileo descent module.

The following sections describe the convective design problem and an experimental study which provided heat-transfer data in a form convenient for thermal design analysis and mission performance predictions.

The Convective Design Problem

The state of the art of thermal design for transient heat-transfer applications is based on the use of nodal

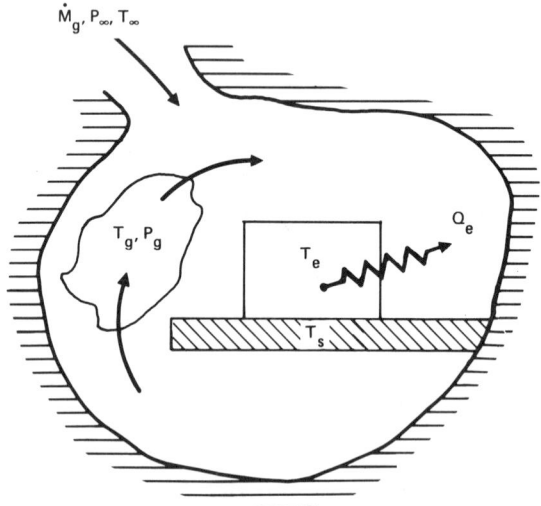

Fig. 3 Schematic representation of vented enclosure containing equipment, heat sink, and gas nodes. Pressure, P_∞, and temperature, T_∞, increase with time.

(discrete element) heat balance equations implicit in a variety of heat-transfer codes. Such equations relate the rate of change of nodal internal energy to energy transport rates due to convection, conduction, radiation, work, electrical dissipation, etc. The Galileo descent module thermal model is based on three nodal categories: 1) power dissipating nodes (e.g., black boxes), 2) heat sink nodes (e.g., equipment shelf), and 3) gas nodes. Consider the simple system shown in Fig. 3. The nodal heat balance equation for a power dissipating node may be written as

$$C_e \frac{dT_e}{dt} = (hA)_{e,g} (T_g - T_e) + K_{e,s} (T_s - T_e)$$
$$+ \sum_j (A\mathcal{F})_{e,j} \sigma(T_j^4 - T_e^4) + Q_e \qquad (1)$$

Equation (1) may be applied to heat sink nodes also by stipulating $Q_e = 0$, i.e., no energy source.

The gas node equation, based on fluid bulk temperature, is written as

$$\frac{d}{dt} (M_g c_v T_g) = (\dot{M}\tilde{H})_{g,in} - (\dot{M}\tilde{H})_{g,out}$$
$$+ \sum_j (hA)_{j,g} (T_j - T_g) \qquad (2)$$

where account is taken of the variable gas mass and enthalpy change due to mass transfer across the system boundaries. The gas node equation can be cast in a form more convenient for computer codes by assuming the ideal gas relations

$$(pV)_g = (MRT)_g \quad \tilde{H}_{g,in} = c_p T_\infty \quad \gamma = c_p/c_v \qquad (3)$$

and observing that $\dot{M}_{g,out} = 0$ and $p_g = p_\infty$ in the present application. (Pressure equality is required by the thin wall design.) With these simplifications, the gas node equation may be written as

$$C_g \frac{dT_g}{dt} = K_p^* (\gamma T_\infty - T_g) + \sum_j (hA)_{j,g} (T_j - T_g) \qquad (4)$$

where

C_g = apparent gas heat capacity

$$= \gamma p_g V_g T_\infty \Big/ J (\gamma - 1) T_g^2 \qquad (5)$$

$\dot{K_p}$ = compressive heating conductance

$$= V_g (dp_g/dt) \Big/ T_g (\gamma - 1) J \qquad (6)$$

The nodal equations may be integrated numerically when all of the heat capacities, conductances, interchange factors, and convective coefficients are known as functions of time and/or temperature and pressure. In the Galileo descent module, the only factors unknown to the thermal designer are the convective "film" coefficients coupling the interior gas with the dissipating and heat sink nodal surfaces.

Experimental Approach

An experimental study was initiated to obtain data on convective heat transfer in an enclosure exposed to compressive heating. Test models and procedures were evolved to meet requirements for well-defined initial and boundary conditions, instrumentation, and data processing based on nodal formulations, budget and schedule limitations, and safety. These requirements all suggested a series of tests incorporating the following features:

1) Geometrical simplicity of source and sink heat-transfer surfaces.

2) Use of helium (instead of a hydrogen/helium mixture) as the convective medium.

3) An initial condition based on steady-state natural convection or gas conduction in an enclosure at one atmosphere (i.e., a "hot start").

4) Use of guard heaters and insulation to control boundary conditions.

5) Redundant instrumentation to measure transient temperatures and pressures.

6) A schedule of test runs in which rate of pressurization was a parameter.

This approach stressed analytical tractability of data interpretation rather than a high degree of simulation of Galileo descent module geometry and descent pressure profiles. Descriptions of test equipment, procedures, and data processing are given below followed by a presentation of test results.

Test Equipment

The test equipment included two different electrically heated models, a pressure vessel, a surge tank, a supply of high pressure helium, copper-constantan (Type T) thermocouples, pressure transducers, power supplies, manual pressure control, and a variety of readout instruments. Fig. 4 shows, schematically, the several articles of test equipment and recording equipment.

The first test model created a vertical annular volume (9.75 in. i.d. x 12.25 in. o.d. x 14 in.) heated at the inner cylinder with the outer cylinder acting as a heat sink. The annular bases were enclosed by "adiabatic" surfaces (balsa wood plugs, 8 lb/ft^3). The electrically heated inner surface was a rolled and welded aluminum cylinder (9.75 in. o.d. x 0.090 in. wall x 12 in.) mounted on a vented balsa wood core. A 1-1/2 in. diam axial manifold in the balsa core transmitted pressure through twenty 1/8 in. diam radial vents machined through both the aluminum cylinder and the core. Ceramic spacers, mounted to the balsa core, maintained a 1-1/4 in. radial clearance during model assembly and testing. The annular bases were not pressure tight insofar as thermocouple leads were conducted along both the outer cylinder and under the balsa core to the axial manifold (clearance for leads was estimated to be less than 0.062 in.); additional penetrations included a pressure tap through the lower base and a valved discharge vent through the upper base. The nominal arrangement of surfaces and dimensions is shown in Fig. 5a. The primary instrumentation for the annular test model consisted of eight thermocouples on the heated inner cylinder, eight thermocouples suspended in the gas volume on nylon cord, four thermocouples on the gas side of the outer cylinder, four thermocouples embedded in the balsa core, a pressure tap, and heater current and voltage measurements. Secondary instrumentation included limited redundancy for primary thermocouples; a second pressure tap in the pipe from the surge tank; thermocouples mounted inside the surge tank, on the pressure vessel exterior, and in the laboratory air; and guard heater power measurements. A total of 50 channels was used to record temperatures, pressures, and heater power while testing this configuration.

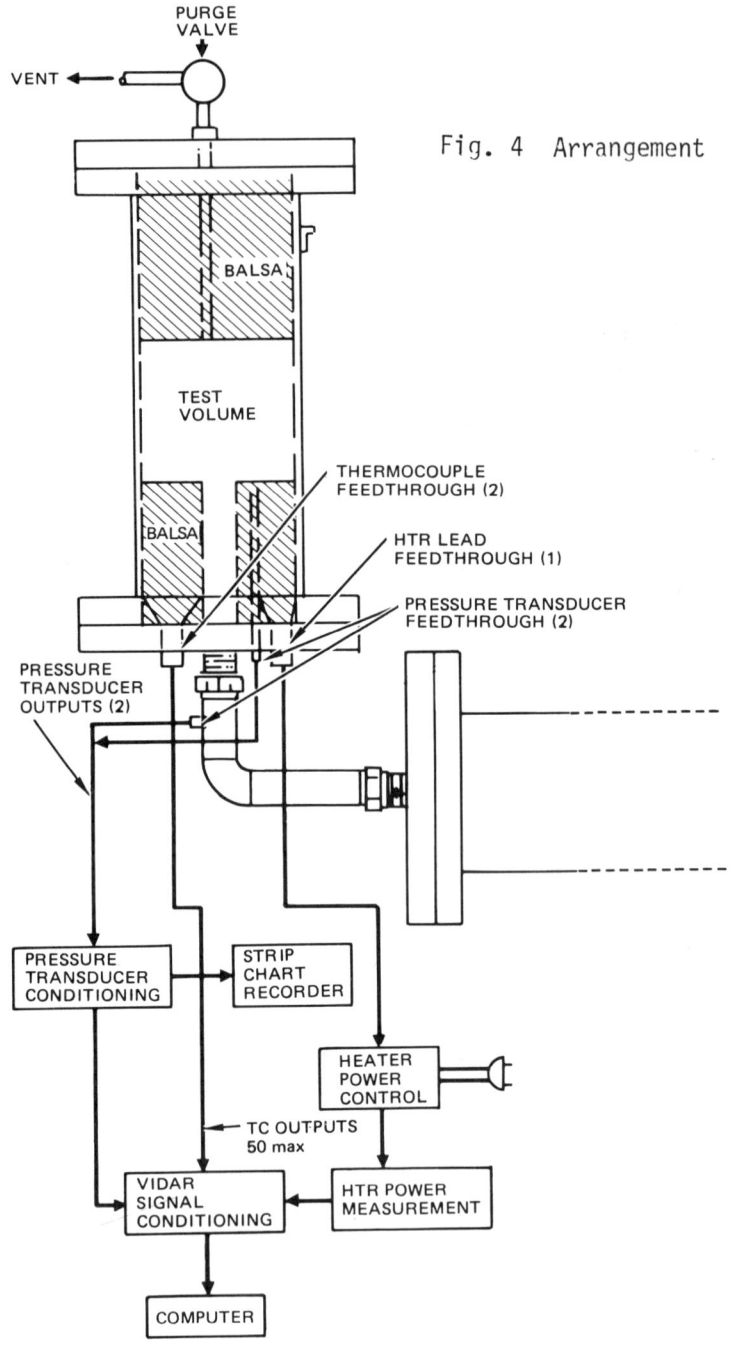

Fig. 4 Arrangement

of equipment used in convection/compressive heating tests.

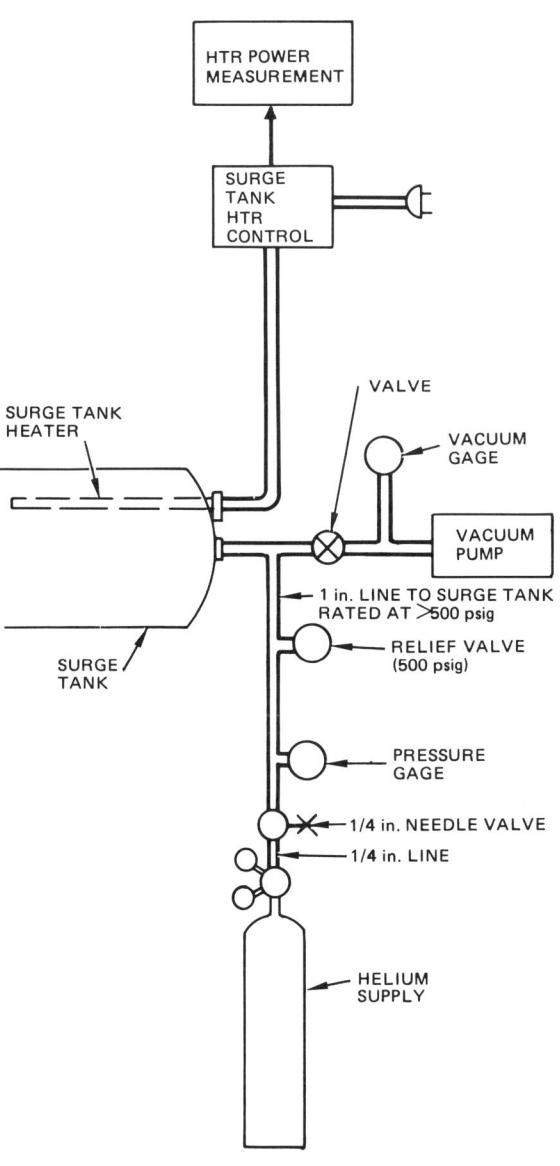

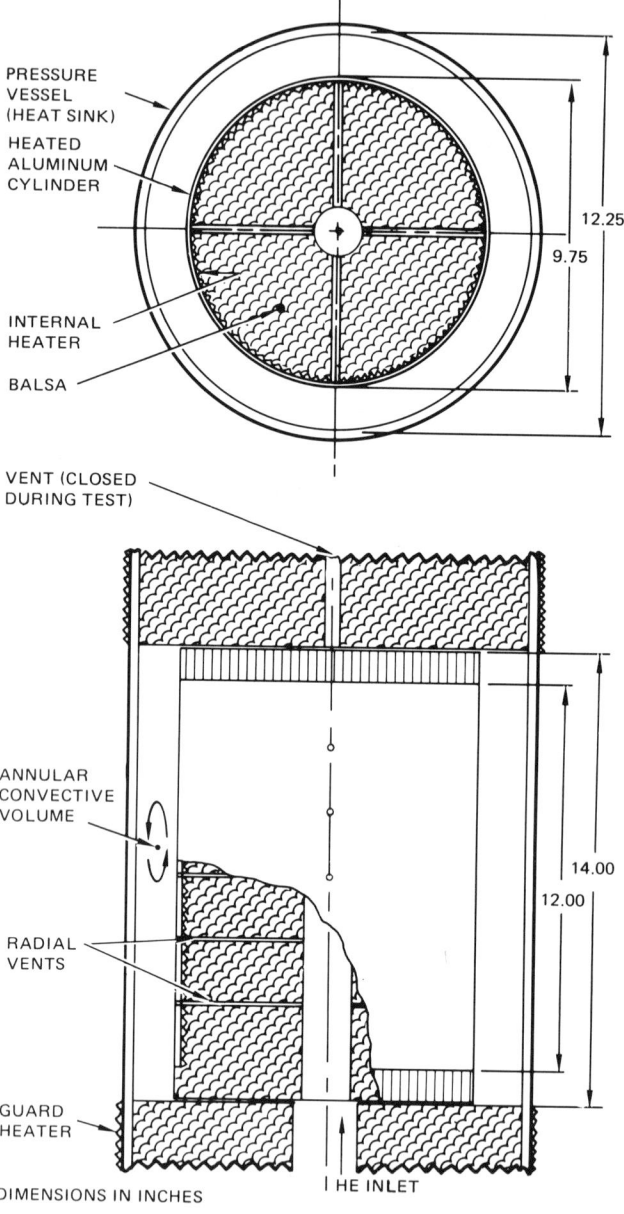

a) **VERTICAL ANNULUS**

Fig. 5 Test model schematic.

The second model was designed to measure convective heat transfer in a three-dimensional gap (6 in. x 6 in. x 1 in.); the gap was created by an electrically heated aluminum plate (6 in. x 6 in. x 0.125 in.) directly opposed by a parallel aluminum plate insulated at the back surface. The bottom of the gap was closed by an adiabatic surface while the top and vertical ends of the gap were open. Physically, the model consisted of two parallel gaps separated by central plates using a common heater; the outboard adiabatic plates were bonded to balsa wood while the bottom of the gap was bare balsa wood. The double gap created a symmetrical arrangement which provided redundancy and simplified the heat balance formulations. The double gap was retained in a slotted balsa cylinder (9.4 in. in diam x 6 in.) which, in turn, was placed in a horizontal cylindrical volume (12.25 in. in diam x 8.8 in.); the horizontal cylinder consisted of the pressure vessel walls with 12.25 in. diam balsa plugs to create adiabatic vertical ends. The principal features of the gap configuration are shown in Fig. 5b. The primary instrumentation consisted of a single thermocouple mounted at the gas side area centroid of

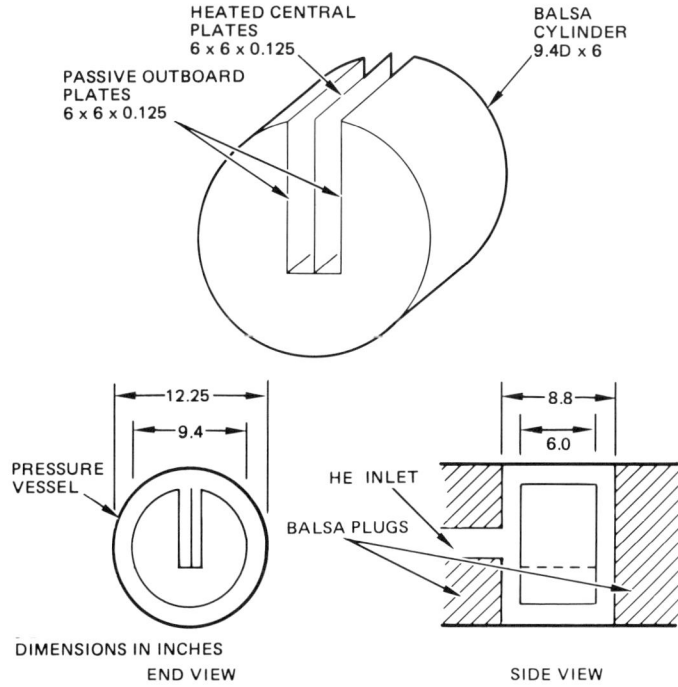

b) **RECTANGULAR GAP**

Fig. 5 (cont.) Test model schematic.

each aluminum plate, a single thermocouple suspended at the volume centroid of each gap, and a pressure tap through one of the 12.25 in. diam balsa plugs. Secondary instrumentation included redundant thermocouples at most of the primary locations, six thermocouples suspended in the annular volume and end volumes surrounding the slotted balsa core, and two thermocouples embedded in the balsa core, plus the surge tank, pressure vessel, and laboratory air instrumentation used with the first model. Once again, 50 channels were used to monitor data during test runs of this configuration.

The test chamber and surge tank were pressure vessels fabricated from 0.25 in. mild steel (ASTM A455-77A). The test cylinder internal dimensions were 12.25 in. in diam x 34 in.; the surge tank dimensions were approximately 12.25 in. in diam x 24 in. The surge tank contained four baffles and was connected to the test chamber by a 2-1/2 in. diam x 18 in. pipe. The inlet flange of the test vessel contained penetrations for model heater power, two thermocouple feedthroughs (12 channels each), and a pressure tap; the outlet flange had an axial vent (1/2 in. diam). The surge tank had penetrations in the nonflanged end for the helium supply (1 in.) and a Cal-Rod heater (1 kW). The test vessel had electric heater blankets bonded to the outer surface adjacent to each flange; each heater covered a cylindrical area 12.75 in. in diam x 7 in. Temperature measurements of the test chamber were made at four internal locations (90 deg apart at the midlength) and eight external locations (four directly opposite the internal thermocouples and two each at planes 11 in. removed from the midlength); a single thermocouple was mounted on the exterior of each flange. The surge tank had four internal thermocouples (two on baffles, two on cylindrical surfaces) during shakedown tests (three thermocouples were reassigned to the test models during convection tests); a fifth thermocouple was suspended in the exit pipe approximately 4 in. downstream from the surge tank.

All temperature measurements were made with commercially fabricated 40 gage copper-constantan (Type T) thermocouples. Surface measuring thermocouples were mounted using pressure sensitive aluminum foil tape; each junction was covered by a rectangular piece of foil tape (1 in. x 2 in. x 0.005 in.). Gas measuring thermocouples in the test volumes were tied to lengths of flat woven nylon cord which were stretched between wooden supports bonded or screwed into noncalorimetric elements of the models; these bare junctions were suspended at the midpoints (tolerance ± 0.063 in.) of all gaps. Pressure measurements were made using two strain gage type pressure transducers (Dynisco, 0-1000 psig).

A manual feedback control system was used to regulate the rate of pressurization (dp/dt) during a test run. The desired test chamber pressure was preplotted on a strip chart and compared, visually, to the pressure transducer output after signal processing by a strip chart recorder. Pressure rate adjustments were made using a 1/4 in. needle valve in the piping connecting the helium supply bottles to the surge tank. In most test runs it was possible to maintain the pressure rate to within ±20% of the desired value.

Data acquisition was provided by a system used for production level testing of spacecraft in the Hughes Aircraft Company Space Simulation Laboratory. Fifty channels of analog data (temperatures, pressures, model heater current and voltage) were supplied to a signal conditioning system ("VIDAR") which scanned the data set at 100 scans/h. Digital data were transmitted to a Honeywell 516 computer for preprocessing and subsequent transfer to a Prime 300 computer for data reduction and analysis. The system was virtually noise-free, and temperature fluctuations between scans were less than $0.75°F$.

Test Runs, Procedures, and Typical Results

Prior to the actual conduct of the experiment, three runs were scheduled for the vertical annulus and four runs for the rectangular gap. The vertical annulus test runs were carried out as planned; the runs were made at constant compression rates of 30, 15, and 6 psi/min with an internal electrical heat load of 100 W (0.112 W/in.^{2}).

The rectangular gap test runs did not proceed according to plan because the model was damaged during the enclosed tests and damage was not discovered until the model was removed from the pressure vessel at the end of the fourth run. The model was repaired, but it was possible to reschedule only one additional test. Scrutiny of the test data showed that the passive plates broke away from the balsa slot near the end of the first test run made at a compression rate of 30 psi/min with an electrical heat load of 17 W (0.236 W/in.^{2}).

This run was repeated after the model was repaired. The three runs made with the damaged model (i.e., skewed gap model) were made at compression rates of 15 and 6 psi/min, each with heat loads of 17 W, and at 30 psi with an electrical load of 40 W (0.556 W/in.^{2}). Data from the damaged model runs provided insight into the influence of heat load on convection in a skewed gap. The three vertical annulus runs, two rectangular gap runs, and three skewed gap runs are summarized in

Table 1. In addition to these runs, a number of diagnostic runs were made with both annular and gap geometries to establish the character and magnitude of heat losses in the test system. Diagnostic tests included steady-state observations at constant pressure, transient response of calorimetric elements to a step increase in heater power at constant pressure, and compressional heating with no electrical heat loads.

The procedure for all nondiagnostic test runs was as follows:

1) **Test system purge** — A vacuum roughing pump removed air from the test system and lines following which a helium bleed was carried out for 5 min and the system was repressurized to 2-5 psig.

2) **Thermal equilibrium** — Test model and guard heaters were energized and temperatures monitored until model surfaces reached a steady state (approximately 200°F and 150°F for the annular and gap models, respectively). Thermal equilibrium was reached in 90-150 min.

3) **Pressurization** — Data acquisition was started approximately 2.5 min (four data scans) prior to pressurization to record initial conditions at constant pressure. Following four data scans the fiberbatt blanket was placed over the test vessel and helium was introduced into the test system at a constant compression rate until a test vessel pressure in the range 425-450 psig was reached. Data acquisition continued for an additional four data scans at constant pressure to verify there were no leaks in the system.

Table 1 Run summary

Geometry	dp/dt, psi/min	Q_h, W	A_h, in.^{2}
Vertical annulus	30	100	896
	15	100	896
	6	100	896
Rectangular gap	30	17	72
	30	17	72
Skewed gap	15	17	72
	6	17	72
	30	40	72

4) <u>Depressurization and cooling</u> — The high-pressure helium was vented to atmosphere and helium was bled through the system for 5-10 min to hasten cooling; the fiberbatt blanket was removed and the vent closed to retain test system pressure at 2-5 psig. Heater power was maintained if a second run was scheduled the same day.

Typical temperature and pressurization histories are shown in Fig. 6. In both the annular and gap tests, convective development was observed through the fourth to sixth data scans (approximately 3-4 min); at the end of that time, convection appeared to be fully established, and all temperature histories became monotonic. Data associated with "transient convection" were not used in the correlations developed below; the results shown represent quasisteady convection associated with compressional heating.

Data Processing

In both annular and gap tests, data reduction relations were developed to complement the nodal formulation shown in the discussion of the convective design problem. That is, geometric mean temperatures were used to obtain instantaneous values of the gas, heated surface, pressure vessel, and balsa support core. For example, in the annular model eight thermocouples in the annular volume were reduced to a single nodal gas temperature. Similarly, eight thermocouples defined the mean value of the heated annular surface while four each were used for the balsa and pressure vessel. The smaller size of the gap model allowed gas, heater, and balsa mean temperatures to be defined with two thermocouples each.

Potential difficulties in obtaining reliable mean temperatures and derivatives were avoided by using orthogonal polynomials[3] to obtain a quadratic least square fit through five successive mean temperatures. Gram-Chebyshef polynomials yield smoothed temperatures and derivatives from the simple expressions

$$\bar{T}(t_3) = [-3T(t_1) + 12T(t_2) + 17T(t_3) + 12T(t_4) - 3T(t_5)] \div 35$$

$$\frac{d}{dt}\bar{T}(t_3) = [-2T(t_1) - T(t_2) + T(t_4) + 2T(t_5)] \div 10\Delta t$$

where $\bar{T}(t_3)$ represents the smoothed value of a mean temperature, such as T_g, at the midpoint, $t = t_3$, of five successive data scans made at equal intervals ($\Delta t = 0.01$ h). The same smoothing formulation was used to obtain the pressure derivative $d\bar{p}(t_3)/dt$, required in the compressive heating conductance.

The annular convection model had three calorimetric elements for use as nodes to obtain two convective coefficients between the gas and the enclosing cylindrical surfaces; the calorimetric elements were 1) a heater node, 2) a gas node, and 3) a pressure vessel node. The data reduction used the heater and gas nodes; the pressure vessel heat balance was not reliable because losses to the laboratory environment could not be controlled. The heater and gas node equations for

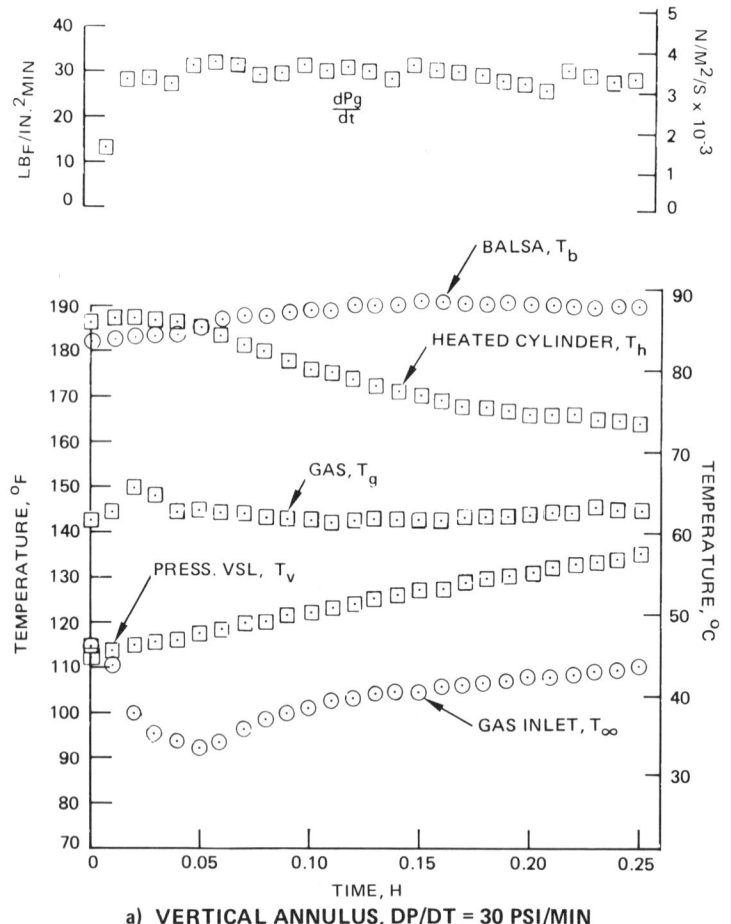

a) VERTICAL ANNULUS, DP/DT = 30 PSI/MIN

Fig. 6 Typical smoothed mean value temperatures.

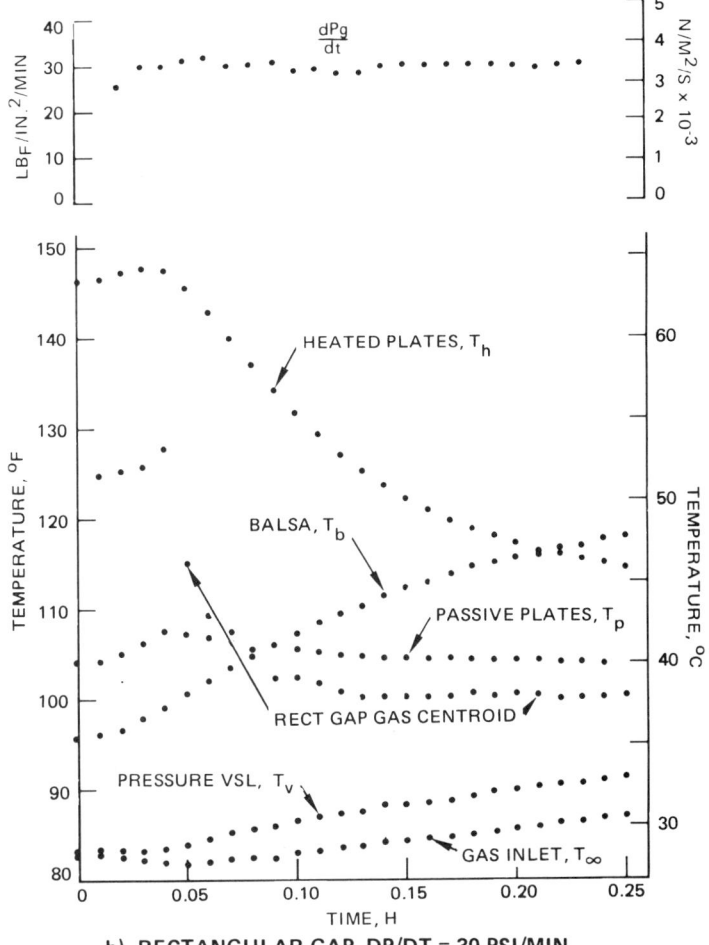

b) RECTANGULAR GAP, DP/DT = 30 PSI/MIN

Fig. 6 (cont.) Typical smoothed mean value temperatures.

convective coefficients follow:

$$H_{hg} = \frac{C_h \dot{T}_h - Q_h - K_{hb}(T_b - T_h) - \tilde{\tilde{R}}_{hv}(T_v^4 - T_h^4) - \tilde{K}_{hx}(T_x - T_h)}{(T_g - T_h)} \quad (7)$$

$$H_{gv} = \frac{C_g \dot{T}_g - H_{hg}(T_h - T_g) - K_{\dot{p}}(\gamma T_\infty - T_g)}{(T_v - T_g)} \quad (8)$$

Both steady-state and transient diagnostic test runs were made to evaluate the heat loss to the balsa, $K_{hb}(T_b - T_h)$, and the axial (end) loss, $K_{hx}(T_x - T_h)$. The balsa loss was due to simple conduction in the steady state but increased 4.3X during pressurization; the convective nature of the balsa heat loss was attributed to the twenty radial holes drilled through the heater and balsa core. The numerical values of the several coefficients are:

$$C_h = 1.02, \quad C_g = 0.162 \; pT_\infty/T_g^2 \quad (Btu/°R)$$

$$K_{hb} = 0.26 \; T_b^{0.598}, \quad K_{hx} = 11.0 \; (Btu/h°R)$$

$$\tilde{R}_{hv} = 2.41 \times 10^{-10} \quad (Btu/h\text{-}°R^4)$$

$$K_p° = 5.81 \; \dot{p}/T_g \quad (Btu/h\text{-}°R)$$

$$\gamma = 5/3$$

All temperatures are in degrees Rankine, pressures in psi, pressure rate in psi/min. The axial loss temperature difference was modeled as $(T_x - T_h) = (T_{\pm 4} - T_{mid})$, where T_{mid} was the mean value of four temperatures at the heater midheight and $T_{\pm 4}$ the mean value of two temperatures 4 in. above and two at 4 in. below the midheight. The heater and pressure vessel areas implicit in H_{hg} and H_{gv} were 2.55 ft² and 3.74 ft², respectively.

Several nodal arrangements were examined for the rectangular gap tests before a mathematical model that satisfied the data was found. The intuitive nodal distribution included a heater node, a passive plate node, and a bulk gas node. This model was unsuccessful because 1) the gas was strongly nonisothermal throughout the enclosure, and this made it impossible to evaluate the various energy transport terms based on a single bulk temperature; 2) diagnostic tests ($Q_h = 0$) suggested the improbable condition that passive plate-to-balsa conductance was greater than the passive plate-to-gas convection coefficient. The data reduction model with the fewest deficiencies postulated convective transport from a heated vertical plate opposed by a vertical adiabatic surface; for this model, the passive plate temperature was used for the bulk temperature of the gas in the gap. This simple model required a single nodal equation for the heater to obtain the convection coefficient:

$$H_{hg} = \frac{C_h \dot{T}_h - Q_h - \tilde{R}_{hp}(T_p^4 - T_h^4) - \tilde{R}_{hv}(T_v^4 - T_h^4)}{(T_g - T_h)} \quad (9)$$

with $T_g = T_p$. The heat capacity and radiation factors had the values $C_h = 0.221$ Btu/°R, $\tilde{R}_{hp} = 4.4 \times 10^{-12}$, and $\tilde{R}_{hv} = 4.5 \times 10^{-11}$ Btu/h-°R⁴. The heater area (two sides) implicit in H_{hg} was 0.5 ft².

Equations (7, 8, and 9) were programmed for calculation by the computer used for data acquisition. Also programmed in the computer were temperature dependent properties of helium and constants occurring in the Rayleigh and Nusselt moduli:

$$Ra = (g \beta \rho^2 c_p/\mu k) (L^3 \Delta T)$$

$$Nu = h L/k$$

Neither the characteristic length, L, nor the characteristic temperature difference, ΔT, were known in advance of the data processing.

Heat-Transfer Correlations

Correlations were sought which would be useful to the thermal designer of an enclosure more complex than the relatively simple annulus and rectangular gap. From a design point of view it was established that the characteristic length, L, should represent the mean separation distance between a surface node and a volumetric gas node; the characteristic temperature difference, ΔT, should be proportional to the rate of compressive heating, $Q_{\dot{p}} = K_{\dot{p}} (T_\infty - T_g)$, and the input from the heated surface to the gas, $Q_{hg} = H_{hg} (T_h - T_g)$. Several L and ΔT possibilities were investigated, but the correlations shown below are based on the following values:

Vertical annulus:

$$L = V_{annulus} \div (A_h + A_v) = 0.0556 \text{ ft} \qquad (10)$$

$$\Delta T^* = Q_g/(k_g A_g/L) \qquad (11)$$

$$Q_g = K_{\dot{p}} (\gamma T_\infty - T_g) + H_{hg} (T_h - T_g) \qquad (12)$$

$$A_g = (A_h + A_v)/2 \qquad (13)$$

Rectangular gap:

$$L = V_{gap} \div (A_h + A_w) = 0.0417 \text{ ft} \tag{14}$$

ΔT^* and Q_g per (11) and (12), respectively

$$A_g = (A_h + A_w)/2 \tag{15}$$

The length, L, represents a hydraulic radius which can be identified for a variety of enclosures; the temperature difference, ΔT^*, is an ostensible potential which can be evaluated by an iterative process during a design analysis.

Data processing of the annular test runs resulted in slightly different values of film coefficient at the heated inner cylinder and the cool outer cylinder ($h_{hg} < h_{gv}$). This observation suggested the mean gas temperature, T_g, may have been in error by up to 5°F. The problem was circumvented by defining an effective thermal conductivity for the convective medium based on a pseudosteady state:

$$Q = [2\pi \ell k_{eff}/\ln(R_v/R_h)](T_h - T_v) = H_{hg}(T_h - T_g)$$
$$= H_{gv}(T_g - T_v) \tag{16}$$

$$k_{eff} = \ln(R_v/R_h)/2\pi\ell \Big/ (1/H_{gv} + 1/H_{hg}) \tag{17}$$

Equations (16) and (17) are correct only if $C_g \dot{T}_g$ is very small and compressive heating is much less than the heat load from the heated surface; a test of these conditions for the highest compression rate (dp/dt = 30 psi/min) showed the following:

$$C_g \dot{T}_g = O(1) \text{ Btu/h}$$

$$Q_p^{\cdot} \approx 100 \text{ Btu/h}$$

$$Q_h \approx 710 \text{ Btu/h}$$

These magnitudes indicate that Eqs. (16) and (17) are reasonable. That is, the magnitude of compressive heating is relatively small but its influence on convective circulation is not negligible. The use of Eq. (17) allows a single value of film coefficient for both surfaces:

$$\bar{h} = k_{eff}/L \tag{18}$$

or the separate film coefficients may be recovered as

$$h_{hg} = 2\pi \ell \, k_{eff}/A_h \, \ln(R_g/R_h) \tag{19}$$

$$h_{gv} = 2\pi \ell \, k_{eff}/A_v \, \ln(R_v/R_g) \tag{20}$$

where R_g represents the radius corresponding to the gas node (say, $R_g = R_h + L$ or $R_g = \sqrt{R_h R_v}$).

Equations (10) and (18) were used to define the Nusselt modulus for the annular volume

$$Nu = \bar{h}L/k \tag{21}$$

Equations (10) - (13) completed the definition of the Rayleigh modulus:

$$Ra^* = g\beta\rho^2 L^3 c_p \Delta T^*/\mu k \tag{22}$$

The reduced data for the three test runs with the vertical annulus are shown in Fig. 7a. A linear regression of these data yielded the correlation

$$Nu = 0.171 \, (Ra^*)^{0.303} \tag{23}$$

$$1.2 \times 10^3 \leq Ra^* \leq 3.5 \times 10^5$$

$$6 \leq dp/dt \leq 30 \text{ psi/min}$$

Although there were slight differences among the three runs, there is no indication that rate of pressurization influenced the correlation.

The two runs made with the rectangular gap had the same compression rate, $dp/dt = 30$ psi/min; the data are shown in Fig. 7b. The first run was made before model damage occurred; the second run was made a week later after the model had been repaired. The values of L and ΔT^* given in Eqs. (15) and (12)

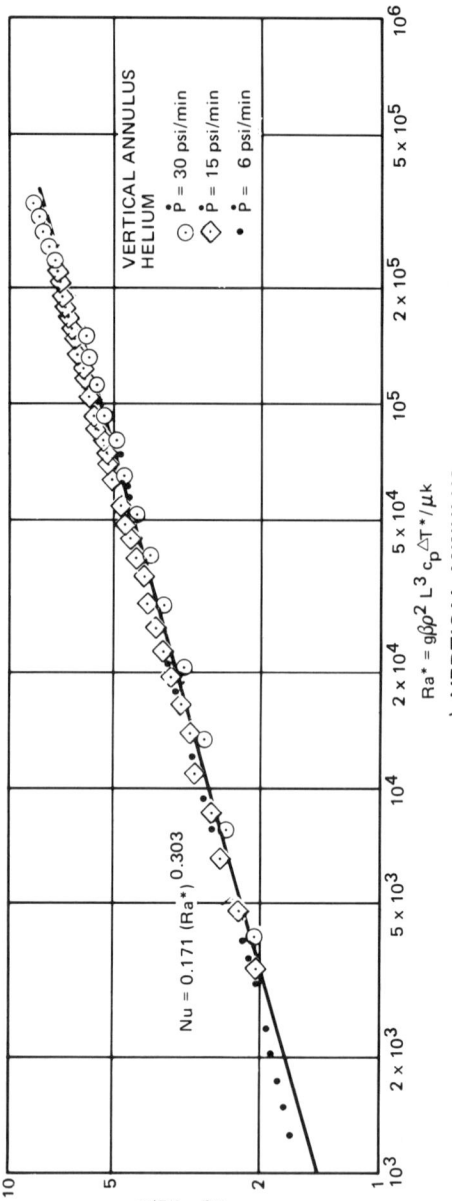

Fig. 7 Experimental results in terms of Nu vs Ra*.

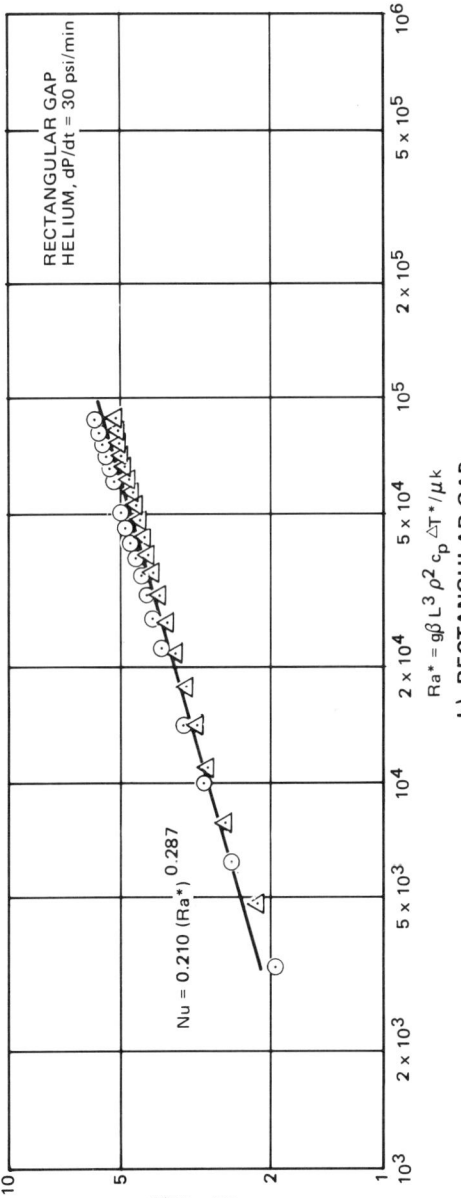

Fig. 7 (cont.) Experimental results in terms of Nu vs Ra*.

yielded the following correlations:

First run: $\text{Nu} = 0.224\,(\text{Ra}^*)^{0.276}$
Second run: $\text{Nu} = 0.196\,(\text{Ra}^*)^{0.297}$

$5 \times 10^3 \leq \text{Ra}^* \leq 1 \times 10^5$

The difference between the correlations is less than 11% over the range of Rayleigh modulus. A linear regression of both runs gives

$$\text{Nu} = 0.210\,(\text{Ra}^*)^{0.287} \qquad (24)$$

$$5 \times 10^3 \leq \text{Ra}^* \leq 1 \times 10^5$$

$$dp/dt = 30 \text{ psi/min}$$

The results of the annular convection tests and examination of reduced data for the skewed gap tests (see below) suggests the correlation is suitable for the pressurization range

$$6 \leq dp/dt \leq 30 \text{ psi/min}$$

Data from the skewed gap runs was reduced in terms of Nusselt and Rayleigh moduli based on values of L and ΔT^* used for the rectangular gap. These runs were of interest because they encompassed the full range of pressure rates and two values of heater load at the active surface. A linear regression of these data showed the same consistency and small spread exhibited by Eqs. (23) and (24):

Skewed gap: $\text{Nu} = 0.487\,(\text{Ra}^*)^{0.206}$ \qquad (25)

$$1.2 \times 10^3 \leq \text{Ra}^* \leq 2 \times 10^5$$

$$6 \leq dp/dt \leq 30 \text{ psi/min}$$

Equation (25) has limited utility for design applications because it is based on inappropriate values of L and ΔT^*. However, the data reduction exercise was useful because 1) it showed that pressurization rate has a negligible influence on heat transfer from the heated surface and 2) the geometrical differences between the rectangular and skewed gaps created less than a $\pm 10\%$ difference in Nusselt modulus over the range $10^4 < \text{Ra}^* < 10^5$. Although these skewed gap observations provide insight into the convective process, the remaining discussion of applications will be confined to the regular geometries.

Discussion of Results

The experimental study described above was motivated by a need to obtain engineering thermal design correlations rather than as an investigation of thermal and fluid dynamic fundamentals in enclosures. The nature of the instrumentation did not provide any information on the fluid dynamics associated with compressive heating in an enclosure. Specifically, there are no data to establish whether the convective circulation was laminar or turbulent; neither is it possible to determine whether the "strength" of the compressive heating rate or the circulation induced by the change of momentum in the enclosure was the dominant effect in the heat-transfer process. It is believed that flow visualization experiments will be required before a rational fluid dynamical model can be developed for compressive heating phenomena.

Very little extrapolation of fluid dynamic effects can be made from the compressive heating experiments of Means and Ulrich[2] or from analytical or experimental studies of free convection in enclosures such as those of Bishop,[4] Flack,[5] Newell and Schmidt,[6] or MacGregor and Emery.[7] For example, the experiments of Means and Ulrich[2] were based on air injection at sonic velocities into simple tanks (e.g., a spherical tank, SCUBA equipment, an oxygen tank) with no heated interior surfaces. They treated the process as forced convection and obtained the following heat transfer correlation:

$$Nu = 22 \, [RePr/(L/D) \, (D/d)^2]^{0.635}$$

where Re is Reynolds modulus based on tank diameter D and nozzle throat velocity; Pr is Prandtl modulus; d and L are nozzle throat diameter and tank length, respectively. Alternatively, the free convection results reported in Refs. 4-7 apply to uncluttered enclosures wherein specified isothermal surfaces led to $\Delta T = T$ (wall 1) $- T$ (wall 2) for the Rayleigh modulus rather than the ostensible potential, ΔT^*, used here.

In the absence of direct flow observations or supportive studies based on comparable boundary conditions it is not possible to draw rigorous conclusions on the thermal and fluid processes; however, it is possible to draw several inferences from superficial features of the data and the correlations on the probable nature of the flow:

1) The correlation coefficients for all of the data sets exceeded 0.99. This suggests that the circulation is fully developed (quasisteady) natural convection and not influenced

by forced convection, i.e., there is little improvement to be gained by introducing Reynolds numbers in the present correlations.

2) Convection appears to be fully established at low values of Rayleigh modulus (Ra* = 1.2×10^3) and there is no remarkable change in slope corresponding to transition from laminar to turbulent circulation. (Some of the test runs showed a slight increase in slope at Ra* = 10^4.)

3) The Rayleigh coefficients and exponents in the present correlations do not provide any insight into the nature of convection in the annular and gap geometries. Experimental and analytical studies of enclosed convection[4-7] show that laminar circulation may lead to exponents in the range 0.25-0.37 with coefficients of 0.117-0.37. These studies also suggest that laminar flow persists for values of conventional Rayleigh modulus as high as 4×10^6.

4) The similarity of the exponents and coefficients in Eqs. (23) and (24), for two dissimilar configurations, suggests that the heat-transfer mechanism is a weak function of geometry.

These inferences are useful for applying the annular and gap correlations to the thermal analysis and design of the more complex geometry in the Galileo descent module.

Application to Design

The annular and rectangular correlations given above will be used to predict film coefficients within the Galileo descent module during its descent into the Jupiter atmosphere. Equation (23) (annular correlation) will be used to find surface-to-gas coefficients at the interior surface of the insulation layer, at the equipment shelf, and at "black box" surfaces facing the insulation across gaps of 1-2 in. Equation (24) (rectangular gap) will be used for gaplike volumes between adjacent "black boxes." Hydraulic radius based on volume-to-area ratio will be used as the characteristic length for each configuration.

The transport properties of the Jupiter atmosphere are easily predicted by Wilke's formulas for nonpolar gases.[8] Preliminary calculations based on a hydrogen-rich atmosphere and a conservative estimate of hydraulic radius indicate that during descent Rayleigh modulus, Ra*, will exceed the upper limit of the range of test runs, Ra* = 3.5×10^5. Equations (23) and (24) will be used for Ra* greater than the test range on the basis that the correlations apply to the same extent as conventional free convection.[4,5]

Summary and Conclusion

An experimental investigation was conducted to determine the influence of compressive heating on convective heat transfer at vertical surfaces in an enclosure. Two configurations were examined:

1) A vertical annular volume with an electrically heated inner cylindrical surface and a cool outer cylindrical surface. The ends of the annular volume were adiabatic.

2) A rectangular parallelopiped volume with a square, electrically heated vertical surface opposite an adiabatic surface. The base of the volume was adiabatic and the top and sides of the volume were open to a larger enclosed volume.

The heat-transfer models were designed to emphasize analytical tractability in the data reduction procedures. Both the vertical annulus and the rectangular gap configurations were tested using helium as the convective medium undergoing compression. The configurations were tested at three pressurization rates: dp/dt = 6, 15, and 30 psi/min.

The gas volumes and heated surfaces were modeled as simple calorimetric elements to develop algorithms for processing temperature data into heat-transfer coefficients. A number of diagnostic test runs were made with different initial and/or boundary conditions to evaluate heat losses and establish the repeatability of the temperature data.

After obtaining convective heat-transfer coefficients, the results were cast in nondimensional form with Nusselt and Rayleigh moduli as the dependent and independent variables, respectively. This free convection formulation provided strong correlations for both configurations when the Rayleigh temperature difference, ΔT^*, was based on the total heat input to the gas volume (convection from the heated surface plus compressive heating). The characteristic length appearing in the Nusselt and Rayleigh moduli was taken as a hydraulic radius based on the ratio of gas volume to the sum of vertical surface areas enclosing the volume. The correlations for the two configurations are given in the text as Eqs.(23) and (24). The test results spanned a Rayleigh modulus range $1.2 \times 10^3 \leq Ra^* \leq 3.5 \times 10^5$ over which the Nusselt modulus increased from 1.5 to 8.2.

The Nusselt-Rayleigh data show virtually no influence of compression rate for the range $6 \leq dp/dt \leq 30$ psi/min. Other significant features of these data are 1) the similarity of the two correlations for two dissimilar configurations, and 2) the continuity of the data and correlations to Rayleigh moduli as low as 1.2×10^3. The first observation suggests that the convective process is a weak function of geometry and that the correlation can therefore be used to predict heat transfer with compressive heating for enclosures somewhat different from the test models. Finally, from the appearance of convection at low Rayleigh modulus, it is concluded that the compression effects are important in initiating convective circulation.

The study reported here has provided an empirical basis for the thermal design of the Galileo descent module. A more complete understanding of the fluid dynamics and heat transfer associated with compressive heating requires additional studies, both experimental and analytical. Flow visualization experiments are needed to investigate early transient effects and to identify convection patterns and parameters that induce laminar and turbulent circulation. Analytical studies that would expand the range of compression rates, Rayleigh moduli, and Prandtl numbers beyond the values that can be studied experimentally would be useful. While practical examples of free convection with compressive heating are not common, additional studies of these phenomena would add to the present understanding of convective heat transfer in enclosures.

Acknowledgments

This study was sponsored by NASA Ames Research Center under the Galileo Probe Program (Contract NAS 2-10000). The author gratefully acknowledges the assistance of the following Hughes Aircraft Company personnel: E.J. Gatz and D. Lombardi, who fabricated the heat-transfer models and conducted the tests, and R.L. Berlinski and J.F. de la Vega, who developed the data acquisition techniques and data processing computer program.

References

[1] Gautier, D., et al., "The Helium Abundance of Jupiter," Bulletin of the American Astronomical Society, Vol. 11, No. 3, 1979, p. 589.

[2] Means, J.D., and Ulrich, R.D., "Transient Convective Heat Transfer During and After Gas Injection into Containers,"

Journal of Heat Transfer, ASME Transactions, Series C, Vol. 97, No. 2, May 1975, pp. 282-287.

[3] Hildebrand, F.B., Introduction to Numerical Analysis, McGraw-Hill, New York, 1956, Chap. 7.

[4] Bishop, E.H., et al., "Heat Transfer by Natural Convection between Concentric Spheres," International Journal of Heat Mass Transfer, Vol. 9, No. 7, July 1966, pp. 649-661.

[5] Flack, R.D., et al., "The Measurement of Natural Convection Heat Transfer in Triangular Enclosures," Journal of Heat Transfer, ASME Transactions, Series C, Vol. 101, No. 4, Nov. 1979, pp. 648-654.

[6] Newell, M.E., and Schmidt, F.W., "Heat Transfer by Laminar Convection within Rectangular Enclosures," Journal of Heat Transfer, ASME Transactions, Series C, Vol. 92, No. 1, 1970, pp. 159-168.

[7] MacGregor, R.K., and Emery, A.F., "Free Convection through Vertical Plane Layers — Moderate and High Prandtl Number Fluids," Journal of Heat Transfer, ASME Transactions, Series C, Vol. 91, No. 3, 1969, pp. 391-403.

[8] Ibele, W., "Thermophysical Properties," Section 2 of Handbook of Heat Transfer, edited by W. M. Rohsenow and J.P. Harnett, McGraw-Hill, New York, 1973

THERMAL ANALYSIS OF A MULTIPURPOSE FURNACE FOR MATERIAL PROCESSING IN SPACE

Gilbert S. Karp,* James N. Holsen,† and Stephen H. Miesner††
McDonnell Douglas Astronautics Company, St. Louis, Mo.

Abstract

A multipurpose furnace has been designed to demonstrate the potential of material processing in outer space. This paper describes the furnace and accessory package designed to meet the requirements of directional solidification. The methodology for producing high gradients for crystal growing is explored in terms of the impact of devices used for mechanical implementation of these requirements such as heat pipes and gas injection. The use of auxiliary heaters is discussed as a means of sharpening the axial gradients.

Nomenclature

A	= area
a	= aperture radius
C_0, C_1, C_2	= constants
E	= black body emissive power
$F_{i,j}$	= view factor of node (area) j relative to node (area) i
G	= irradiance
G_L	= axial gradient on liquid side of interface
G_S	= axial gradient on solid side of interface
J_i	= radiosity of node i
k	= thermal conductivity

Presented as Paper 90-1514 at the AIAA 15th Thermophysics Conference, Showmass, Colo., July 14-16, 1980. Copyright © American Institute of Aeronautics and Astronautics, Inc., 1980. All rights reserved.

*Lead Engineer.
†Technical Specialist.
††Engineer.

q	=	heat transfer rate
r	=	radius
R	=	radius of curvature
T	=	temperature
x	=	axial distance along sample
α	=	absorptance
ε	=	emittance
σ	=	Stefan-Boltzmann constant
τ	=	transmittance

Subscripts

A	=	ampule
C	=	cold
H	=	hot
i, j	=	surface i, j respectively
L	=	liquid
M	=	muffle
S	=	specimen, solid

Introduction

NASA has initiated a major materials processing in space program to demonstrate the feasibility of manufacturing new materials in outer space. As a subcontractor to TRW, Inc. McDonnell Douglas Astronautics Company - St. Louis Division designed a multipurpose material processing furnace and ancillary equipment to support this program. Experiments in directional solidification and isothermal and gradient freeze processing will be accommodated through the use of interior accessory packages specifically designed to meet the requirements of each type of experiment. This paper will discuss the analytically derived thermal characteristics of the furnace and its application to directional solidification experiments typical of those planned for the first Space Shuttle flight on which the furnace will be carried.

The near zero gravitational environment of space, with the absence or near absence of convection currents due to concentration and temperature gradients, offers the opportunity of growing more perfect crystals and more finely dispersed alloys than may be possible on Earth. Crystalline samples may approach the ultimate in magnetic or semiconductor properties. Alloys are expected to be stronger than those produced on Earth or to process other unusual properties. The lead tin telluride and mercury cadmium telluride samples selected for analysis are both used as infrared detectors.

Processing in space is expected to eliminate the necessity for probing and sampling crystals produced by present methods to identify those smaller sections which may have suitable semiconductor characteristics.

Directional solidification (DS) processing is designed for the production of highly uniform crystalline solids from a melt. Furnaces for DS processing must provide a uniformly high temperature environment in one end and a low temperature in the other end. By adjusting the hot and cold end temperatures, the desired thermal gradient is created in the sample and the isotherm corresponding to the melt/solid interface is positioned at a location in the gradient development zone where radial gradients are minimal. Crystallization proceeds as the experiment sample, generally encased in a quartz or alumina ampule, is slowly withdrawn into the cold end. High thermal gradients, the specific values depending upon the translation rate (the rate at which the ampule is withdrawn into the cold end of the furnace), are required to avoid constitutional supercooling, a phenomenon whereby liquid/solid equilibrium conditions are created in the melt in front of the desired solidification interface. A planar interface is desirable to ensure uniform crystal growth. Processing in the low-gravity environment of space minimizes the thermal convection currents that distort the interface in ground-based experiments.

Every experiment sample has its particular requirements. Hot end furnace temperatures are limited by the softening point of the quartz ampules encapsulating the sample, or by the vapor pressure of the sample. Cold end temperatures are determined by the necessity to reduce the temperature below some transition point in the solidified sample. Hot end temperatures as high as 1200 - 1400°C and cold end temperature as low as 50 - 100°C are required for the DS experiments scheduled for the first Shuttle flight. Cartridge translation rates may vary from 0.07 to 75 cm/h. Sophisticated control algorithms are required to maintain temperature uniformity and to position the interface at the desired location. Thermal analysis therefore guides furnace design and predicts the optimum processing conditions for each experiment.

Furnace Design Concepts

The furnace mechanical design has been shaped by the following requirements:
1) Operating temperature range 50 - 1600°C
2) Sample cartridge size envelope: length 2.5 - 25 cm, diameter 1.2 - 3.2 cm

3) Maximum system steady state power $\leq$ 3 Kw
4) Maximum system transient power $\leq$ 4 Kw
5) Heated cavity heat-up time $\leq$ 1 hour
6) Environment helium/vacuum
7) Furnace coolant F21
8) Coolant temperature range 0 - 40°C
9) Process modes: directional solidification, isothermal processing, gradient freeze processing
10) Heater input voltage 0 - 26 Vdc
11) Maximum nominal current 40 A

Out of these requirements, the present design concept of multiple accessory packages has been developed. Accessory packages can be inserted into the basic furnace to specialize its function for a particular experiment or generic class of experiments.

The multipurpose furnace assembly is illustrated in Fig. 1. The basic furnace consists of a multizone heater, multilayer insulation and structure support members. The accessory package concept for directionaly solidification is described in detail. The main heater is a one piece construction with three independently controlled heater zones. Only heater zones 2 and 3 are operational during solidification processing. Heat loss to the furnace shell, rear flange and forward flange is minimized by four multilayer foil insulation assemblies. In operation, a vacuum environment exists in the space between the furnace shell and the accessory package muffle. The multilayer insulation in this space will be thermally shorted with helium from an external gas system during furnace cooldown/quench operations. Freon 21 is circulated through coolant tubes attached to the furnace shell, rear flange and forward flange.

The design for the directional solidification accessory package was shaped by two experiment requirements:

1) Maximize the sample temperature gradient.
2) Produce a planar or near planar melt/solid interface.

In order to produce these effects, the general accessory package design characteristics are a near-step change in boundary temperatures and an insulated zone to allow for the development of a planar melt/solid interface.

The first concept explored consisted of two adjacent heat pipes separated by an insulated barrier. This concept is an outgrowth of several years of work at the McDonnell

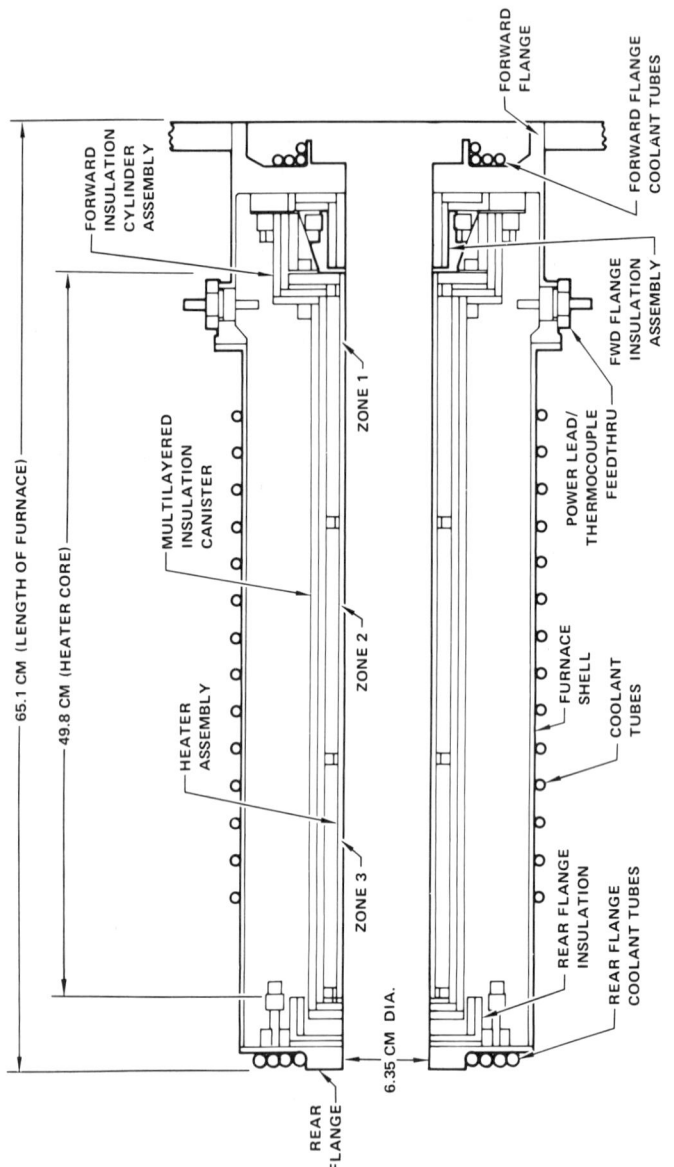

Fig. 1 Multipurpose furnace assembly.

Research Laboratory[1] using sodium heat pipes for directional solidification. However, heat pipes have two major shortcomings:

1) Commercially available heat pipes are currently limited to approximately 900°C for vacuum application. Higher temperature heat pipes have been developed but are not commercially available.

2) Heat transfer on the cold end of the furnace from the cartridge to the cold end heat pipe is radiation limited and therefore will impact the maximum sample gradient achievable.

Early studies indicated that substituting a thick wall metallic liner for the hot end heat pipe results in little or no degradation in thermal performance. Eliminating the heat pipe in the hot zone simplifies the accessory package design and improves the furnace heatup characteristics. Early thermal analyses also revealed that higher axial temperature gradients can be obtained using helium gas as a convective cooling agent. The lowest cold end heat pipe temperature that can be realized is limited by the fact that it is a temperature integrating device and will therefore be severely affected by heat transfer within the heated cavity. A recirculation system was incorporated into the furnace system design to supply helium for convective cooling of the cartridge cold end.

Our present concept for the directional solidification accessory package is illustrated in Fig. 2. It consists of a gas tight, metallic muffle containing an insulation baffle to separate the heated cavity into hot and cold zones. The insulation baffle also provides a quasiadiabatic region for developing a large axial temperature gradient and locating a planar or near planar liquid/solid interface temperature isotherm.

The hot and cold end cartridge temperatures are controlled separately. Zones 2 and 3 of the heater, illustrated in Fig. 1, are controlled to obtain the desired hot end cartridge temperature. To enhance and steepen the axial gradient, a tickler heater adjacent to the insulation baffle and furnace hot zone supplies heat selectively to the hot end of the cartridge at a point near the gradient development zone.

The zone 1 heater is inactive for directional solidification processing, but will be at a significantly higher tem-

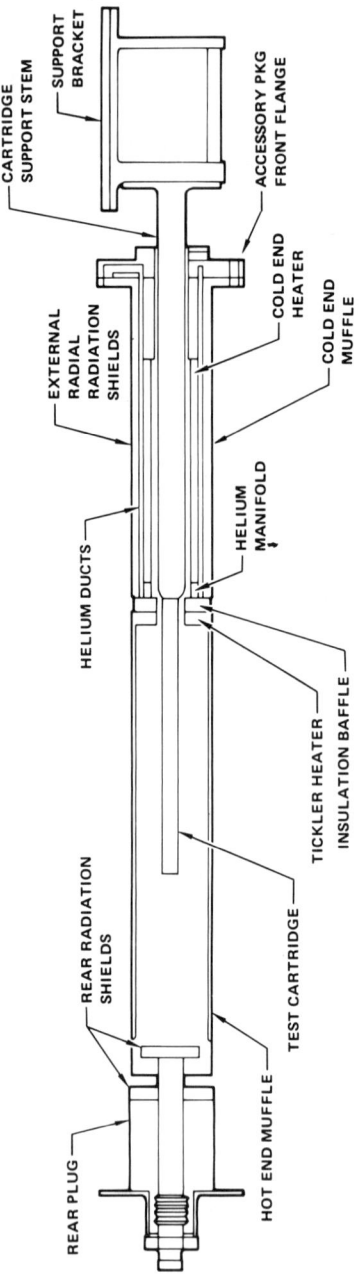

Fig. 2 Directional solidification accessory package.

perature than the desired cold end cartridge temperature because of conduction from the hot end heater zones. Thus, multilayer insulation is located on the external side of the cold end muffle to isolate the cold end muffle from the Zone heater. The cold end muffle is as thin as possible to minimize the heat losses to the forward flange. Inside are the helium ducts and a temperature controlled cold end modulating heater. Helium enters the front flange of the accessory package at a flow rate of 0.25 g/s and flows through an annular duct to a manifold near the insulation baffle. The manifold directs the stream of helium coolant onto the cartridge surface, enhancing the cold end heat transfer. A cold end modulating heater controls the temperature of the helium before it impinges on the cartridge.

As illustrated in Fig. 2 the support bracket is bolted to a slide mechanism that provides extremely accurate translational movement of the test cartridge in and out of the furnace heated cavity.

In the flight experiment cartridges, the specimen is either encapsulated in a quartz ampule or is contained in an alumina encapsulent with an outer inconel safety enclosure. Two flight experiment cartridges are analyzed in this paper. These cartridges are quartz encapsulated specimens 25 cm in length with an o.d. of 1.2 cm.

Theoretical Gradient Analysis

The earliest design concept for the production of large axial gradients employed two adjacent heat pipes to produce a near-step change in the wall temperature of the furnace heated cavity. This geometry can be idealized by considering radiation boundary conditions only and neglecting axial conduction in the sample. The results of this analysis serve a useful purpose of delimit the performance characteristics of directional solidification cartridges. Fig. 3 shows the idealized furnace heated cavity. The sample here is treated as having a totally transparent body and negligible conductivity.

A heat balance on a differential area (dA) on the sample centerline takes the following form:

$$F_{dA,A_H}(T_{dA}^4 - T_H^4) + F_{dA,A_C}(T_{dA}^4 - T_C^4) = 0 \tag{1}$$

Note that $F_{dA,A_H} + F_{dA,A_C} = 1.0$, so that Eq. (1) can be solved for T_{dA}:

$$T_{dA} = [F_{dA,A_H}(T_H^4 - T_C^4) + T_C^4]^{0.25} \qquad (2)$$

The axial centerline gradient is:

$$\frac{dT_{dA}}{dx} = \frac{1}{4}[F_{dA,A_H}(T_H^4 - T_C^4) + T_C^4]^{-0.75}(T_H^4 - T_C^4)\frac{dF_{dA,A_H}}{dx} \qquad (3)$$

Using Nusselt's method the view factor F_{dA,A_H} can be shown to be equal to:

$$F_{dA,A_H} = F_{dA,A_O} = 1/2 - X/2\sqrt{X^2 + a^2} \qquad (4)$$

Differentiating the above gives the following:

$$\frac{dF_{dA,A_H}}{dx} = \frac{X^2(X^2 + a^2)^{-0.5} - (X^2 + a^2)^{0.5}}{2(X^2 + a^2)} \qquad (5)$$

Combining Eq. (5) with Eq. (3) results in:

$$\frac{dT_{dA}}{dx} = \frac{1}{8}\left[\frac{X^2(X^2 + a^2)^{-0.5} - (X^2 + a^2)^{0.5}}{(X^2 + a^2)}\right]$$

$$\cdot\left[\frac{(X^2 + a^2)^{0.5} - X}{2(X^2 + a^2)^{0.5}}(T_H^4 - T_C^4) + T_C^4\right]^{-0.25}(T_H^4 - T_C^4) \qquad (6)$$

The maximum value of dT_{dA}/dx occurs at $x \simeq 0.0$, so that Eq. (6) can be further simplified:

$$\left(\frac{dT_{dA}}{dx}\right)_{max} \simeq \frac{1}{8a}[\frac{1}{2}(T_H^4 - T_C^4)]^{-0.75}(T_H^4 - T_C^4) \qquad (7)$$

and for the limit case where $T_H >>> T_C$, Eq. (7) can be further simplified:

$$\left(\frac{dT_{dA}}{dx}\right)_{max} \simeq \frac{-0.21 T_H}{a} \qquad (8)$$

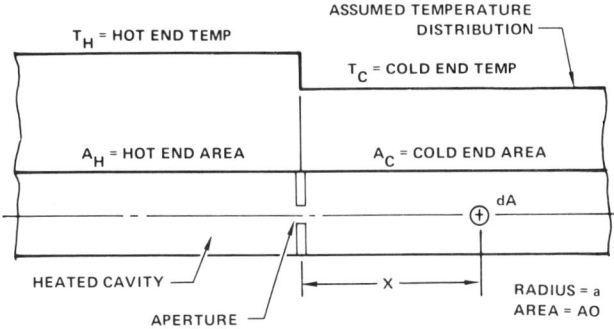

Fig. 3 Idealized heated cavity geometry assumed for gradient limit case caluclation.

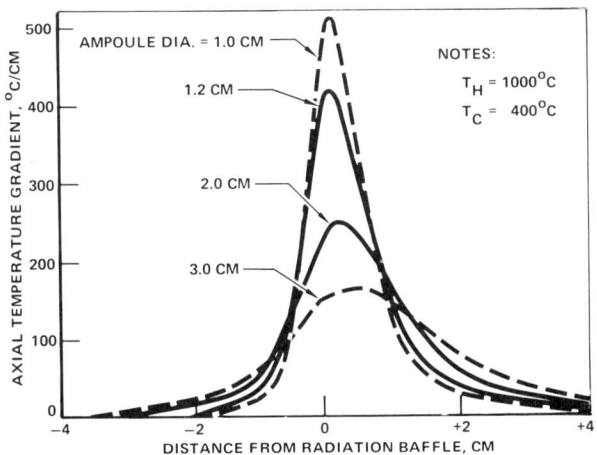

Fig. 4 Theoretical axial gradient distribution (radiation only).

Fig. 4 shows the axial gradient distribution derived from Eq. (6) for several difference cartridge diameters. This figure illustrates the general characteristics of the gradient distribution: a narrow gradient zone, the width of which increases with increasing diameter, and a decreasing gradient with increasing diameter.

Thermal Analysis Math Model

The characteristics of real materials involving conductive, convective, and radiative effects are not amenable to a closed form mathematical solution. Consequently, we have created a finite difference math model of the furnace heated

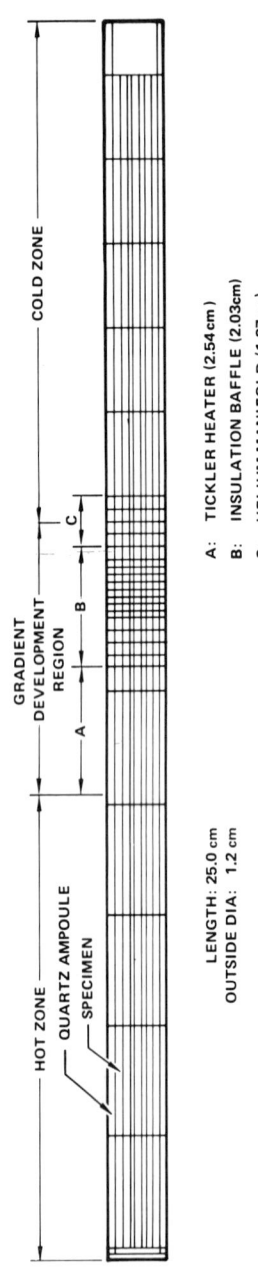

LENGTH: 25.0 cm
OUTSIDE DIA: 1.2 cm

A: TICKLER HEATER (2.54cm)
B: INSULATION BAFFLE (2.03cm)
C: HELIUM MANIFOLD (1.27cm)

Fig. 5 Cartridge nodal arrangement.

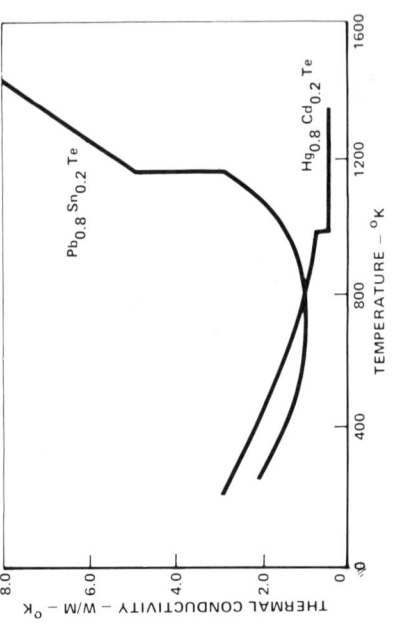

Fig. 7 Assumed thermal conductivities for PbSnTe and HgCdTe.

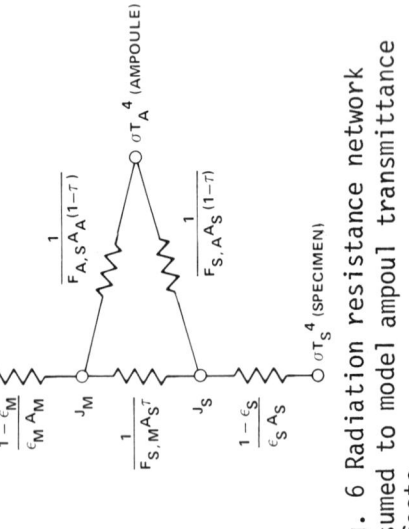

Fig. 6 Radiation resistance network assumed to model ampoul transmittance effects.

cavity in order to study the gradient distribution within several selected experimental samples.

The thermal math model consists of a set of elementary volumes, each of which can be assumed to be at a uniform temperature. The heat balance equations have been formulated in a backward difference form and the resulting matrix solved by the Gauss-Siedel method. Material thermophysical properties are input into the thermal math model as a function of temperature.

The cartridge, drawn to scale and showing the nodal distribution, is illustrated in Fig. 5. The nodes are relatively small in the vicinity of the insulated baffle to improve analytical accuracy, since this is the region where the liquid/solid interface occurs. Where thermal mapping is not as critical, the cartridge nodes are larger to reduce the model size.

Radiation heat transfer is particularly important in the analysis of the directional solidification accessory package. Radiation heat transfer through the multilayer radiation shields in the furnace is a major factor affecting power requirements. Within the furnace heated cavity, the radiation heat transfer to and from the test cartridge plays a major role in creating the required thermal gradients. The close spacing between the elements of the multilayer insulation permits them to be treated as semi-infinite concentric cylinders. We have approximated the shield-to-shield radiant transfer by the following expression:

$$q_{ij} = \frac{A_i \sigma (T_i^4 - T_y^4)}{\frac{1}{F_{ij}} + (\frac{1}{\varepsilon_i} - 1) + \frac{A_i}{A_j}(\frac{1}{\varepsilon_j} - 1)} \qquad (9)$$

Conductive effects in the multilayer insulation have been developed from experimental data. Radiation heat transfer in the furnace heated cavity is treated as diffuse radiation in an enclosure of gray surfaces.

To allow for the variation of surface emittance and transmittance with temperature within the heated cavity, a series of additional nodes has been included which represent the radiosity of each real surface node to which it is assigned. The radiosity (J) is defined as the total amount of energy per unit area leaving a node.

In the case of the quartz encapsulent, the transmittance of the quartz-to-infrared radiation is significant at temperatures greater than approximately 200°C, and the transparency of the quartz allows radiation cross-coupling within the gradient development zone, diminishing the axial gradient potential. Consequently, this effect is included in the model.

If one assumes the internal reflectance of the quartz is negligible and that Kirchoff's identify applies (i.e., $\alpha + \tau = \varepsilon + \tau = 1$), then an analog resistance network can be developed representing the radiation heat transfer between the muffle, the quartz ampule and the specimen. The muffle radiosity (J_M) is:

$$J_M = \varepsilon_M \sigma T_M^4 + (1 - \varepsilon_M) G_M$$

where G_M = muffle irradiance = net energy per unit area incident on the muffle. The net amount of energy leaving the muffle is:

$$q_{M-net} = A_M(J_M - G_M) \qquad (10)$$

Combining these two equations results in:

$$q_{M-net} = G_M A_M (\sigma T_M^4 - J_M^4)/(1 - \varepsilon_M) \qquad (11)$$

By the same reasoning, the net amount of energy leaving the specimen is:

$$q_{S-net} = \varepsilon_S A_S (\sigma T_S^4 - J_S)/(1 - \varepsilon_S) \qquad (12)$$

The energy radiated by the muffle, transmitted through the quartz, and incident on the specimen is:

$$J_M A_M F_{M,S} \tau$$

Also, the energy leaving the specimen and incident on the muffle is:

$$J_S A_S F_{S,M} \tau$$

Therefore, the net heat flow from the muffle to the specimen is:

$$(J_M - J_S) A_S F_{S,M} \tau \qquad (13)$$

Finally, the net heat flow between the quartz and the muffle is the difference between the energy leaving the quartz ampule incident on the muffle ($\varepsilon_A F_{AM} A_A \sigma T_A^4$) and that which originates from the muffle and is absorbed (not transmitted) by the quartz ampule ($\varepsilon_A J_M, F_{M,A} A_M$), so:

$$q_{A,M-net} = \varepsilon_A A_A F_{A,M} \sigma T_A^4 - \varepsilon_A J_M F_{M,A} A_M$$

$$= (1 - \tau) A_A F_{A,M} (\sigma T_A^4 - J_M) \quad (14)$$

By the same token the net heat flow between the quartz ampule and the sample is:

$$q_{A,S-net} = (1 - \tau) A_S F_{S,A} (\sigma T_A^4 - J_S) \quad (15)$$

Eqs. (10-15) can be collectively represented by the analog resistance network shown in Fig. 6.

Analysis Results

Two examples have been analyzed to illustrate the relationship between the thermophysical properties of the sample and its optimal operating conditions (lead tin telluride and mercury cadmium telluride). The first material we have chosen to report on is lead tin telluride ($Pb_{0.8}Sn_{0.2}Te$).

The assumed thermal conductivity for this material is shown in Fig. 7.[2,3] Large radial gradients distort the crystal structure causing defects. Limiting the cold end temperature to 400°C limits radial gradients in the specimen in the helium impingement region. The specific thermal requirements to be met for lead tin telluride are:

1) Minimum cartridge temperature = 400°C
2) Maximum temperature = 1200°C
3) Melt/solid interface temperature = 890°C
4) Axial gradient = 100 - 400°C/cm--determined in the liquid phase at the melt/solid interface
5) Radial gradient $\leq$ 0.45°C/cm
6) No axial temperature inversion (i.e., temperature must increase/decrease monotonically spatially).

To demonstrate the optimization process we have assumed a static sample. Four basic variables are examined: 1) tickler heater temperature; 2) sample hot end temperature; 3) centerline radius of curvature at the melt/solid interface; and 4) axial location of melt/solid interface. The tickler heater temperature was set at the highest temperature possible that would not cause a temperature inversion.

The location of the melt/solid interface is determined by a linear interpolation of the nodal temperature distribution. The gradient distribution can then be determined on both sides of the melt/solid interface by a backward or forward difference method derived from a Taylor series expansion of temperature at the interface.

The radius of curvature at the melt/solid interface was chosen as the primary variable in lieu of a local radial gradient parameter. This was done because a reversal of sign of the radius of curvature will indicate the presence of a planar interface. Also accurately defining a radial gradient at the interface is difficult due to the discontinuity in the thermal conductivity at the melt temperature. The radius of curvature (R) is equal to:

$$R = (1 + (dx/dr)^2)^{3/2} / (d^2x/dr^2) \qquad (16)$$

In order to calculate the radius of curvature at the centerline of the melt/solid interface an RSS fit to a second order polynomial is assumed of the following form:

$$X = C_0 + C_1 r + C_2 r^2$$

Note that $C_1 = 0$, so,

$$R = 0.5/C_2 \qquad (17)$$

Fig. 8 shows the variation in the calculated axial gradient with respect to variations in the sample hot end temperature. Figs. 9 and 10 show the variation of radius of curvature and the axial location of the melt/solid interface with respect to the sample hot end temperature. The asymptote shown on Fig. 9 represents an infinite radius of curvature of a planar interface. Fig. 10 illustrates that adjusting the sample hot end temperature also causes the melt/solid interface to move.

Adjustment of the cold end temperature results in an infinite number of operating points possessing a planar melt/solid interface. Fig. 11 shows the locus of such operating points. The characteristic temperature and gradient distributions for the case of a planar interface with a 400°C cold end temperature are shown in Fig. 12 and Fig. 13. Referenc-

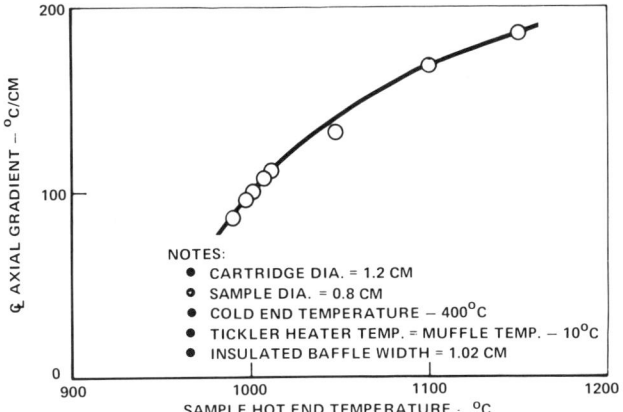

Fig. 8 Influence of sample hot end temperature on sample centerline axial gradient ($Pb_{0.8}Sn_{0.2}Te$).

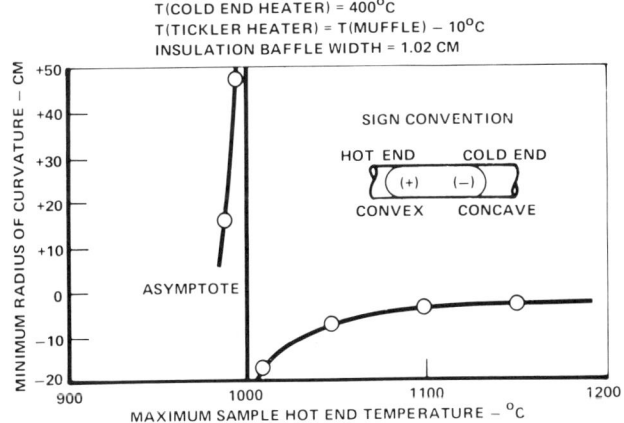

Fig. 9 Influence of sample hot end temperature on melt/solid interface radius of curvature.

ing Fig. 4, the maximum axial gradient predicted for these same boundary temperatures is 420°C/cm in the liquid.

An analysis of the directional solidification of mercury cadmium telluride, $Hg_{0.8}Cd_{0.2}Te$, yields an interesting comparison with lead tin telluride. As shown in Fig. 7, mercury cadmium telluride possesses a low thermal conductivity and, in contrast to lead tin telluride, the thermal conductivity in the liquid is less than that in the solid. The high vapor

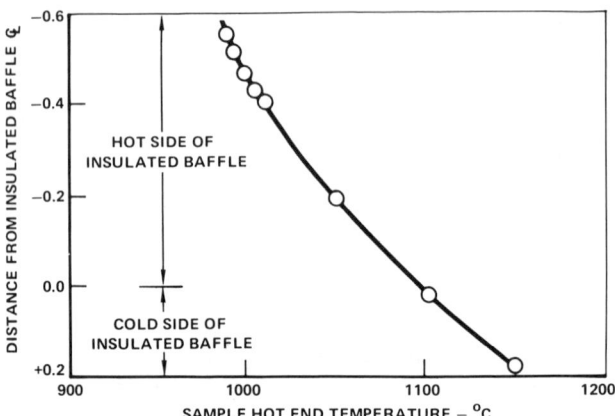

Fig. 10 Axial location of melt/solid interface relative to insulated baffle width C_L ($Pb_{0.8}Sn_{0.2}Te$).

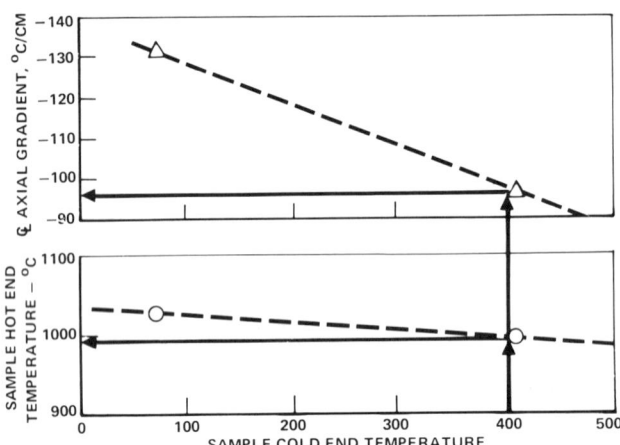

Fig. 11 Locus of optimum operating points (based on planar melt/solid interface) -- $Pb_{0.8}Sn_{0.2}Te$.

pressure of mercury in equilibrium with mercury cadmium telluride at elevated temperatures imposes a practical limit of approximately 1000°C for processing this material in the commonly used fused silica ampule.

The specific thermal requirements to be met for this material are listed below:

1) Maximum cartridge temperature = 1000°C
2) Melt/solid interface temperature = 705°C

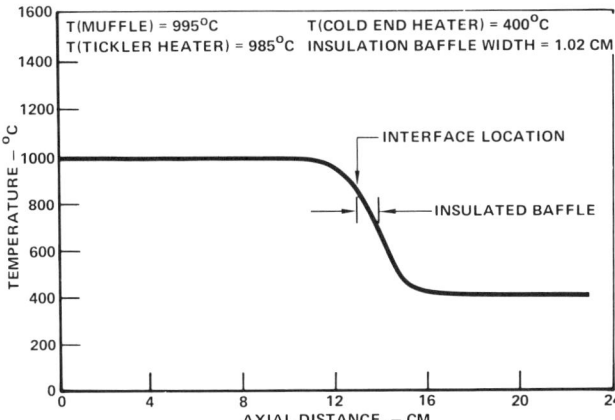

Fig. 12 Centerline temperature profile optimized for minimum radial gradient ($Pb_{0.8}Sn_{0.2}Te$).

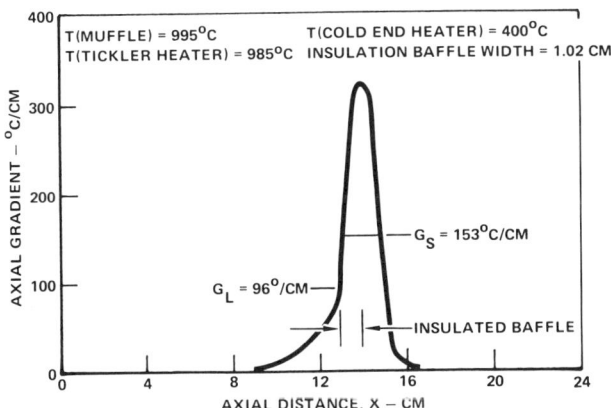

Fig. 13 Centerline axial temperature gradient: optimized for minimum radial gradient ($Pb_{0.8}Sn_{0.2}Te$).

3) Axial gradient $\geq 400-600°C/cm$ (determined in liquid phase at melt/solid interface)
4) Radial gradient $\leq 0.35°C/cm$
5) No axial temperature inversion

With the furnace hot end temperature fixed at 1000°C, the effect of varying the cavity cold end temperature is shown in Fig. 14. The maximum axial temperature gradient in the sample is obtained with the lowest cold end temperature, but this condition also yields the highest radial gradient

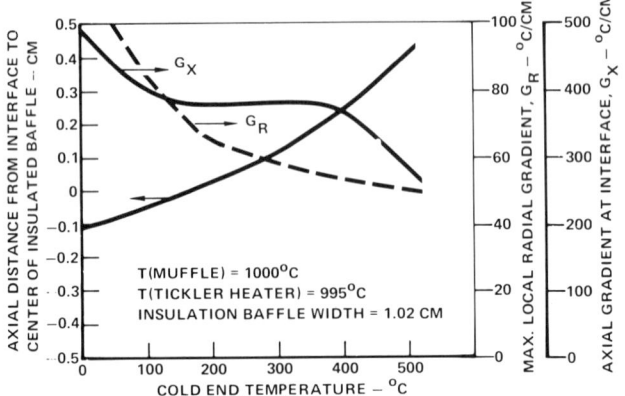

Fig. 14 Melt/solid interface location and temperature gradients for $Hg_{0.8}Cd_{0.2}Te$.

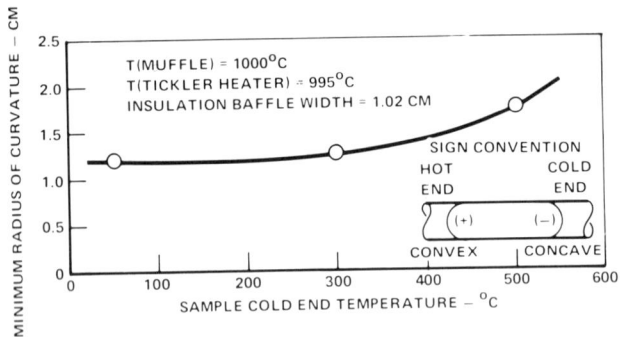

Fig. 15 Influence of sample cold end temperature on melt/solid interface radius of curvature ($Hg_{0.8}Cd_{0.2}Te$).

(low radius of curvature) at the melt/solid interface (see Fig. 15). A cold end temperature of 500°C may be identified as an upper limit beyond which the axial gradient decreases with little decrease in radial gradient. The positive sign for the distance parameter signifies moving towards the cold end of the insulated baffle. Note that increasing the cold end temperature has the effect of moving the melt/solid interface towards the cold end of the sample and, conversely, decreasing the sample cold end temperature moves the interface towards the hot end of the sample.

Steeper axial gradients can be obtained by accepting larger radial gradients (smaller radii of curvature), as shown in Fig. 14. For mercury cadmium telluride it was

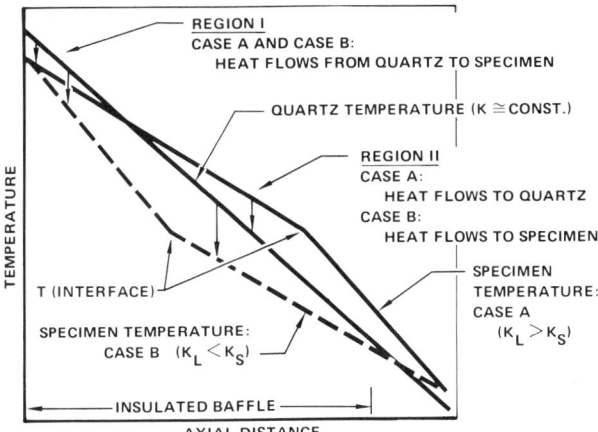

Fig. 16 Variation of gradient shape in melt/solid zone showing effect of specimen thermal conductivity.

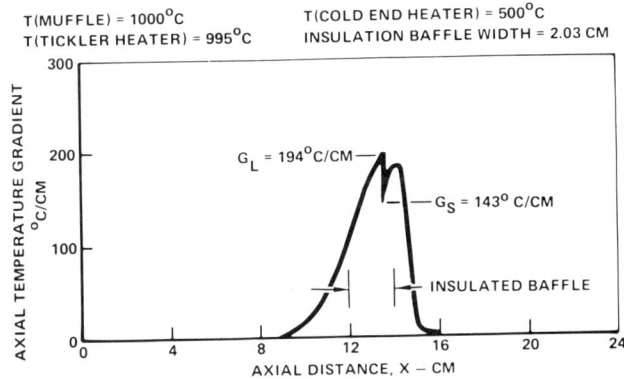

Fig. 17 Centerline axial temperature gradient: optimized for minimum radial gradient ($Hg_{0.8}Cd_{0.2}Te$).

impossible to identify a cold end temperature that yielded a planar interface. An explanation of this is shown schematically in Fig. 16. As an approximation, the thermal conductivity of the quartz encapsulent is assumed constant. Then the temperature distribution in the quartz through the melt region is approximately linear, as shown in Fig. 16. In the mercury cadmium telluride sample, however, due to the thermal conductivity being lower in the liquid than in the solid, the gradient is relatively steeper on the liquid side of the melt/solid interface. Approaching the interface from the hot (liquid) side, it is evident that temperatures in encap-

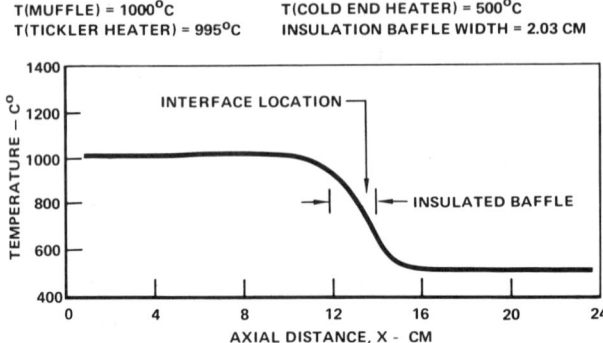

Fig. 18 Centerline temperature profile optimized for minimum radial gradient ($Hg_{0.8}Cd_{0.2}Te$)

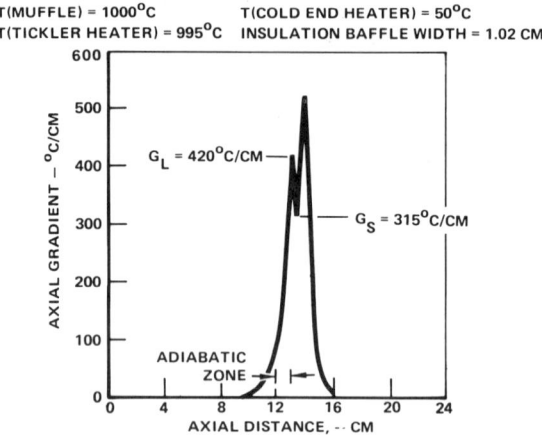

Fig. 19 Centerline axial temperature gradient for 50°C cold end temperature ($Hg_{0.8}Cd_{0.2}Te$).

sulent and sample diverge, with the largest differential existing at the interface itself. For these cases (Case B), radial gradients will always be present in the liquid, with the gradients being largest near the interface. For samples in which the thermal conductivity in the liquid exceeds that in the solid (Case A), it is in general possible to identify hot end processing temperatures for which the quartz adjacent to the melt/solid interface is at or very close to the interface temperature, yielding a planar interface. The lead tin telluride sample, previously discussed is an example of a Case A system. The temperature dependence of the thermal conductivity of quartz, i.e. lower at low temperatures, contributes to the ease of locating a condition yielding a planar interface for Case A type samples.

Some increase in radius of curvature is obtained by using a longer radiation baffle between the hot and cold ends of the furnace. With the cold end heater temperature at 500°C, the radius of curvature is 1.91 cm with a 2 cm baffle as opposed to 1.75 cm with a 1 cm baffle. At the same time the axial gradient is reduced from 280 to 194°C/cm. Figs. 17 and 18 show the temperature gradient and temperature distribution, respectively, for this case.

The effect of the variation of the cold end temperature can also be shown. Fig. 19 shows that with 1 cm baffle, reducing the cold end temperature from 500°C to 50°C results in reducing the interface radius of curvature from 1.75 cm to 1.2 cm.

Conclusions

The furnace design concept described in this paper has been shown to provide the large axial temperature gradients required by the experimenters. It has also been demonstrated that by adjusting the furnace hot and cold end temperatures that the melt/solid interface can be moved to a position within the sample that minimizes the interface radius of curvature. The magnitude of the radius of curvature, i.e. the degree of planarity is limited by the intrinsic thermal conductivity of the sample. The impact of the potential lack of planarity will be a point of investigation in the experimental portion of this program. Lastly, the idealized analysis presented here is useful as a guideline to understand the relationship between the sample diameter, furnace hot and cold end temperatures and the resultant axial gradient.

Acknowledgment

This work has been performed under TRW subcontract No. 70161YR9S.

References

[1] Summers, C. J. et al, "Growth and Characterization of $Hg_{1-x}Cd_xSe$ Alloys", presented to the American Conference on Crystal Growth, 16-19 July 1978, Washington, D.C.

[2] Fedorov, V. I. and Machuev, V. I., "The Thermal Conductivity of PbTe, SnTe, and GeTe in the Solid and Liquid Phases", Soviet Physics-Solid State, Vol. 2, No. 5, Nov. 1969.

[3] Touloukian, Y. S. et al, "Thermal Conductivity-Nonmetallic Solids - Thermophysical Property of Matter", Vol. 2, Plenum Press, New York, 1970.

Author Index for Volume 78

Alario, J. 305
Al-Astrabadi, F. R. 266
Altgilbers, L. L. 61
Basiulis, A. 324
Behnia, M. 110
Bobco, R. P. 487
Chin, J. H. 249
Cogley, A. C. 49
Ferguson, R. E. 61
Fernandes, R. 92
Fleischman, G. L. 324,375
Fote, A. A. 467
Francis, J. 92
Funai, A. I. 25
Giles, G. E., Jr. 152
Hall, D. F. 467
Harwell, W. 357
Holsen, J. N. 516
Hwangbo, H. 452
Karp, G. S. 516
Kelly, W. H. 452
Kemme, J. E. 345
Kirkpatrick, J. R. 152
Kunitomo, T. 3
Kosson, R. 305
Li, W. O. 130
Loose, J. D. 375
Martin, K. A. 202
McLean, W. J. 130
Meyer, R. 402
Miesner, S. H. 516
Minning, C. P. 324
O'Callaghan, P. W. 266
Ollendorf, S. 357
Prenger, F. C., Jr. 345
Probert, S. D. 266
Reddy, J. N 92
Schlitt, R. 402
Schneider, G. E. 179
Schuster, J. P. 130
Scollon, T. R., Jr. 375
Sheffield, J. W. 285
Shafey, H.M. 3
Sharma, A. 49
Veziroglu, T. N. 285
Viskanta, R. 110
Williams, A. 285
Wood, R. F. 152
Wright, J. P. 429
Wu, S. T. 61
Yovanovich, M. M. 179,202
Zien, T. F. 229

RAYMOND H. FOGLER LIBRARY
DATE DUE